Digital Signal Processing: A Breadth-First Approach

River Publishers Series in Signal, Image and Speech Processing

Volume 1

The "River Publishers Series in Signal, Image and Speech Processing" is a series of comprehensive academic and professional books which focus on all aspects of the theory and practice of signal processing. Books published in the series include research monographs, edited volumes, handbooks and textbooks. The books provide professionals, researchers, educators, and advanced students in the field with an invaluable insight into the latest research and developments.

Topics covered in the series include, but are by no means restricted to the following:

- Signal Processing Systems
- Digital Signal Processing
- Image Processing
- Signal Theory
- Stochastic Processes
- Detection and Estimation
- Pattern Recognition
- Optical Signal Processing
- Multi-dimensional Signal Processing
- Communication Signal Processing
- Biomedical Signal Processing
- Acoustic and Vibration Signal Processing
- Data Processing
- Remote Sensing
- Signal Processing Technology
- Speech Processing
- Radar Signal Processing

For a list of other books in this series, visit www.riverpublishers.com

Digital Signal Processing: A Breadth-First Approach

Muhammad N. Khan
The University of Lahore
Pakistan

Syed K. Hasnain
Swedish College of Engineering and Technology
Pakistan

Mohsin Jamil
National University of Sciences and Technology
Pakistan

Published, sold and distributed by:
River Publishers
Niels Jernes Vej 10
9220 Aalborg Ø
Denmark

River Publishers
Lange Geer 44
2611 PW Delft
The Netherlands

Tel.: +45369953197
www.riverpublishers.com

ISBN: 978-87-93379-40-4 (Hardback)
978-87-93379-39-8 (Ebook)

I dedicate this book to my loving family: Atika, M. Hamdaan Khan,
Ayaan Ahmad Khan, and Imaan Khan and to my parents:
Habibullah Khan and Saira Khan.

Contents

Preface

With the rate at which technology is advancing and the level of research being conducted in various fields, particularly Digital Signal Processing (DSP), it is vital to keep up with the times.

By the grace of almighty Allah, We are able to complete my book in the form of Fourth edition, which we feel it is complete course book for undergraduate students. It has been our prime aim to streamline the flow of the book by connecting the numerical problems with the theory in a manner which will be most beneficial to the student. We wish to thank all our students and colleagues in suggesting improvements for this book. The organization of the book is as follows:

Chapter 1 includes the introduction of Digital Signal Processing, with a brief history of DSP, classification of signals and application of DSP signal.

Chapter 2 is devoted entirely to the characterization and analysis of linear time invariant continuous and discrete time signals and systems, block diagram representation of discrete time systems, which can be left if the students have already gone through the course of Signals and Systems in previous semesters.

Chapter 3 is devoted to impulse response, convolution sum, application of convolution, properties of convolution, and different methods and techniques of finding convolution, correlation and its properties, application, and correlation coefficients.

Chapter 4 plays an important role in signal processing applications which brushes up the Z-transform, its properties which has been introduced are used to solve the problem, three cases of inverse Z-Transform, classical method for solution of difference equations with different forcing functions has also been included.

Chapter 5 describes the classical method of solution of difference equation. Classical method for solution of difference equations with different forcing functions has also been included.

Chapter 6 covers Fourier synthesis, discrete time Fourier transform (DTFT), discrete Fourier transform (DFT), and its properties and finally fast Fourier transform (FFT) of radix-2 by two methods, decimation in time fast Fourier transform (DIT-FFT) and decimation in frequency fast Fourier transform (DIF-FFT).

Chapter 7 focuses on the realization structures of the FIR and IIR digital filter using interconnection of basic building block, few basic structures of the filter has been included to give basic concepts of it such as direct form-1, direct form-2, cascade, parallel and lattice, and lattice-ladder form.

Chapter 8 gives a basic knowledge about filter types used in DSP, criteria for selecting digital filters, design steps, advantage and disadvantage of Finite Impulse Response (FIR) and Infinite Impulse Response (IIR) filters.

Chapter 9 is devoted to FIR filter design consideration. A simple approach of designing filter coefficients has been described including FIR low pass, high pass, band pass, and band stop filters using window methods have been analyzed in detail for all type of filters non-causal and causal results of filter coefficient calculation using Matlab has also been included.

Chapter 10 is developed for IIR Filter design consideration of analog filters; step by step method of designing the digital IIR filter employing Butterworth and Chebyshev approximation. Bilinear transformation, Impulse invariance, and Pole placement methods for calculation of coefficients has been described using the help of MATLAB.

Chapter 11 is finite word length effects in digital filters, which we think should be inducted at undergraduate level, covering fixed point and floating point numbers, quantization noise, etc. Although this chapter is not included in the course at undergraduate level, it has been included here to provide basic knowledge regarding this chapter.

Chapter 12 provides a 245 review questions and 85 multiple choice questions to test the knowledge of students.

Chapter 13 includes sample test papers for practicing by the students; these questions can be given to students as assignment work.

Audience

This textbook is for a first course on DSP. It can be used in both computer science and electrical engineering departments. In terms of programming languages, the book assumes only that the student may have basic experience with MATLAB or C language. Although this book is more precise and analytical than many other introductory DSP texts, it uses mathematical concepts that are taught in higher secondary school. We have made a deliberate effort to avoid using most advanced calculus, probability, or stochastic process concepts (although we've included some basic and homework problems for students with this advanced background). The book is, therefore, appropriate for undergraduate courses. It should also be useful to practitioners in the telecommunications industry.

Unique about This Textbook

The subject of DSP is enormously complex, involving many concepts, probabilities, and signal processing that are woven together in an intricate manner. To cope with this scope and complexity, many DSP texts are often organized around the "numerical examples" of a communication system. With such organization, students can see through the complexity of DSP they learn about the distinct concepts and protocols in one part of the communication system while seeing the big picture of how all parts fit together. From a pedagogical perspective, our personal experience has been that such approach indeed works well.

Special Features for Students and Instructors

MATLAB includes several Signal Processing features and is an important tool for illustrating many of the field's applications. The use of MATLAB has been linked to some aspects of this book to assist students in their understanding and to give them confidence in the subject.

MATLAB is not a pre-requisite for this book. Its working is described in sections where it is utilized. For further specifics the help documentation is available online from Mathworks (http://www.mathworks.com), which is easy to use and contains many examples. Our experience has shown that signal processing students completely unfamiliar with MATLAB are able to use MATLAB within a week or two of exposure to tutorial exercises.

Every attempt has been made to ensure the accuracy of all material of the book. Readers are highly welcomed for a positive criticism and comments. Any suggestions or error reporting can be sent to dr.nasirkhan@ucp.edu.pk

One Final Note: We'd Love to Hear from You

We encourage students and instructors to e-mail us with any comments they might have about our book. It's been wonderful for us to hear from so many instructors and students from around the world about our first international edition. We also encourage instructors to send us new homework problems (and solutions) that would complement the current homework problems. We also encourage instructors and students to create new MATLAB programs that illustrate the concepts in this book. If you have any topic that you think would be appropriate for this text, please submit it to us. So, as the saying goes, "Keep those cards and letters coming!" Seriously, please *do* continue to send us interesting URLs, point out typos, disagree with any of our claims, and tell us what works and what doesn't work. Tell us what you think should or shouldn't be included in the next edition.

Acknowledgments

Since we began writing this book in 2013, many people have given us invaluable help and have been influential in shaping our thoughts on how to best organize. We want to say A BIG THANKS to everyone who has helped us in drafting this book. Our special thanks to:

- Dr. Hammad Omer from COMSATS, Islamabad who has helped me to in thorough checking of the manuscript of the book for the International edition.
- Dr. Mohsin Jamil, Deputy Head of Department, School of Manufacturing and Mechanical Engineering, National University of Science and Technology, Islamabad for keeping my moral up to keep writing and compiling such engineering books for undergraduate students.
- Prof. Dr. Mansoor-U-Zafar Dawood from King Abdul Aziz University, Kingdom of Saudi Arabia for giving the manuscript a final reading for the International edition.
- Prof. Dr. Jameel Ahmad, Chairman, Department of Electrical Engineering, HITEC University Islamabad for checking the manuscript for this edition.
- Dr. Ishtiaq Ahmad, Assistant Professor, Department of Electrical Engineering, The University of Lahore, Lahore for checking the manuscript for this edition.
- Ms Tarbia Iftikhar, my student of the University of Lahore at graduate level, for helping me in writing and editing my book.
- Dr. Kamran Ezdi, Assistant Professor, Electrical Engineering Department, University of Central Punjab, Lahore for technical support during preparation of this book.
- Dr. Ghulam Abbas, Assistant Professor, Electrical Engineering Department, The University of Lahore, Lahore for technical support during preparation of this book.
- Mr. Farhan Abbas Jaffery from OGDCL and Mr. Tajammul Ahsan Rasheed from SNGPL for designing the cover of this edition of the book.

- Mr. Qaiser Mahmood for typing this book for first and second edition with great zeal and enthusiasm.

We also want to thank the entire publisher team, who has done an absolutely outstanding job on this International edition. Finally, most special thanks go to the editor. This book would not be what it is (and may well not have been at all) without their graceful management, constant encouragement, nearly infinite patience, good humor, and perseverance.

Muhammad N. Khan
S. K. Hasnain

List of Figures

List of Tables

List of Abbreviations

ACF	Auto-Correlation Function
ADC	A/D (Analog to Digital Converter)
BZT	Bilinear Z-Transform
CAD	Computer Aided Design
CCF	Cross Correlation Function
CT	Continues Time
CTFT	CT Fourier Transforms
DAC	D/C (Digital to Analog Converter)
DFS	Discrete Fourier Series
DIF	Decomposition in Frequency
DIT	Decomposition in-Time
DSP	Digital Signal Processing
DT	Discrete Time
DTFS	DT Fourier Series
DTFT	Discrete Time Fourier Transforms
ECG	Electrocardiogram
FFT	Fast Fourier Transform
FIR	Finite Impulse Response
IC	Integrated Circuit
IDTFT	Inverse DTFT
IIR	Infinite Impulse Response
LTI	Linear Time Invariant
MRI	Magnetic Resonance Imaging
PDF	Probability Density Function
RADAR	RAdio Detection And Ranging
RC	Resistance Capacitance
RMS	Root Mean Square
ROC	Region of Convergence
SNR	Signal-to-Noise Ratio
SONAR	SOund Navigation And Ranging
VLSI	Very Large Scale Integration

1

Introduction

This chapter covers fundamental concepts of analog signal processing and digital signal processing, history of digital signal processing, basic definition, advantages of the signal processing in basic systems, basic blocks, key operation, and classification of signals along with applications of digital signal processing.

1.1 Concept of Signal Processing

Signal Processing is basically the analysis, interpretation, and manipulation of signals. It is the manipulation of the basic nature of a signal to get the desired shaping of the signal at the output. It is concerned with the representation, transformation, and manipulation of signals and the information they contain. Signal processing can be grouped into two classes:

- Analog Signal Processing
- Digital Signal Processing

1.1.1 Analog Signal Processing

It is the analysis of analog signals through analog means. In analog signal processing, continuous time signals are processed. Different types of analog signals are processed through low-pass filters, high-pass filters, band pass filters, and band stop filters to get the desired shaping of the input signal.

1.1.2 Digital Signal Processing

Digital signal processing is the numerical processing of signals on a digital computer or some other data processing machine. Digital signal processors (DSPs) take real-world signals like voice, audio, video, temperature, pressure, or position that have been digitized and then mathematically manipulate them. For example, a digital system such as a digital computer takes a signal in discrete-time sequence form and converts it into a discrete time output sequence.

The environment is full of signals that we sense; examples including the sound, temperature and light.

In case of sound we use our ears to convert into electrical signal in to our brain. We then analyze properties such as frequency, amplitude, and phase to categorize the sound and determine its direction. We may recognize it as music, speech or noise of a machine.

In case of temperature our nerves are exposed through skin will send signals to the brain. The example includes in this case is witch on of a heating or opening a window.

In case of light, our eyes focus the image into the retina, which converts it into electrical signal to send to the brain. Our brain analyses the color, shape, intensity, etc.

The processing which apply to the signals is carried out by the digital computer and is thus called digital signal processing (DSP).

Digital signal processing involves the extraction of information from signals which in turn, depends upon the type of signal and the nature of information it carries. In the case of a calculator, the application of different operators on a set of values comes under DSP.

It covers the mathematics, the algorithms, and the techniques used to influence and control signals after they have been converted into digital form. Visual images, recognition and generation of speech, compression of data for storage and transmission are some of its applications.

1.2 Roots of DSP

Owing to the high level of research conducted, the subject of DSP has developed very rapidly over the last few decades. This rapid development has been a result of significant advances in digital computer technology and IC fabrication techniques.

Since the advent of computers in the 1960s, time and money have been invested in incorporating DSP in all the sections of engineering. The efforts were first made in four key areas, namely

1. Radar and Sonar (where national security was at risk)
2. Oil Exploration (where large amounts of money could be made)
3. Space Exploration (where the data are irreplaceable)
4. Medical Imaging (where lives could be saved)

Digital Signal Processing	Communication Theory
	Numerical Analysis
	Probability and Statistics
	Analog Signal Processing
	Decision Theory
	Digital Electronics
	Analog Electronics

Figure 1.1 Fuzzy and overlapping boundaries of DSP.

Digital Signal Processing has ties to many other areas of science and engineering as shown in Figure 1.1. Hence, for a thorough knowledge of the subject, it is vital to have some level of exposure to these other fields.

1.3 Advantages of DSP

What is it about DSP that makes it so popular? The answer to it is not short and clear. There are many advantages in using digital technique for general purpose signal processing.

1. Digital programmable systems allow flexibility. DSP programs can be configured by simply making alterations in our program. Reconfiguration of an analog system usually implies a redesign of the hardware.
2. Digital signal processing systems exhibit high accuracy.
3. DSP programs can be stored on magnetic media (disk) without any loss in signal. As a consequence, the signals become portable and can be processed off-line in a remote laboratory.
4. Processing in DSP reduces the cost by time-sharing of the processor among a number of signals.
5. Digital circuits are less sensitive to tolerance of a component value.
6. The implementation of highly sophisticated signal processing algorithms is made possible with DSP. It is very difficult to perform precise mathematical operations on signals in the analog form.

1.4 Basic Blocks of Signal Processing System

Most signals are analog in nature. For us to apply DSP on these signals, it is vital to efficiently bring these signals to the digital realm. Thus, there is a need for an interface between the analog and the digital signal processor as shown in Figure 1.2.

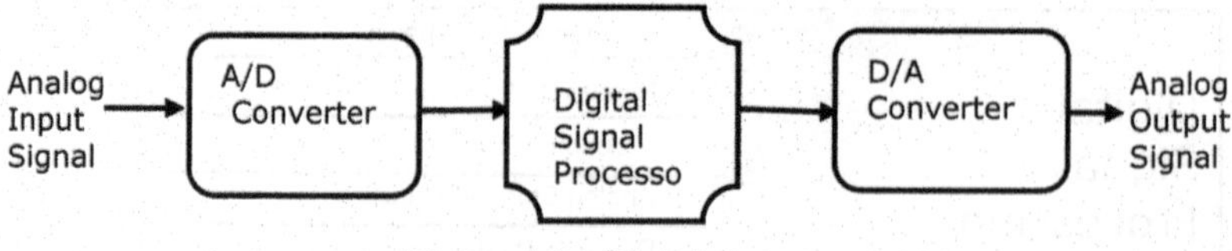

Figure 1.2 Digital system.

The interface that performs this operation is called the analog-to-digital (A/D) converter. The digital output is inputted to a digital processor as per our requirements.

1.5 DSP Key Operations

There are five main principles in DSP operations that need to be studied to familiarize with the field of DSP, which are as under:

- Convolution,
- Correlation,
- Filtering,
- Transformation, and
- Modulation.

1.6 Classification of Signals

The characteristic attributes of a specific signal determine which methods to employ in processing a signal or analyzing the response of a system to a signal. There are techniques that apply only to specific families of signals.

1.6.1 Continuous-Time versus Discrete-Time Signals

Continuous-time (or analog) signals exist for the continuous interval (a, b) where a can be $-\infty$ and b can be ∞. Discrete-time signals exist only for certain specific instances of time. These time instants may not be equidistant, but in practice, they are usually taken at equally spaced intervals for ease of calculations and mathematical tractability.

1.6.2 Continuous-Valued versus Discrete-Valued Signals

Continuous-time or discrete-time signals may give values that are continuous or discrete. If a signal takes on all possible values on a finite or an infinite range, it is said to be a continuous-valued signal. A discrete-time signal having a set of discrete values is called a digital signal.

1.6.3 Deterministic versus Random Signals

A deterministic signal is one that can be uniquely described by an explicit mathematical expression, a table of data, or a well-defined rule. Signals that cannot be described to any reasonable degree of accuracy by explicit mathematical formulas are of little practical use. Some examples of deterministic signals are:

$$x(t) = bt. \tag{1.1}$$

The above equation represents a ramp signal whose amplitude increases linearly with time and the slope is b.

$$x(n) = A \sin \omega n. \tag{1.2}$$

In the above case, the amplitude varies sinusoidally with time and has maximum amplitude of A. For such signals, it can be seen that the amplitude can be predicted in advance for any time instant. Hence, the signals represented above are deterministic signals.

A non-deterministic signal, on the other hand, is one whose occurrence is always random in nature. The absence of a relationship in these signals implies that they evolve over time in an unpredictable manner; such signals are also called random signals.

A typical example of non-deterministic signals is thermal noise generated in an electric circuit. Such a signal has probabilistic behavior.

1.6.4 Multi-Channel and Multi-Dimensional Signals

A signal is expressed by a function of one or more independent variables. In the case of electrocardiograms, 3 lead and 12 lead ECG are often used. This results in 3 channel and 12 channel signals. If the signal is a function of a single independent variable, the signal is called a one-dimensional signal.

One example of a two-dimensional signal is a picture, since the intensity or the brightness $I(x, y)$ at each point is a function of two-independent variables. Furthermore, since the brightness is a function of time, it may be represented as $I(x, y, t)$. Hence, the TV picture may be treated as a three-dimensional signal. The color TV picture may be described by three intensity function $I(x, y, t)$, $I(x, y, t)$, and $I(x, y, t)$ corresponding to the brightness of the three principle colors (red, green, and blue) as a function of time. The color TV picture is, thus, a three channel, three-dimensional signal, which can be represented by the vector.

$$I(x, y, t) = \begin{bmatrix} I_{\mathrm{r}}(x, y, t) \\ I_{\mathrm{g}}(x, y, t) \\ I_{\mathrm{b}}(x, y, t) \end{bmatrix}$$

1.7 Application of DSP

High-resolution spectral analysis has created various application areas in DSP. It requires a high-speed processor to implement the Fast Fourier Transform (FFT).

New applications are being added to DSP all the time. The sound production in home theatre systems employs DSP. Digital Computers are used to process the images of Mars sent back to Earth by the Mars pathfinder at the National Aeronautics and Space Administration (NASA). Any area where information is handled in digital form or controlled by a digital processor is working on the principles of DSP.

Even the special effects in movies such as Pan's Labyrinth, Harry Potter, and the Lord of the rings are created using special purpose digital computers and softwares. The generation of the cartoon characters and the lighting and shading effects in computer animation movies such as Shrek and Wall-E have all been carried out digitally.

Signaling tone generation and detection, frequency band shifting, filtering to remove power line hum etc., are all implemented by DSP. Further application areas are discussed below.

1.7.1 Telecommunications

Let us consider the three examples of multiplexing, compression, and echo control present in the telephone network.

1.7.1.1 Multiplexing

Here, audio signals are converted into a stream of serial digital data by the process of DSP. The property of bits to be easily intertwined and later separated allows telephone conversations to be transmitted on a single channel.

The financial advantage of transmitting digitally is enormous. Digital logic gates are far cheaper than wires and analog switches.

1.7.1.2 Compression

A voice signal is digitized at 8000 samples/s. Much of this digital information is superfluous. For this reason, several DSP algorithms have been developed to convert digitized voice signals into data streams that require fewer bits per second. These are called data compression algorithms. Matching of uncompressed algorithms is used to restore the signal to its original form.

1.7.1.3 Echo control

Echoes are a common problem faced in long distance telephone connections. With increasing distances, the echo becomes even more noticeable and irritating. It is particularly objectionable for intercontinental communication, where the delay can be several hundred milliseconds.

Digital signal process tackles this problem by measuring the returned signal and generating an appropriate anti-signal to cancel the offending echo. This technique allows speakerphone users to hear and speak at the same time without fighting audio feedback (squealing). In this way, environmental noise can also be reduced by cancelling it with digitally generated anti-noise signal.

1.7.2 Audio Signal Processing

The area of Speech Signal Processing has been revolutionized by the introduction of DSP.

1.7.2.1 Speech generation

Speech generation and recognition are used for the communication between humans and machines. Generated speech systems provide digital recording and vocal tract simulation.

1.7.2.2 Speech recognition

Digital signal processing approaches the problem of voice recognition in two steps: Feature extraction followed by feature matching.

1.7.3 Echo Location

A common method of obtaining information about a remote object is to bounce a wave of it. For example, radar operates by transmitting pulses of radio waves and examining the received signal for echoes from aircraft.

In sonar applications, submarines, and other submerged objects are detected by transmitting sound waves through the water. Geophysicists have long probed the earth by setting off explosions and listening for the echoes from deeply buried layers of rock. While these applications have a common thread, each has its own specific problems and needs. DSP has produced revolutionary changes in all three areas.

1.7.3.1 Radar (RAdio Detection And Ranging)

For a few microseconds, a radio transmitter produces a pulse of radio frequency energy. This pulse is inputted to a highly directional antenna, where the resulting radio wave propagates away at the speed of light. The aircraft in the path of this wave will reflect a small portion of the energy back toward a receiving antenna, situated near the transmission site. The elapsed time between the transmitted pulse and the received echo are used in the distance calculation. The direction to the object is known, as we know which direction the directional antenna was facing when the echo was received.

Digital signal processing has revolutionized radar in three areas, all of which relate to this basic problem. First, pulse compression can be carried out by DSP after it is received, providing better distance determination without reducing the operating range. Second, DSP can filter the received signal to decrease the noise. This increases the range, without degrading the distance determination. Third, DSP enables the rapid selection and generation of different pulse shapes and lengths. Among other things, this allows the pulse to be optimized for a particular detection problem. Much of this is done at a sampling rate comparable to the radio frequency used, as high as several hundred megahertz. When it comes to the radar, DSP is as much about high-speed hardware design as it is about algorithms.

1.7.3.2 Sonar (SOund Navigation And Ranging)

In *active sonar*, sound pulses transmitted into the water are between 2 and 40 kHz, and the resulting echoes are detected and analyzed. Some uses of active sonar are: the detection and localization of undersea bodies, for navigation, communication, and mapping the sea floor, with a maximum operating range of 10–100 km.

Passive sonar covers listening to underwater sounds, such as: natural turbulence, marine life, and mechanical sounds from submarines and surface vessels.

No energy is emitted in passive sonar and is, therefore, ideal for covert operations. You want to detect the enemy, without him detecting you.

The most significant application of passive sonar is in military surveillance systems that detect and track submarines. The frequencies utilized by passive sonar are typically lower than those of active sonar because less absorption occurs as they propagate through water. Detection ranges can be thousands of kilometers.

Rather than just a single channel, sonar systems usually employ extensive arrays of transmitting and receiving elements. The sonar system can steer the emitted pulse to the desired location by properly controlling and mixing the signals in these elements, and determine the direction the echoes are received from. To handle these multiple channels, sonar systems require the same massive DSP computing power as those employed in radars.

1.7.3.3 Reflection seismology

Today, the primary method for locating petroleum and mineral deposits is the reflection seismic method. Ideally, a sound pulse sent into the ground produces a single echo for each boundary layer the pulse passes through.

Each echo returning to the surface must pass through all the other boundary layers above from where it is originated. This can result in the echo bouncing between layers, giving rise to echoes that is being detected at the surface. These secondary echoes can make the detected signal very complicated and difficult to interpret.

Since the 1960s, DSP has been widely used to isolate the primary echoes from the secondary echoes in reflection seismograms. DSP allows oil exploration at difficult locations, such as under the ocean.

1.7.4 Image Processing

Images are signals with special characteristics. While most signals are a measure of a parameter over time, images are a measure of a parameter over space (distance). They contain a great deal of information. More than 10 MB can be required to store one second of television video. This is more than a thousand times greater than for a similar length voice signal. The final judge of quality is often a subjective human evaluation, rather than being an objective criterion. These special characteristics have made image processing a distinct subgroup within DSP.

1.7.4.1 Medical

Since 19th century, medical X-ray imaging was limited by four problems. First, overlapping structures in the body hide behind each other creating problems with visibility. For example, portions of the heart might not be visible behind the ribs. Second, it would not always be possible to distinguish between similar tissues. For example, it may be possible to discern bone from soft tissue, but distinguishing a tumor from the liver would not be as straightforward. Third, X-ray images show anatomy, the body's structure, and not physiology, the body's operation. The X-ray image of a living person would look exactly like the X-ray image of a dead person. Finally, X-ray exposure can also cause cancer, requiring it to be used sparingly and only with proper justification.

The last three X-ray problems have been solved by the use of penetrating energy other than X-rays, such as radio and sound waves. Magnetic resonance imaging (MRI) uses magnetic fields in conjunction with radio waves to probe the interior of the human body. This resonance results in the emission of a secondary radio wave, detected with an antenna placed near the body. Information about the localized region in resonance can be obtained from the strength and other characteristics of this detected signal.

With the adjustment of the magnetic field, the resonance region can scan throughout the body, mapping the internal structure. Just as in computed tomography, this information is usually presented as images. Besides providing excellent discrimination between different types of soft tissue, MRI can provide information about physiology, such as blood flow through arteries. It relies totally on DSP techniques and could not be implemented without them.

1.7.4.2 Space

With images taken from unmanned satellites and space exploration vehicles, the feed received is frequently of the lowest quality. DSP can improve the quality of images taken under extremely unfavorable conditions in several ways: brightness and contrast adjustment, edge detection, noise reduction, focus adjustment, motion blur reduction, etc. It is due to these merits that DSP proves ideal in this application as well.

2

Signals and Systems (Continuous and Discrete)

This chapter presents the basic foundation of signal and systems in discrete-time (DT). We introduced important types of signals with their properties and operations. Major topics of signals and systems have been introduced in this chapter. This chapter covers: Different continuous-time (CT) signals, concepts of frequency in CT signals, processing of analog-to-digital (A/D) and digital-to-analog (D/A) conversion, sampling theorem, quantization error, DT signals, concepts of frequency in DT signals, simple manipulation of DT signals, classification of DT signals, energy and power signals, DT systems, block diagram representation of DT systems, classification of DT systems, and Problems and solutions.

2.1 Introduction

In this modern age of microelectronics, signals, and systems play vital roles. A function of one or more independent variables which contains some information is called a signal. In other words, a signal can be defined as a varying phenomenon, which can be measured. Signals could be varied with respect to time or space. More suitable examples of signals include sounds, temperature, a voltage, and an image of video camera. Signals can be thought as either CT or DT. Signals normally occurring in nature (e.g., speech) are continuous in time as well as amplitude. Such signals are called CT signals. DT signals have values defined only at discrete instants of time. These time instants need not be equidistant, but in practice, they are usually taken at equally spaced intervals for computational convenience and mathematical tractability. If amplitude of DT signal is also made discrete through process of quantization or rounding off, then this becomes a Digital Signal. Digital signal processing (DSP) is concerned with digital processing of signals.

2.2 CT Signals

An analog signal has infinite variety of values with the varying time and continuous changes (e.g., smoothly) over time. CT signals are often denoted by $x(t)$. Such a signal is often called an analog signal, but a better term is continuous signal. The following are few **CT** signals for positive values of time (i.e., $t \geq 0$). The values of these signals are given below for $t \geq 0$ for a causal input

$$x_1(t) = 1 \tag{2.1}$$

$$x_2(t) = t \tag{2.2}$$

$$x_3(t) = t^2 \tag{2.3}$$

$$x_4(t) = e^{-t} \tag{2.4}$$

$$x_5(t) = \cos(\omega t + \theta). \tag{2.5}$$

2.2.1 Unit Impulse Function

The first specific signal we discuss is the unit impulse as is given in Equation (2.1). The unit impulse is a building block signal used for creating more complex signals as well as an effective signal for determining the time and frequency domain characteristics of certain classes of systems. The unit impulse has a magnitude of ∞, pulse width or time duration of 0 and area of

1. Following are the simple examples of such signals:

$$\delta(t) = \begin{cases} \infty, & t = 0 \\ 0, & t \neq 0 \end{cases}, \tag{2.6}$$

and Equation (2.6) is also constrained to satisfy the identity as

$$\int_{-\infty}^{+\infty} \delta(t)\mathrm{d}t = 1 \tag{2.7}$$

2.2.2 Step Function

The step function is commonly used to test the response time of a system. The unit step response is desirable because the signal varies from zero magnitude value to a finite value theoretically at zero time. The most often used unit step function is described as a function having magnitude of 1 occurring at time equal to and greater than zero.

$$u(t) = \begin{cases} 0, & t < 0 \\ 1, & t \geq 0 \end{cases}. \tag{2.8}$$

2.2.2.1 Properties of unit step function

Properties describe the continuous as well as sampled impulse and step function. Following are the important properties of the Unit step function

1. $u(t)\delta(t-a) = \delta(t-a)$
2. $\frac{du(t)}{dt} = \delta(t)$ and then, $u(t) = \int_{t=-\infty}^{\infty} \delta(t)dt$
3. The step and impulse response are related by derivatives; the impulse represents the instantaneous rate of change of the step function and accordingly the step function is equal to the integral of impulse function.

2.2.3 Ramp Function

The ramp function is uniformly increasing time domain signal of a constant slope k. The ramp function is commonly used as a test signal after step function. The signal is designate as $r(t)$. The ramp function is described as a function having a magnitude of t at $t \geq 0$.

$$r(t) = \begin{cases} 0, & t < 0 \\ t, & t \geq 0 \end{cases}. \tag{2.9}$$

2.2.4 Parabolic Function

The parabolic function is not uniformly increasing with respect to time and having a slope k. The signal is designated as t^2.

$$t^2 = \begin{cases} 0, & t < 0 \\ t^2, & t \geq 0 \end{cases}. \tag{2.10}$$

2.2.5 Exponential Function

The exponential function is increasing or decreasing exponentially. Because of stability issue, we use most frequently the decreasing exponential function. The signal is designate as e^{-t}. The unit exponential function is described as having a magnitude of 1 at zero time and exponentially decaying or rising for time greater than zero.

$$e^{-t} = \begin{cases} 1, & t < 0 \\ e^{-t}, & t > 0 \end{cases}. \tag{2.11}$$

2.2.6 Sinusoidal Function

The sinusoidal function is time domain signal. The signal is designated as $\sin \omega t$. This signal is used to find out the steady-state response of a system.

2.3 Concept of Frequency: Continuous Time Signals

A simple harmonic oscillation is given by

$$x(t) = A\cos(\omega t + \theta) \quad \text{for} \quad -\infty < t < \infty, \tag{2.12}$$

Where $x(t)$ is a CT analog signal, A is the amplitude, ω is the frequency in radian per second, and θ is the phase.

2.3.1 Periodic and Aperiodic Signals

A periodic signal is that type of signal which has a finite pattern and repeat with a repetition period of T. In other words, a CT signal is called periodic if it exhibits. The smallest value of period T, which satisfies Equation (2.12) is called fundamental period and is denoted by T_o.

The CT sinusoids are characterized by the following properties:

(1) Periodic functions are assumed to exist for all time. In Equation (2.12), we can eliminate the limit of t.
(2) Aperiodic function can be written with period nT, where n is an integer. Hence for a periodic function, $x(t) = x(t+T) = x(t+nT)$ with n be any integer.
(3) We define the fundamental period T_o as the minimum value of the period $T > 0$ that satisfies $x(t) = x(t+T)$.

Example 2.1

Determine the fundamental period and periodicity of the following sinusoids.

(a) $x(t) = \sin \pi t$
(b) $x(t) = \sin \sqrt{2}\,\pi t$.

Solution

(a) The fundamental period T is given as $T = \frac{2\pi}{\omega} = \frac{2\pi}{\pi} = 2, x(t) = \sin \pi t$ is a periodic signal.

$$\begin{aligned} x(t+T) &= \sin \pi(t+T) = \sin \pi(\ t+2\ ) \\ &= \sin \pi t\ \ \cos 2\pi + \cos \pi\, t \sin\ 2\pi = \sin \pi t. \end{aligned}$$

(b) The fundamental period T is given as $T = \frac{2\pi}{\Omega} = \frac{2\pi}{\pi\sqrt{2}} = \frac{2}{\sqrt{2}} = \sqrt{2} = 1.414,\, x(t) = \sin \pi\sqrt{2}t$ is a periodic signal

$$\begin{aligned} x(t+T) &= \sin\ \sqrt{2}\pi t = \sin\ \sqrt{2}\pi(\ t+\sqrt{2}\ ) \\ &= \sin \sqrt{2}\pi t \cos 2\pi + \cos \sqrt{2}\pi t \sin\ 2\ \pi = \sin\ \sqrt{2}\pi t. \end{aligned}$$

Example 2.2

Assume $x_1(t)$ and $x_2(t)$ are periodic signals with period T_1 and T_2, respectively. Under what conditions the sum $x(t) = x_1(t) + x_2(t)$ is periodic. What will be period of $x(t)$, if it is periodic?

Solution

Given that $x_1(t)$ and $x_2(t)$ are periodic signals with period T_1 and T_2, respectively.

Thus $x_1(t)$ and $x_2(t)$ may be written as
$x_1(t) = x_1\left(t+T_1\right) = x_1\left(t+mT_1\right)$, where m is an integer;
$x_2(t) = x_2\left(t+T_2\right) = x_2\left(t+nT_2\right)$, where m is an integer.
Now, if T_1 and T_2 are such that $mT_1 = nT_2 = T$. Then,
$x\left(t+T\right) = x_1\left(t+T_1\right) + x_2\left(t+T_2\right)$;
$x\left(t+T\right) = x_1(t) + x_2(t)$, i.e., $x(t)$ is periodic in this case.
Therefore, condition of $x(t)$ to be periodic is $\frac{T_1}{T_2} = \frac{n}{m}$ is a rational number.

Example 2.3

The sinusoidal signal $x(t) = 10\cos(200t + \pi/2)$ is passed through a square-law device defined by the input–output relation. Using the trigonometric identity $\cos^2\theta = \frac{1}{2}(\cos 2\theta + 1)$.

(a) Specify the DC component.
(b) Specify the amplitude and fundamental frequency of the sinusoidal component in the output $y(t)$.

Solution

$y(t) = x^2(t) = \left[10\cos\left(200t + \frac{\pi}{2}\right)\right]^2 \Rightarrow y(t) = \frac{100}{2}\left[1 + \cos\left(400t + \pi\right)\right] \Rightarrow$

(a) The DC component is 50.

(b) The amplitude is 50 and fundamental frequency is $\frac{200}{\pi}$.

Example 2.4

Consider the following analog sinusoidal signal $x(t) = 5\sin(100)\Pi t) = 5$. Sketch the signal $x_a(t)$ for $0 \leq t \leq 60$ ms

Solution

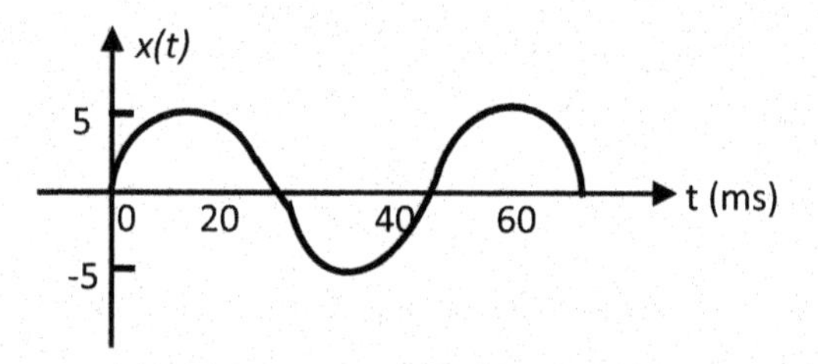

2.4 DT Signals

To emphasize DT nature, a **DT** signal is denoted as $x[n]$, instead of $x[n]$. A DT signal $x[n]$ is a function of an independent variable n, which is an integer. A DT signal is not defined at instants between two successive samples. **DT** signals are defined only at discrete values of time. The values of the signals are given below for $n \geq 0$ for a causal input:

$$x_1[n] = 1, \tag{2.13}$$

$$x_2[n] = n, \tag{2.14}$$

$$x_3[n] = n^2, \tag{2.15}$$

$$x_4[n] = a^n, \quad \text{and} \tag{2.16}$$

$$x_5[n] = \cos\omega n. \tag{2.17}$$

2.4.1 CT versus DT Signals

The values of a **CT** or **DT** signal may be continuous or discrete. If a signal takes on all possible values on a finite or infinite range, it is said to be a continuous valued signal.

If the signal takes on values from finite set of possible values it is said to be a discrete-valued signal. A **DT** signal having a set of discrete values is called a digital signal.

2.4.2 Unit Impulse

The first specific signal we discuss is the unit impulse. The unit impulse is a building block signal used for creating more complex signals as well as an effective signal for determining the time and frequency domain characteristics of certain classes of systems. The unit impulse has a magnitude of ∞, pulse width or time duration of zero and area 1.

$$\delta\left[n\right] = \begin{cases} \infty, & n = 0 \\ 0, & n \neq 0 \end{cases}. \tag{2.18}$$

2.4.3 Unit Step Function

The unit step function, designated as *u*(*n*). The unit step function is commonly used to test the response time of a system. The unit step response is desirable because the signal goes from zero magnitude value to a finite value theoretically zero time. The unit step function is described as a signal having magnitude of 1 and occurring at time equal to and greater than zero.

$$u\left[n\right] = \begin{cases} 0, & n < 0 \\ 1, & n \geq 0 \end{cases}. \tag{2.19}$$

2.4.4 Ramp Function

The ramp function is u uniformly increasing time to domain signal of a constant slope *k*. The ramp function is commonly used as a test signal after step function. The signal is designate as $r[n]$.

$$r[n] = \begin{cases} 0, & n < 0 \\ n, & n \geq 0 \end{cases}. \tag{2.20}$$

2.4.5 Parabolic Function

The parabolic function is not uniformly increasing time to domain signal of a slope *k*. The signal is designate as n^2.

$$n^2 = \begin{cases} 0, & n < 0 \\ n^2, & n \geq 0 \end{cases}. \tag{2.21}$$

2.4.6 Exponential Function

The exponential function used may be increasing exponential or decreasing exponential. Here because of stability we use most of the time the decreasing exponential. Here the signal is designate as e^{-n}. The unit exponential function is described as has a magnitude of 1 and occurring at time zero.

$$e^{-n} = \begin{cases} 1, & n < 0 \\ e^{-n}, & n > 0 \end{cases}. \tag{2.20}$$

2.4.7 Sinusoidal Function

The sinusoidal function is time domain signal. The signal is designate as $\cos(\omega n)$ or $\sin \omega n$. This signal is used to find out the steady-state response.

2.4.8 Concept of Frequency: DT Signals

A **DT** sinusoidal signal may be expressed as:

$$x[n] = A\cos[\omega n + \theta], + < n < \infty, \tag{2.21}$$

where $x[n]$ is a **DT** analog signal, A is the amplitude, ω is the frequency in radian per second and θ is the phase.

The DT sinusoids are characterized by the following properties:

(a) A DT sinusoid is periodic only if its frequency f is a rational number. By definition, DT signal $x[n]$ is periodic with period $N(N > 0)$ if and only if $x[n+N] = x[n]$ for all n. The smallest value of N is called the fundamental period.

$$x[n+N] = x[n] \text{ or } \cos[2\pi f(n+N) + \theta] = \cos[2\pi fn + \theta]. \tag{2.22}$$

The relationship in Equation (2.22) is true if and only if there exists an integer k such that $2\pi fN = 2k\pi$ or $f = \frac{k}{N}$. A **DT** signal is periodic only if its frequency f can be expressed as the ratio of two integer (i.e., f is rational).

(b) Discrete-time sinusoids whose frequencies are separated by an integer multiple of 2π are identical.

$$\cos\left[(\omega + 2\pi)\, n + \theta\right] = \cos\left[\omega n + 2\pi n + \theta\right] = \cos\left[\omega n + \theta\right]. \tag{2.23}$$

As a result all sequences

$$x_k\left[n\right] = A\cos\left[\omega_k n + \theta\right]; \qquad k = 0, 1, 2, \ldots \tag{2.24}$$

where $\omega_k = \omega + 2k\pi;\ -\pi < \omega < \pi$ are identical.

(c) The highest rate of oscillation in a DT sinusoid is attained when

$$\omega = \pi \;(\text{or}\; \omega = -\pi) \;\text{or equivalent}\; f = 0.5 \;\text{or}\; f = -0.5,\; x\left[n\right] = \cos\omega n,$$

where the frequency varies from 0 to π. To simplify the argument, we take the value of $\omega = 0, \pi/8,\ \pi/4,\ \pi/2, \pi$ corresponding to $f = 0$, 1/16, 1/8, 1/4, 1/2 which results in periodic sequence having period $N = \infty$, 16, 8, 4, 2. It is to be noted that period of sinusoidal decreases as the frequency increasing. The rate of oscillation increases as the frequency increases.

Frequency range of a **DT** sinusoids is finite with duration 2π. Usually the range of $0 \leq \omega \leq 2\pi$ or $-\pi \leq \omega \leq \pi$ or $-1/2 \leq f \leq 1/2$ is chosen which is also called the fundamental range.

Example 2.5

Determine a relationship between the sampling interval and signal period that ensures periodicity in a sampled sinusoid.

Solution

We determine that the digital frequency

$$\omega = 2\pi \frac{f}{F_s} \quad \text{or} \quad \frac{f}{F_s} = \frac{\omega}{2\pi}.$$

Example 2.6

A sinusoidal signal $x(n)$ having an amplitude 4 V and a frequency of 1000 Hz, is sampled at 125 μs, and begins at sample $n = 3$. Express $x(n)$ mathematically $x(t) = A\,(\sin \Omega_0 t + \varphi)$ using digital frequency notation.

Solution

If the sampling period is 125 μs = 125 $\times$ 10^{-3}, then the sampling frequency is: $\frac{1}{T_s} = \frac{1}{125\times10^{-6}} = 8000$ Hz.

Remembering that

$$\omega = 2\pi \frac{f}{F_s}$$

$$\omega = 2\pi \frac{1000}{8000} \text{ radians } = 0.784 \text{ radians.}$$

The sinusoid part of our wave form can be expressed as sin(0.784n) and scaled by 4, the delayed unit step function $u(n - 3)$ can be used to activates the sinusoid at $n = 3$ (or three times 125 or 375 μs). The expression for $x(n)$ becomes $x(n) = 4n(n - 3)\sin(0.784n)$.

Example 2.7

Compute the fundamental period (N_p) and determine the periodicity of the signal. (a) $\cos 0.01\pi n$; (b) $\cos\left(\pi\frac{30\,n}{105}\right)$.

Solution

(a) $f = \frac{0.01\pi}{2\pi} = \frac{1}{200} \Rightarrow$ periodic with $N_p = 200$
$\cos 0.01\,\pi(n + 200) = \cos 0.01\,\pi n \cdot \cos 2\,\pi - \sin\,0.01\,\pi n \cdot \sin\,2\,\pi =$ $\cos\,2\,\pi$. It is a periodic signal.

(b) $f = \frac{30\pi}{105}\left(\frac{1}{2\pi}\right) = \frac{1}{7} \Rightarrow$ periodic with $N_p = 7$
$\cos\left(\pi\frac{30n}{105}\right) = \cos\left(\pi\frac{30}{105}(n + 7)\right) = \cos\left(\pi\frac{30\,n}{105} + \pi\frac{210}{105}\right)$
$\cos\left(\pi\frac{30\,n}{105} + 2\pi\right) = \cos\left(\pi\frac{30n}{105}\right)\cos 2\pi - \sin\left(\pi\frac{30n}{105}\right)\sin 2\pi =$
$\cos\left(\pi\frac{30\,n}{105}\right)$.

2.5 Time-Domain and Frequency-Domain

The signals introduced so far were all function of time; we measured the signal's amplitude at different time instants, thus monitoring the signal with the passage of time. A graph representation of these results is known as time domain representation of the signal.

The time-domain representation is very useful for many applications such as measuring the average value of a signal or determining when the amplitude of the signal exceeds certain limits.

There are certain applications for which other representation is more useful, i.e., the frequency-domain representation. Sine waves occur commonly in nature and are the building blocks of many other waveforms.

The frequency-domain representation tells us the amplitudes, and frequencies, of the constituent sinusoidal waves present in the signal being measured. This representation is also known as the spectrum of the signal.

2.6 A/D and D/A Conversion

Most signals of practical interest, such as speech, biological signals, seismic signals, radar signals, sonar signals, and various communication signals such as audio and video signals, are analog. To process analog signals by digital means, it is first necessary to convert them into digital form, i.e., to convert them to a sequence of numbers having finite precision. This procedure is called A/D conversion, and the corresponding devices are called analog-to-digital converters (ADCs).

2.6.1 Processing Steps for A/D Conversion

2.6.1.1 Sample and hold

This is the conversion of a CT signal into a **DT** signal obtained by taking samples of the CT signal at **DT** instants. Thus, if $X_{\mathrm{a}}(t)$ is the input to the sample, the output is $X_{\mathrm{a}}[nT] = X[n]$, where *T* is called the sampling interval. For the ease of convenience *T* is suppressed. In sample and hold operation the previous value is held, using a zero-order hold, till the next value comes. It is most frequently used in data converters. The process of obtaining signal values from a continuous signal at regular time interval is basically known as *sampling*. The result is being represented by a sequence of numbers and the sample and hold circuit is one of the basics circuits used in the sampling procedure. After getting sampled, we get the **DT** signal rather than digital signal. A lot of details about the sampling and the sampling theorem can be found in DSP books. We are skipping that much detail from this book.

2.6.1.2 Quantization

After getting sampled of a continuous signal, we get sequence of numbers, which can still get any values on a continuous range of values. This is the conversion of **DT** continuous-valued signal into a **DT**, discrete-valued (digital) signal. The value of each signal sample is represented by a value selected from a finite set of possible values. After the discretizing of time variable, we now have to discretize the amplitude variable as well. This discretizing of amplitude variable is called quantization. We can assume a range of sequence for the quantization. Then we start doing the quantization of the discrete valued signal. The difference between the un-quantized sample $x[n]$ and the quantized output $x_{\mathrm{q}}[n]$ is called the quantization error.

Mathematically, we can view the error signal as random signal with a uniform probability density function (pdf). For this reason, the quantization

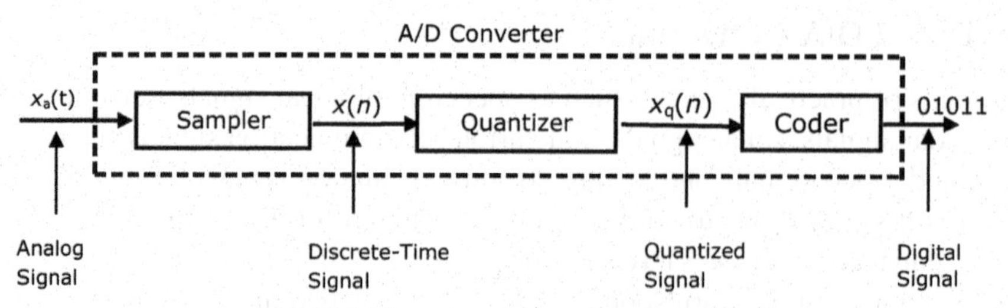

Figure 2.1 Basic blocks of an A/D converter.

error is often also referred to as the quantization noise. A more rigorous mathematical justification for the treatment of the quantization error as uniformly distributed white noise is provided by Window's Quantization Theorem but we are not to discuss that here any further.

2.6.1.3 Coding

In the coding process, each discrete value $x_q[n]$ is presented by a b-bit binary sequence. An A/D converter is modeled as a sampler, quantizer followed by a coder as shown in Figure 2.1.

In practice, the A/D conversion is performed by a single device that takes $x_q[n]$ and produces a binary coded number. The operation of sampling and quantization can be performed in either order, in practice; sampling is always performed before quantization.

In many cases, it is desirable to have a signal in analog form such as in speech signal processing because sequence of samples cannot be understood. So the conversion is required from D/A. All D/A converters connect the dots to make an analog signal by performing some kind of interpolation such as linear interpolation and quadratic interpolation. The process of interpolation can be found in other books.

2.6.2 Sampling of Analog Signals

There are many ways to sample an analog signal. Discussion is limited here to periodic or uniform sampling, which is the type of sampling used most often in practice. The relation is given below:

$$x[n] = x_a(nT)\,;; \quad ussn < \infty, \tag{2.25}$$

where $x[n]$ is the **DT** signal obtained by taking samples of the analog signal $x_a(nT)$ every T seconds. This procedure is explained in Figure 2.2. The time interval T between successive samples is called the sampling period or sample

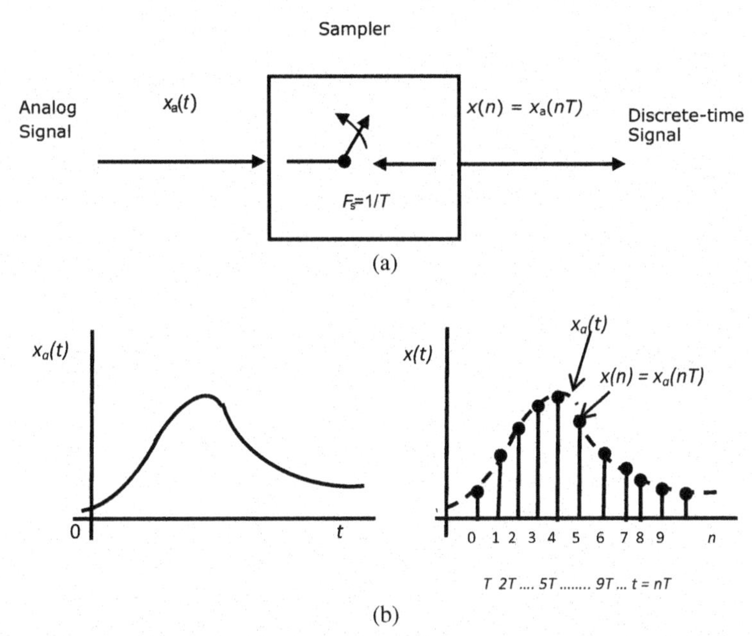

Figure 2.2 (a) Sampler and (b) Periodic sampling of an analog signal.

interval and its reciprocal $\frac{1}{T} = F_s$ is called the sampling rate (samples per second) or the sampling frequency (in hertz).

Periodic sampling establishes a relationship between the time variables t and n of CT and **DT** signals, respectively. Indeed, these variables are linearly related through the sampling period T or, equivalently, through the sampling rate $F_s = \frac{1}{T}$ or $t = nT = \frac{n}{F_s}$.

There exists a relationship between the frequency variable F (or Ω) for analog signals and the frequency variable f (or ω) for **DT** signals. To establish this relationship, consider an analog sinusoidal signal of the form

$$x_a(t) = A\cos(2\pi Ft + \theta) \tag{2.26}$$

when sampled periodically at a rate $F_s = \frac{1}{T}$ samples per second, yields:

$$x_a\,(nT) = A\cos(2\pi nFt + \theta)$$

$$x_a(n) = A\cos\left(\frac{2\pi nF}{F_s} + \theta\right). \tag{2.27}$$

Comparing the Equations (2.26) and (2.27), we note that the frequency variables F and f are linearly related as

$$f = \frac{F}{F_s} = \frac{T}{T_p} \text{ or } \omega = \Omega T \tag{2.28}$$

where f is named as the normalized frequency, it can be used to determine the frequency F in hertz only if the sampling frequency F_s is known.

Example 2.8

The **DT** sequence

$$x(n) = \cos\left(\frac{\pi}{2}\right) n \quad -\infty < n < \infty$$

was obtained by sampling analog signal

$$x_a(t) = \cos(\Omega t) \quad -\infty < t < \infty$$

at a sampling rate of 1000 samples/s. What are two possible values of Ω that could have resulted in the sequence $x(n)$?

Solution

The DT sequence is given as $x(n) = \cos\left(\frac{\pi}{2}\right) n$.

results by sampling the CT signal $x_a(t) = \cos(\Omega t)$

$$\omega = \frac{\Omega}{\mathrm{F}} \text{ or } \Omega = \omega F_s \quad \text{or} \quad \Omega = \left(\frac{\pi}{2}\right) 1000 = 500\pi$$

or possibly $\Omega = \left(2\pi + \frac{\pi}{2}\right) 1000 = 2500\pi$

Example 2.9

The implications of frequency conversion can be considered by the analog sinusoidal signal. Consider a signal $x(t) = \cos(4000\pi)\mathrm{t}$, which is sampled at a rate $F_s = 4000$ Hz and $F_s = 2000$ Hz. Show that recovery of the signal is possible when the analog signal is sampled at correct frequency.

Solution

The corresponding **DT** signal at $F_s = 4000$ Hz is

$$x(n) = \cos 2\pi\left(\frac{2000}{4000}\right) n = \cos 2\pi(1/2)n$$

$$x(n) = \cos 2\pi(1/2)n.$$

Using Equation (2.27) it follows that digital frequency $f = 1/2$. Since only the frequency component at 4000 Hz is present in the sampled signal, the analog signal we can recover is $y_a(t) = \cos 4000\pi t$.

The corresponding **DT** signal at $F_s = 2000$ Hz is

$$x(n) = \cos 2\pi \left(\frac{2000}{2000}\right) n = \cos 2\pi(1)n$$

$$x(n) = \cos 2\pi(1)n.$$

From Equation (2.27) it follows that the digital frequency $f = 1$ cycle/sample. Since only the frequency component at 2000 Hz is present in the sampled signal, the analog signal we can recover is $y_a(t) = x(F_s t) = \cos 2000\pi t$. This example shows that by taking a wrong selection of the sampling frequency, the original signal cannot be recovered.

2.7 The Sampling Theorem

The sampling theorem indicates that a continuous signal can be properly sampled, only if it does not contain frequency components above one-half of the sampling rate.

Figures given below indicate several sinusoids before and after digitization. The continuous line represents the analog signal entering the ADC, while the square markers are the digital signal leaving the ADC. The analog signal is a constant DC value as shown in Figure 2.3(a), a cosine wave of zero frequency. Since the analog signal is a series of straight lines between each of the samples, all of the information needed to reconstruct the analog signal is contained in the digital data. According to definition, this is proper sampling.

The sine wave shown in Figure 2.3(b), a 90 cycle/second sine wave being sampled at 1000 samples/s, it has a frequency of 0.09 of the sampling rate. Expressed in another way, this results in only 11.1 samples per sine wave cycle.

This situation is more complicated than the previous case, because the analog signal cannot be reconstructed by simply drawing straight lines between the data points. Do these samples properly represent the analog signal? The answer is yes, because no other sinusoid, or combination of sinusoids, will produce this pattern of samples. These samples correspond to only one analog signal, and, therefore, the analog signal can be exactly reconstructed. Again, it is an instance of proper sampling.

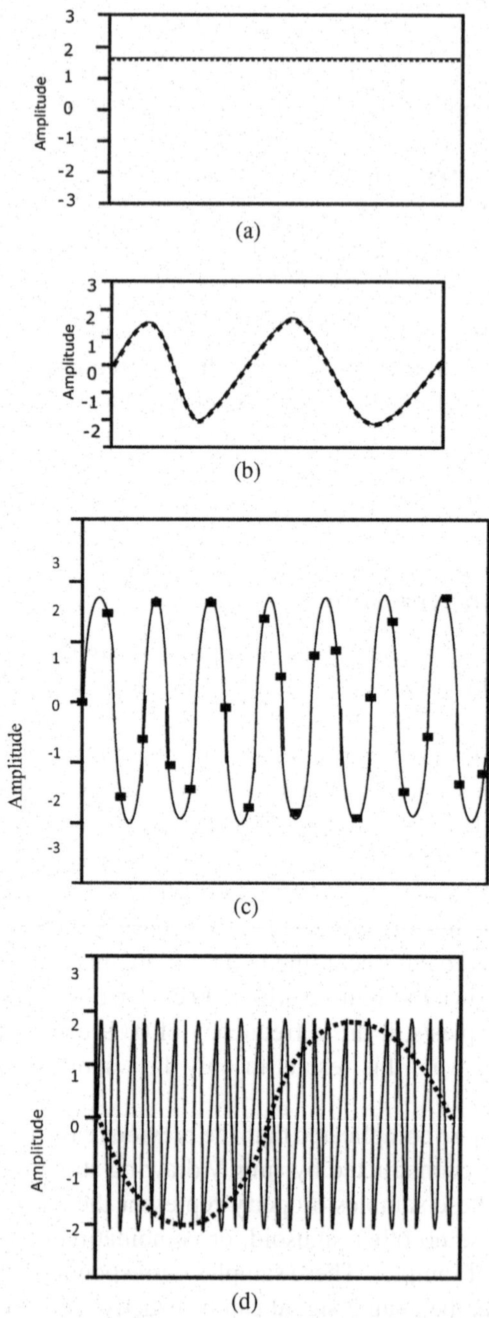

Figure 2.3 Time (or sample) number: (a) Analog Frequency = 0.0 (i.e., DC). (b) Analog frequency = 0.09 of sample rate. (c) Analog Frequency = 0.31 of sampling rate. (d) Analog Frequency = 0.95 of sampling rate.

In Figure 2.3(c), a 310 cycle/second sine wave being sampled at 1000 samples/second, the situation is made more difficult by increasing the sine wave's frequency to 0.31 of the sampling rate, this result in only 3.2 samples per sine wave cycle.

Here the samples are so sparse that they don't even appear to follow the general trend of the analog signal. Do these samples properly represent the analog waveform? Again, the answer is yes, and for exactly the same reason. The samples are a unique representation of the analog signal.

All of the information needed to reconstruct the continuous waveform is contained in the digital data. Obviously, it must be more sophisticated than just drawing straight lines between the data points. As strange as it seems, this is the proper sampling according to our definition.

In Figure 2.3(d), a 950 cycle/second sine wave being sampled at 1000 samples/s, the analog frequency is pushed even higher to 0.95 of the sampling rate, with a mere 1.05 samples per sine wave cycle. Do these samples properly represent the data? No, they do not. The samples represent a different sine wave from the one contained in the analog signal. In particular, the original sine wave of 0.95 frequencies misrepresents itself as a sine wave of 0.05 frequencies in the digital signal.

This phenomenon of sinusoids changing frequency during sampling is called aliasing. Just as a criminal might take on an assumed name or identity (an alias), the sinusoid assumes another frequency that is not its own. Since the digital data is no longer uniquely related to a particular analog signal, an unambiguous reconstruction is impossible. There is nothing in the sampled data to suggest that the original analog signal had a frequency of 0.95 rather than 0.05.

The sine wave has hidden its true identity completely; the perfect crime has been committed! According to our definition, this is an example of improper sampling.

This line of reasoning leads to a milestone in DSP, the sampling theorem. This is called the Shannon sampling theorem, or the Nyquist sampling theorem.

For instance, a sampling rate of 2,000 samples/s requires the analog signal to be composed of frequencies below 1000 cycles/s. If frequencies above this limit are present in the signal, they will be aliased to frequencies between 0 and 1000 cycles/s, combining with whatever information that was legitimately there.

Two terms are widely used when discussing the sampling theorem: the Nyquist frequency and the Nyquist rate. Unfortunately, their meaning is not standardized. There are different terms used in analog to digital conversions and has been given here as Nyquistrate ($F_N < \frac{F_s}{2}$), Shannon Theorem ($F_{max} < \frac{F_s}{2}$), and Folding Frequency ($F_{fold} < \frac{F_s}{2}$).

To understand these, consider an analog signal composed of frequencies between DC and 3 kHz. To properly digitize this signal it must be sampled at 6,000 samples/s (6 kHz) or higher. Suppose if we choose to sample at 8,000 samples/s (8 kHz), the allowing frequencies between DC and 4 kHz to be properly represented. In this situation there are four important frequencies:

(1) The highest frequency in the signal, 3 kHz;
(2) Twice of this frequency, 6 kHz;
(3) The sampling rate should be 8 kHz; and
(4) One-half the sampling rate, 4 kHz.

Example 2.10

Right or wrong frequency conversion can be considered by the analog sinusoidal signal. The two signals of different frequencies $x_1(t) = \cos(2\pi 10)t$ and $x_2(t) = \cos(2\pi 50)t$ are sampled at a rate $F_s = 40$ Hz. Show that one signal which has not been sampled correctly is alias of the other signal, which has been sampled properly.

Solution

The corresponding **DT** signal at $F_s = 40$ Hz is

$$x_1(n) = \cos 2\pi \left(\frac{10}{40}\right) n = \cos\left(\frac{\pi}{2}\right) n \quad \text{and}$$

$$x_2(n) = \cos 2\pi \left(\frac{50}{40}\right) n = \cos\left(\frac{5\pi}{2}\right) n$$

$$\cos\left(\frac{5\pi}{2}\right) n = \cos\left(2\pi n + \left(\frac{\pi}{2}\right) n\right) = \cos\left(\frac{\pi}{2}\right) n.$$

Hence $x_1(n) = x_2(n)$, thus the sinusoidal signals are identical and consequently indistinguishable. If we are given the sampled value generated by $\cos\left(\frac{\pi}{2}\right) n$, there is an ambiguity as to whether these sampled values correspond to $x_1(t)$ or $x_2(t)$, when the two are sampled at a rate $F_s = 40$ Hz. We say that the frequency $F_2 = 50$ Hz is an alias of frequency F_1 of 10 Hz at the sampling rate of 40 samples/s. It is important to note that F_2 is not the only alias of F_1. In fact at the sampling rate of 40 samples/second, the frequency $F_3 = 90$ Hz is an alias of F_1, so the frequency $F_4 = 130$ Hz and so on. All the sinusoids $\cos 2\pi(F_1 + 40k)t,\ k = 1, 2, 3\ldots$ sampled at 40 samples/s yield identical values, consequently they are all alias of $F_1 = 10$ Hz.

2.8 Quantization Error

It is already discussed that the process of converting a **DT** continuous-amplitude signal into a digital signal by expressing each sample value as a finite (instead of infinite) number of digits, is called **quantization**. Accuracy of the signal representation is directly proportional to how many discrete levels are allowed to represent the magnitude of the signal.

The error introduced in representing the continuous-valued signal by a finite set of discrete value levels is called **quantization error** or **quantization noise.** Quantization error is the difference between the actual value and the quantized value.

Generally number of bits determines the resolution and the degree of quantization error or noise. It is common practice in signal processing to describe the ratio of the largest undistorted signal to the smallest undistorted signal that system can process. The ratio is called the **dynamic range**.

The noise voltage and the maximum signal-to-noise ratio (SNR) are both important measures of quantization error. They are related with the following expression.

$$V_{\text{noise (rms)}} = \frac{V_{\text{full scale}}\,(0.289)}{2^n} \tag{2.29}$$

$$\gamma_{\text{dB}} = 6.02n + 1.76, \tag{2.30}$$

where γ is the SNR in dBs and n is the number of bits.

It is possible to reduce level by increasing the number of bits. However, a large number of bits dictate higher cost, more storage space, and also longer processing time. Also, high resolution ADCs are slower. Thus there is a compromise.

The quantization process is illustrated with an example. Let us consider the **DT** signal,

$$x(n) = \begin{cases} 0.9^n, & n \geq 0 \\ 0, & n < 0 \end{cases}. \tag{2.31}$$

The signal in Equation (2.31) was obtained by sampling the analog exponential signal $x_a(t) = 0.9^t,\ t \geq 0$ with a sampling frequency $F_s = 1$ Hz (as shown in Figure 2.4(a)), the quantization is done by rounding, although it

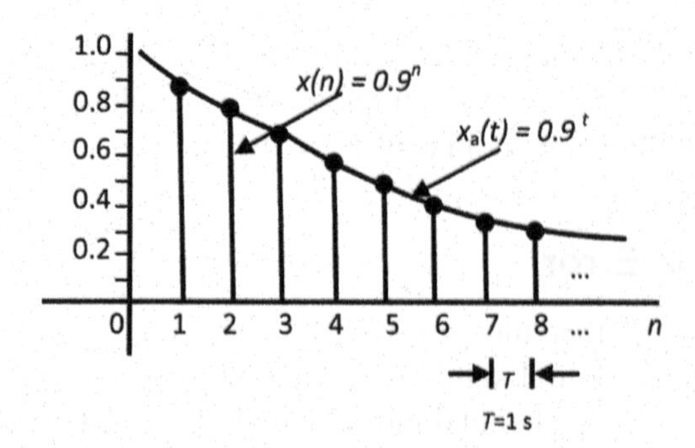

Figure 2.4 (a) Sampling with frequency of 1 Hz.

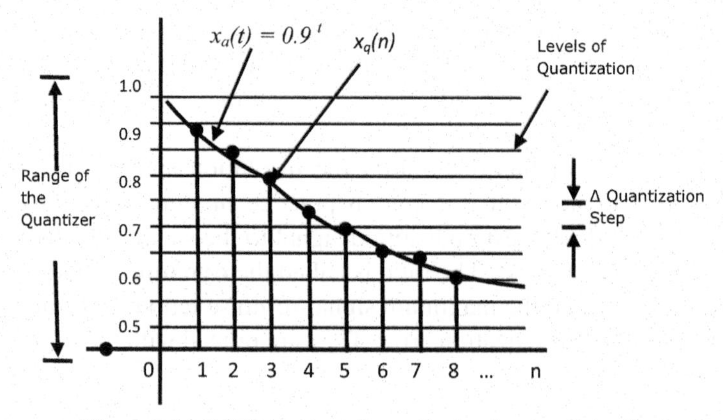

Figure 2.4 (b) Illustration of quantization by rounding off.

is easy to treat truncation. The rounding process is graphically illustrated in Figure 2.4(b).

It is to be noted that if $x_{\min}$ and $x_{\max}$ represent the minimum and maximum value of $x(n)$ and L is the number of quantization level, then

$$\Delta = \frac{x_{\max} - x_{\min}}{L-1} = \frac{\text{range}}{L-1}, \tag{2.32}$$

where $x_{\max} - x_{\min}$ is called the range of the signal, where L is the quantization level.

Example 2.11

Determine the root mean square noise (rms) quantization noise voltage for 8 and 12 bit systems when the signal to noise ratio is from 0 to 5 V. Find also the maximum signal to noise ratio for a 12-bit DSP system.

Solution

Applying Equation (2.32)

$$\Delta = \frac{x_{\max} - x_{\min}}{L-1} = \frac{\text{range}}{L-1}$$

$V_{\text{noise (rms)}} = \frac{(5\,\text{V})\,(0.289)}{2^8} = 5.64\,\text{mv}$ and $V_{\text{noise (rms)}} = \frac{(5\,\text{V})\,(0.289)}{2^{12}} = 353\,\mu\text{v}.$

Applying Equation (2.30)

$$\gamma\,(\text{dB}) = 6.02\,n + 1.76$$

$$\gamma = 6.02\,(12) + 1.76 \ = 74\text{ dB}$$

2.9 Further about DT Signals

A **DT** signal is not defined at instants between two successive samples. **DT** signals are defined only at discrete values of time. There are few **DT** signals which are much used in DSP, the way they are represented and types of them are mentioned below

2.9.1 Representing DT Signal

A **DT** signal *x*(*n*) is a function of an independent variable, i.e., an integer. A **DT** signal is not defined at instants between two successive samples. The following methods are in use to illustrate the digital signals

$$x(n) = \left\{\ldots, 2,\ 1,\ -2,\ -2,\ 3,\ \underset{\uparrow}{2},\ 2,\ -2,\ 1, \ldots\right\}$$

2.9.1.1 Graphical representation (Figure 2.5)

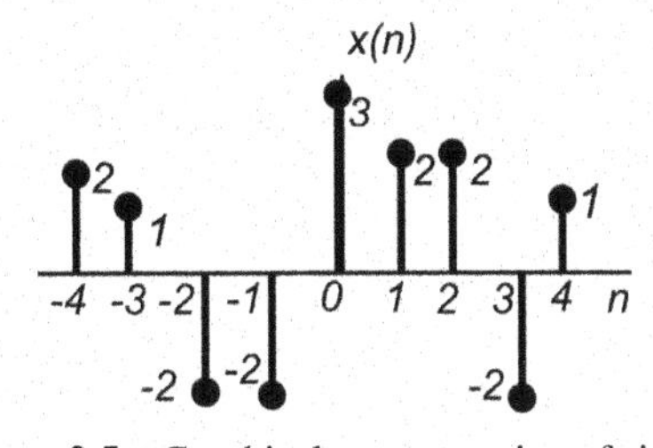

Figure 2.5 Graphical representation of signal.

2.9.1.2 Functional representation

$$x(n) = \begin{cases} 3 & \text{for } n = 0 \\ 1 & \text{for } n = -3,\ 4 \\ 2 & \text{for } n = 1,\ 2\ -4 \\ -2 & n = -1,\ -2,\ 3 \\ 0 & \text{elsewhere.} \end{cases}$$

2.9.1.3 Sequence representation

$$x(n) = \{\ldots, 2,\ 1, -2,\ -2,\ \underset{\uparrow}{3},\ 2,\ 2,\ -2,\ 1, \ldots\}$$

2.9.1.4 Tabular representation

n	-4	-3	-2	-1	0	1	2	3	$4\ldots$
$x(n)$	2	1	-2	-2	3	2	2	-2	1

2.10 Simple Manipulations

There are few operations which are required to be handled at different stages in digital signal processing. Any signal $y(n)$ on which operation has to be performed can be easily understood and calculated by substituting the different values of n in the original signal and getting the new values of $y(n)$. The following examples are given to elaborate this point.

2.10.1 Reflection/Folding/Flipping

The modification of time base is to replace the independent variable n by $-n$. The result of this of operation is a folding or a reflection of the signal about origin.

If $x(n) = \{-1, 1, 2, 2, 1, 1\}$, to find $x(-\underline{n})$, replacing $y(n) = x(-n)$, and substituting the value of n such as $\{1, 1, 2, 2, 1, -1\}$, because these are the values of n for which the original signal $x(n)$ exists (Figure 2.6).

2.10.2 Shifting (Advance and Delayed)

A signal $x(n)$ may be shifted in time by replacing independent variable n by $n - k$, where k is an integer. If k is a positive integer, the time shift results in a delay of the signal by k units of time. It means the new signal shifts to the right by k amount. If k is a negative integer, the time shift results in an advance of the signal by $|k|$ units in time. It means the new signal shifts to the left by k amount (Figure 2.7).

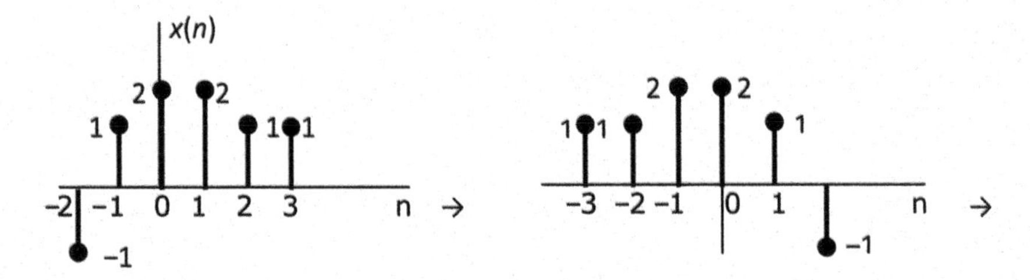

Figure 2.6 (a) Signal $x(n)$ and (b) Reflected signal $x(-n)$.

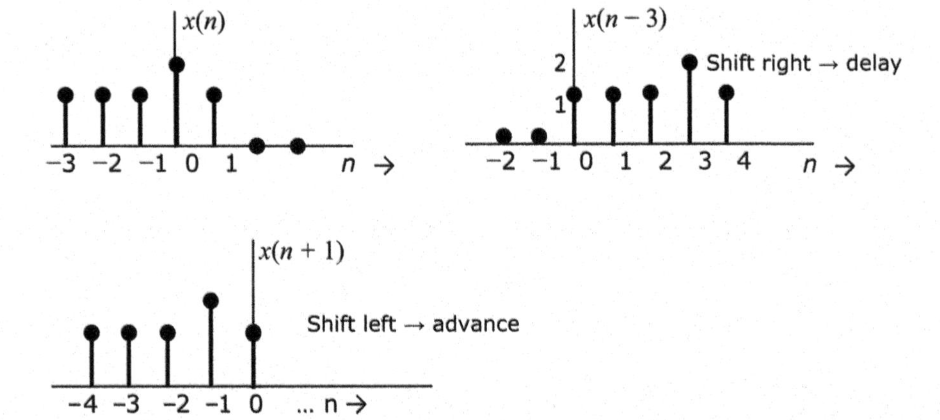

Figure 2.7 Original signal $x(n)$, delayed by 3 units and advanced by 1 unit.

2.10.3 Scaling (Time and Magnitude)

(i) Time scaling or down sampling:

In this case n is replaced by cn where c is an integer,

(ii) Magnitude Scaling:

Magnitude scaling is to multiply numbers with the every value of signals.

$$y(n) = Ax(n); \; -\infty < n < \infty.$$

Example 2.12

Consider signal $x(n) = \{\ldots 0, 2, 1, 2, -2, -2, 1, 0 \ldots\}$

Find (a) $x(2n)$

(b) $4x(n)$

Solution

(a) Let $y(n) = x(2n)$

$$y(0) = x(0) = 0; y(1) = x(2) = 1; y(2) = x(4) = -2;$$
$$y(3) = x(6) = 1;$$
$$y(-1) = x(-2) = 0;$$
$$y(-2) = x(-4) = 0$$
$$y(n) = x(2n) = \{\ldots 0, 1, -2, 1 \ldots\}.$$

The value of y gives the location and the value of $x(2n)$ gives its magnitude.
(b) $y(n) = 4x(n) = \{\ldots 0, 8, 4, 8, -8, -8, 4, 0 \ldots\}$.

2.10.4 Addition and Multiplication

The sum of two signals $x_1(n)$ and $x_2(n)$ is a signal $y(n)$, whose value at any instant is equal to sum of values of these signals at that instant,

$$y(n) = x_1(n) + x_2(n); -\infty < n < \infty.$$

The product of two signals is similarly defined on a sample-to-sample basis as

$$y(n) = x_1(n)x_2(n); -\infty < n < \infty.$$

Example 2.13

Consider the length – 7 sequences defined for $-3 \leq n \leq 3$:

$$x(n) = \{3, -2, 0, 1, 4, 5, 2\}$$
$$y(n) = \{0, 7, 1, -3, 4, 9, -2\}$$
$$w(n) = \{-5, 4, 3, 6, -5, 0, 1\}$$

Generate the following sequence:

(a) $u(n) = x(n) + y(n)$
(b) $v(n) = x(n) \cdot w(n)$
(c) $s(n) = y(n) - w(n)$
(d) $r(n) = 4.5y(n)$

Solution

(a) $u(n) = x(n) + y(n) = \{3, 5, 1, -2, 8, 14, 0\}$
(b) $v(n) = x(n) \cdot w(n) = \{-15, -8, 0, 6, -20, 0, 2\}$
(c) $s(n) = y(n) - w(n) = \{5, 3, -2, -9, 9, 9, -3\}$
(d) $r(n) = 4.5y(n) = \{0, 31.5, 4.5, -13.5, 19, 40.5, -9\}$.

Example 2.14

A **DT** signal $x[n]$ is shown in Figure. Sketch and label carefully each of the following signals.

(i) $x[n-1]\delta[n-3]$
(ii) $\frac{1}{2}x[n] + \frac{1}{2}(-1)^n x[n]$
(iii) $x[n^2]$

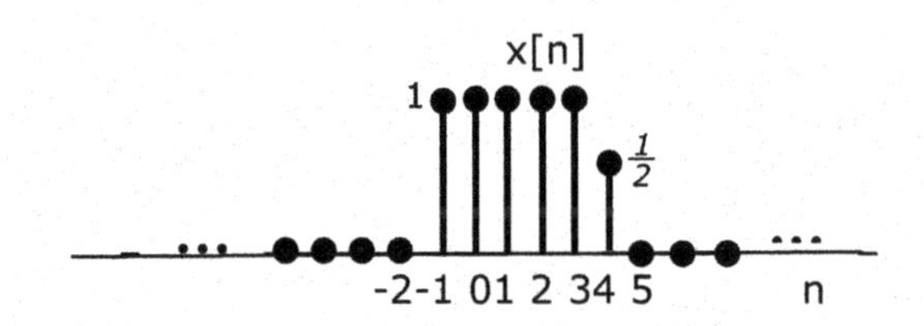

Solution

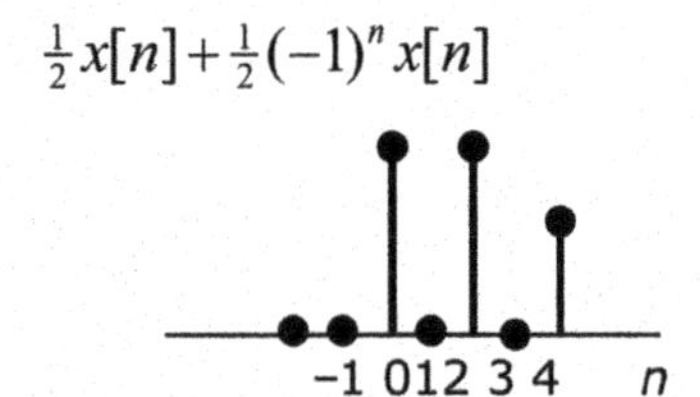

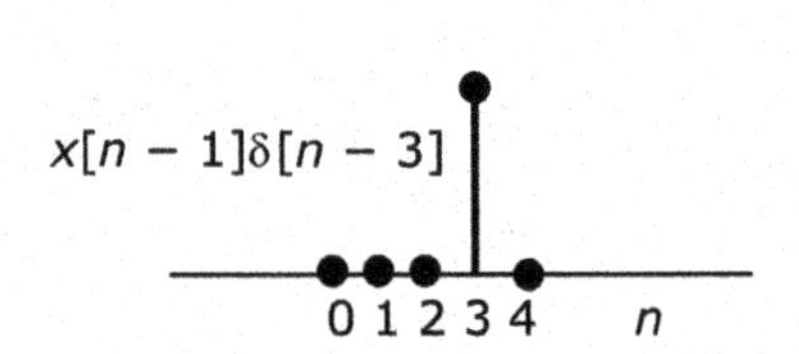

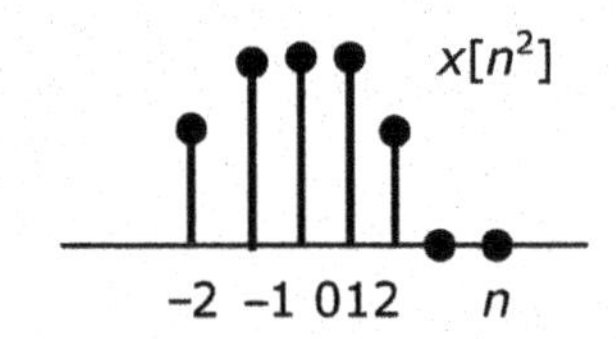

Example 2.15

Let $x[n]$ and $y[n]$ be given in Figures, respectively.

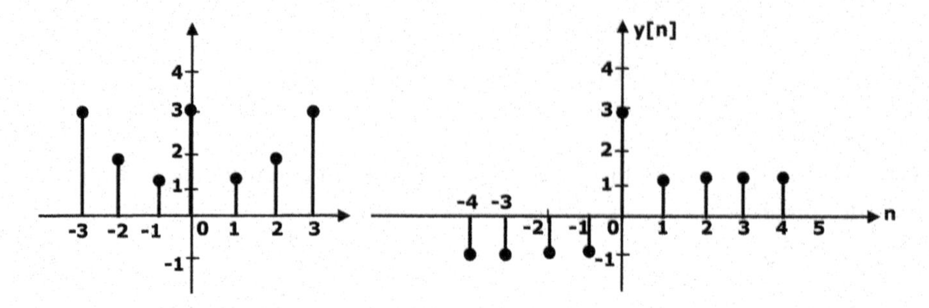

Carefully sketch the following signals.

(a) $y[2-2n]$
(b) $x[n-2]+y[n+2]$
(c) $x[2n]+y[n-4]$
(d) $x[n+2]y[n-2]$

Solution

(a) $y[2-2n]$

$$y[2-2n] = \begin{cases} 1, & n = 0, -1 \\ -1, & n = 2, 3 \\ 3, & n = 1 \end{cases}$$

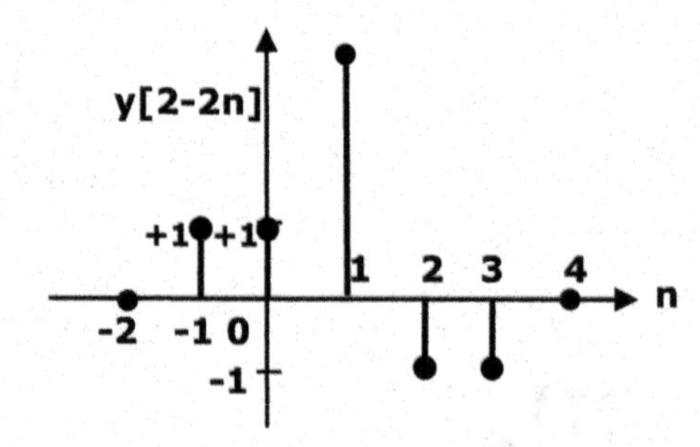

(b) $x[n-2]+y[n+2]$

$$x[n-2] = \begin{cases} 1, & n = 1,\ 3 \\ 2, & n = 0,\ 4 \\ 3, & n = -1, 2, 5 \\ 0, & n = \text{rest} \end{cases}$$

$$y[n+2] = \begin{cases} 1, & n = -1, 0, 1, 2 \\ -1, & n = -3, -4, -5, -6 \\ 3, & n = -2 \end{cases}$$

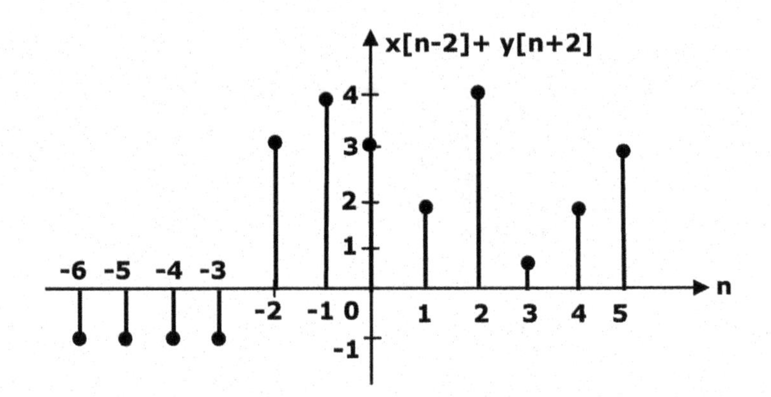

(c) $x[2n] + y[n-4]$

$$x[2n] = \begin{cases} 2, & n = \pm 1 \\ 0, & n = 3 \end{cases}$$

$$y[n-4] = \begin{cases} 1, & n = 5, 6, 7, 8 \\ -1, & n = 0, 1, 2, 3 \\ 3, & n = 4 \end{cases}$$

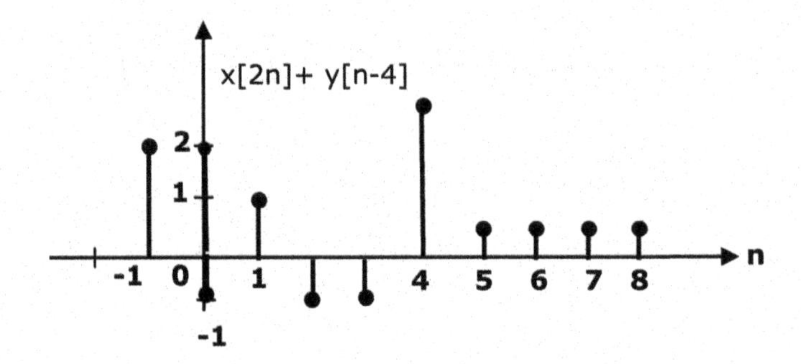

(d) $x[n+2]y[n-2]$

$$x[n+2] = \begin{cases} 1, & n = -3, \ -1 \\ 2, & n = -4, 0 \\ 3, & n = -5, \ -2, 1 \end{cases}$$

$$y[n-2] = \begin{cases} 1, & n = 3, 4, 5, 6 \\ -1, & n = 1, 0, -1, -2 \\ 3, & n = \ 2 \end{cases}$$

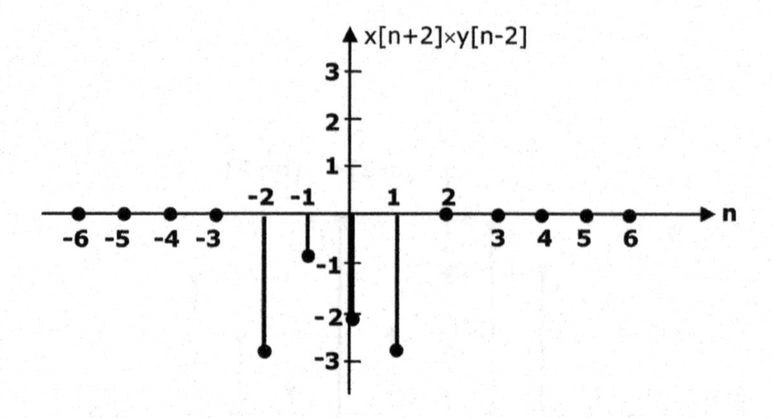

2.10.5 Even and Odd Signals

An even signal is that type of signal which exhibits symmetry in time domain. This type of signal is identical about the origin. Mathematically a signal $x(n)$ is called to be an even (symmetric) if $x(n) = x(-n)$ and odd (anti-symmetric) if $x(n) = -x(-n)$ (Figure 2.8).

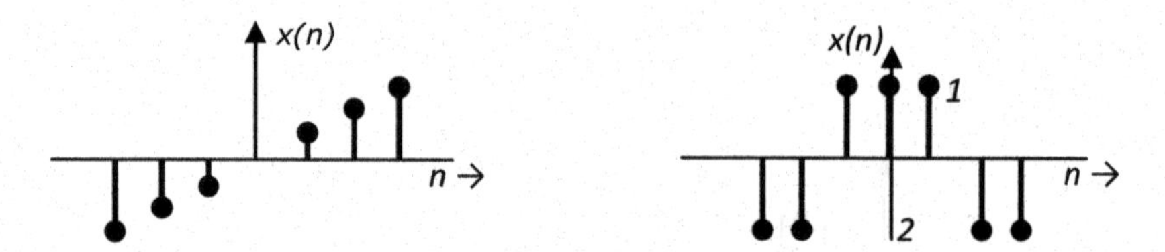

Figure 2.8 (a) Odd signal; (b) Even signal.

If a signal is neither even nor odd it can be decomposed into its odd $x_{\mathrm{o}}(n)$ and even $x_{\mathrm{e}}(n)$ components. The following important relationship is used to find the odd and even components. Using this $x(n) = x_{\mathrm{e}}(n) + x_{\mathrm{o}}(n)$ relationship the same original signal can be recovered.

$$x_{\mathrm{e}}(n) = \frac{1}{2}\left[x(n) + x(-n)\right] \tag{2.33}$$

$$x_{\mathrm{o}}(n) = \frac{1}{2}\left[x(n) - x(-n)\right]. \tag{2.34}$$

Example 2.16

A **DT** signal $x(n)$ is shown in Figure. Sketch and label carefully each of the following signals.

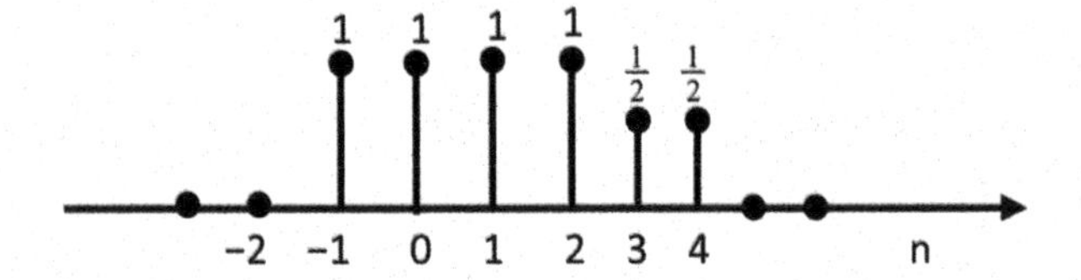

(a) Even part of $x(n)$
(b) Odd part of $x(n)$

Solution

$x(n) = \{\ldots 0, 1, \underset{\uparrow}{1}, 1, 1, \frac{1}{2}, \frac{1}{2}, 0 \ldots\}$

(a)

$$
\begin{aligned}
x_e(n) &= \frac{x(n) + x(-n)}{2}, \; x(-n) \\
&= \left\{\ldots 0, \frac{1}{2}, \frac{1}{2}, 1, 1, \underset{\uparrow}{1}, 1, 0, 0, 0, \ldots\right\} \\
&= \left\{\ldots 0, \frac{1}{4}, \frac{1}{4}, \frac{1}{2}, 1, \underset{\uparrow}{1}, 1, \frac{1}{2}, \frac{1}{4}, \frac{1}{4}, 0, \ldots\right\}
\end{aligned}
$$

(b)

$$
\begin{aligned}
x_o(n) &= \frac{x(n) + x(-n)}{2}, \\
&= \left\{\ldots 0, -\frac{1}{4}, -\frac{1}{4}, -\frac{1}{2}, 0, \underset{\uparrow}{0}, 0, \frac{1}{2}, \frac{1}{4}, \frac{1}{4}, 0, \ldots\right\},
\end{aligned}
$$

2.11 Energy and Power Signals

Signals may be classified as energy and power signals. Since we often think of a signal as a function of varying amplitude through time, it seems to reason that a good measurement of the strength of a signal would be area under the curve. However, this area may have negative part. The negative part does not have less strength than a positive signal of the same size. This suggests squaring the signal or taking it absolute value, then finding the area under the curve. It turns out that what we call the energy of a signal is the area under squared signal. The energy signal is one which has finite energy and zero average power. However, there are some signals which can neither be classified as neither energy signals nor power signals.

Table 2.1 A comparison of energy and power signal

	Energy Signal	Power Signal
1	Total normalized energy is finite and non-zero	Total normalized average power is finite and non-zero
2	The energy is obtained by $E = \sum_{n=-\infty}^{\infty} \vert x(n)\vert^2$	The average power is obtained by $P = \lim_{N\to\infty} \frac{1}{N} \sum_{n=-N/2}^{N/2} \vert x(n)\vert^2$
3	Non-periodic signals are energy signals	Practically periodic signals are power signals
4	These signals are time limited	These signals can exist over infinite time
5	Power of energy signal is zero	Energy of power signal is infinite

The energy signal is one which has finite energy and zero average power. Hence $x(n)$ is an energy signal, if $0 < E < \infty$ and $P = 0$. Where E is the energy and P is the power of the signal $x(n)$.

The power signal is one which has finite average power and infinite energy. Hence $x(n)$ is an power signal, if $0 < P < \infty$ and $E = \infty$.

The signal power P is equal to the mean square value of $x(n)$. Table 2.1 shows a comparison of energy and power signal.

However if the signal does not satisfy any of the above conditions, then it is neither energy nor power signal. For **DT** signals the area under the squared signal makes no sense, so we have to us another energy definition. We define energy as the sum of the squared magnitude of the samples.

For **DT** signals, we define energy as the sum of the squared magnitude of the samples.

$$E = \sum_{n=-\infty}^{\infty} |x(n)|^2 \tag{2.35}$$

In **DT** $x(n)$ is an energy signal, if $0 < E < \infty$ and $P = 0$. Where E is the energy and P is the power of the signal $x(n)$.

Squared values of $x(n)$ can be applied to both complex-values signal and real signals. Energy of signal may be finite or infinite. If E is finite then $x(n)$ is an energy signal.

The power in a **DT** signal $x(n)$ is given as:

$$P = \lim_{N\to\infty} \frac{1}{N} \sum_{n=-N/2}^{N/2} |x(n)|^2 \tag{2.36}$$

Example 2.17
Compute the energy of the length-N sequence

$$x(n) = \cos\left(\frac{2\,\pi kn}{N}\right) \quad 0 \le n \le N-1$$

Solution 2.17
$x(n) = \cos\left(\frac{2\,\pi kn}{N}\right)$ $0 \le n \le N-1$ $\cos 2\theta = 2\cos^2\theta - 1 \Rightarrow \cos^2\theta = \frac{1+\cos 2\theta}{2}$

$$E = \sum_{n=0}^{N-1} |x(n)|^2 = \sum_{n=0}^{N-1} \cos^2(2\,\pi\,k\,n/N) = \frac{1}{2}\sum_{n=0}^{N-1}(1+\cos(4\pi kn/N))$$
$$= \frac{N}{2} + \frac{1}{2}\sum_{n=0}^{N-1}\cos(4\pi kn/N)$$

$\sum_{n=0}^{N-1} a^n = \frac{1-a^N}{1-a}$ (1) and $e^{j\theta} = \cos\theta + j\sin\theta$ $\sum_{n=0}^{N-1} e^{j4\,\pi\,k\,n/N} = \frac{e^{j4\,\pi\,k}-1}{e^{j4\,\pi\,k\,/N}-1} = 0$ it follows that $E = \frac{N}{2}$.

Example 2.18
The angular frequency ω of the sinusoidal signal $x[n] = A\cos(\omega n + \varphi)$ satisfies the condition for $x[n]$ to be periodic. Determine the average power of $x[n]$.

Solution 2.18
Average power, $P = \frac{1}{N}\sum_{n=0}^{N-1} x^2(n)$, here $N = \frac{2\pi}{\omega}$

Then, $P = \frac{\omega}{2\pi}\sum_{n=0}^{\left(\frac{2\pi}{\omega}-1\right)} A^2\{\cos(\omega n + \varphi) + 1\}^2$
$P = \frac{A^2\omega}{4\pi}\{1 + \cos 2\varphi + 1 + \cos[2(2\pi - \omega) + \varphi]\}$
$P = \frac{A^2\omega}{4\pi}\{2 + \cos 2\varphi + \cos(2\varphi - 2\omega)\}$

2.12 Systems

A system may be defined as a set of elements or functional block which are connected together and produce an output in response to the input signal. The response or output of the system depends upon transfer function of the system.

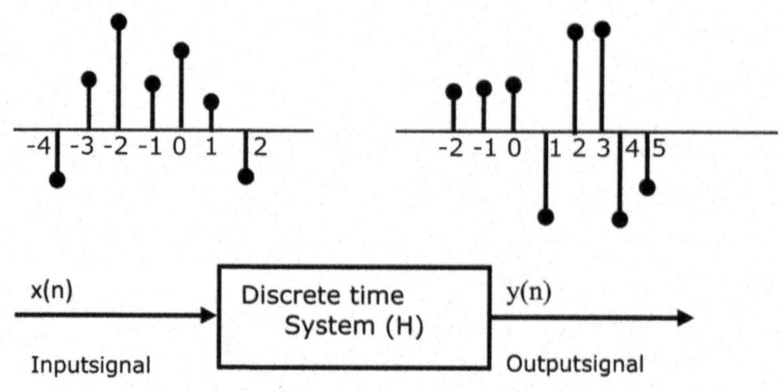

Figure 2.9 Representation of a discrete time system.

Mathematical functional relationship between and output may be written as $y(t) = f(x,t)$.

2.12.1 DT Systems

A **DT** system is a device that operates on **DT** signals (input), according to some rules, to produce another **DT** signal-(output or response) of the systems (Figure 2.9).

2.13 System's Representation

Discrete time system can be representation by two ways:

(a) Difference Equations
(b) Block diagram

2.13.1 Symbol used for DT Systems

The operation of a **DT** system may be described simply by drawing a block schematic. We use different building block to form a complete schematic.

(a) Adder
(b) Constant Multiplier
(c) Signal Multiplier
(d) Unit delay
(e) Unit advance

2.13.2 An Adder

Addition operation is memory-less (Figure 2.10(a)).

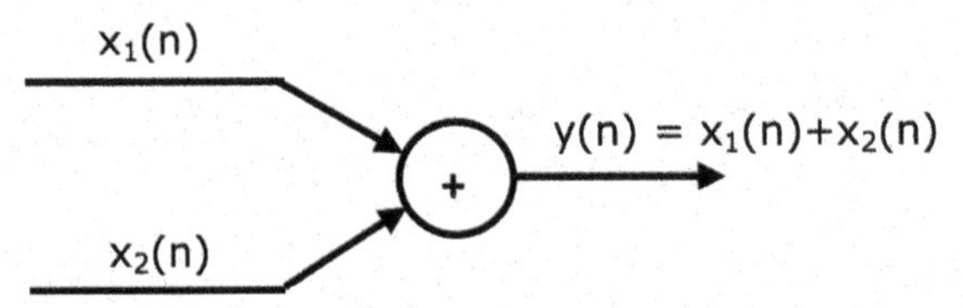

Figure 2.10 (a) Graphical representation of adder.

2.13.3 A Constant Multiplier

Simply represent by applying a scale factor on input. It is also memory-less (Figure 2.10(b)).

x(n) c y(n) = cx(n)

Figure 2.10 (b) Graphical representation of a constant multiplier.

2.13.4 A Signalmultiplier

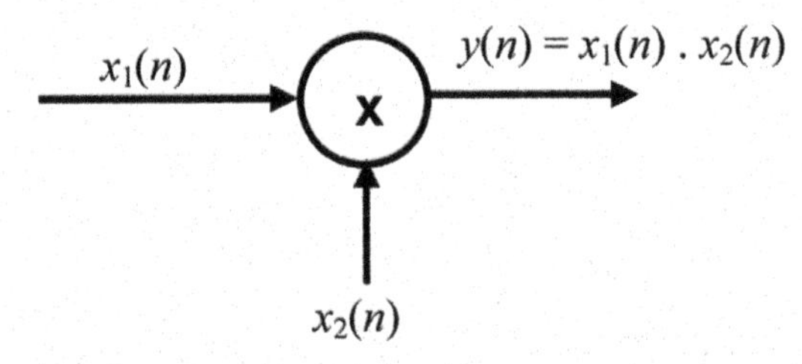

Figure 2.10 (c) Graphical representation of a multiplier.

2.13.5 Unit Delay Element

It is the system that simply delays the signal passing through it by one sample. If input signal is $x(n)$, the output is $x(n-1)$. The sample $x(n-1)$ is stored in memory at **DT** $n-1$ and can be recalled at time n to form $y(n) = x(n-1)$ (Figure 10(d)).

The symbol of unit delay $= z^{-1}$

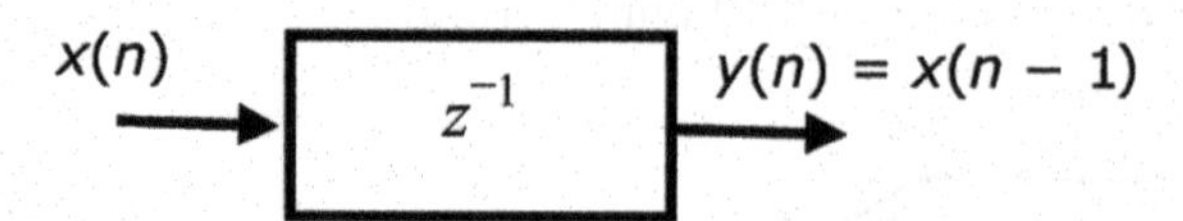

Figure 2.10 (d) Unit delay element.

2.13.6 Unit Advanced Element

A unit advance moves the input $x(n)$ ahead by one sample in time to yield $x(n + 1)$. This advancement is denoted by z (Figure 10(e)).

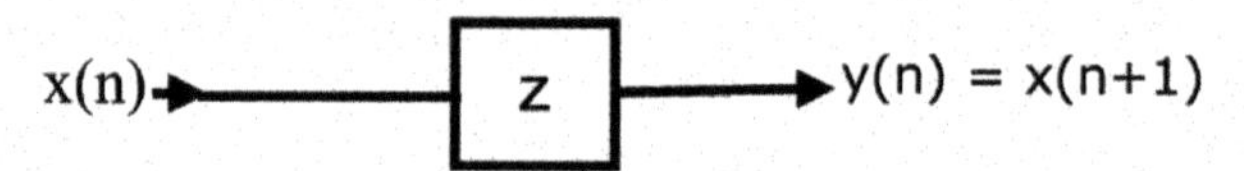

Figure 2.10 (e) Unit delay element.

Example 2.19

Sketch the block diagram

$y(n) = 1/2\, y\,(n - 1) + 1/4\, x(n) + 1/4\, x(n - 1)$ where $x(n)$ is the input $y(n)$ is the output.

Solution

$$y(n) = 0.25\, x(n) + 0.25\, x(n - 1) + 0.5\, y(n - 1)$$

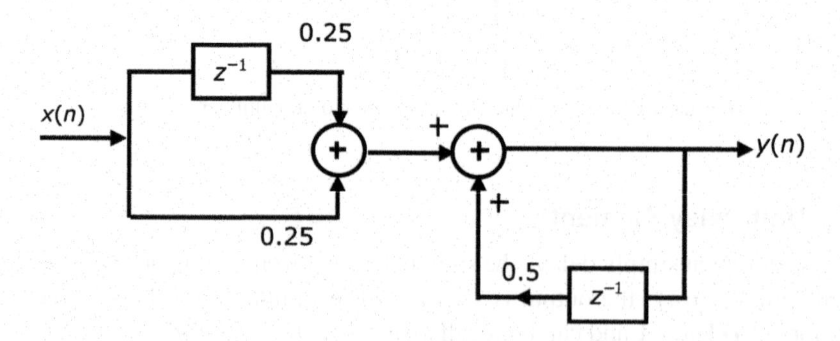

or

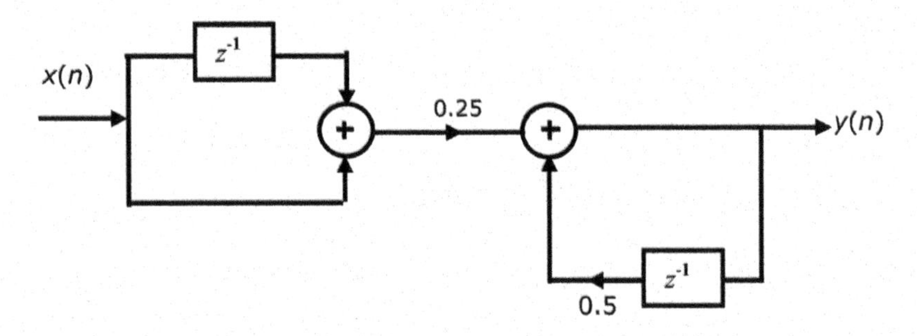

Example 2.20
A **DT** system is realized by the structure shown in Figure. Determine a realized for its inverse system, that is, the system which produces $x(n)$ as an output when $y(n)$ is used as an input.

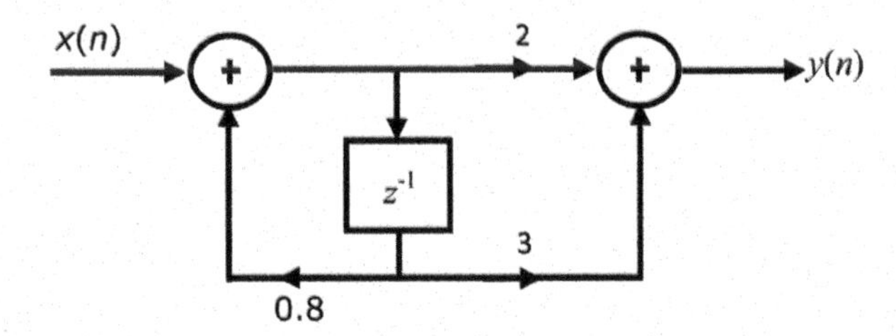

Solution
The inverse system is characterized by the difference equation

$$x(n) = -1.5x(n-1) + \frac{1}{2}y(n) - 0.4y(n-1)$$

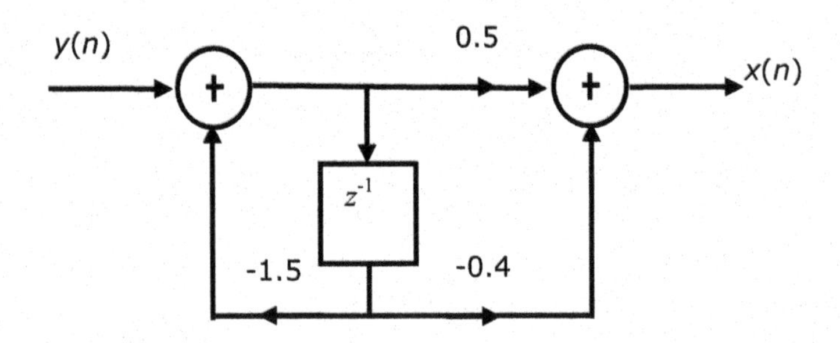

2.14 System's Classification

In designing the systems, the general characteristic of the systems have to be considered. A number of properties or categories that can be used to develop general characteristics of the system are explained here.

2.14.1 Static versus Dynamic Systems

A **DT** system is called static or memory-less if its output at any instant n depends at most on the input sample at the same time, but not on past or future samples of the input. In any other case, the system is said to be dynamic or to have memory.

If the output of a system at time n is completely determined by the input samples in the interval from $n - N$ to n ($N \geq 0$), the system is said to have memory of duration N. If $N = 0$, the system is static. If $0 < N < \infty$, the system is said to have finite memory, whereas if $N = \infty$, the system is said to have infinite memory.

The systems described by the following input–output equations

$$y(n) = ax(n) \tag{2.37}$$

$$y(n) = nx(n) + bx^3(n) \tag{2.38}$$

are both static or memory-less. Note that there is no need to have stored any of the past inputs or outputs in order to compute the present output. On the other hand, the systems described by the following input–output relations are dynamic systems or systems with memory.

$$y(n) = x(n) + 3x(n-1) \tag{2.39}$$

$$y(n) = \sum_{k=0}^{n} x(n-k) \tag{2.40}$$

$$y(n) = \sum_{k=0}^{\infty} x(n-k). \tag{2.41}$$

The systems described by Equations (2.39) and (2.40) have finite memory, whereas the system described by Equation (2.41) has infinite memory.

2.14.2 Time-Invariant versus Time-Variant System

The system is said to be time invariant if its input–output characteristics do not change with time.

2.14.2.1 Method to workout for time-invariant and time-variant system

Consider the following system

$$y(n) = H[x(n)]. \tag{2.42}$$

If the input signal is delayed by k unit in the function only and again delayed by k unit to overall system irrespective that it is a function or not. If $y(n, k) = y(n - k)$ it means that the characteristics of system do not change with time, and the system is time invariant.

$$x(n) \xrightarrow{H} y(n) \quad \text{relaxed}$$

$$x(n-k) \xrightarrow{H} y(n-k) \quad \text{shift.} \tag{2.43}$$

$y(n, k)$ means the delay is to be given in function only, while $y(n-k)$ means that where ever n is existing it has to be replaced by $n - k$ unit. If $y(n, k) = y(n - k)$ then system is called time invariant.

Example 2.21

Determine if the shown in Figure 2.11 are time invariant or time variant

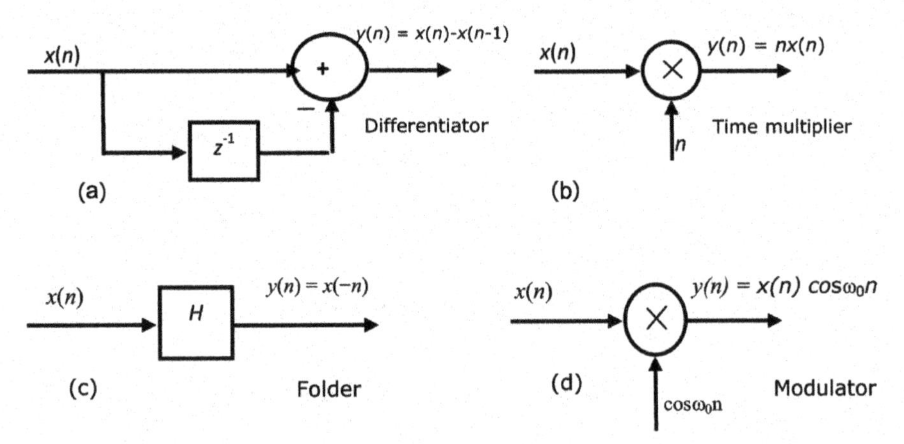

Figure 2.11 Time-invariant (a) and some time-variant systems (b)–(d).

Solution

(a) This system is described by the input–output equations

$$y(n) = H[x(n)] = x(n) - x(n-1). \tag{1}$$

Now if the input is delayed by k units in time and applied to the system, it is clear from the block diagram that the output will be

$$y(n,k) = x(n-k) - x(n-k-1) \tag{2}$$

On the other hand, from Equation (1) we note that if we delay $y(n)$ by k units in time, we obtain

$$y(n-k) = x(n-k) - x(n-k-1) \tag{3}$$

Since the right-hand sides of Equations (2) and (3) are identical, it follows that $y(n, k) = y(n - k)$. Therefore, the system is time invariant.

(b) The input–out equation for this system is

$$y(n) = H[x(n)] = nx(n). \tag{4}$$

The response of this system to $x(n - k)$ is

$$y(n,k) = nx(n-k). \tag{5}$$

Now if we delay $y(n)$ in Equation (4) by k units in time, we obtain

$$y(n-k) = (n-k)x(n-k) = nx(n-k) - kx(n-k). \tag{6}$$

This system is time variant, since $y(n, k) \neq y(n - k)$.

(c) This system is described by the input–output relation

$$y(n) = H[x(n)] = x(-n) \tag{7}$$

The response of this system to

$$\begin{aligned} x(n-k) \text{ is } y(n,k) &= H[x(n-k)] \\ &= x(-n-k) \end{aligned} \tag{8}$$

Now, if we delay the output $y(n)$, as given by Equation (7), by k units in time, the result will be

$$y(n-k) = x(-n+k) \tag{9}$$

Since $y(n, k) \neq y(n - k)$, the system is time variant.

(d) The input–output equation for this system is

$$y(n) = x(n)\cos\omega_0 n \tag{10}$$

The response of this system to $x(n - k)$ is $y(n,k) = x(n-k)\cos\omega_0 n$

$$\tag{11}$$

If the expression in Equation (10) is delayed by k units and the result is compared to Equation (11), it is evident that the system is time variant.

Example 2.22

Check the **DT** system for time-invariance which is described by the following difference equation.

$$y(n) = 4nx(n)$$

Solution 2.22

The response to a delayed input is

$$y(n, k) = 4nx(n - k).$$

The delayed response will be

$$y(n - k) = 4(n - k)x(n - k)$$

it is clear that both responses are not equal, i.e.,

$$y(n, k) \neq y(n - k)$$

Therefore, the given **DT** system $y(n) = 4nx(n)$ is not time invariant. It is time varying system.

Example 2.23

Check the **DT** system for time-invariance which is described by the following difference equation.

$$y(n) = ax(n - 1) + bx(n - 2)$$

Solution 2.23

The response to a delayed input is

$$y(n, k) = ax(n - k - 1) + bx(n - k - 2).$$

The delayed response will be

$$y(n - k) = ax((n - k) - 1) + bx((n - k) - 2).$$

It is clear that both responses are not equal, i.e.,

$$y(n, k) = y(n - k)$$

Therefore, the given **DT** system $y(n) = ax(n - 1) + bx(n - 2)$ is time invariant.

2.14.3 Linear versus Non-linear System

2.14.3.1 Linear system

A linear system is that which satisfies the properties of superposition theorem. The response of system to a weighted sum of signal is equal to the corresponding weighted sum of the response of each individual input signals.

$$H[a_1x_1(n) + a_2x_2(n)] = a_1H[x_1(n)] + a_2H[x_2(n)] \tag{2.23}$$

2.14.3.2 Non-linear system

If a system produces a nonzero output with a zero input, the system may be either non-relaxed or non-linear. If a relaxed system does not satisfy the superposition principle it is non-linear (Figure 2.12).

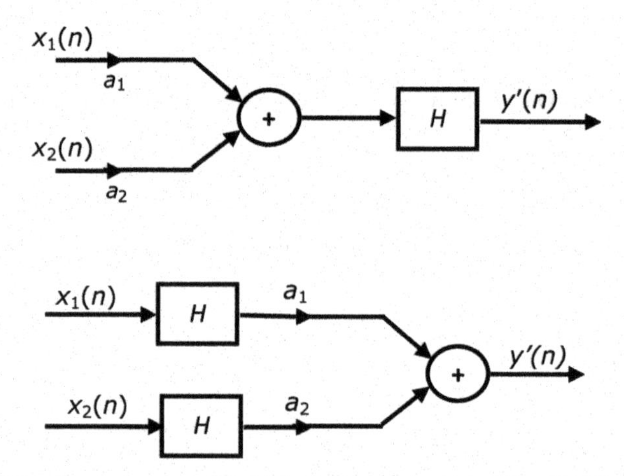

Figure 2.12 Graphical representation of the superposition principle, H is linear if and only if $y(n) = y'(n)$.

Example 2.24

Determine if the systems described by the following input–output equations are linear or non-linear.

(a) $y(n) = nx(n)$ **(b)** $y(n) = x(n^2)$ **(c)** $y(n) = x^2(n)$
(d) $y(n) = Ax(n) + B$ **(e)** $y(n) = e^{x(n)}$

Solution 2.24

(a) For two input sequences $x_1(n)$ and $x_2(n)$, the corresponding outputs are

$$y_1(n) = nx_1(n) \quad (1)$$
$$y_2(n) = nx_2(n)$$

A linear combination of the two input sequence results in the output

$$y_3(n) = H[a_1x_1(n) + a_2x_2(n)] = n[a_1x_1(n) + a_2x_2(n)] \quad (2)$$
$$= a_1nx_1(n) + a_2nx_2(n)$$

On the other hand, a linear combination of the two outputs in Equation (2) results in the output

$$a_1y_1(n) + a_2y_2(n) = a_1nx_1(n) + a_2nx_2(n) \quad (3)$$

Since the right-hand sides of Equations (2) and (3) are identical, the system is linear.

(b) As in part (a), we find the response of the system to two separate input signals $x_1(n)$ and $x_2(n)$. The result is

$$y_1(n) = x_1(n^2) \quad (4)$$

$$y_2(n) = x_2(n^2)$$

The output of the system to a linear combination of $x_1(n)$ and $x_2(n)$ is

$$y_3(n) = H[a_1x_1(n) + a_2x_2(n)] = a_1x_1(n^2) + a_2x_2(n^2) \quad (5)$$

Finally, a linear combination of the two outputs in (2.21) yields

$$a_1y_1(n) + a_2y_2(n) = a_1x_1(n^2) + a_2x_2(n^2) \quad (6)$$

By comparing Equation (5) with Equation (6), we conclude that the system is linear.

(c) The output of the system is the square of the input. (Electronic devices that have such an input–output characteristic and are called square-law devices) From our previous discussion it is clear that such a system is memory-less.

To illustrate that this system is nonlinear, the responses of the system to two separate input signals are

$$y_1(n) = x_1^2(n); y_2(n) = x_2^2(n) \quad (7)$$

The response of the system to a linear combination of these two input signals is

$$\begin{aligned} y_3(n) &= H\left[a_1x_1(n) + a_2x_2(n)\right] \\ &= \left[a_1x_1(n) + a_2x_2(n)\right]^2 \\ &= a_1^2x_1^2(n) + 2\,a_1\,a_2\,x_1(n)\,x_2(n) + a_2^2x_2^2(n). \end{aligned} \tag{8}$$

On the other hand, if the system is linear, it would produce a linear combination of the two outputs in Equation (7), namely,

$$a_1y_1(n) + a_2y_2(n) = a_1x_1^2(n) + a_2x_2^2(n) \tag{9}$$

Since the actual output of the system, as given by Equation (8), is not equal to Equation (9), the system is non-linear.

(d) Assuming that the system is excited by $x_1(n)$ and $x_2(n)$ separately, we obtain the corresponding outputs

$$y_1(n) = Ax_1(n) + B \tag{10}$$

$$y_2(n) = Ax_2(n) + B$$

A linear combination of $x_1(n)$ and $x_2(n)$ produces the output

$$\begin{aligned} y_3(n) &= H[a_1x_1(n) + a_2x_2(n)] \\ &= A[a_1x_1(n) + a_2x_2(n)] + B \\ &= Aa_1x_1(n) + a_2Ax_2(n) + B. \end{aligned} \tag{11}$$

On the other hand, if the system were linear, its output to the linear combination of $x_1(n)$ and $x_2(n)$ would be a linear combination of $y_1(n)$ and $y_2(n)$, that is,

$$a_1y_1(n) + a_2y_2(n) = a_1Ax_1(n) + a_1B + a_2Ax_2(n) + a_2B. \tag{12}$$

Clearly, Equations (11) and (12) are different and hence the system fails to satisfy the linearity test.

The reason that this system fails to satisfy the linearity test is not that the system is nonlinear (in fact, the system is described by a linear equation) but it is the presence of the constant B. Consequently, the output depends on both the input excitation and on the parameter $B \neq 0$, Hence, for $B \neq 0$, the system is not relaxed. If we set $B = 0$, the system is now relaxed and the linearity test is satisfied.

(e) Note that the system described by the input–output equation

$$y(n) = e^{x(n)}$$

is relaxed. If $x(n) = 0$, we find that $y(n) = 1$. This is an indication that the system is non-linear. This, in fact, is the conclusion reached when the linearity test, as described above, is applied.

Example 2.25

Determine if the system described by the following input–output equations is linear or non-linear.

$$y(n) = x^2(n) - x(n-1)x(n+1)$$

Solution 2.25

For two input sequences $x_1(n)$ and $x_2(n)$, the corresponding outputs are $y_1(n)$ and $y_2(n)$

$$\begin{aligned} y_1(n) &= x_1^2(n) - x_1(n-1)x_1(n+1) \\ y_2(n) &= x_2^2(n) - x_2(n-1)x_2(n+1) \qquad (1) \\ a_1y_1(n) + a_2y_2(n) &= [a_1x_1^2(n) - x_1(n-1)a_1x_1(n+1)] \\ &\quad + [a_2x_2^2(n) - x_2(n-1)a_2x_2(n+1)]. \end{aligned}$$

A linear combination of the two input sequence result in the output

$$\begin{aligned} y_3(n) &= H[a_1x_1(n) + a_2x_2(n)] \\ &= [(a_1x_1(n) + a_2x_2(n)]^2 - [(a_1x_1(n-1) \\ &\quad a_2x_2(n-1)] + [(a_1x_1(n+1)a_2x_2(n+1)] \qquad (2) \\ &= a_1^2[(x_1^2(n) + x_1(n-1)x_1(n+1)] + a_2^2[x_2^2(n) \\ &\quad - x_2(n-1)x_2(n+1)] + a_1a_2[2x_1(n)x_2(n) \\ &\quad - x_1(n-1)x_2(n+1) - x_1(n+1)x_2(n-1)]. \end{aligned}$$

Since the right-hand sides of Equations (1) and (2) are not identical, the system is non-linear.

Example 2.26

A **DT** system is represented by following difference equation in which $x(n)$ is input $y(n)$ is output. Determine if the system described by the input–output equations is linear or non-linear.

$$y(n) = 5\,y^2(n-1) - nx(n) + 4x(n-1) - 2x(n+1)$$

Solution 2.26

The given expression is $y(n) = 5\,y^2\,(n-1) - nx(n) + 4x(n-1) - 2x(n+1)$

It may be noted that real condition for linearity is $H[ax(n)] = aH[x(n)]$

$$H[ax(n)] = ay(n) = 5a^2y^2(n-1) - anx(n) + 4ax(n-1) - 2ax(n+1)$$

$$aH[x(n)] = a[y(n)] = 5ay^2(n-1) - an\,x(n) + 4ax(n-1) - 2ax(n+1)$$

from above it is clear that $H[ax(n)] \neq aH[x(n)]$. System is non-linear.

2.14.4 Causal versus Non-Causal System

A system is said to be causal if the output of the system at any time n depends only on present and past inputs [i.e., $x(n)$, $x(n-1)$, $x(n-2)\ldots$], but does not depends on future inputs [$x(n+1)$, $x(n+2)\ldots$].

$$y(n) = F[x(n), x(n-1), x(n-2)\ldots]. \tag{2.24}$$

If a system does not satisfy this definition, it is called non-causal.

Example 2.27

Determine if the systems described by the following input–output equations are causal or non-causal.

(a) $y(n) = x(n) - x(n-1)$
(b) $y(n) = \sum_{k=-\infty}^{n} x(k)$
(c) $y(n) = ax(n)$
(d) $y(n) = x(n) + 3x(n+10)$
(e) $y(n) = x(n^2)$
(f) $y(n) = x(2n)$
(g) $y(n) = x(-n)$

Solution 2.27

The systems described in parts (a), (b), and (c) are clearly causal, since the output depends only on the present and past inputs. On the other hand, the systems in parts (d), (e), and (f) are clearly non-casual, since the output depends on future values of the input.

The system in (g) is also non-causal, as we note by selecting, for example, $n = -1$, which yields $y(1) = x(1)$. Thus, the output at $n = -1$ depends on the input at $n = 1$, which is two units of time into the future.

Example 2.28

A **DT** system is represented by following difference equation in which $x(i)$ is input $y(n)$ is output. Determine if the system described by the input–output equations is causal or non-causal.

$$y(n) = 10y^2(n-1) - nx(n) + 4x(n-1) - 5x(n+1).$$

Solution 2.28

It may be noted that the required condition for causality is that the output of a causal system must be dependent only on the present and past values of the input. From the given equation, it is obvious that output $y(n)$ is dependent on future sample value $x(n + 1)$.

2.14.5 Stable versus Un-Stable System

An arbitrary relaxed system is said to be bounded input-bounded output (BIBO) stable if and only if every bounded input produces a bounded output.

By definition, a signal $x(n)$ is bounded if there exists a member M such that

$$jx(n)j \leq M \quad \text{for all } n. \tag{2.25}$$

Hence a system is BIBO stable if, for a number R

$$jy(n)j \leq R \quad \text{for all } x(n) \tag{2.26}$$

If for some bounded input sequence $x(n)$, the output is unbounded (infinite), the system is unstable. A system is stable if the output remains bounded for any bounded input.

Example 2.29

Check whether the system $y(n) = ax^2(n)$ is BIBO stable or not.

Solution 2.29

(a) The given expression is $y(n) = ax^2(n)$

If $x(n) = \delta(n)$, then $y(n) = h(n)$. Thus the impulse response is given by $h(n) = a\delta^2(n)$ when $n = 0$, $h(0) = a\delta^2(0) = a$ and when $n = 1$, $h(1) = \delta^2(1) = 0$.

In general we have $h(n) = a$ when $n = 0$, $h(n) = 0$ when $n \neq 0$ we know the necessary and sufficient condition for BIBO stability is expressed as

$\sum_{n=0}^{\infty} |h(k)| < \infty$. Here we have $\sum_{n=0}^{\infty} | h(k)| = | h(0)| + | h(1)| + | h(2)|$ $+ \ldots | h(k)| + \ldots = |a|$.

Therefore, we conclude that the given system is BIBO stable only if $\alpha < \infty$.

2.15 Problems and Solutions

Problem 2.1

Determine which of the following sinusoids are periodic and compute their fundamental period.

$$\text{(a) } \cos 3\pi n; \quad \text{(b) } \sin 3n; \quad \text{and} \quad \text{(c) } \sin\left(\pi\frac{62n}{8}\right)$$

Solution 2.1

(a) $f = \frac{3\pi}{2\pi} = \frac{3}{2} \Rightarrow$ periodic with $N_p = 2$
$\cos 3\pi(n + N) = \cos 3\pi(n + 2)$
$= \cos 3\pi n \cos 2\pi - \sin 3\pi n \sin 2\pi = \cos 3\pi n$

(b) $f = \frac{3}{2\pi} \Rightarrow$ non-periodic

(c) $f = \frac{62\pi}{8}\left(\frac{1}{2\pi}\right) = \frac{31}{8} \Rightarrow$ periodic with $N_p = 8$
$\sin\left(\pi\frac{62n}{8}\right) = \sin\left(\pi\frac{62}{8}(n+8)\right) = \sin\left(\pi\frac{62n}{8}\right)$
$\Rightarrow$ periodic with $N_p = 8$.

Problem 2.2

Determine whether or not each of the following signals are non-periodic. In case a signal is periodic, specify its fundamental period.

(a) $x_a(t) = 3\cos(5t + \pi/4)$
(b) $x(n) = 3\cos(5n + \pi/4)$
(c) $x(n) = 2\exp[j(n/5 - \pi)]$
(d) $x(n) = \cos(n/8)\cos(\pi n/8)$
(e) $x(n) = \cos(\pi n/3) - \sin(\pi n/4) + 3\cos(\pi n/4 + \pi/3)$

Solution 2.2

(a) Periodic with period $T_p = 2\pi/5$
(b) $f = 5/2\pi \Rightarrow$ non-periodic

(c) $f = 1/10\pi \Rightarrow$ non-periodic
(d) cos(n/8) is non-periodic; cos(πn/8) in periodic. Their product is no-periodic
(e) cos(πn/3) is periodic with period N_p = 6
sin (πn/4) is periodic with period N_p = 8

$$\cos\left(\frac{\pi n}{4} + \frac{\pi}{3}\right) \text{ is periodic with period } N_p = 8.$$

Therefore, $x(n)$ is periodic with period N_p = 96 (96 = least common multiple of 6, 8, 8).

Problem 2.3

Consider the following analog sinusoidal signal $x_a(t) = 3\sin(100\pi t)$

(a) Sketch the signal $x_a(t)$ for $0 \leq$ to $\leq$30 ms.
(b) The signal $x_a(t)$ is sampled with a sampling rate F_s = 300 samples. Determine the frequency of the **DT** signal $x(n) = x_a(nT)$, $T = 1/F_s$, and show that it is periodic.
(c) Compute the sample values in one period of $x(n)$. Sketch $x(n)$ on the same diagram with $x_a(t)$. What is the period of the **DT** signal in milli-seconds?
(d) Can you find a sampling rate F_s such that the signal $x(n)$ reaches its peak value of 3? What is the minimum F_s suitable for this task?

Solution 2.3

(a)

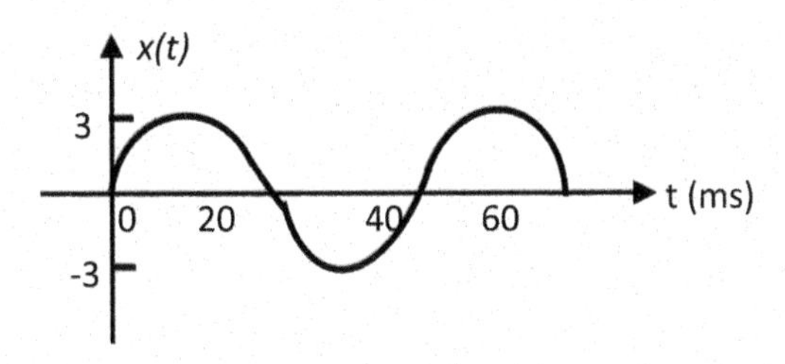

(b)

$$x(n) = x_a(nT) = x_a(n/F_s) = 3\sin\left(\frac{\pi}{3}n\right) \Rightarrow f = \frac{1}{2\pi}\left(\frac{\pi}{3}\right) = \frac{1}{6}, \; N_p = 6$$

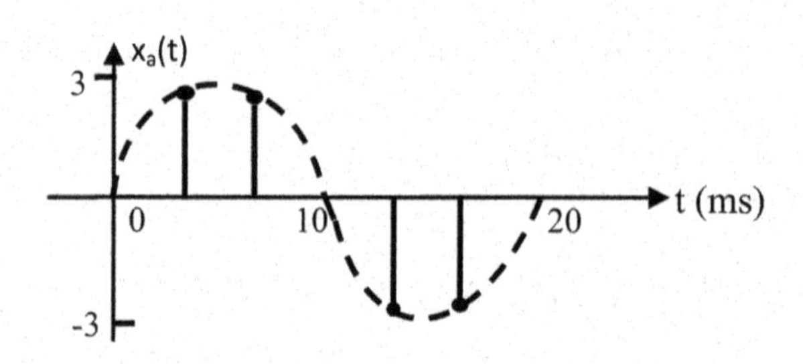

$$x(n) = \left\{0, \frac{3\sqrt{3}}{2}, \frac{3\sqrt{3}}{2}, 0 - \frac{3\sqrt{3}}{2}, -\frac{3\sqrt{3}}{2}\right\}, \quad N_\text{p} = 6$$

(d)

$$\text{Yes.} \quad x(1) = 3 = 3\sin\left(\frac{100\pi}{F_\text{s}}\right) \Rightarrow F_\text{s} = 200 \text{ samples/s.}$$

Problem 2.4

A CT sinusoid x_a(t) with fundamental period $T_\text{p} = 1/F_\text{s}$ is sampled at a rate $F_\text{s} = 1/T$ to produce a **DT** sinusoid $x(n) = x_\text{a}(nT)$.

(a) Show that $x(n)$ is periodic if $T/T_\text{p} = k/N$ (i.e., T/T_p is a rational number).
(b) If $x(n)$ is periodic, what is the fundamental period T_p in seconds?
(c) Explain the statement: $x(n)$ is periodic if its fundamental period T_p in seconds, is equal to an integer number of periods of $x_\text{a}(t)$.

Solution 2.4

(a) $x(n) = A\cos(2\pi F_\text{o} n/F_\text{s} + \theta) = A\cos[2\pi(T/T_\text{p})n + \theta]$
But $T/T_\text{p} = f \Rightarrow x(n)$ is periodic if f is rational
(b) If $x(n)$ is periodic, then $f = k/N$ where N is the period.
Then, $T_\text{discrete} = \left(\frac{k}{f}T\right) = k\left(\frac{T_\text{p}}{T}\right)T = kT_\text{p}$. Thus, it takes k periods (kT_p) of the analog signal to make1 period (T_d) of the discrete signal.
(c) $T_\text{d} = kT_\text{p} \Rightarrow NT = kT_\text{p} \Rightarrow f = k/N = T/T_\text{p} \Rightarrow f$ is rational $\Rightarrow x(n)$ is periodic.

Problem 2.5

An analog signal contains frequency up to 10 kHz.

(a) What range of sampling frequencies will allow exact reconstruction of this signal from its sample?
(b) Suppose that we sample this signal with a sampling frequency $F_s = 8$ kHz. Examine what will happen to the frequency $F_1 = 5$ kHz
(c) Repeat part (b) for a frequency $F_2 = 9$ kHz.

Solution 2.5

(a) $F_{max} = 10$ kHz $\Rightarrow F_s = 2F_{max} = 20$ kHz
(b) For $F_s = 8$ kHz, $F_{fold} = F_s/2 = 4$ kHz $\Rightarrow$ 5 kHz will alias to 3 kHz.
(c) $F = 9$ kHz will alias to 1 kHz.

Problem 2.6

An analog electrocardiogram (ECG) signal contains useful frequencies up to 100 Hz.

(a) What is the Nyquist rate for this signal?
(b) Suppose that we sample this signal at a rate of 250 samples/s. What is the highest frequency that can be represented uniquely at this sampling rate?

Solution 2.6

(a) $F_{max} = 100$ Hz, $F_s = 2F_{max} = 200$ Hz
(b) $F_{fold} = \frac{F_s}{2} = 125$ Hz.

Problem 2.7

An analog signal $x_a(t) = \sin(480\pi t) + 3\sin(720\pi t)$ is sampled 600 times per second.

(a) Determine the Nyquist sampling rate for $x_a(t)$.
(b) Determine the folding frequency.
(c) What are the frequencies, in radians, in the resulting **DT** signal $x(n)$?
(d) If $x(n)$ is passed through an ideal D/A converter, what is the reconstructed signal $y_a(t)$?

Solution 2.7

(a) $F_{max} = 360$ Hz, $F_N = 2F_{max} = 720$ Hz
(b) $F_{fold} = \frac{F_s}{2} = 300$ Hz .
(c) $x(n) = x_a(nT) = x_a(n/F_s) = \sin(480\pi n/600) + 3\sin(720\pi n/600)$

$x(n) = \sin(4\pi n/5) - 3\sin(4\pi n/5) = -2\sin(4\pi n/5)$. Therefore, $\omega = 4\pi/5$.

(d) $y_a(t) = x(F_s t) = -2\sin(480\pi t)$.

Problem 2.8

A digital communication link caries binary-coded word representing samples of an input signal $x_a(t) = 3\cos 600\pi t + 2\cos 1800\pi t$. The link is operated at 10,000 bits/s and each input sample is quantized into 1024 different voltage levels.

(a) What is the sampling frequency and the folding frequency?
(b) What is the Nyquist rate for the signal $x_a(t)$?
(c) What are the frequencies in the resulting **DT** signal $x(n)$?
(d) What is the resolution Δ?

Solution 2.8

(a) Number of bits/sample = $\log_2 1024 = 10$
F_s = [10,000 bits/s]/[10 bits/sample]
= 1000 sample/s. ; $F_{fold} = 500$ Hz

(b) $F_{max} = 1800\pi/2\pi = 900$ Hz; $F_N = 2F_{max} = 1800$ Hz

(c) $f_1 = \frac{600\pi}{2\pi}\left(\frac{1}{F_s}\right) = 0.3;\ f_2 = \frac{1800\pi}{2\pi}\left(\frac{1}{F_s}\right) = 0.9$
But $f_2 = 0.9 > 0.5 \Rightarrow f_2 = 0.1$
Hence, $x(n) = 3\cos[(2\pi)(0.3)n] + 2\cos[(2\pi)(0.1)n]$

(d) $\Delta = \frac{x_{max}-x_{min}}{L-1} = \frac{5-(-5)}{1023} = \frac{10}{1023}$.

Problem 2.9

Consider the simple signal processing shown Figure P 2.9. The sampling periods of the A/D and D/A converters are T = 5 ms and T' = 1 ms, respectively. Determine the output $y_a(t)$ of the system, if the input is

$$x_a(t) = 3\cos 100\pi t + 2\sin 250\pi t \ (t \text{ in seconds})$$

The post-filter removes any frequency component above $F_s/2$.

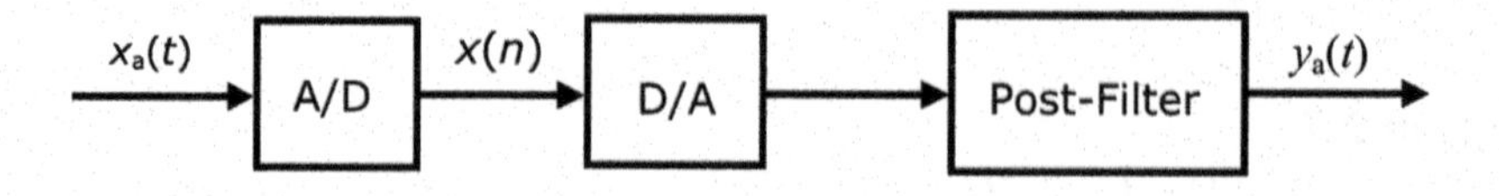

Solution 2.9

$$\begin{aligned} x(n) &= x_a(nT) = 3\cos\left(\tfrac{100\pi n}{200}\right) + 2\sin\left(\tfrac{250\pi n}{200}\right) \\ &= \cos(\pi n/2) - 2\sin(3\pi n/4) \end{aligned}$$

$$T^{-1} = \frac{1}{1000} \quad y_a(t) = x\left(\tfrac{1}{T}\right) = 3\cos\left(\tfrac{\pi}{2}1000t\right) - 2\sin\left(\tfrac{3\pi}{4}1000t\right)$$

$$y_a(t) = 3\cos(500\pi t) - 2\sin(750\pi t)$$

Problem 2.10

Consider the analog signal $x_a(t) = 3\cos 2000\pi t + 5\sin 6000\pi t + 10\cos 12,000\pi t$

(a) Assume now that the sampling rate F_s = 12000 samples/s. What is the **DT** signal obtained after sampling?

(b) What is the analog signal $y_a(t)$ we can reconstruct from the samples if we use ideal interpolation?

Solution 2.10

(a) The frequencies existing in the analog signal are

$$F_1 = 1 \text{ kHz}, \ F_2 = 3 \text{ kHz}, \text{ and } F_3 = 6 \text{ kHz},$$

Thus $F_{max} = 6$ kHz , and according to the sampling theorem,

$$F_s \geq 2F_{max} = 12 \text{ kHz (sampled correctly)}$$

The Nyquist rate is $F_N = 12$ kHz.

Since we have chosen $F_s = 12$ kHz, the folding frequency is $\frac{F_s}{2} = 6$ kHz and this is the maximum frequency that can be represented uniquely by the sampled signal.

The three frequencies F_1, F_2, and F_3 are below or equal to the folding frequency and they will not be changed by the aliasing effect.

From (2.7) it follows that three digital frequencies $f_1 = \frac{1}{12}$, $f_2 = \frac{3}{12}$, and $f_3 = \frac{6}{12}$.

Again using (2.7) we obtain

$$\begin{aligned} x(n) &= x_a(nT) = x_a\left(\frac{n}{F_s}\right) \\ x(n) &= 3\cos 2\pi\left(\tfrac{1}{12}\right)n + 5\sin 2\pi\left(\tfrac{3}{12}\right)n + 10\cos 2\pi\left(\tfrac{6}{12}\right)n \end{aligned}$$

which are in agreement with the result obtained by (2.7) above.

The frequency F_1, F_2, and $F_3 \leq F_s/2$ and thus it is not affected by aliasing.

(b) Since all the frequency components at 1 kHz, 3 kHz, and 6 kHz are present in the sampled signal, the analog signal we can recover is

$$x_a(t) = x(F_s t) = 3\cos 2000\pi t + 5\sin 6000\pi t \ + \ 10\cos 12,000\pi t$$

which is obviously not different from the original signal $x_a(t)$.

Problem 2.11

The **DT** signal $x(n) = 6.35\cos(\pi/10)n$ is quantized with a resolution

(a) $\Delta = 0.1$
(b) $\Delta = 0.02.$

How many bits are required in the A/D converter in eachcase?

Solution 2.11

(a) Range $= x_{\max} - x_{\min} = 12.7$

$$L = 1 + \frac{\text{range}}{\Delta} = 127 + 1 = 128 \Rightarrow \log_2 128 = 7 \text{ bits}$$

(b) $L = 1 + \frac{12.7}{0.02} \ = \ 636 \Rightarrow \ \log_2 636 \Rightarrow 10$ bit A/D.

Problem 2.12

How many bits are required for the storage of a seismic signal if the sampling rate is F_s = 20 samples/s and we use an 8-bit A/D converter? What is the maximum frequency that can be present in the resulting digital seismic signal?

Solution 2.12

$R = (20\text{ sample/s}) \times (8\text{ bits/sample}) = 160\text{ bits/s}.$
$F_{\text{fold}} = \frac{F_s}{2} = 10\text{ Hz}.$

Problem 2.13

A **DT** signal $x(n)$ is defined as

$$x(n) = \begin{cases} 1 + \frac{n}{3} & -3 \leq n \leq -1 \\ 1, & 0 \leq n \leq 3 \\ 0, & \text{elsewhere} \end{cases}$$

(a) Determine its values and sketch the signal $x(n)$.

(b) Sketch the signals that result if we:

(1) First fold $x(n)$ and then delay the resulting signal by four sample.
(2) First delay $x(n)$ by four samples and then fold the resulting signal.

(c) Sketch the signal $x(-n+4)$.
(d) Compare the results in parts (b) and (c) and derive a rule for obtaining the signal $x(-n+k)$ from $x(n)$.

Solution 2.13

(a) $x(n) = \{\ldots\ 0,\ \frac{1}{3},\ \frac{2}{3},\ \underset{\uparrow}{1},\ 1,\ 1, 1,\ 0 \ldots\}$

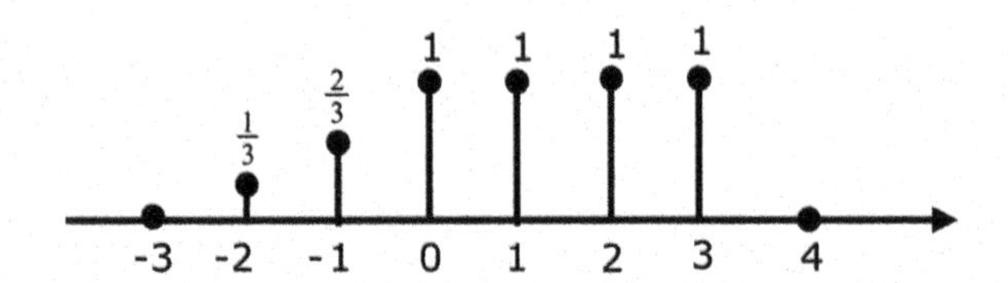

(b) After folding $x(n)$ we have

$$x(-n) = \{\ldots 0, 1, 1, 1, \underset{\uparrow}{1}, \tfrac{2}{3}, \tfrac{1}{3}, 0 \ldots\}$$

After delaying the folded signal by four samples, we have $x(-n+4) = \{\ldots \underset{\uparrow}{0}, 1, 1, 1, 1, \frac{2}{3}, \frac{1}{3}, 0 \ldots\}$ on the other hand, if we delay $x(n)$ by 4 samples we have $x(n-4) = \{\ldots 0, \underset{\uparrow}{0}, 0, \frac{1}{3}, \frac{2}{3}, 1, 1, 1, 1, \ldots\}$ now, if we fold $x(n-4)$ we have
$x(-n-4) = \{\ldots, 0, \underset{\uparrow}{1}, 1, 1, 1, \frac{2}{3}, \frac{1}{3}, 0, 0, \ldots\}$

(c) $x(-n+4) = \{\ldots \underset{\uparrow}{0}, 1, 1, 1, 1, \frac{2}{3}, \frac{1}{3}, 0 \ldots\}$

(d) To obtain $x(-n+k)$, first we fold $x(n)$, this yields $x(-n)$; then, we shift $x(-n)$ by k samples to the right if $k > 0$, or k samples to the left if $k < 0$.

Problem 2.14

A **DT** signal $x(n)$ is shown in Figure.

Sketch and label carefully each of the following signals.

(a) $x(n-2)$, (b) $x(4-n)$, (c) $x(n+2)$, (d) $x(n)u(2-n)$, (e) $x(n-1)\delta(n-3)$, (f) $x(n^2)$, (g) even part of $x(n)$, (h) odd part of $x(n)$

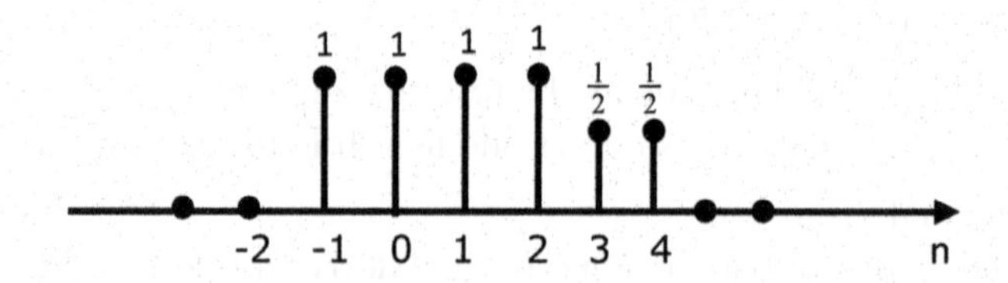

Solution 2.14

$$x(n) = \{\ldots 0, 1, \underset{\uparrow}{1}, 1, 1, \tfrac{1}{2}, \tfrac{1}{2}, 0 \ldots\}$$

(a) $x(n-2) = \{\ldots 0, \underset{\uparrow}{0}, 1, 1, 1, 1, \frac{1}{2}, \frac{1}{2}, 0 \ldots\}$

(b) $x(4-n) = \{\ldots 0, \underset{\uparrow}{\frac{1}{2}}, \frac{1}{2}, 1, 1, 1, 1, 0, \ldots\}$

(c) $x(n+2) = \{\ldots 0, 1, \mathbf{1}, 1, \underset{\uparrow}{1}, \frac{1}{2}, \frac{1}{2}, 0 \ldots\}$

(d) $x(n)u(2-n) = \{\ldots 0, 1, \underset{\uparrow}{1}, 1, 1, 0, 0, \ldots\}$

(e) $x(n-1)\delta(n-3) = \{\ldots 0, \underset{\uparrow}{0}, 0, 0, 1, 0, \ldots\}$

(f) $x(n^2) = \{\ldots 0, x(4), x(1), x(0), x(1), x(4), 0, \ldots\}$
$x(n^2) = \{\ldots 0, \frac{1}{2}, 1, \underset{\uparrow}{1}, 1, \frac{1}{2}, 0 \ldots\}$

(g) $x_e(n) = \frac{x(n)+x(-n)}{2}$, $x(-n) = \{\ldots 0, \frac{1}{2}, \frac{1}{2}, 1, 1, \underset{\uparrow}{1}, 1, 0, 0, 0, \ldots\}$

$x_e(n) = \{\ldots 0, \frac{1}{4}, \frac{1}{4}, \frac{1}{2}, 1, \underset{\uparrow}{1}, 1, \frac{1}{2}, \frac{1}{4}, \frac{1}{4}, 0, \ldots\}$

(h) $x_o(n) = \frac{x(n)-x(-n)}{2}$,
$x_o(n) = \{\ldots 0, -\frac{1}{4}, -\frac{1}{4}, -\frac{1}{2}, 0, \underset{\uparrow}{0}, 0, \frac{1}{2}, \frac{1}{4}, \frac{1}{4}, 0, \ldots\}$

Problem 2.15

Show that
(a) $\delta(n) = u(n) - u(n-1)$
(b) $u(n) = \sum_{k=0}^{\infty} \delta(k) = \sum_{k=-\infty}^{n} \delta(n-k)$

Solution 2.15

(a)

$$u(n) - u(n-1) = \begin{cases} 0, & n > 0 \\ 1, & n = 0 \\ 0, & n < 0 \end{cases} = \delta(n)$$

(b)

$$\sum_{k=-\infty}^{n} \delta(k) = \begin{cases} 0, & n < 0 \\ 1, & n \geq 0 \end{cases} = u(n)\ ; \sum_{k=0}^{\infty} \delta(n-k) = \begin{cases} 0, & n < 0 \\ 1, & n \geq 0 \end{cases}$$

Problem 2.16

Show that any signal can be decomposed into an even and an odd component. Is the decomposition unique? Illustrate your arguments using the signal

$$x(n) = \{2, 3, \underset{\uparrow}{4}, 5, 6\}$$

Solution 2.16

Let $x_e(n) = \frac{1}{2}[x(n) + x(-n)], x_0(n) = \frac{1}{2}[x(n) - x(-n)]$. Since $x_e(-n) = x_e(n)$ and $x_0(-n) = -x_0(n)$, it follow that $x(n) = x_e(n) + x_0(n)$

The decomposition is unique.

For $x(n) = \{2, 3, \underset{\uparrow}{4}, 5, 6\}$, we have

$$x_e(n) = \{4, 4, \underset{\uparrow}{4}, 4, 4\} \quad \text{and} \quad x_0(n) = \{-2, -1, \underset{\uparrow}{0}, 1, 2\}.$$

Problem 2.17

Determine the energy of the following sequence:

$$x(n) = \left(\tfrac{1}{2}\right)^n \quad \text{for } n \geq 0$$
$$x(n) = 0 \quad \text{for } n \leq 0$$

Solution 2.17

We know that for a **DT** signal, the energy is expressed as

$$\mathrm{E} = \sum_{n=-\infty}^{\infty} |x(n)|^2$$

$E = \sum_{n=-\infty}^{\infty} \left(\frac{1}{2}\right)^2$. Therefore summing the infinite series $E = 2$.

Problem 2.18

Show that the energy (power) of a real-valued energy (power) signal is equal to the sum of the energies (powers) of its even and odd components.

Solution 2.18

$$\sum_{n=-\infty}^{\infty} x_{\mathrm{e}}(n)\, x_{\mathrm{o}}(n) = 0$$

$$\sum_{n=-\infty}^{\infty} x_{\mathrm{e}}(n)\, x_{\mathrm{o}}(n) = \sum_{m=-\infty}^{\infty} x_{\mathrm{e}}(-m)\, x_{\mathrm{o}}(-m) = \sum_{m=-\infty}^{\infty} x_{\mathrm{e}}(m)\, x_{\mathrm{o}}(m)$$

$$= -\sum_{n=-\infty}^{\infty} x_{\mathrm{e}}(n)\, x_{\mathrm{o}}(n) \Rightarrow \sum_{n=-\infty}^{\infty} x_{\mathrm{e}}(n)\, x_{\mathrm{o}}(n) = 0$$

Then,

$$\sum_{n=-\infty}^{\infty} x^2(n) = \sum_{n=-\infty}^{\infty} [x_{\mathrm{e}}(n) + x_{\mathrm{o}}(n)]^2$$

$$= \sum_{n=-\infty}^{\infty} x_{\mathrm{e}}^2(n) + \sum_{n=-\infty}^{\infty} x_{\mathrm{o}}^2(n) + 2\sum_{n=-\infty}^{\infty} x_{\mathrm{e}}(n)\, x_{\mathrm{o}}(n)$$

$$= E_{\mathrm{e}} + E_{\mathrm{o}}$$

Problem 2.19

Consider the system

$$y(n) = H[x(n)] = x(n^2)$$

(a) Determine if the system is time invariant.
(b) To clarify the result in part (a) assume that the signal

$$x(n) = \begin{cases} 1, & 0 \leq n \leq 3 \\ 0, & \text{elsewhere} \end{cases}$$

is applied into the system.

(1) Sketch the signal $x(n)$.
(2) Determine and sketch the signal $y(n) = H[x(n)]$.
(3) Sketch the signal $y_2(n) = y(n-2)$.
(4) Determine and sketch the signal $x_2(n) = x(n-2)$.
(5) Determine and sketch the signal $y_2(n) = H[x_2(n)]$.
(6) Compare the signal $y_2(n)$ and $y(n-2)$. What is your conclusion?

(c) Repeat part (b) for the system $y(n) = x(n) - x(n-1)$.

Can you use this result to make any statement about the time invariance of this system? Why?

Solution 2.19

(a) Yes, the system is time invariant.
Proof: If $y(n,k) = x(n-k)^2$
$y(n-k) = x[(n-k)^2]$
$y(n-k) = x[(n^2 + k^2 - 2nk) = y(n,k)$

(b) (1) $x(n) = \{0, \underset{\uparrow}{1}, 1, 1, 1, 0, \ldots\}$

(2) $y(n) = x(n^2)\{\ldots 0, 1, \underset{\uparrow}{1}, 1, 0, \ldots\}$

(3) $y(n-2) = \{...0, \underset{\uparrow}{0}, 1, 1, 1, 0, \ldots\}$

(4) $x(n-2) = \{\ldots \underset{\uparrow}{0}, 0, 1, 1, 1, 1, \ldots\}$

(5) $y_2(n) = H[x(n-2)] = \{\ldots 0, 1, 0, \underset{\uparrow}{0}, 0, 1, 0, \ldots\}$

(6) $y_2(n) \neq y(n-2) \Rightarrow$ system is time variant

(c) (1) $x(n) = \{\underset{\uparrow}{1}, 1, 1, 1\}$

(2) $y(n) = \{\underset{\uparrow}{1}\, 0, 0, 0, -1, 0\}$

(3) $y(n-2) = \{\underset{\uparrow}{0}, 0, 1, 0, 0, 0, -1, 0\}$

(4) $x(n-2) = \{\ldots \underset{\uparrow}{0}, 0, 1, 1, 1, 1, 0 \ldots\}$

(5) $y_2(n) = \{\ldots \underset{\uparrow}{0}, 0, 1, 0, 0, 0, -1, \ldots\}$

(6) $y_2(n) = y(n-2) \Rightarrow$ The system is time invariant but this example does not constitute a proof.

Problem 2.20

A **DT** system can be

(1) Static or dynamic
(2) Linear or non-linear
(3) Time invariant or time varying
(4) Causal or no causal
(5) Stable or unstable

Examine the following systems with respect to the properties above.

(a) $y(n) = \cos[x(n)]$
(b) $y(n) = \sum_{k=-\infty}^{n+1} x(k)$
(c) $y(n) = x(n)\cos(\omega_o n)$
(d) $y(n) = x(-n+2)$
(e) $y(n) = |x(n)|$

(f) $y(n) = x(n)u(n)$
(g) $y(n) = x(n) + nx(n+1)$
(h) $y(n) = x(2n)$
(i) $y(n) = \begin{cases} x(n), & \text{if } x(n) \geq 0 \\ 0, & \text{if } x(n) < 0 \end{cases}$
(j) $y(n) = x(-n)$

Solution 2.20

(a) Static, non-linear, time invariant, causal, and stable.
(b) Dynamic, linear, time invariant, non-causal, and unstable. The latter is easily proved.
For the bounded input $x(k) = u(k)$, the output becomes

$$y(n) = \sum_{k=-\infty}^{n+1} u(k) = \begin{cases} 0, & n < 1 \\ n+2, & n \geq -1 \end{cases}$$

since $y(n) \rightarrow \infty$ as $n \rightarrow \infty$, the system is unstable.
(c) Static, linear, time variant, causal, and stable.
(d) Dynamic, linear, time invariant,non-causal, and stable.
(e) Same answers as in (e).
(f) Static, linear, time invariant, causal, and stable.
(g) Dynamic, linear, time variant, non-causal, and unstable.
Note that the bounded input $x(n) = u(n)$ produces an unbounded output.
(h) Dynamic, linear, time variant, non-causal, and stable.
(i) Static, nonlinear, time invariant, causal, and stable.
(j) Same answer as in (d).

Problem 2.21

Two **DT** systems H_1 and H_2 are connected in cascade to form a new system H as shown in Figure. Prove or disprove the following statements.

(a) If H_1 and H_2 are linear, then H is linear (i.e., the cascade connection of two linear systems is linear).
(b) If H_1 and H_2 are time invariant, then H is time invariant.
(c) If H_1 and H_2 are causal, then H is causal.
(d) If H_1 and H_2 are linear and time invariant, the same holds for H.
(e) If H_1 and H_2 are linear and time invariant, then interchanging their order does not change the system H.
(f) As in part (e) except that H_1, H_2 are now time varying. (Hint: Use an example.)

(g) If H_1 and H_2 are non-linear, then H is non-linear.
(h) If H_1 and H_2 are stable, then H is stable.
(i) Show by an example that the inverse of parts (c) and (h) do not hold in general.

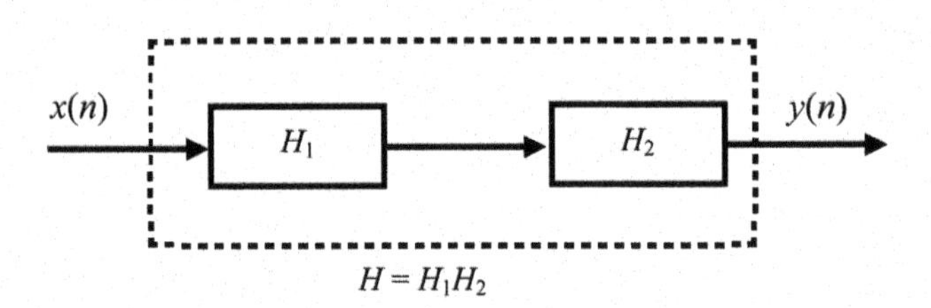

Solution 2.21

(a) True.
If $v_1(n) = H_1[x_1(n)]$ and $v_2(n) = H_1[x_2(n)]$, then $a_1x_1(n) + a_2x_2(n)$ yields
$a_1v_1(n) + a_2v_2(n)$ by the linearity property of H.
Similarly, if $y_1(n) = H_2[v_1(n)]$ and $y_2(n) = H_2[v_2(n)]$,
Then $b_1v_1(n) + b_2v_2(n) \rightarrow y(n) = b_1y_1(n) + b_2y_2(n)$ by the linearity property of H_2.
Since $v_1(n) = H_1[x_1(n)]$ and $v_2(n) = H_2[x_2(n)]$, it follows that $A_1x_1(n)+A_2x_2(n)$ yields the output $A_1H[x_1(n)]+A_2H[x_1(n)]$, where $H = H_1H_2$. Hence H is linear.

(b) True
For H_1, if $x(n) \rightarrow v(n)$ and $x(n-k) \rightarrow v(n-k)$ for H_2, if $v(n) \rightarrow y(n)$,
Then $v(n-k) \rightarrow y(n-k)$ Hence, for H_1H_2, $x(n) \rightarrow y(n)$ and $x(n-k) \rightarrow y(n-k)$
Therefore $H = H_1H_2$ is time invariant.

(c) True
H_1 is causal $\Rightarrow v(n)$ depends only on $x(k)$ for $k \leq n$
H_2 is causal $\Rightarrow y(n)$ depends only on $v(k)$ for $k \leq n$
Therefore, $y(n)$ depends only on $x(k)$ for $k = n$. Hence, H is causal.

(d) True Combine (a) and (b)

(e) True This follows from $h_1(n) * h_2(n) = h_2(n) * h_1(n)$

(f) False
For example, consider H_1: $y(n) = n\,x(n)$ and H_2: $y(n) = nx(n+1)$. Then,
$H_2[H_1[\delta(n)]] = H_2(0) = 0$
$H_1[H_2[\delta(n)]] = H_1[\delta(n+1)] = -\delta(n+1) \neq 0$

(g) False
For example, consider H_1: $y(n) = x(n) + b$ and H_2: $y(n) = x(n) - b$, where $b \neq 0$. Then,
$H[x(n)] = H_2\,[H_1(x(n)]] = H_2[x(n) + b] = x(n)$. Hence H is linear.

(h) True
H_1 is stable $\Rightarrow v(n)$ is bounded if $x(n)$ is bounded H_2 is stable $\Rightarrow y(n)$ is bounded Hence, $y(n)$ is bounded if $x(n)$ is bounded $\Rightarrow H = H_1 H_2$ is stable.

(i) Inverse of (c)
H_1 and for H_2 are non-causal $\Rightarrow H$ is non-causal
For example: H_1: $y(n) = x(n + 1)$, H_2: $y(n) = x(n - 2) \Rightarrow H$: $y(n) = x(n - 1)$, which is causal. Hence, the inverse of (c) is False.
Inverse of (h)
H_1 and for H_2 is unstable, implies H is unstable.
For example, H_1: $y(n) = e^{x(n)}$, stable and H_2: $y(n) = \ln[x(n)]$, which is unstable.
But H: $y(n) = x(n)$, which is stable. Hence, the inverse of (h) is false.

Problem 2.22

Show that the necessary and sufficient condition for a relaxed LTI system to be BIBO stable is

$$\sum_{n=-\infty}^{\infty} |h(n)| \leq M_{\mathrm{h}} < \infty$$

for some constant M_h.

Solution 2.22

A system is BIBO stable if and only if a bounded input producer a bounded output.

$$y(n) = \sum_k h(k)x(n-k)$$

$$|y(n)| = \sum_k |h(k)||x(n-k)| \leq M_x \sum_k |h(k)|,$$

where $|x(n-k)| = M_x$. Therefore, $|y(n)| < \infty$

$$\text{for all } n, \text{ if and only if} \sum_k |h(k)| < \infty.$$

Problem 2.23

Show that:

(a) A relaxed linear system is causal if and only if for any input $x(n)$ such that

$$x(n) = 0 \text{ for } n < n_o \Rightarrow y(n) = 0, \text{ for } n < n_o$$

(b) A relaxed LTI system is causal if and only if $h(n) = 0$, for $n < 0$

Solution 2.23

(a) A system is causal ⇔ the output becomes nonzero after the input become non-zero. Hence, $x(n) = 0$ for $n < n_0 \Rightarrow y(n) = 0$ for $n < n_o$.

(b)

$$y(n) = \sum_{k=-\infty}^{n} h(k)x(n-k), \quad \text{where } x(n) = 0 \text{ for } n < 0.$$

$$\text{If } h(k) = 0 \text{ for } k < 0, \text{ Then } y(n) = \sum_{k=0}^{n} h(k)x(n-k)$$

and hence, $y(n) = 0$ for $n < 0$. On the other hand, if $y(n) = 0$ for $n < 0$, then

$$\sum_{k=-\infty}^{n} h(k)x(n-k) \Rightarrow h(k) = 0, \quad k < 0$$

Problem 2.24

The **DT** system

$$y(n) = ny(n-1) + x(n), n \geq 0 \text{ is at rest [i.e., } y(-1) = 0].$$

Check if the system is linear, time invariant, and BIBO stable.

Solution 2.24

If $H[a_1y_1(n) + a_2y_2(n)] = a_1H[y_1(n)] + a_2H[y_2(n)]$, the system is linear

$H[a_1y_1(n) + a_2y_2(n)] = ny_1(n-1) + x_1(n) + ny_2(n-1) + x_2(n)$

And $a_1H[y_1(n)] + a_2H[y_2(n)]$, produces $ny_1(n-1) + x_1(n) + ny_2(n-1) + x_2(n)$.

Hence, the system is linear.

$y(n,k) = (n)y(n-k-1) + x(n-k)$

$y(n-k) = (n-k)y(n-k-1) + x(n-k)$

$y(n,k) \neq y(n-k)$.

Hence, the system is time variant.

If $x(n) = u(n)$ [a initial step function] then $|x(n)| = 1$. But for this bounded input, the output is

$y(0) = 1,\ y(1) = 1 + 1 = 2,\ y(2) = 2 \times 2 + 1 = 5, \ldots$ which is unbounded.

Hence the system is unstable.

Problem 2.25

Determine whether the following signals derived from $x(n)$ are periodic. If they are periodic, find the fundamental period.

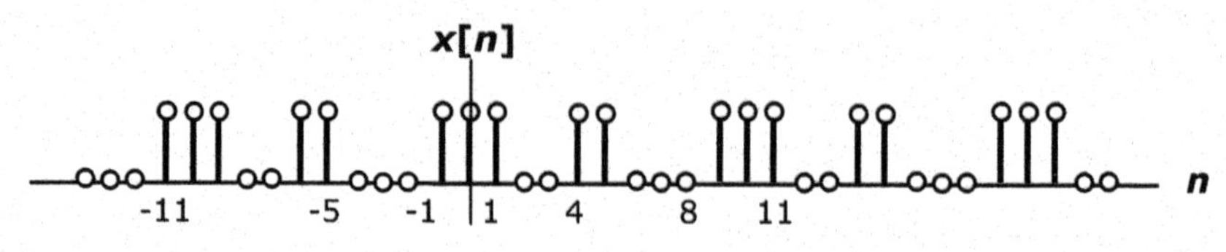

Discrete time signal $x(n)$

(a) $x[n] = (-1)^n$
(b) $x[n] = (-1)^{n^2}$
(c) $x[n]$ depicted in Figure
(d) $x[n] = \cos(2n)$
(e) $x[n] = \cos(2\pi n)$

Solution 2.25

(a) $x[n] = (-1)^n$

For a discrete signal to be periodic,

$x[n] = x[n + N]$

For all integers N, n, where N is a positive integer.

Here, $N = 2$ (fundamental period)

(b) $x[n] = (-1)^{n^2}$

$x[n]$ periodic signal and fundamental period is 1.

(c) For $x[n]$ to be periodic,

$x[n] = x(n + 10)$ The fundamental period is $N = 10$.

(d) Here $x[n] = \cos(2n) \neq x[n + N] = x[2(n + N)]$ for any positive integer N and, thus $x[n]$ is non-periodic

(e) $x[n] = \cos(2\pi n)$

$x[n] = x[n + 1]$ for any n and the fundamental period is $N = 1$

Problem 2.26

Consider the analog signal $x_a(t) = 3\cos 100\pi t$

(a) Determine the minimum required sampling rate to avoid aliasing.
(b) Suppose that the signal is sampled at the rate $F_s = 200$ Hz. What is the **DT** signal obtained after sampling?
(c) Suppose that the signal is sampled at the rate $F_s = 75$ Hz. What is the **DT** signal obtained after sampling?
(d) What is the frequency $F < F_s/2$ of a sinusoid that yields samples identical to those obtained in part (c)?

Solution 2.26

(a) The frequency of the analog signal is $F = 50$ Hz. Hence the minimum sampling rate required to avoid aliasing is $F_s = 100$ Hz.
(b) The signal is sampled at $F_s = 200$ Hz, the **DT** signal is

$$x(n) = 3\cos\frac{100\pi}{200}n = 3\cos\frac{\pi}{2}n$$

(c) If the signal is sampled $F_s = 75$ Hz , the **DT** signals is

$$x(n) = 3\cos\tfrac{100\pi}{75}n \;= 3\cos\tfrac{4\pi}{3}n$$
$$x(n) = 3\cos\left(2\pi - \tfrac{2\pi}{3}\right)n \;= 3\cos\tfrac{2\pi}{3}n$$

(d) For the sampling rate of $F_s = 75$ Hz , we have

$$F = fF_s = 75f$$

The frequency of the sinusoid in part (c) is $f = \frac{1}{3}$. Hence $F = 25$ Hz.

Clearly, the sinusoidal signal $y_a(t) = x(F_s t) = 3\cos 2\pi Ft = 3\cos 50\pi t$ sampled at $F_s = 75$ samples/s yields identical samples. Hence $F = 50$ Hz is an alias of $F = 25$ Hz for the sampling rate $F_s = 75$ Hz.

Problem 2.27

Consider the analog signal $x_a(t) = 3\cos 2000\pi t + 5\sin 6000\pi t + 10\cos 12{,}000\pi t$

(a) What is the Nyquist rate for this signal?
(b) Assume the sampling rate $F_s = 5000$ samples/s. What is the **DT** signal obtained after sampling?
(c) What is the analog signal $y_a(t)$ we can reconstruct from the samples if we use ideal interpolation?

Solution 2.27

(a) The frequencies existing in the analog signal are

$$F_1 = 1\ \text{kHz}, F_2 = 3\ \text{kHz}, \text{ and } F_3 = 6\ \text{kHz},$$

Thus $F_{\text{max}} = 6$ kHz, and according to the sampling theorem,

$$F_s \geq 2F_{\text{max}} = 12\ \text{kHz}$$

The Nyquist rate is

$$F_N = 12\ \text{kHz}$$

(b) Since we have chosen $F_s = 5$ kHz, the folding frequency is

$$\frac{F_s}{2} = 2.5\ \text{kHz}$$

and this is the maximum frequency that can be represented uniquely by the sampled signal.

From $F_k = F_0 + kF_s$, we have $F_0 = F_k - kF_s$. $F_1 = 1$ kHz, the other two frequencies F_2 and F_3 are above the folding frequency and they will be changed by the aliasing effect.

Indeed, $F_2' = F_2 - F_s = -2$ kHz and $F_3' = F_3 - F_s = 1$ kHz

From (2.7) it follows that three digital frequencies $f_1 = \frac{1}{5}$, $f_2 = -\frac{2}{5}$, and $f_3 = \frac{1}{5}$.

Again using (2.7) we obtain

$$\begin{aligned}
x_n = x_a(nT) &= x_a\left(\frac{n}{F_s}\right) \\
&= 3\cos 2\pi\left(\frac{1}{5}\right)n + 5\sin 2\pi\left(\frac{3}{5}\right)n + 10\cos 2\pi\left(\frac{6}{5}\right)n \\
&= 3\cos 2\pi\left(\frac{1}{5}\right)n + 5\sin 2\pi\left(1-\frac{1}{5}\right)n \\
&\quad + 10\cos 2\pi\left(1+\frac{1}{5}\right)n \\
&= 3\cos 2\pi\left(\frac{1}{5}\right)n + 5\sin 2\pi\left(-\frac{2}{5}\right)n + 10\cos 2\pi\left(\frac{1}{5}\right)n
\end{aligned}$$

Finally, we obtain

$$x(n) = 13\cos 2\pi(\tfrac{1}{5})n - 5\sin 2\pi(\tfrac{2}{5})n$$

which are in agreement with the result obtained by (2.7)
Thus F_0 can be obtained by subtracting from F_k an integer multiple of F_s such that $F_s/2 \leq F_0 \leq F_s/2$.
The frequency F_1 is less than $F_s/2$ and thus it is not affected by aliasing.

(c) Since only the frequency components at 1 and 2 kHz are present in the sampled signal, the analog signal we can recover is

$$y_a(t) = y(F_s t) = 13\cos 2000\pi t - 5\sin 4000\pi t$$

which is obviously different from the original signal $x_a(t)$. This distortion of the original analog signal was caused by the aliasing effect, due to the low sampling rate used.

Problem 2.28

Categorize the following signals as an energy or power signal, and find the energy and power of the signal.

$$\text{(a)}\quad x[n] = \begin{cases} n, & 0 \leq n \leq 5 \\ 10 - n, & 5 \leq n \leq 10 \\ 0, & \text{otherwise} \end{cases}$$

$$\text{(b)}\quad x[n] = \begin{cases} \cos \pi n & -4 \leq n \leq 4 \\ 0, & \text{otherwise} \end{cases}$$

Solution 2.28

(a) $E = \sum_{n=-¥}^{¥} x^2[n] = \sum_{n=0}^{5} n^2 + \sum_{n=5}^{10} (10-n)^2 = 110$
Since $0 < E < \infty$, $x[n]$ is an energy signal

$$P = \frac{1}{N}\sum_{n=0}^{N-1} x^2[n] = 0$$

(b) $E = \sum_{n=-4}^{4} \cos^2(\pi n) = \sum_{n=-4}^{4} \left(\frac{1+\cos 2\pi n}{2}\right) = 4$
Since $0 < E < \infty$, $x[n]$ is an energy signal

$$P = \frac{1}{N}\sum_{n=0}^{N-1} x^2[n],$$

where N is the fundamental period here, $N = 2$

$$p = \frac{1}{2}\sum_{n=0}^{1} \cos^2(\pi\, n) = \frac{\cos^2 \pi}{2} = 0$$

$x(n)$ cannot be a power signal.

3

Convolution and Correlation

This chapter covers: Introduction to convolution, Impulse Response, Convolution sum and its general formula, Properties of convolution, Applications of convolution, Convolution description and the various methods of calculating convolution, Introduction to correlation and its general formula, Properties of correlation, Applications of correlation, Analysis of cross-correlation, Cross-correlation coefficients, Auto correlation, Correlation description, and methods of finding correlation, Problems and solutions.

3.1 Introduction

The response of a filter to an impulse is called the impulse response of a filter. In other words, when the input to a filter is a unit impulse function, the output of the filter is the unit impulse response. The difference equation for a digital filter can be used to calculate the impulse response for the filter, and is normally designated by $h(n)$. In digital convolution, this response is used to calculate the output for a general input. Convolution is an essential tool for understanding DSP and will be analyzed here. For one, it provides an alternative to the difference equation used in filter implementation. In addition, it accounts for the creation of spectral copies by sampling and by aliasing errors that sampling can introduce.

Multiplication of the signal $x(n)$ by a unit impulse at some delay k, [i.e., $\delta(n-k)$], in essence picks out the single value $x(k)$ of the signal $x(n)$ at the delay where the unit impulse is non-zero.

If multiplication of $x(k)$ over all possible delays is repeated over the range $-\infty < k < \infty$, and all the product sequences are summed, it will be a sequence that is equal to the sequence $x(n)$.

$$x(n) = \sum_{k=-\infty}^{\infty} x(k)\delta(n-k). \tag{3.1}$$

The right side of Equation (3.1) is the summation of an infinite number of unit sample sequences $\delta(n-k)$ and has an amplitude value $x(k)$. Thus, the right hand side of Equation (3.1) gives the resolution or decomposition of input signal $x(n)$ sample sequences.

Example 3.1

Consider the special case of a finite duration sequence given as
$x(n) = \left\{-2,\ \underset{\uparrow}{4},\ 0,\ 3\right\}$ Resolve the sequence $x(n)$ into a sum of weighted impulse response

Solution 3.1

Since the sequence $x(n)$ is non-zero for the time instants $n = -1, 0, 1, 2$, we need impulses at delay $k = -1, 0, 1, 2$

$$x(n) = \sum_{k=-\infty}^{\infty} x(k)\delta(n-k).$$

Using the above equation, we get
$x(n) = 2\delta(n+1) + 4\delta(n) + 0\delta(n-1) + 3\delta(n-2).$

3.2 The Convolution Sum

If the response of linear-time invariance (LTI) system to the unit sample sequence $\delta(n)$ is denoted as $h(n)$ (Figure 3.1).

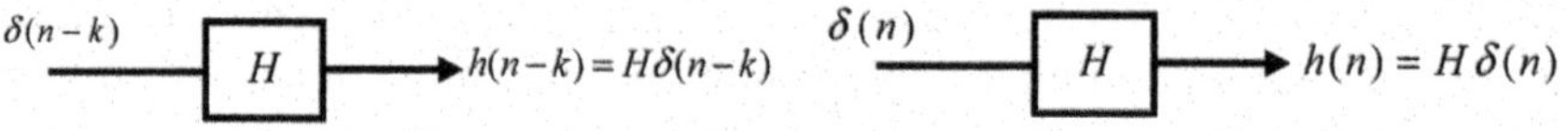

Figure 3.1 A LTI system with and without delay.

Consider a system $h(n) = H[\delta(n)]$. If a delay is applied, then by the time-invariance property, the response of the system to the delayed unit sample sequence $\delta(n-k)$ is
$h(n-k) = H[\delta(n-k)]$
For each impulse function $[\delta(n-k)]$ that forms part of the input

$$x(n) = \sum_{k=-\infty}^{\infty} x(k)\,\delta\,(n-k) \tag{3.2}$$

for a digital filter, the output is an impulse response $h(n-k)$, i.e., every $\delta(n-k)$ in Equation (3.2) becomes $h(n-k)$ as it passes through the filter. The sample $x(k)$ provides the weighting of each impulse function. For example, an input sample $x(5) = 4$, corresponds to a system input $4\delta(n-5)$ and gives a filter output $4h(n-5)$. Thus the total output $y(n)$ due to input $x(n)$ is the sum of all of the weighted impulse response or

$$y(n) = x(n) \otimes h(n) = \sum_{k=-\infty}^{\infty} x(k)\, h(n-k). \tag{3.3}$$

The response $y(n)$ of the LTI system as a function of the input signal $x(n)$ and the unit sample (impulse) response $h(n)$ is called a Convolution sum. The input $x(n)$ is convolved with the impulse response $h(n)$ to yield the output.

If the two samples to be convoluted are of finite sequences which is often the case the above equation becomes

$$y(n) = \sum_{k=0}^{N-1} x(k)\, h(n-k). \tag{3.4}$$

Example 3.2

Consider the discrete-time signals depicted in the Figures 3.2 below. Evaluate the convolution sums indicated below.

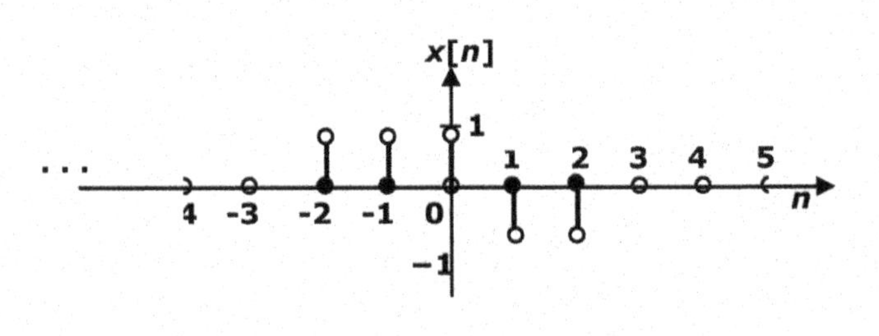

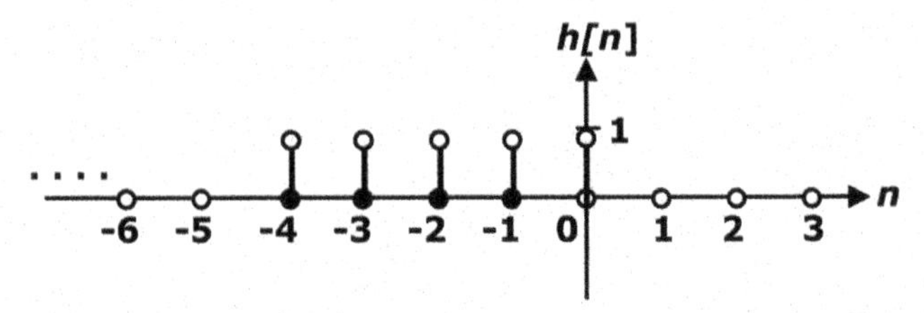

Figure 3.2 Discrete time signals for convolution.

Solution 3.2

$$y[n] = x[n] \otimes h[n] \quad \text{or} \quad y[n] = \sum_{k=-\infty}^{\infty} x[k]\,\delta[n-k]$$

$$y(n) = \ldots + x[-2]\,\delta[n+2] + x[-1]\,\delta[n+1] + x[0]\,\delta[n] + x[1]\,\delta[n-1] + x[2]\,\delta[n-2] + \ldots = 3$$

3.3 Properties of Convolution

3.3.1 Commutative Law

$$x(n) \otimes h(n) = h(n) \otimes x(n), \tag{3.5}$$

Where $\otimes$ is used as convolution sign.

Convolution satisfies the commutative law (Figure 3.3).

Figure 3.3 Commutative law.

3.3.2 Associative Law

$$[x(n) \otimes h_1(n)] \otimes h_2(n) = x(n) \otimes [h_1(n) \otimes h_2(n)] \tag{3.6}$$

Convolution operation also fulfils the associative law (Figure 3.4).

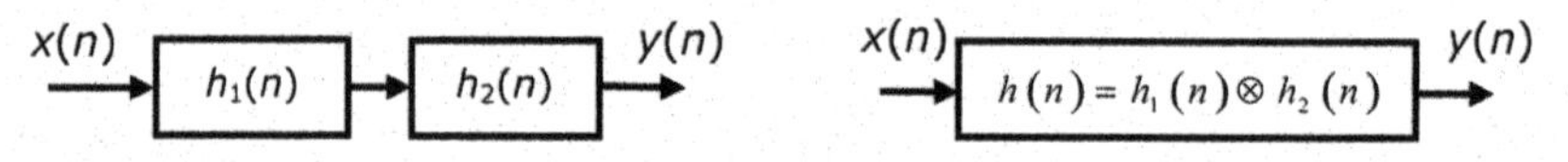

Figure 3.4 Associative law.

3.3.3 Distributive Law

The third property satisfied by the convolution operation is the distributive law (Figure 3.5).

$$x(n) \otimes [h_1(n) + h_2(n)] = x(n) \otimes h_1(n) + x(n) \otimes h_2(n) \tag{3.7}$$

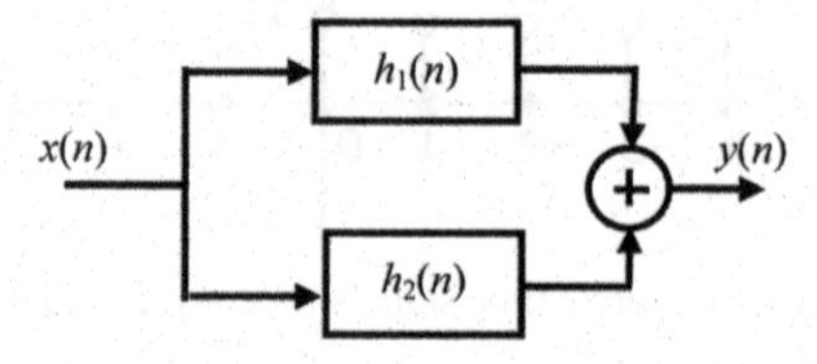

Figure 3.5 Distributive law.

Example 3.3

Determine the expression for the impulse response of the LTI system shown in the figure.

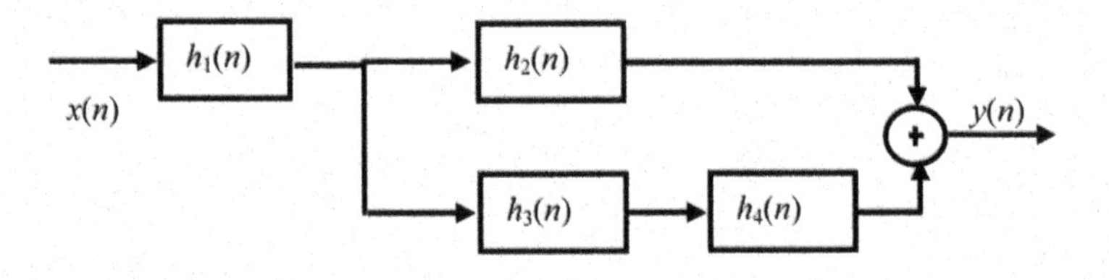

Solution 3.3

$$y(n) = x(n) \otimes [h_1(n) \otimes h_2(n)] + x(n) \otimes [h_1(n) \otimes h_2(n) \otimes h_4(n)]$$

$$y(n) = [\{h_1(n) \otimes h_2(n)\} + \{h_1(n) \otimes h_2(n) \otimes h_4(n)\}]x(n)$$

$$h(n) = \{h_1(n) \otimes h_2(n)\} + \{h_1(n) \otimes h_2(n) \otimes h_4(n)\}$$

3.4 Application of Convolution

3.4.1

Convolution and related operations are found in many applications of engineering and mathematics.

(1) In electrical engineering, the convolution of one function (the input signal) with a second function (the impulse response) gives the output of a LTI system. At any given moment, the output is an accumulated effect of all the prior values of the input function, with the most recent values typically having the most influence (expressed as a multiplicative factor). The impulse response function provides that factor as a function of the elapsed time since each input value occurred.

(2) In DSP and image processing applications, the entire input function is often available for computing every sample of the output function. In that case, the constraint that each output is the effect of only prior inputs can be relaxed.

(3) Convolution amplifies or attenuates each frequency component of the input independently of the other components.

(4) In statistics, as noted above, a weighted moving average is a convolution.

(5) In probability theory, the probability distribution of the sum of two independent random variables is the convolution of their individual distributions.

(6) In optics, many kinds of "blur" are described by convolutions. A shadow (e.g., the shadow on the table when you hold your hand between the table and a light source) is the convolution of the shape of the light source that is casting the shadow and the object whose shadow is being cast. An out-of-focus photograph is the convolution of the sharp image with the shape of the iris diaphragm. The photographic term for this is Bokeh.
(7) Similarly, in digital image processing, convolutional filtering plays an important role in many important algorithms in edge detection and related processes.
(8) In linear acoustics, an echo is the convolution of the original sound with a function representing the various objects that are reflecting it.
(9) In artificial reverberation (DSP and pro audio), convolution is used to map the impulse response of a real room on a digital audio signal (see previous and next point for additional information).
(10) In time-resolved fluorescence spectroscopy, the excitation signal can be treated as a chain of delta pulses, and the measured fluorescence is a sum of exponential decays from each delta pulse.
(11) In radiotherapy treatment planning systems, most part of all modern codes of calculation applies a convolution-superposition algorithm.
(12) In physics, wherever there is a linear system with a "superposition principle", a convolution operation makes an appearance.
(13) In geographic information systems, the result of a kernel estimate of the intensity function of the point pattern is the convolution of the isotropic Gaussian kernel of a standard deviation with point masses at each of the data points. (Diggle 1995) cited by documentation of the "Kernel Smoothed Intensity of Point Pattern" of the SDA4PP QGIS plugin.
(14) In computational fluid dynamics, the large eddy simulation (LES) turbulence model uses the convolution operation to lower the range of length scales necessary in computation thereby reducing computational cost.

3.4.2

One of the most widely used applications of convolution in DSP is that of digital filtering. A digital filter is similar to a digital signal in that it is also a sequence of numbers. Filtering can be used to remove high-frequency noise (rapid fluctuations) from a signal and thus can be used to smooth a signal. Figures given below show how a contaminated signal can been recovered? A simple way to do this is by the process of averaging which has been

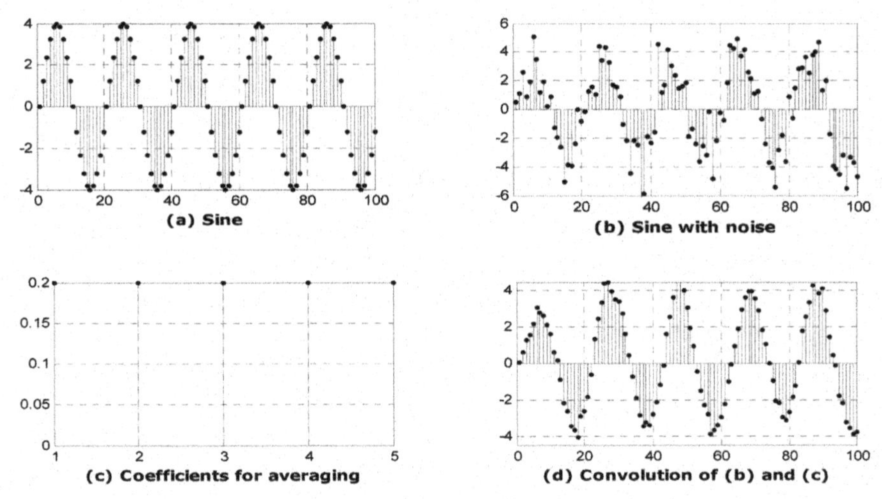

Figure 3.6 (a, b) Sample of a sine wave and sample of a sine wave with noise added to it, and (c, d) Coefficients used for averaging the noisy sine-wave signal and result of convolution between the coefficients and the noisy sine wave.

simulated in Matlab and shown here for understanding one of its applications (Figure 3.6).

3.5 Methods of Calculating Convolution

Before we enter into the rule of convolution it is advisable to learn the technique of convolution here, which is very helpful in some of the example and problems given in this chapter.

3.5.1 Convolution of Delta Function with Delta Function

It is easy to evaluate convolution of $\delta(n)$ and $\delta(n)$ is equal to $\delta(n)$, because both signal existing at $n = 0$. Convolution of $\delta(n)$ and $\delta(n-1)$ is equal to $\delta(n-1)$. Convolution of $\delta(n)$ and $\delta(n-4)$ is equal to $\delta(n-4)$. But the convolution of $\delta(n-1)$ and $\delta(n-4)$ is equal to $\delta(n-5)$. The convolution of $\delta(n+1)$ and $\delta(n-4)$ is equal to $\delta(n-3)$. Similarly the convolution of $\delta(n-1)$ and $\delta(n+5)$ is equal to $\delta(n+4)$.

It means the inner values of the integer are added up such as $\delta(n)$ with $\delta(n)$ its value is equal to $\delta(n)[0+0=0]$, and $\delta(n-4)$ with $\delta(n+1)$ is equal to $\delta(n-3)[1+-3]$, therefore, the convolution of $\delta(n+4)$ and $\delta(n-1)$ is equal to $\delta(n + 3)[4 + -1]$. Likewise we can generate the rule of thumb for delta function convolved with delta function, that is $\delta(n \pm j)$ with $\delta(n \pm k)$ is equal to $\delta(n \pm j \pm k)$.

3.5.2 Convolution of Delta Function with Step Function

All these methods have been verified using Z-transform method. The convolution of $\delta(n)$ and $u(n)$ is equal to $u(n)$. The convolution of $\delta(n)$ and $u(n - 1)$ is equal to $u(n - 1)$. The convolution of $\delta(n)$ and $u(n - 5)$ is equal to $u(n - 5)$. Similarly, now we convolute of $\delta(n - 2)$ and $u(n - 5)$ is equal to $u(n - 7)$. The convolution of $\delta(n - 1)$ and $u(n - 2)$ is equal to $u(n - 3)$. The convolution of $\delta(n - 4)$ and $u(n - 1)$ is equal to $u(n - 5)$.

It means the final answer comes in $u(n)$ format, inner values of the integer are added up such as $\delta(n)$ with $u(n)$ its value is equal to $u(n)$, $\delta(n + 1)$ with $u(n + 1)$ is equal to $u(n + 2)$, and $\delta(n - 4)$ with $\delta(n + 1)$ is equal to $u(n - 3)$, therefore, the convolution of $\delta(n + 4)$ and $u(n - 1)$ is equal to $u(n + 3)$. Like this we can generate the rule of thumb for delta function convolved with step function, that is $\delta(n \pm j)$ with $u(n \pm k)$ is equal to $u(n \pm j \pm k)$.

3.5.3 Convolution of Step Function with Step Function

There could be no rule established in case of convolution of $u(n)$ with $u(n)$. Now the convolution of $u(n)$ and $u(n)$ is equal to $(n+1)u(n)$. The convolution of $u(n)$ and $u(n - 1)$ is equal to $nu(n)$. The convolution of $u(n)$ and $u(n - 2)$ is equal to $(n - 1)u(n) + \delta(n)$. The convolution of $u(n)$ and $u(n + 1)$ is equal to $u(n)(n + 2)$. Similarly the convolution of $u(n)$ and $u(n - 2)$ is equal to $(n - 1)u(n) + \delta(n)$. The convolution of $u(n - 2)$ and $u(n - 1)$ is equal to $nu(n) + 2u(n) + 2\delta(n)$.

The rule of thumb could not be made for convolution of step function with step function. Students are advised to solve such problem accordingly. The proof of all these convolutions is easier to calculate by using Z-transform approach.

Example 3.4

Consider the interconnection of LTI system as shown in the figure.

1. Express the overall impulse response in terms of $h_1(n)$, $h_2(n)$, $h_3(n)$, and $h_4(n)$.
2. Determine $y(n)$ when $x(n) = u(n - 1)$, first find out $h(n)$.

$$
\begin{aligned}
h_1(n) &= \delta(n - 1)\\
h_2(n) &= \delta(n - 2)\\
h_3(n) &= \delta(n - 3)\\
h_4(n) &= \delta(n + 4)
\end{aligned}
$$

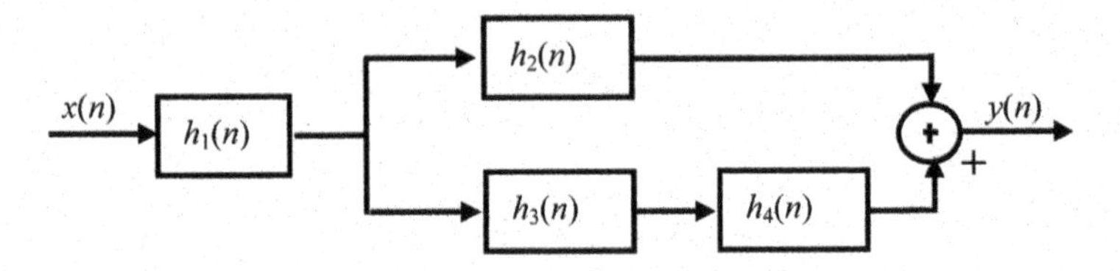

Solution 3.4

$$
\begin{aligned}
h(n) &= h_1(n) \otimes [h_2(n) + \{ h_3 \otimes (n)\, h_4(n)\}]\\
h_3(n) \otimes h_4(n) &= \delta(n - 3)\, \delta(n) = \delta(n - 3)\\
h_2(n) + [\, h_3(n) \otimes h_4(n)] &= [\delta(n - 2)]+[\delta(n + 1)]\\
h(n) &= \delta(n - 1) \otimes [\delta(n - 2)]+[\delta(n + 1)]\\
h(n) &= \delta(n - 1) \otimes \delta(n - 2) + \delta(n - 1) \otimes \delta(n + 1)\\
&= \delta(n - 3) + \delta(n - 3) = 2\delta(n - 3)\\
y(n) &= h(n) \otimes x(n) = h(n) \otimes u(n - 1) = 2\delta(n - 3) \otimes u(n - 1)\\
&= 2u(n - 4).
\end{aligned}
$$

It is worth noting at this stage that convolution in the time domain is not the multiplication of two signals, but is in fact the process where the four operations of flipping, shifting, multiplication, and addition are required. There are two formats the question may be presented in this section. In the first case the input $x(n)$ and the impulse response $h(n)$ is described in terms of the function format and the second case in which both input $x(n)$ and impulse response $h(n)$ are provided as a sequence format. Here, both methods of calculation are covered, i.e., function and the sequence format.

A table for summing the finite and infinite series is given here for the students to apply in calculating the function format in closed form in Table 3.1.

3.5.4 Linear Convolution: Function Format

When the functions for input $x(n)$ and impulse $h(n)$ are given the analytical formula of convolution is employed.

$$
y(n) = \sum_{k=-\infty}^{\infty} x(k)h(n - k).
$$

Table 3.1 Closed form expression for some commonly used series in convolution

Finite Series	Infinite Series
$\sum_{n=0}^{N-1} a^n = \frac{1-a^N}{1-a}$	$\sum_{n=0}^{\infty} a^n = \frac{1}{1-a} \quad \lvert a \rvert < 1$
$\sum_{n=0}^{N-1} n = \frac{1}{2}N(N-1)$	$\sum_{n=0}^{\infty} n.a^n = \frac{a}{(1-a)^2} \quad \lvert a \rvert < 1$
$\sum_{n=0}^{N-1} na^n = \frac{(N-1)a^{N+1} - Na^N + a}{(1-a)^2}$	$\sum_{n=0}^{\infty} n^2 = \frac{1}{6}N(N-1)(2N-1)$

The steps for calculating are summarised below:

(i) The given $x(n)$ or the input function is written in terms of $x(k)$.
(ii) The given $h(n)$ or the impulse response function is written in terms of $h(k)$.
(iii) The convolution formula indicates that either sequence can be reversed.
(iv) It is advisable that the simpler looking function be reversed.
(v) The limits of the summation is changed accordingly after seeing the function and $x(k)$, $h(n - k)$ or $h(k)$, $x(n - k)$ sign is dropped.
(vi) Once we obtain both functions after combining them through the given tabular series, it is written in closed form.

Example 3.5

Derive a closed-form expression for the convolution of $x(n)$ and $h(n)$ where

$$x(n) = \left(\frac{1}{2}\right)^n u(n)$$

$$h(n) = \left(\frac{1}{4}\right)^n u(n)$$

Solution 3.5

The generalized formula for convolution is

$$y(n) = \sum_{k=-\infty}^{\infty} x(k)h(n-k)$$

Because both sequences are infinite in length,
$x(n)$ is changed into $x(k)$

$$x(k) = \left(\frac{1}{4}\right)^k u(k)$$

$h(n)$ is changed into $h(n-k)$

$$h(n-k)=\left(\frac{1}{2}\right)^{n}u(n-k).$$

Substituting $x(k)$ and $h(n-k)$ into the convolution sum, we have

$$y(n)=\sum_{k=-\infty}^{\infty}\left(\frac{1}{4}\right)^{k}u(k)\left(\frac{1}{2}\right)^{n-k}u(n-k)$$

Due to the step $u(k)$ in the first function, the lower limit on the sum may be changed to $k=0$, and the upper limit may be changed to $k=n$. Thus, for $n\geq 0$ the convolution sum becomes

$$y(n)=\sum_{k=0}^{n}\left(\frac{1}{4}\right)^{k}\left(\frac{1}{2}\right)^{n-k}=\left(\frac{1}{2}\right)^{n}\sum_{k=0}^{n}\left(\frac{2}{4}\right)^{k}=\left(\frac{1}{2}\right)^{n}\sum_{k=0}^{n}\left(\frac{1}{2}\right)^{k}\quad n\geq 0.$$

Using the geometric series to evaluate the sum, we have

$$y(n)=\left(\tfrac{1}{2}\right)^{n}\frac{1-\left(\frac{1}{2}\right)^{n+1}}{1-\frac{1}{2}}$$
$$y(n)=2\left(\tfrac{1}{2}\right)^{n}\left[1-\tfrac{1}{2}\left(\tfrac{1}{2}\right)^{n}\right]=2\left(\tfrac{1}{2}\right)^{n}-\left(\tfrac{1}{4}\right)^{n}\quad n\geq 0$$

OR

The generalized formula for convolution is

$$y(n)=\sum_{k=-\infty}^{\infty}h(k)x(n-k).$$

Because both sequences are finite in length,
$h(n)$ is changed into $h(n-k)$

$$h(n-k)=\left(\frac{1}{4}\right)^{n-k}u(n-k)$$

$x(n)$ is changed into $x(k)$

$$x(k)=\left(\frac{1}{2}\right)^{k}u(k).$$

Substituting $x(k)$ and $h(n-k)$ into the convolution sum, we have

$$y(n) = \sum_{k=-\infty}^{\infty} \left(\frac{1}{4}\right)^{n-k} u(n-k) \left(\frac{1}{2}\right)^{k} u(k).$$

Due to the step $u(k)$ in the first function, the lower limit on the sum may be changed to $k = 0$, and the upper limit may be changed to $k = n$. Thus, for $n \geq 0$ the convolution sum becomes

$$y(n) = \sum_{k=0}^{n} \left(\frac{1}{2}\right)^{k} \left(\frac{1}{4}\right)^{n-k} = \left(\frac{1}{4}\right)^{n} \sum_{k=0}^{n} \left(\frac{4}{2}\right)^{k} = \left(\frac{1}{2}\right)^{n} \sum_{k=0}^{n} (2)^{k} \quad n \geq 0.$$

Using the geometric series to evaluate the sum, we have

$$y(n) = \left(\tfrac{1}{4}\right)^{n} \tfrac{1-2^{n+1}}{1-2}$$
$$y(n) = \left(\tfrac{1}{4}\right)^{n} \left[(2)^{n+1} - 1\right] = \left(\tfrac{1}{4}\right)^{n} [2((2)^{n} - 1] = 2\left(\tfrac{1}{2}\right)^{n} - \left(\tfrac{1}{4}\right)^{n} \quad n \geq 0.$$

Example 3.6

Derive a closed-form expression for the convolution of $x(n)$ and $h(n)$ where

$$x(n) = \left(\frac{1}{4}\right)^{n-2} u(n) \quad h(n) = \left(\frac{1}{3}\right)^{n} u(n-3)$$

Solution 3.6

The generalized formula for convolution is

$$y(n) = \sum_{k=-\infty}^{\infty} x(k)h(n-k).$$

Because both sequences are infinite in length,
$x(n)$ is changed into $x(k)$

$$x(k) = \left(\frac{1}{4}\right)^{k-2} u(k)$$

$h(n)$ is changed into $h(n-k)$

$$h(n-k) = \left(\frac{1}{3}\right)^{n} u(n-k-3).$$

Substituting $x(k)$ and $h(n-k)$ into the convolution sum, we have

$$y(n) = \sum_{k=-\infty}^{\infty} \left(\frac{1}{4}\right)^{k-2} u(k) \left(\frac{1}{3}\right)^{n-k} u(n-k-3).$$

Due to the step $u(k)$ in the first function, the lower limit on the sum may be changed to $k = 0$, and the upper limit may be changed to $k = n - 3$. Thus, for $n \geq 3$ the convolution sum becomes

$$y(n) = \sum_{k=0}^{n-3} \left(\frac{1}{4}\right)^{k-2} \left(\frac{1}{3}\right)^{n-k} = 4^2 \left(\frac{1}{3}\right)^{n} \sum_{k=0}^{n-3} \left(\frac{3}{4}\right)^{k} \quad n \geq 3.$$

Using the geometric series to evaluate the sum, we have

Example 3.7

Convolve $x(n) = (0.5)^n u(n)$ with a ramp $h(n) = nu(n)$

Solution 3.7

The convolution of $x(n)$ with $h(n)$ is $y(n) = x(n) \otimes h(n)$

$= \sum_{k=-\infty}^{\infty} x(k)h(n-k).$

Because both sequences are infinite in length

$x(n)$ is changed into $x(k)$, $x(k) = (0.5)^k u(k)$

$h(n)$ is changed into $h(n-k)$, $h(n-k) = (n-k)u(n-k)$

$$= \sum_{k=-\infty}^{\infty} [(0.5)^k u(k)][(n-k)u(n-k)].$$

Due to the step $u(k)$ in the first function, the lower limit on the sum may be changed to $k = 0$, and the upper limit may be changed to $k = n$, $n > 0$. Thus the convolution sum becomes:

$$y(n) = \sum_{k=0}^{n} (n-k)(0.5)^k \; n \geq 0 \text{ or } y(n) = n\sum_{k=0}^{n} (0.5)^k - \sum_{k=0}^{n} k(0.5)^k \; n \geq 0.$$

Using the series given in Table 3.1, we have $\sum_{n=0}^{N-1} na^n = \frac{(N-1)a^{N+1} - Na^N + a}{(1-a)^2}$.

$$y(n) = n\frac{1-(0.5)^{n+1}}{1-0.5} - \frac{n(0.5)^{n+2} - (n+1)(0.5)^{n+1} + 0.5}{(1-0.5)^2}$$

$$y(n) = n\frac{1-(0.5)^{n+1}}{0.5} - \frac{n(0.5)^{n+2}-(n+1)(0.5)^{n+1}+0.5}{0.25}$$

$$= 2\,n[1-(0.5)^{n+1}] - 4\,[n\,(0.5)^{n+2}-(n+1)(0.5)^{n+1}+0.5] \quad n \geq 0$$

$$y(n) = 2\,n[1-0.5(0.5)^{n}] - 4\,[0.25\,n\,(0.5)^{n}-(n+1)\{0.5(0.5)^{n}+0.5] \quad n \geq 0$$

$$y(n) = 2\,n - n(0.5)^{n} - 4[0.25\,n\,(0.5)^{n} - 0.5n(0.5)^{n} - 0.5(0.5)^{n}+0.5] \quad n \geq 0$$

$$y(n) = 2\,n - n(0.5)^{n} - \;n\,(0.5)^{n} + 2n(0.5)^{n} + 2(0.5)^{n} - 2 \;\; n \geq 0$$

which may be simplified to

$$y(n) = [2n - 2 + 2(0.5)^{n}] \quad n \geq 0.$$

3.5.5 Linear Convolution: Sequence Format

There are several methods of evaluating convolution of a sequence format; ultimately the different methods lead to the same solution. The methods can be classified as follows:

(a) Graphical
(b) Analytical
(c) Matrix
(d) Overlap and Add

3.5.5.1 Linear convolution by graphical method

When presented in sequence, the first operation is done only once, the rest three are done repeatedly till the overlapping of the signals exists either to the left or to the right. If the overlapping occurs in both directions the shifting has to be done in sequence in both directions, i.e., for all possible time shifts $-\infty < n < \infty$.

The graphical method is included here to give a complete understanding of how the convolution in the time domain is not only the multiplication of two signals, but is a complete process consisting of four operations.

These four steps are listed below:

(i) Folding: Fold $h(k)$ about $k = 0$ to obtain $h(-k)$
(ii) Shifting: Shift $h(-k)$ by n_0 to the right (left) if n is positive (negative), to obtain $h(n_0 - k)$.
(iii) Multiplication: Multiply $x(k)$ by $h(n_0 - k)$ to obtain the product sequence $y_o(k)x(k)\;h(n_0 - k)$.
(iv) Summation: Sum all the values of the product sequence $y_{n0}(k)$ to obtain the values of the output at time $n = n_0$.

Example given next is included here to help make an understanding on how the graphical convolution process is carried out.

Example 3.8

Find the convolution $y(n)$ of the system given in Figure.

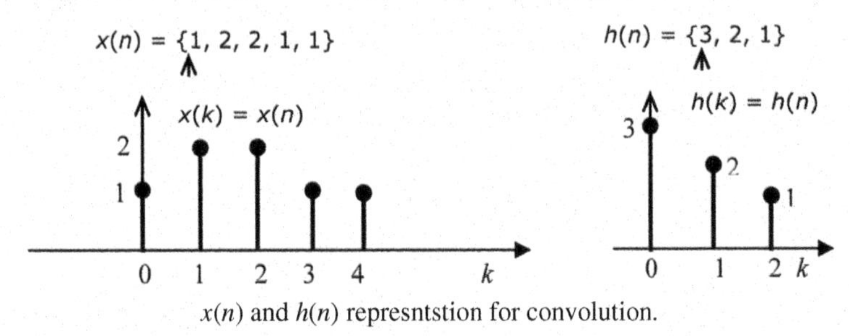

$x(n)$ and $h(n)$ represntstion for convolution.

Solution 3.8

The function is not given but its impulse response is $h(n)$. The convolution of the two sequences $h(n)$ and $x(n)$ is represented as $y(n) = h(n) \times x(n)$.

We now proceed to applying the four steps of convolution. We begin by selecting the simpler sequence from the two and then fold it. In this example, we have selected $h(k)$.

Fold $h(k)$

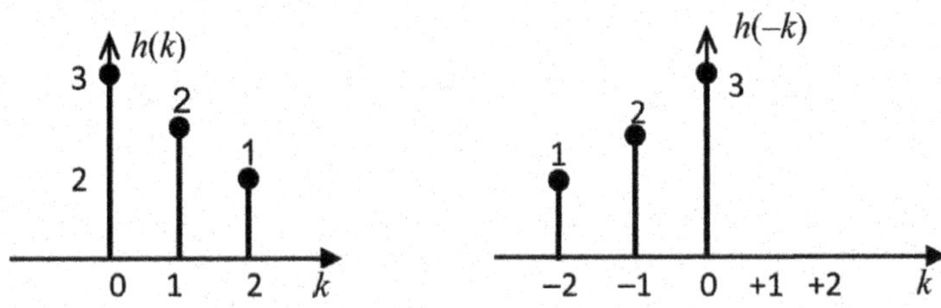

Shifting of the folded signal is the second step, but in this step first time instant (no shifting) gives us a value so step 3 is carried out ($n_0 = 0$).

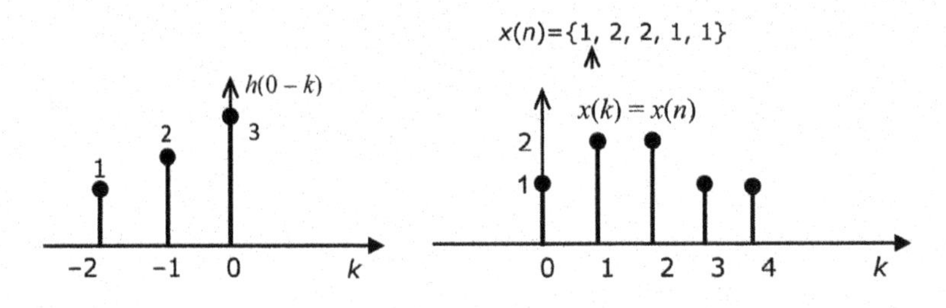

Multiplication and summation of the two signals $h(-k)$ and $x(n)$ for the corresponding discrete-time is done according accordingly as below,

$$y(0) = x(0)\,h(0) = (1)(3) = 3.$$

Now returning to the second step of convolution, the signal $h(-k)$ is shifted to the right by 1 unit of discrete time ($n_0 = 1$).

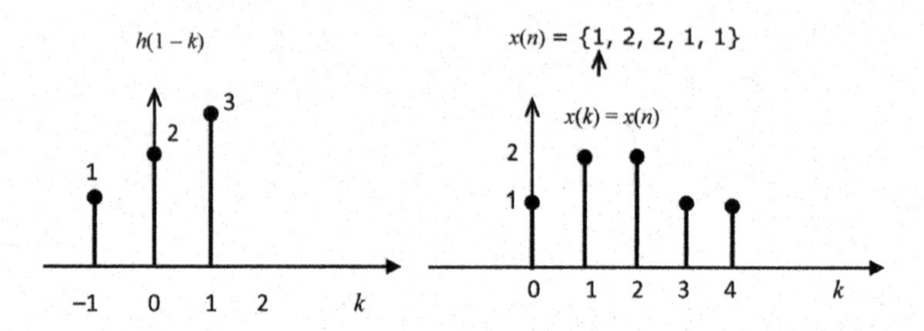

Multiplication and summation of the two signals $h(1-k)$ and $x(n)$ for the corresponding discrete-time values is done accordingly as below,

$$y(1) = x(0)h(0) + x(1)h(1) = (1)(2) + (2)(3) = 8.$$

Now returning to the second step of convolution, the signal $h(-k)$ is shifted to the right by 2 units of discrete-time ($n_0 = 2$).

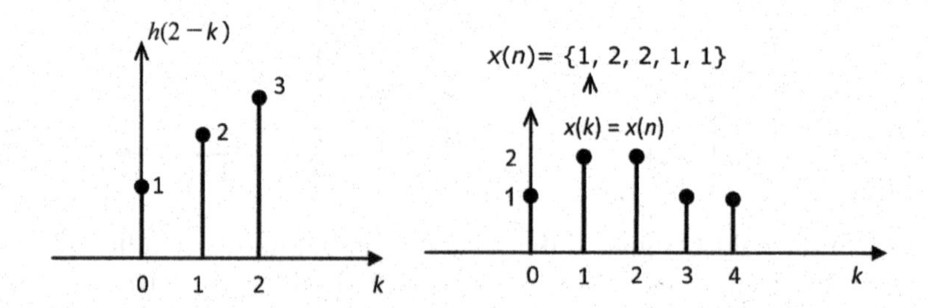

Multiplication and summation of the two signals $h(1-k)$ and $x(n)$ for the corresponding discrete-time values is done according to the formula given below

$$y(2) = x(0)h(0) + x(1)h(1) + x(2)h(2)$$
$$y(2) = (1)(1) + (2)(2) + (2)(3) = 11.$$

Now returning to the second step of convolution, the signal $h(-k)$ is shifted to the right by 3 units of discrete-time ($n_0 = 3$).

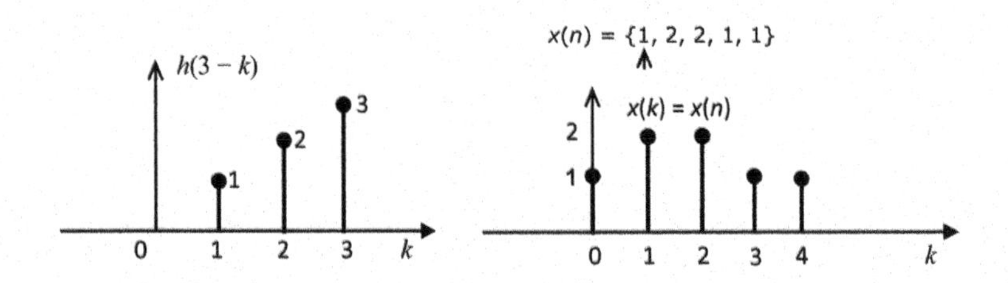

Multiplication and summation of the two signals $h(1 - k)$ and $x(n)$ for the corresponding discrete time values is done according to the formula given below,

$$y(3) = x(1)h(1) + x(2)h(2) + x(3)h(3)$$
$$y(3) = 2x1 + 2x2 + 3x1 = 9.$$

Now returning to the second step of convolution, the signal $h(-k)$ is shifted to the right by 4 units of discrete-time ($n_0 = 4$).

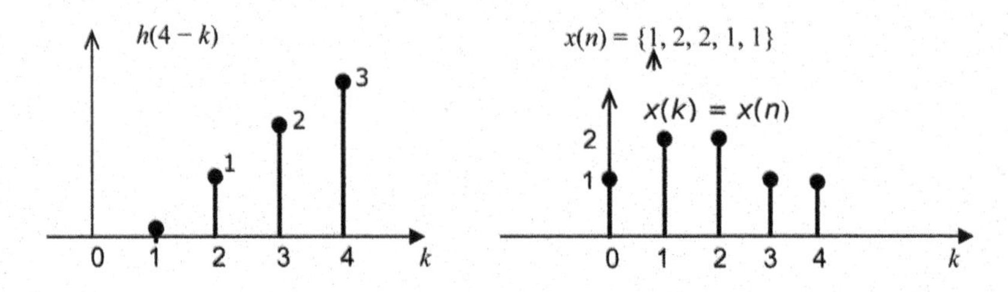

Multiplication and summation of the two signals $h(1 - k)$ and $x(n)$ for the corresponding discrete-time values is done according to the formula given below,

$$y(4) = x(2)h(2) + x(3)h(3) + x(4)h(4)$$
$$y(4) = 2 \times 1 + 2 \times 2 + 1 \times 3 = 9.$$

Now returning to the second step of convolution, the signal $h(-k)$ is shifted to the right by 5 units of discrete-time ($n_0 = 5$).

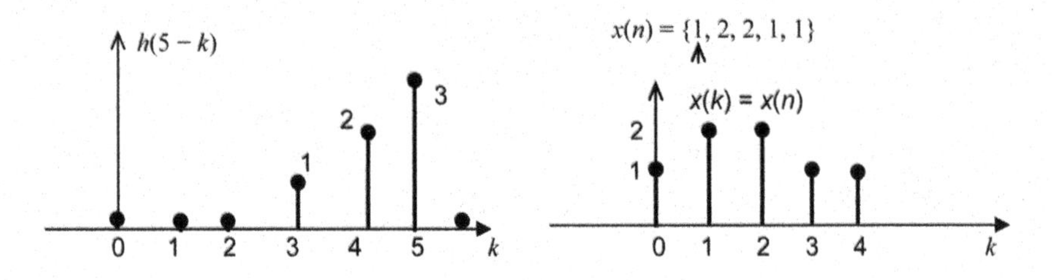

Multiplication and summation

$$y(5) = x(3)h(3) + x(4)h(4)$$
$$y(5) = 1 \times 1 + 1 \times 2 = 3.$$

Now returning to the second step of convolution, the signal $h(-k)$ is shifted to the right by 6 Units of discrete-time ($n_0 = 6$).

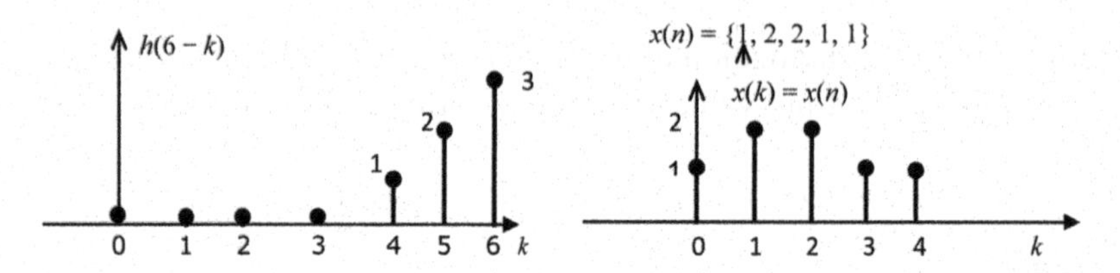

Multiplication and summation of the two signals $h(1 - k)$ and $x(n)$ for the corresponding discrete-time values is done according to the formula given below,

$$y(6) = x(4)\ h(4)$$
$$y(6) = x(4)h(4) = (1)(1) = 1.$$

The entire response of the system for $-\infty < n < \infty$ is shown in the sequence form and plotted in the figure below

$$y(n) = \{ 3, \underset{\uparrow}{8}, 11, 9, 7, 3, 1\}$$

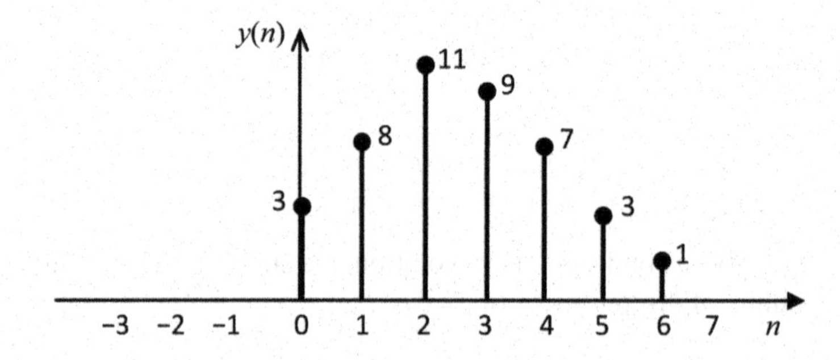

3.5.5.2 Linear convolution by analytical method

The analytical method is adopted when calculating the convolution of two signals in the sequence form in the time domain. We will see how adjustments are made to the limits of the value $y(n)$. This direct formula method is given in Equations (3.2) and (3.3).

$$y(n) = \sum_{k=-\infty}^{\infty} x(k)h(n-k); \quad y(n) = \sum_{k=0}^{N-1} x(k)h(n-k).$$

Example 3.9

The example previously solved by the graphical method is solved here using the analytical method. Find the convolution of the two sequences given below.

$$x(n) = \{\underset{\uparrow}{1},\ 2,\ 2,\ 1,\ 1\}; \quad h(n) = \{\underset{\uparrow}{3},\ 2,\ 1\}.$$

Solution 3.9

In the given sequence, both the values of $x(n)$ and $h(n)$ start from $x(0)$ and $h(0)$. Hence, the lower limit will start from $k = 0$. The final sequence value of $x(n)$ is four and that of $h(n)$ is two, therefore, the upper limit of the summation formula will be the summation of $4 + 2 = 6$.

$$y(n) = \sum_{k=0}^{6} x(k)h(n-k) \quad \text{OR} \quad y(0) = \sum_{k=0}^{6} x(k)\ h(0-k).$$

$$y(0) = x(0)\ h(0) + x(1)\ h(-1) + x(2)\ h(-2) + x(3)h(-3) + x(4)h(-4) + x(5)\ h(-5) + x(6)\ h(-6).$$

The values of $h(-1)$ to $h(-6)$ are not written because the sequence values for these bits are not given (i.e., are zeros). Therefore, the expression for $y(0)$ reduces to

$$y(0) = x(0)\ h(0) = (1)(3) = \mathbf{3}$$

$$y(1) = \sum_{k=0}^{6} x(k)\ h(1-k).$$

$$y(1) = x(0)\ h(1) + x(1)\ h(0) + x(2)\ h(-1) + x(3)\ h(-2) + x(4)\ h(-3) + x(5)\ h(-4) + x(6)\ h(-5)$$

The values of $h(-1)$ to $h(-5)$ are not written because the sequence values for these bits are not given (i.e., are zeros). Therefore the expression for $y(1)$ reduces to

$$y(1) = x(0)h(1) + x(1)h(0) = (1)(2) + (2)(3) = \mathbf{8}$$

$$y(2) = \sum_{k=0}^{6} x(k)\ h(2-k)$$

$$y(2) = x(0)\,h(2) + x(1)\,h(1) + x(2)\,h(0) + x(3)\,h(-1) + x(4)\,h(-2) + x(5)\,h(-3) + x(6)\,h(-4).$$

The values of $h(-1)$ to $h(-4)$, and $x(5)$ to $x(6)$ are not written because the sequence values for these bits are not given (i.e., are zeros). Therefore the expression for $y(2)$ reduces to

$$y(2) = x(0)h(2) + x(1)h(1) + x(2)h(0)$$

$$= (1)(1) + (2)(2) + (2)(3) = \mathbf{11}$$

$$y(3) = \sum_{k=0}^{6} x(k)\,h(3-k)$$

$$y(3) = x(0)h(3) + x(1)h(2) + x(2)h(1) + x(3)h(0) + x(4)h(-1) + x(5)h(-2) + x(6)h(-3)$$

The values of $h(-1)$ to $h(-3)$, $h(3)$, and from $x(5)$ to $x(6)$ are not written because the sequence values for these bits are not given (i.e., are zeros). Therefore, the expression for $y(3)$ reduces to

$$y(3) = x(1)h(2) + x(2)\,h(1) + x(3)h(0) = (2)(1) + (2)(2) + (1)(3) = \mathbf{9}$$

$$y(4) = \sum_{k=0}^{6} x(k)\,h(4-k)$$

$$y(4) = x(0)\,h(4) + x(1)\,h(3) + x(2)\,h(2) + x(3)\,h(1) + x(4)h(0) + x(5)\,h(-1) + x(6)\,h(-2)$$

The values of $h(-1)$ to $h(-2)$, $h(3)$ to $h(4)$ and from $x(5)$ to $x(6)$ are not written because the sequence values for these bits are not given (i.e., are zeros). Therefore, the expression for $y(4)$ reduces to

$$y(4) = x(2)h(2) + x(3)h(1) + x(4)h(0) = (2)(1) + (1)(2) + (1)(3) = \mathbf{7}$$

$$y(5) = \sum_{k=0}^{6} x(k)\,h(5-k)$$

$$y(5) = x(0)\,h(5) + x(1)\,h(4) + x(2)\,h(3) + x(3)\,h(2) + x(4)h(1) + x(5)\,h(0) + x(6)\,h(-1)$$

The values of $h(-1)$, $h(3)$ to $h(6)$ and from $x(5)$ to $x(6)$ are not written because the sequence values for these bits are not given (i.e., are zeros). Therefore, the expression for $y(5)$ reduces to

$$\begin{aligned} y(5) &= x(0)h(5) + x(1)h(4) + x(2)h(3) + x(3)h(2) + x(4)h(1) \\ &\quad + x(5)h(0) \\ &= x(3)h(2) + x(4)h(1) \; = (1)\,(1) + (1)(2) = \mathbf{3} \end{aligned}$$

$$y(6) = \sum_{k=0}^{6} x(k)\; h(6-k)$$

$$\begin{aligned} y(6) = x(0)\; h(6) + x(1)\; h(5) + x(2)\; h(4) + x(3)\; h(3) + x(4)\; h(2) \\ + x(5)\; h(1) + x(6)h(0). \end{aligned}$$

The value of $h(3)$ to $h(6)$ and from $x(5)$ to $x(6)$ are not written because the sequence values for these bits are not given (i.e., are zeros). Therefore, the expression for $y(6)$ reduces to

$$y(6) = \; x(4)h(2) = (1)(1) = \mathbf{1}.$$

So the entire response of the system for $-\infty < n < \infty$ is

$$y(n) \; = \; \left\{3,\; 8,\; \underset{\uparrow}{11},\; 9,\; 7,\; 3,\; 1\right\}.$$

3.5.5.3 Linear convolution by matrix method

The matrix method is a quick way of numerically calculating the convolution of two signals as opposed to the earlier two methods discussed. How the limits of the final result $y(n)$ is adjusted is evident from the examples.

Example 3.10

Find the linear convolution of the two sequences given below, using matrix method

$$x(n) = \{\underset{\uparrow}{1},\; 2,\; 2,\; 1,\; 1\} \quad h(n) = \{\underset{\uparrow}{3},\;\; 2,\;\; 1\}$$

Solution 3.10

Maximum size of the bit in this example is 5, i.e., called N, $N - 1$ zeros padding has to be done in either bit for linear convolution. In this example, $N - 1 = 5 - 1 = 4$ zeros padding is required. To keep the size of the matrix small, zeros have been added to $h(n)$.

$$\begin{bmatrix} & & & h(n) & & & \\ 3 & 0 & 0 & 0 & 0 & 1 & 2 \\ 2 & 3 & 0 & 0 & 0 & 0 & 1 \\ 1 & 2 & 3 & 0 & 0 & 0 & 0 \\ 0 & 1 & 2 & 3 & 0 & 0 & 0 \\ 0 & 0 & 1 & 2 & 3 & 0 & 0 \\ 0 & 0 & 0 & 1 & 2 & 3 & 0 \\ 0 & 0 & 0 & 0 & 1 & 2 & 3 \end{bmatrix} \begin{bmatrix} x(n) \\ 1 \\ 2 \\ 2 \\ 1 \\ 1 \\ 0 \\ 0 \end{bmatrix} = \begin{bmatrix} y(n) \\ 3 \\ 8 \\ 11 \\ 9 \\ 7 \\ 3 \\ 1 \end{bmatrix}$$

$$y(n) = \{\underset{\uparrow}{3}, 8, 11, 9, 7, 3, 1\}.$$

Example 3.11

Find the linear convolution of the two sequences given below, using the matrix method

$$x(n) = \{\underset{\uparrow}{1}, 2, 1\} \qquad h(n) = \{\underset{\uparrow}{1}, 1\}$$

Solution 3.11

Maximum size of the bit in this example is 3, i.e., called N, $N - 1$ zeros padding has to be done in either bit for linear convolution. In this example, $N - 1 = 3 - 1 = 2$ zeros padding is required. To keep the size of the matrix small, zeros have been added to $h(n)$.

$$\begin{bmatrix} & h(n) & & \\ 1 & 0 & 0 & 1 \\ 1 & 1 & 0 & 0 \\ 0 & 1 & 1 & 0 \\ 0 & 0 & 1 & 1 \end{bmatrix} \begin{bmatrix} x(n) \\ 1 \\ 2 \\ 1 \\ 0 \end{bmatrix} = \begin{bmatrix} y(n) \\ 1 \\ 3 \\ 3 \\ 1 \end{bmatrix}$$

$$y(n) = \{\underset{\uparrow}{1}, 3, 3, 1\}.$$

3.5.5.4 Linear convolution by overlap and add method

The overlap and add method is the fastest and simplest method to calculating the results of convolution. How the limits of the final result $y(n)$ is adjusted can be seen in the following examples. The method of calculation is as given here. Write any sequence on top and the second one the left of it. The sequence which has been placed on left must be written in column format. Multiplying by each number the top sequence line, draw a slant line and then add up the magnitude values, gives us the final value of convolution. It can be easily very well understood from the following examples.

Example 3.12

Perform the linear convolution of the two sequences below in the time domain using the Overlap and Add method.

$$x(n) = \{\underset{\uparrow}{1}, 2, 2, 1, 1\} \qquad h(n) = \{\underset{\uparrow}{3}, 2, 1\}$$

Solution 3.12

In the given sequences both the values of $x(n)$ and $h(n)$ start at $x(0)$ and $h(0)$. Hence, the lower limit will start from $k = 0$. The final sequence value of $x(n)$ is four and that of $h(n)$ is two, therefore, the upper limit of the summation formula will be the summation of $4 + 2 = 6$.

$$\begin{aligned} y(n) &= x(n) \times h(n)\ y(0) &= 3 \\ y(1) &= 8,\ y(2) &= 11 \\ y(3) &= 9,\ y(4) &= 7 \\ y(5) &= 3,\ y(6) &= 1 \end{aligned}$$

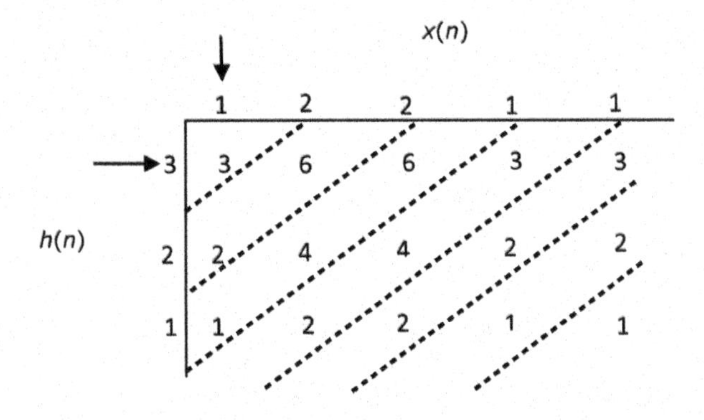

One sequence $x(n) = \{1, 2, 2, 1, 1\}$ is written here on the top of the table, while the other sequence $h(n) = \{3, 2, 1\}$ is placed on the left hand side of the table in column format, each $h(n)$ value is multiplied with the $x(n)$ values and written in the table. Then finally it is added as in the above example. The correctness of the linear convolution can be checked and verified by using the formula which indicates that the multiplication of summation of both sequences is equal to the summation of the output $y(n)$.

$$y(n) = \sum h(k)\, x(n-k)$$

$$\sum y(n) = \sum\sum h(k)x(n-k) = \sum h(k) \sum_{n=-\infty}^{+\infty} x(n-k)$$

$$= \left(\sum h(k)\right)\left(\sum x(k)\right)$$

$$\sum y(n) = 42, \quad \sum h(k) = 6 \sum x(k) = 7.$$

The four different methods for calculation of linear convolution ultimately lead to the same solution as described above.

Example 3.13

Perform the linear convolution of the two sequences below in the time domain using the Overlap and Add method.

$$x(n) = \{\underset{\uparrow}{1},\ 2,\ 1\} \qquad h(n) = \{\underset{\uparrow}{1},\ 1\}$$

Solution 3.13

In the given sequences, both the values of $x(n)$ and $h(n)$ start at $x(0)$ and $h(0)$. Hence, the lower limit will start from $k = 0$. The final sequence value of $x(n)$ is 3 and that of $h(n)$ is 2, therefore, the final answer of the will be $3 + 2 - 1 = 4$.

$$\begin{aligned} y(n) &= x(n) \times h(n)\ y(0) \ = \ 1 \\ y(1) &= 3, \\ y(2) &= 3, \\ y(3) &= 1 \end{aligned}$$

$$y(n) = \{\underset{\uparrow}{1}, 3, 3, 1\}$$

$$\sum y(n) = \sum\sum h(k)x(n-k) = \sum h(k) \sum_{n=-\infty}^{+\infty} x(n-k)$$

$$= \left(\sum h(k)\right)\left(\sum x(k)\right)$$

$$\sum y(n) = 8, \sum h(k) = 2 \sum x(k) = 4.$$

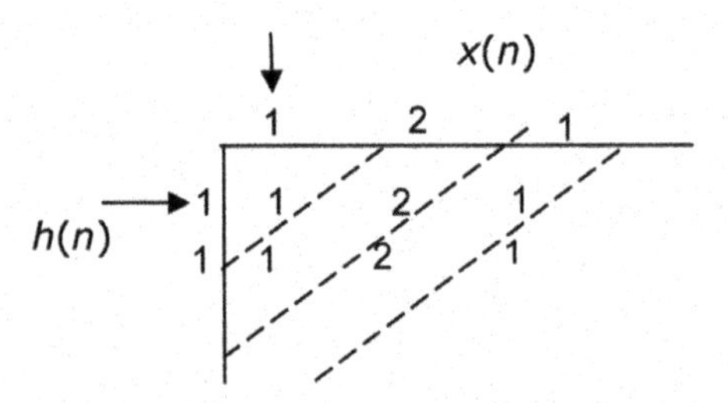

3.5.6 Circular Convolution

The easiest method of evaluating circular convolution is the matrix method, in which no additional zero padding is required as it was in the case of linear convolution and the sizes of resultant sequence bits is the same as larger sequence.

Example 3.14

Find the circular convolution of the two sequences given below, using the matrix method

$$x(n) = \{\underset{\uparrow}{1},\ 2,\ 2,\ 1,\ 1\} \qquad h(n) = \{\underset{\uparrow}{3},\ 2,\ 1\}$$

Solution 3.14

Here maximum size of the sample in this example is 5, therefore, only two zeros are required to be padded up in $h(n)$ make a square matrix, not in $x(n)$ as it was the case of linear convolution.

$$\begin{bmatrix} & & x(n) & & \\ 1 & 1 & 1 & 2 & 2 \\ 2 & 1 & 1 & 1 & 2 \\ 2 & 2 & 1 & 1 & 1 \\ 1 & 2 & 2 & 1 & 1 \\ 1 & 1 & 2 & 2 & 1 \end{bmatrix} \begin{bmatrix} h(n) \\ 3 \\ 2 \\ 1 \\ 0 \\ 0 \end{bmatrix} = \begin{bmatrix} y(n) \\ 6 \\ 9 \\ 11 \\ 9 \\ 7 \end{bmatrix}$$

$$y(n) = \{\underset{\uparrow}{6},\ 9,\ 11,\ 9,\ 7\}.$$

Example 3.15

Find the circular convolution of the two sequences given below, using the matrix method

$$x(n) = \{\underset{\uparrow}{1},\ 2,\ 1\} \qquad h(n) = \{\underset{\uparrow}{1},\ 1\}$$

Solution 3.15

Maximum size of the bit in this example is 3, therefore, only one zero is required to be padded up in $h(n)$ to make a square matrix

$$\begin{bmatrix} & h(n) & \\ 1 & 0 & 1 \\ 1 & 1 & 0 \\ 0 & 1 & 1 \end{bmatrix} \begin{bmatrix} x(n) \\ 1 \\ 2 \\ 1 \end{bmatrix} = \begin{bmatrix} y(n) \\ 2 \\ 3 \\ 3 \end{bmatrix} y(n) = \{\underset{\uparrow}{2}, 3, 3\}$$

Because convolution is a commutative process we prove in this example.

$$\begin{bmatrix} & x(n) & \\ 1 & 1 & 2 \\ 2 & 1 & 1 \\ 1 & 2 & 1 \end{bmatrix} \begin{bmatrix} h(n) \\ 1 \\ 1 \\ 0 \end{bmatrix} = \begin{bmatrix} y(n) \\ 2 \\ 3 \\ 3 \end{bmatrix} y(n) = \{\underset{\uparrow}{2}, 3, 3\}$$

3.6 Correlation

Correlation is a relationship that exists between objects, phenomena, or signals and occurs in such a way that it cannot be by chance alone. Unconsciously correlation is being used every day without its realization. How does one recognize his parents, his friends, his car, or his house? Mental images of each of these are already present. When one's eyes look at another person, car or house, his brain attempts to match the incoming image with hundreds (or thousands) of images that are already stored in their memory.

Recognition occurs when the incoming images bears a strong correlation with an image in memory that "best" corresponds to fit or is most similar to it. This process also helps one to distinguish between say, a dog and a cat, a rose and sunflower, or a train and an airplane.

A similar process is used in DSP to measure the similarity between two signals. This process is known as autocorrelation, if the two signals are exactly the same and as cross-correlation, if the two signals are different.

Since correlation measures the similarity between two signals, it is quite useful in identifying a signal by comparing it with a set of known reference signals. For example, the signal from an unknown aircraft can be correlated with a number of reference signals that have been pre-recorded from different types of aircrafts whose identity was known. The reference signal that results in the lowest value of the correlation with the unknown signals is most likely the identity of the unknown aircraft.

Correlation combines the following three operations,
1. Shifting.
2. Multiplication.
3. Addition (also known as accumulation)

3.7 Properties of Correlation

The properties of correlation listed here are critical for the understanding and calculation of cross- and auto-correlation.

(a) The auto-correlation is always an even sequence, i.e.,

$$r_{xx}(j) = r_{xx}(-j) \tag{3.8}$$

(b) The cross-correlation satisfies the following relationship.

$$r_{xy}(j) = r_{yx}(-j) \tag{3.9}$$

(c) Correlation can be performed by using the convolution property shown below

$$r_{xy}(j) = x(n) \otimes y(-n) \text{ and } r_{yx}(j) = y(n) \otimes x(-n) \tag{3.10}$$

The third property of correlation can be easily verified using over lap and method.

3.8 Application of Correlation

Correlation of signals is often encountered in radar, sonar, digital communications, geology, and other areas of science and engineering. To be very specific, let us suppose that we have two sequences $x(n)$ and $y(n)$ that we wish to compare.

In radar and sonar applications $x(n)$ may represent the sampled version of the transmitted signal and $y(n)$ may represent the sampled version of the received signal at the output of the analog-to-digital (A/D) converter. If the target in space is being searched by the radar or sonar, the received signal $y(n)$ consists of a delayed version of the transmitted signal, reflected from the target, and corrupted by the additive noise.

We may represent the received signal sequence as $y(n) = \alpha x(n - \mathrm{D}) + w(n)$, where α is some attenuation factor representing the signal loss involved in the round trip transmission of the signal $x(n)$, D is the round trip delay, which is assumed to be an integer multiple of the sampling interval, and $w(n)$ is the additive noise which has been picked up by the antenna and the noise generated by the electronics components and amplifiers contained in the front end of the receiver. On the other hand, if there is no target in the space searched by the radar and sonar, the received signal $y(n)$ consists of noise alone.

Further application areas include the following:

- Image processing for robotic vision.
- Remote sensing by satellite in which data from different images is compared.
- Radar and sonar systems for range and position finding in which transmitted and reflected waveforms are compared.
- Detection and identification of signals in noise, and control engineering for observing the effect of inputs on outputs.
- Identification of binary code words in pulse code modulation systems using correlation detectors, as an integral part of the ordinary least squares estimation technique.
- Computation of the average power in waveforms, and in many other fields, such as, for example, climatology. Correlation is also an integral part of the process of convolution. The convolution process is essentially the correlation of two data sequences in which one of the sequences has been reversed. Same algorithms may be used to compute correlations and convolutions simply by reversing one of the sequences. The process of convolution gives the output from a system, which filters the input. The spectrum of a recorded signal consists of the convolution of the spectrum of the signal with the spectrum of its window function.

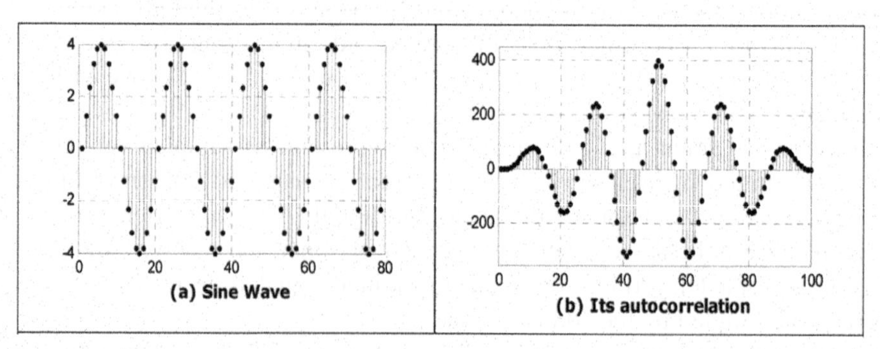

Figure 3.8 A sine wave and its auto-correlation. The auto-correlation of a sine wave is a sinc function.

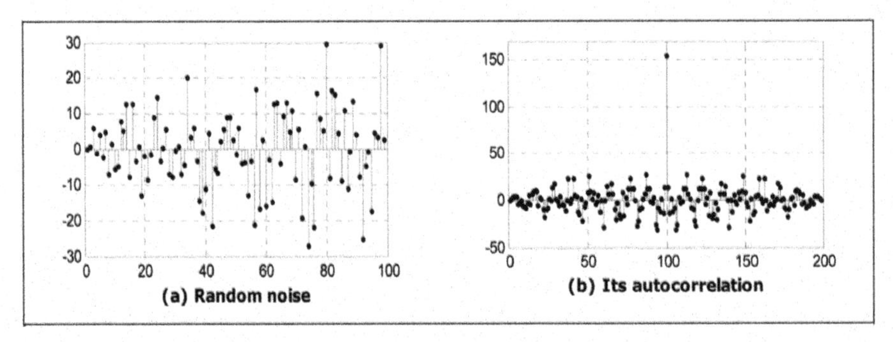

Figure 3.9 Random noise and its auto-correlation: the auto-correlation of random noise is an impulse.

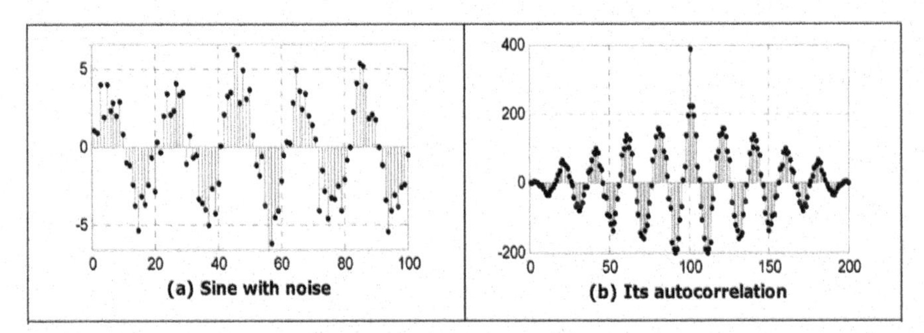

Figure 3.10 Sine wave buried in noise and its auto-correlation. The autocorrelation clearly shows the presence of noise and a periodic signal.

The use of cross-correlation is to detect and estimate periodic signals in noise. Signal buried in noise can be estimated by cross-correlating it with an adjustable template signal. The template is adjusted by trial and error, guided by any foreknowledge, until the cross-correlation function has been maximized. This template is then the estimate of the signal.

The following are the few graphs which show clearly the recovery of actual signal mixed with random noise (Figures 3.8–3.10).

3.9 Types of Correlation

3.9.1 Cross-Correlation

The cross-correlation function (CCF) is a measure of the similarities or shared properties between two signals. Application of CCF's include cross spectral density, detection/recovery of signals buried in noise, the detection of radar return signals, pattern, and delay measurement.

The general formula for cross-correlation $r_{xy}(n)$ between two data sequences $x(n)$ and $y(n)$ each containing N data might therefore be written as

$$r_{xy} = \sum_{n=0}^{N-1} x(n)\, y(n). \tag{3.11}$$

If the two waveforms varied similarly point to point, then a measure of their correlation might be obtained by taking the sum of the products of the corresponding pair of points. This proposal becomes more convincing when the case of two independent and random data sequences is considered.

In this case the sum of the products will tend toward a vanishingly small random number as the number of pairs of points is increased. This is because all numbers, positive and negative, are equally likely to occur so that the product pairs tend to be self-cancelling on summation. By contrast, the existence of a finite sum will indicate a degree of correlation.

A negative sum will indicate negative correlation, i.e. an increase in one variable is associated with a decrease in the other variable.

This definition of cross-correlation, however, produces a result, which depends on the number of sampling points taken. This is corrected for by normalizing the result to the number of points by dividing by N. Alternatively this may be regarded as averaging the sum of products.

Thus, an improved definition presented in which j has been designated as lag between two signals

$$r_{xy}(j) = \frac{1}{N} \sum_{n=0}^{N-1} x(n)\; y(n) \tag{3.12}$$

3.9.2 Auto-Correlation

The auto-correlation function (ACF) involves only one signal and provides information about the structure of the signal or its behaviour in the time domain. It is a special form of CCF and is used in similar applications. It is particularly useful in identifying hidden properties.

$$r_{xx} = \sum_{n=0}^{N-1} x(n)\; x(n) \tag{3.13}$$

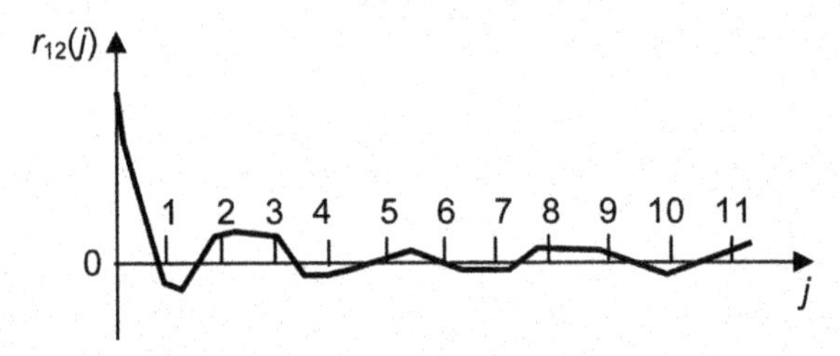

Figure 3.11 Auto-correlation function of a random waveform.

The auto-correlation function has one very useful property that is

$$r_{11}(j) = \frac{1}{N}\sum_{n=0}^{N-1} x^2(n) \;= S. \tag{3.14}$$

Here j has been designated as lag between two signals, if there is no lag j is equal to zero, where S is the normalized energy of the waveform. This provides a method for calculating the energy of a signal. If the waveform is completely random, for example corresponding to that of white, Gaussian noise in an electrical system, then the auto-correlation will have its peak value at zero lag and will reduce to random fluctuation of small magnitude about zero for lags greater than about unity (Figure 3.11). This also constitutes a test for random waveforms.

$$r_{11}(0) \geq r_{11}(j).$$

Example 3.16

The calculation of r_{12} is illustrated using Equation (3.12) in the following example, in which the point numbers in the data sequences are n and the sequences are x_1 and x_2.

n	0	1	2	3	4	5	6	7	8
x_1	4	2	–1	3	–2	–6	–5	4	5
x_2	–4	1	3	7	4	–2	–8	–2	–1

Solution 3.16

$$r_{12} = 4 \times (-4) + 2 \times 1 + (-1) \times 3 + 3 \times 7 + (-2) \times 4 + (-6) \times (-2) + (-5) \times (-8) + 4 \times (-2) + 5 \times (-1)$$

$$r_{12} = \; 34.92$$

3.10 Further Analysis of Cross-Correlation

The definition of correlation needs modification to be used in a better way. The waveforms are clearly highly correlated, even if they are out of phase. The phase difference could, for example, occur because x_1 is the reference signal while x_2 is the delayed output from a circuit. To overcome such phase differences it is necessary to shift, or lag, one of the waveforms with respect to the other.

$$r_{12}(j) = \frac{1}{N} \sum_{n=0}^{N-1} x_1(n)\, x_2(n+J) \tag{3.15}$$

$$r_{12}(-j) = \frac{1}{N} \sum_{n=0}^{N-1} x_1(n)\, x_2(n-J). \tag{3.16}$$

Typically x_2 is shifted to the left to align the waveforms prior to correlation.

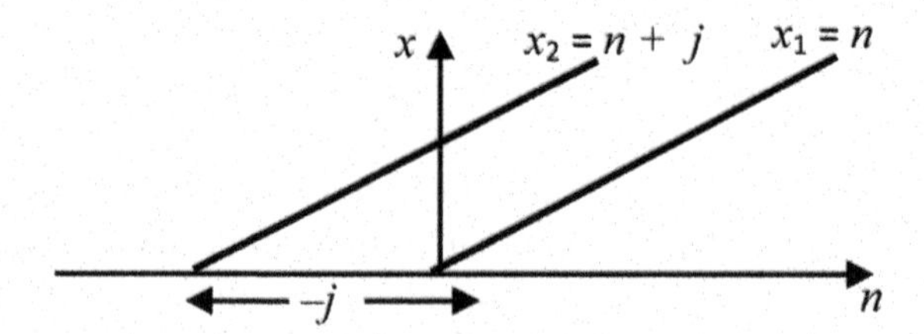

Figure 3.12 Waveform $x_2 = x_1 + j$ shifted j lags to the left of waveform x_1.

As illustrated in Figure 3.12 this is equivalent to changing $x_1(n)$ to $x_2(n+j)$, where j represents the amount of lag which is the number of sampling points by which x_2 has been shifted to the left. An alternative, but equivalent, procedure is to shift x_1 to the right.

Example 3.17

Find the cross-correlation $r_{12}(3)$ of the two sequence.

n	0	1	2	3	4	5	6	7	8
x_1	4	2	–1	3	–2	–6	–5	4	5
x_2	–4	1	3	7	4	–2	–8	–2	–1

Solution 3.17

The second sequence has to be shifted to the left by 3 units

n	0	1	2	3	4	5	6	7	8
x_1	4	2	–1	3	–2	–6	–5	4	5
x_2	7	4	–2	–8	–2	–1			

$$r_{12}(3) = 1/9\,(4 \times 7 + 2 \times 4 + (-1) \times (-2) + 3 \times (-8) + (-2) \times (-2) + (-6) \times (-1))$$
$$r_{12}(3) = 2.667$$

There is another difficulty associated with cross-correlating finite lengths of data. This can be seen in the above example in which $r_{12}(3) = 2.667$ was determined. As x_2 is shifted to the left by 3 units the waveforms no longer overlap and data at the ends of the sequences no longer form pair products. This is known as the end effect.

In the example the number of pairs has dropped from 9 to 6 for a lag of 3. This results in a linear decrease of $r_{12}(j)$ because of increasing j, which leads to debatable values of $r_{12}(j)$. There are two possible solutions available to cater end effect.

One possible solution is to make one of the sequences twice as long as the required length for correlation. This should be achieved by recording more data, or, if one of the sequences were periodic, by repeating the sequence.

Example 3.18

The data values of one pair of waveforms $x_1(n), x_2(n)$ are shown in the tabular form below. Calculate $r_{12}(0)$ and $r_{12}(1)$.

n	0	1	2	3	4	5	6	7	8
$x_1(n)$	0	3	5	5	5	2	0.5	0.25	1
$x_2(n)$	1	2	3	4	1	2	3	4	1

Solution 3.18

For calculating the value of $r_{12}(1)$, it is assumed that the signal is periodic.

$$r_{12}(0) = 1/9\,(0 \times 1 + 3 \times 2 + 5 \times 3 + 5 \times 4 + 5 \times 1 + 2 \times 2 + 0.5 \times 3 + 0.25 \times 4 + 1 \times 1)$$
$$r_{12}(0) = 5.944$$

n	0	1	2	3	4	5	6	7	8
$x_1(n)$	0	3	5	5	5	2	0.5	0.25	1
$x_2(n)$	2	3	4	1	2	3	4	1	1

$$r_{12}(1) = 1/9\,(0 \times 2 + 3 \times 3 + 5 \times 4 + 5 \times 1 + 5 \times 2 + 2 \times 3 + 0.5 \times 4 + 0.25 \times 1 + 1 \times 1)$$
$$r_{12}(1) = 6.0278$$

Example 3.19

The calculation of r_{12} is illustrated using improved definition in the following example, in which the point numbers in the data sequences are the *n*, and the sequences are x_1 and x_2.

n	0	1	2	3	4	5	6	7	8
x_1	4	2	–1	3	–2	–6	–5	4	5
x_2	–4	1	3	7	4	–2	–8	–2	-1

Solution 3.19

$$x(n) = \delta(n + 2) + 3\delta(n - 1) - 4\delta(n - 3)$$
$$r_{12} = 1/9\,(4 \times (-4) + 2 \times 1 + (-1) \times 3 + 3 \times 7 + (-2) \times 4 + (-6) \times (-2) + (-5) \times (-8) + 4 \times (-2) + 5 \times (-1))$$
$$r_{12} = 3.88.$$

Example 3.20

Find the auto-correlation coefficient $\rho_{11}(0)$ of the signal given in tabular form

n	0	1	2	3	4	5	6	7	8	9
$x(n)$	4	3	3	3	2	0	1	2	3	4

Solution 3.20

Cross-correlations $\rho_{11}(0)$ is found as

$$r_{11}(0) = 7.7$$

$$\frac{1}{N}\left\{\sum_{n=0}^{N-1} x_1^2(n)\right\} = \frac{1}{10}\{(77)\} = 7.7$$

$$\rho_{11}(0) = \frac{r_{11}(0)}{7.7} = \frac{7.7}{7.7} = 1.0.$$

Now $\rho_{11}(0)$ demonstrates that auto-correlation of the sample sequence is equal to 1, it means it is highly correlated.

Another possible solution is to add a correction to all computed values. Figure 3.13 shows how $r_{12}(j)$ decreases with j as a result of the end effect.

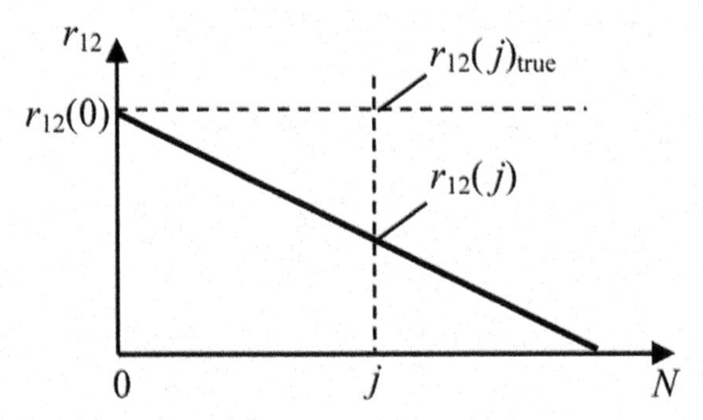

Figure 3.13 The effect of the end-effect on the cross-correlation $r_{12}(j)$.

At $j = 0$, $r_{12}(j) = r_{12}(0)$, which can be computed. At $j = N$, $r_{12}(N) = 0$, because the waveforms no longer overlap. In between, at some lag j, the true value of $r_{12}(j)$ is $r_{12}(j)_{\text{true}}$ while the actual value caused by the end effect is $r_{12}(j)$. Then, from the triangle of Figure 3.13, we developed the following equation.

$$\frac{r_{12}(j)_{\text{true}} - r_{12}(j)}{j} = \frac{r_{12}(0)}{N} \tag{3.17}$$

$$r_{12}(j)_{\text{true}} = r_{12}(j) + \frac{j}{N} r_{12}(0). \tag{3.18}$$

Computed values of the cross-correlation are therefore easily corrected to get $r_{12}(j)_{\text{true}}$ to cater end effects by adding j $r_{12}(0)/N$ to the values of $r_{12}(j)$.

Example 3.21

The data values of one pair of waveforms $x_1(n)$, $x_2(n)$ are shown in the tabular form below. Calculate $r_{12}(0)_{\text{true}}$ and $r_{12}(1)_{\text{true}}$.

n	0	1	2	3	4	5	6	7	8
$x_1(n)$	0	3	5	5	5	2	0.5	0.25	1
$x_2(n)$	1	2	3	4	1	2	3	4	1

Solution 3.21

n	0	1	2	3	4	5	6	7	8
$x_1(n)$	0	3	5	5	5	2	0.5	0.25	1
$x_2(n)$	1	2	3	4	1	2	3	4	1

$$r_{12}(0) = 1/9\ (0 \times 1 + 3 \times 2 + \ 5 \times 3 + 5 \times 4 + 5 \times 1 + 2 \times 2 + 0.5 \times 3 \\ + 0.25 \times 4 + 1 \times 1)$$
$$r_{12}(0) = 5.944$$

For the value of $r_{12}(1)$, now it is assumed that the signal is non-periodic.

n	0	1	2	3	4	5	6	7	8
$x_1(n)$	0	3	5	5	5	2	0.5	0.25	1
$x_2(n)$	2	3	4	1	2	3	4	1	0

$$r_{12}(1) = 1/9\ (0 \times 2 + 3 \times 3 + \ 5 \times 4 + 5 \times 1 + 5 \times 2 + 2 \times 3 + 0.5 \times 4 \\ + 0.25 \times 1 + 1 \times 0)$$
$$r_{12}(1) = 5.8055$$

$$r_{12}(j)_{\text{true}} = r_{12}(j) + \tfrac{j}{N} r_{12}(0)$$
$$r_{12}(1)_{\text{true}} = 5.8055 + \tfrac{j}{N} r_{12}(0)$$
$$r_{12}(1)_{\text{true}} = 5.8055 + \tfrac{1}{9}(5.944) = 6.4659.$$

3.11 Cross-Correlation Coefficient

The cross-correlation coefficients value shows how much the two signal are in correlation with each other. The values are computed according to the above formulae depending on the values of the data. It is often necessary to measure cross-correlations according to the fixed scale between –1 and +1, which is named as cross-correlation coefficients.

This situation can be rectified by normalizing the cross-correlation $r_{12}(j)$ by the factor given in Equation (3.19)

$$\frac{1}{N}\left\{\sum_{n=0}^{N-1} x_1^2(n) \times \sum_{n=0}^{N-1} x_2^2(n)\right\}^{1/2} = \frac{1}{N}\left\{\sum_{n=0}^{N-1} x_1^2(n) \sum_{n=0}^{N-1} x_2^2(n)\right\}^{1/2}. \tag{3.19}$$

The normalized expression for $r_{12}(j)$ then becomes $\rho_{12}(j)$

$$\rho_{12}(j) = \frac{r_{12(j)\text{true}}}{\frac{1}{N}\left[\sum_{n=0}^{N-1} x_1^2(n) \sum_{n=0}^{N-1} x_2^2(n)\right]^{1/2}} \tag{3.20}$$

Its values always lie between –1 and +1. +1 means 100% correlation in the same sense, –1 means 100% correlation in the opposing sense, for example, signals in anti-phase.

A value of 0 signifies zero correlation. This means the signals are completely independent. This would be the case, for example, if one of the waveforms were completely random.

Small value of $\rho_{12}(j)$ indicates very low correlation between two signals. This can be achieved by normalizing the values by an amount depending on the energy content of the data. Consider the two pairs of waveforms $x_1(n)$, $x_2(n)$, and $x_3(n)$, $x_4(n)$. The data values are given in the tabular form below.

n	0	1	2	3	4	5	6	7	8
$x_1(n)$	0	3	5	5	5	2	0.5	0.25	0
$x_2(n)$	1	1	1	1	1	1	0	0	0
$x_3(n)$	0	9	15	15	15	6	1.5	0.75	0
$x_4(n)$	2	2	2	2	2	2	0	0	0

As may be seen from Figure 3.14, if the data points are plotted, waveforms $x_1(n)$ and $x_3(n)$ are alike, differing only in magnitude. The same is true of the pair $x_2(n)$ and $x_4(n)$.

The correlation between $x_1(n)$ and $x_2(n)$ is therefore the same as that between $x_3(n)$ and $x_4(n)$, which is proved in the example given below.

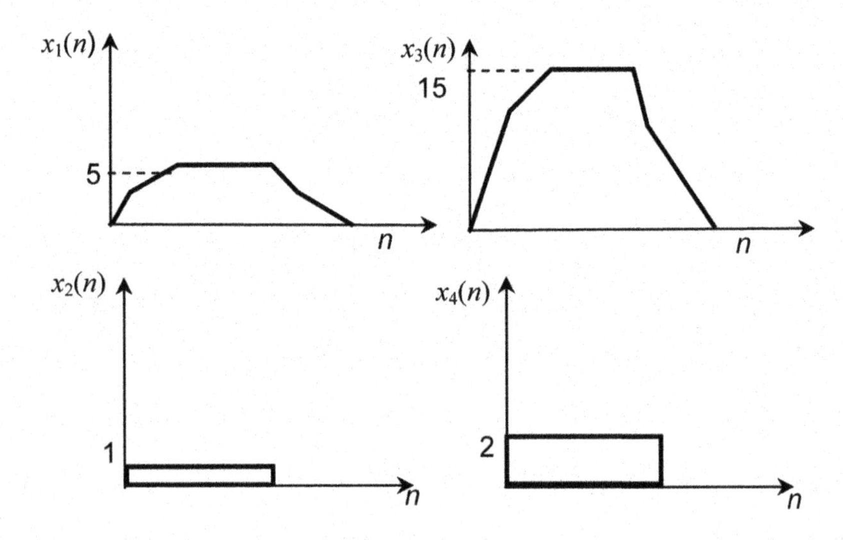

Figure 3.14 Pairs of waveforms $\{x_1(n), x_2(n), x_3(n), x_4(n)\}$ of different magnitudes but equal cross-correlations.

Example 3.22

The data values of two pairs of waveforms $x_1(n)$, $x_2(n)$, and $x_3(n)$, $x_4(n)$ are shown in the table below.

n	0	1	2	3	4	5	6	7	8
$x_1(n)$	0	3	5		5	2	0.5	0.25	0
$x_2(n)$	1	1	1	1	1	1	0	0	0
$x_3(n)$	0	9	15	15	15	6	1.5	0.75	0
$x_4(n)$	2	2	2	2	2	2	0	0	0

Find the correlation between coefficients $\rho_{12}(1)$ and $\rho_{34}(1)$ between $x_1(n)$ and $x_2(n)$ and $x_3(n)$ and $x_4(n)$.

Solution 3.22

The cross-correlations $\rho_{12}(1)$ and $\rho_{34}(1)$ is calculated as follows

$$r_{12}(0) = 1/9\ (0 \times 1 + 3 \times 1 + \ 5 \times 1 + 5 \times 1 + 5 \times 1 + 2 \times 1 + 0.5 \times 0 \ + \ 0.25 \times 0 + 0 \times 0)$$

$$r_{12}(0) = \ 2.22$$

$$r_{34}(0) = 1/9\ (0 \times 2 + 9 \times 2 + 15 \times 2 + 15 \times 2 + 15 \times 2 + 6 \times 2 + 1.5 \times 0 \ + 0.75 \times 0 \ + 0 \times 0)$$

$$r_{34}(0) = \ 13.33$$

$$r_{12}(1) = 1/9\ (0 \times 1 + 3 \times 1 + \ 5 \times 1 + 5 \times 1 + 5 \times 1 + 2 \times 0 + 0.5 \times 0 + 0.25 \times 0 + 0 \times 0)$$

$$r_{12}(1) = \ 2$$

$$r_{34}(1) = 1/9\ (0 \times 2 + 9 \times 2 + 15 \times 2 + 15 \times 2 + 15 \times 2 + 6 \times 0 + 1.5 \times 0 \ + 0.75 \times 0 + 0 \times 0)$$

$$r_{34}(1) = \ 12$$

$$r_{12}(1)_{\text{true}} = r_{12}(1) + \frac{1}{9} r_{12}(0) \qquad r_{12}(1)_{\text{true}} = 2 + \frac{2.22}{9} = 2.246$$

$$r_{34}(1)_{\text{true}} = r_{34}(1) + \frac{1}{9} r_{34}(0) \qquad r_{34}(1)_{\text{true}} = 12 + \frac{13.33}{9} = 13.48.$$

The normalizing factor for $r_{12}(j)$ in the above illustration is introduced below

$$\frac{1}{N}\left\{\sum_{n=0}^{N-1} x_1^2(n) \times \sum_{n=0}^{N-1} x_2^2(n)\right\}^{1/2} = \frac{1}{9}\{(88.31) \times (6)\}^{1/2} = 2.5577$$

and for $r_{34}(j)$ it is

$$\frac{1}{N}\left\{\sum_{n=0}^{N-1} x_3^2(n) \times \sum_{n=0}^{N-1} x_4^2(n)\right\}^{1/2} = \frac{1}{9}\{794.8) \times (24)\}^{1/2} = 15.346.$$

Therefore

$$\rho_{12}(1) = \frac{r_{12}(1)}{2.5577} = \frac{2.246}{2.5577} = 0.8785 \quad \text{and}$$

$$\rho_{34}(1) = \frac{r_{34}(1)}{15.346577} = \frac{13.486}{15.346} = 0.8785.$$

Now $\rho_{12}(1) = \rho_{34}(1)$ which demonstrates that this normalization process indeed allows a comparison of cross-correlations independently of the absolute data values.

Example 3.23

Find the auto correlation coefficient $\rho_{11}(2)$ of the signal given in tabular form

n	0	1	2	3	4	5	6	7	8	9
$x_1(n)$	4	3	3	3	2	0	1	2	3	4

Solution 3.23

n	0	1	2	3	4	5	6	7	8	9
$x_1(n)$	4	3	3	3	2	0	1	2	3	4
$x_2(n)$	3	3	2	0	1	2	3	4	0	0

Cross-correlations $\rho_{11}(2)$ is found as

$$r_{12}(2)_{\text{true}} = r_{12}(2) + \frac{2}{10} r_{12}(0) \quad r_{12}(2)_{\text{true}} = 4 + \frac{2(7.7)}{10} = 5.54.$$

The normalizing factor for $r_{12}(0)$ in the above illustration is calculated as below

$$\frac{1}{N}\left\{\sum_{n=0}^{N-1} x_1^2(n) \times \sum_{n=0}^{N-1} x_2^2(n)\right\}^{1/2} = \frac{1}{10}\{(77) \times (77)\}^{1/2} = 7.7.$$

Therefore

$$\rho_{11}(2) = \frac{r_{11true}(2)}{7.7} = \frac{5.54}{7.7} = 0.7914$$

$\rho_{11}(2)$ demonstrate that auto-correlation of the sample sequence is decreased and is not equal to 1, although even by using the true value of r_{12}.

3.12 Correlation Methods

There are several methods of evaluating cross- and auto-correlation; ultimately the different methods lead to the same solution. Here, these methods can be classified as

(a) Graphical
(b) Analytical
(c) Tabular Shifting
(d) Convolution Property using Overlap and Add

This definition of cross-correlation is further elaborated here which produces results, which is largely used and very similar to Equation (3.15) but this time the value of the cross correlation is not being divided by (number of sample points) N.

The two digital signal sequences $x(n)$ and $y(n)$ each of which has a finite energy. The general formula for the cross-correlation Equation (3.15) thus becomes

$$r_{xy}(j) = \sum_{n=-\infty}^{\infty} x(n)y(n-j) \quad j = 0, \pm 1, \pm 2, \ldots \tag{3.21}$$

$$r_{xy}(j) = \sum_{n=-\infty}^{\infty} x(n+j)y(n) \quad j = 0, \pm 1, \pm 2, \ldots \tag{3.22}$$

The index j is the (time) shift (or lag) parameter and the subscripts xy on the cross correlation sequence $r_{xy}(j)$ indicate the sequence being correlated. The order of the subscripts, with x preceding y, indicates the direction in which one sequence is shifted, relative to other.

To elaborate in Equation (3.21), the sequence $x(\mathrm{i})$ is left unshifted and $y(n)$ is shifted by j units of time, to the right for j positive and to the left for j negative.

Equivalently, in Equation (3.22), the sequence $y(n)$ is left unshifted, and sequence $x(n)$ is shifted by j units of time, to the left for j positive and to the right for j negative. But shifting $x(n)$ to the left by j units relative to $y(n)$ is equivalent to shifting $y(n)$ to the right by j units relative to $x(n)$.

In the reverse the role of $x(n)$ and $y(n)$ in Equations (3.23) and (3.24) and hence reverse the order of the indices xy, we obtain the cross-correlation sequence.

$$r_{yx}(j) = \sum_{n=-\infty}^{\infty} x(n)y(n-j) \quad j = 0, \pm 1, \pm 2, \ldots \tag{3.23}$$

$$r_{yx}(j) = \sum_{n=-\infty}^{\infty} y(n+j)x(n) \quad j = 0, \pm 1, \pm 2, \ldots \tag{3.24}$$

By comparing Equation (3.22) with Equations (3.25) or (3.23) with Equation (3.24), we conclude that

$$r_{xy}(j) = r_{yx}(-j), \tag{3.25}$$

Therefore, $r_{yx}(j)$ is the folded version of $r_{xy}(j)$, where the folding is done with respect to $j = 0$. Hence $r_{yx}(j)$ provides exactly the same information as $r_{xy}(j)$ about the similarity of $x(n)$ to $y(n)$.

In dealing with finite duration sequences, it is customary to express the auto correlation and cross correlation in terms of finite limits on summation. In particular if $x(n)$ and $y(n)$ are causal sequences of length N [i.e., $x(n) = y(n) = 0$ for $n < 0$ and $n \geq N$], the cross correlation and auto correlation sequences may be expressed as

$$r_{xy}(j) = \sum_{n=i}^{N-|K|-1} x(n)\, y(n-j) \quad j = 0, \pm 1, \pm 2, \ldots \tag{3.26}$$

$$r_{xx}(j) = \sum_{n=i}^{N-|K|-1} x(n)\, x(n-j) \quad j = 0, \pm 1, \pm 2, \ldots \tag{3.27}$$

where N is the maximum length of the either sequence and $i = j$, $k = 0$ for $j \geq 0$, and $i = 0$, $k = j$ for $j < 0$.

3.12.1 Correlation by Graphical Method

Example 3.24

Find the correlation $r_{yx}(j)$ of the samples of sequences given in Figure.

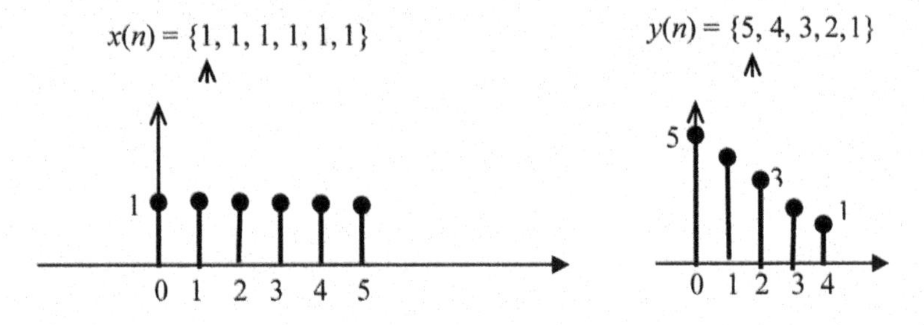

Discrete representation for cross-correlation of $x(n)$ and $y(n)$

Solution 3.24

$$r_{xy}(j) = \sum_{n=-\infty}^{\infty} x(n)\, y(n-j) \quad j = 0, \pm1, \pm2, \ldots$$

Here index j is the time shift (or lag) parameter and the subscripts xy on the cross-correlation sequence $r_{xy}(j)$ indicate the sequence being correlated.

To elaborate the above equation, the sequence $x(n)$ is left unshifted and $y(n)$ is shifted by j units of time, to the right for j positive and to the left for j negative.

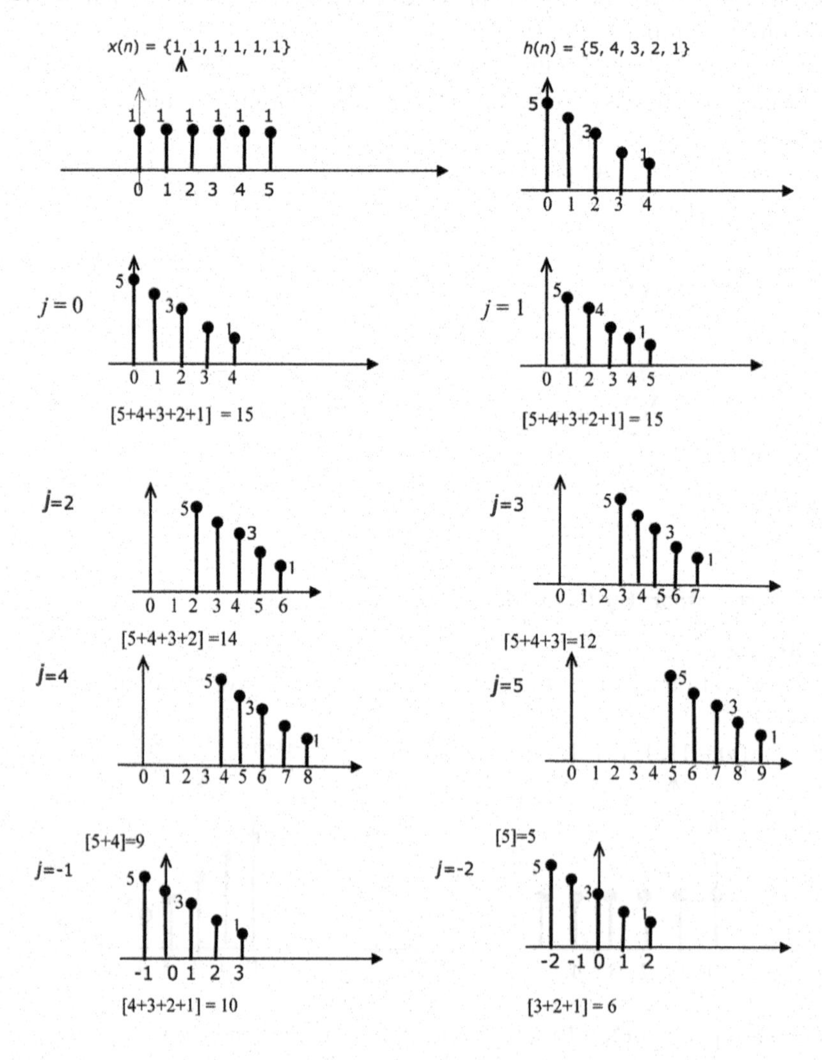

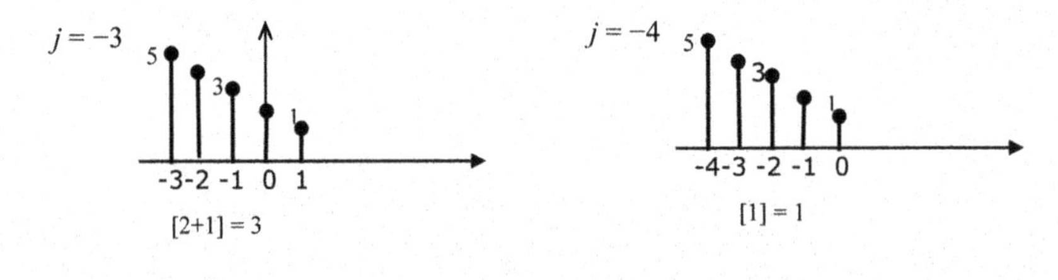

$$r_{xy} = \left\{1,\ 3,\ 6,\ 10,\ 15,\ \underset{\uparrow}{15},\ 14,\ \underset{\uparrow}{12},\ 9,\ 5\right\}$$

3.12.2 Correlation by Analytical Method

Analytical method uses the formula directly. The following examples have been included to explain the procedure.

Example 3.25

Find the cross-correlation of the signal given in tabular form

n	0	1	2	3
$x(n)$	4	3	1	6
$y(n)$	5	2	3	

Solution 3.25

Here in this example the sequence $x(n)$ and $y(n)$ are causal,

$$r_{xy}(j) = \sum_{n=-\infty}^{\infty} x(n)\, y(n-j) \quad j = 0, \pm1, \pm2, \ldots$$

Here the sequence lengths are 4 and 3 and the number of lags necessary is $4 + 3 - 1 = 6$ (that is sample size after correlation)

$$r_{xy}(j) = \sum_{n=0}^{5} x(n) y(n-j) \quad j = 0, \pm1, \pm2, \ldots$$

$$r_{xy}(j) = x(0)\, y(0-j) + x(1)\, y(1-j) + x(2)\, y(2-j) + x(3)\, y(3-j) + x(4)\, y(4-j) + x(5)\, y(5-j)$$

because the value of $x(4)$ and $x(5)$ is given zero in the problem, so deleting the two additional terms.

$$r_{xy}(j) = x(0)\, y(0-j) + x(1)\, y(1-j) + x(2)\, y(2-j) + x(3)\, y(3-j)$$

now substituting the value of 0, ± 1, ± 2, ... in the above expression

$$\begin{aligned}
r_{xy}(0) &= x(0)y(0) + x(1)y(1) + x(2)y(2) + x(3)y(3) \\
&= (4)(5) + (3)(2) + (1)(3) = 29 \\
r_{xy}(1) &= x(0)\,y(-1) + x(1)\,y(0) + x(2)\,y(1) + x(3)\,y(2) \\
&= (3)(5) + (1)(2) + (6)(3) = 35 \\
r_{xy}(-1) &= x(0)\,y(1) + x(1)\,y(2) + x(2)\,y(3) + x(3)\,y(4) \\
&= (4)(2) + (3)(3) = 17 \\
r_{xy}(-2) &= x(0)\,y(2) + x(1)\,y(3) + x(2)\,y(4) + x(3)\,y(5) = (4)(3) = 12 \\
r_{xy}(3) &= x(0)\,y(-3) + x(1)\,y(-2) + x(2)\,y(-1) + x(3)\,y(0) \\
&= (6)(5) = 30 \\
r_{xy}(2) &= x(0)\,y(-2) + x(1)\,y(-1) + x(2)\,y(0) + x(3)\,y(1) \\
&= (1)(5) + (6)(2) = 17
\end{aligned}$$

$$r_{xy}(j) \;=\; \{12,\ 17,\ 29,\ 35,\ 17,\ 30\}.$$

The above example can be solved also using the formula shown below, which is not solved here and left for students.

$$r_{xy}(j) = \sum_{n=i}^{N-|K|-1} X(n)\,Y(n-j) \quad j = 0,\ \pm 1,\ \pm 2, \ldots$$

where N is the maximum length of the either sequence and $i = j$, $k = 0$ for $j \geq 0$, and $i = 0$, $k = j$ for $j < 0$.

3.12.3 Correlation by Tabular Shifting Method

For calculating the linear correlation using tabular form zero padding is selected, if the sequence lengths are N_1 and N_2 and the number of lags necessary is $N_1 + N_2 - 1$ is required. This reveals the general rule for obtaining the linear cross-correlation two periodic sequences of lengths N_1 and N_2: add augmenting zeros to each sequence to make the lengths of each sequence $N_1 + N_2 - 1$.

This may be expressed as adding $N_2 - 1$ zeros to the sequence of length N_1 and adding $N_1 - 1$ zeros to the sequence of length N_2. The following example shows the procedure for length of correlated sequence, zeros padding, repetition of the sequence but does not show the location of zero sequence.

Example 3.26

What will be the cross-correlation length of the signal after zero padding?

n	0	1	2	3
$x(n)$	4	3	1	6
$y(n)$	5	2	3	0

Solution 3.26

Here the sequence lengths are 4 and 3 and the number of lags (size of the correlated sample) necessary is 4 + 3 – 1 = 6. Therefore 2 zeros padding is required in the larger sample.

n	0	1	2	3	4	5
$x(n)$	4	3	1	6	0	0
$y(n)$	5	2	3	0	0	0

If $j < 0$, negative, left shifting is carried out.
If $j > 0$, positive, right shifting is carried out.
This is now demonstrated for the given sequences x and y.

			Sequence				Lag	$r_{xy}(j)$
4	3	1	6	0	0	j		
5	2	3	0	0	0	0	0 no shifting	29
2	3	0	0	0	5	–1	I unit left shifting	17
3	0	0	0	5	2	–2	2 units left shifting	12
0	0	0	5	2	3	3	3 units right shifting	30
0	0	5	2	3	0	2	2 units right shifting	17
0	5	2	3	0	0	1	1 unit right shifting	35
5	2	3	0	0	0		Repetition starts	29

Thus, the required linear cross-correlation of x and y is

$$r_{xy}(j) = \{12, 17, 29, 35, \underset{\uparrow}{17}, 30\}.$$

3.12.4 Correlation by Convolution Property Method

The convolution property method is easiest method to find the linear auto and cross correlation. It uses the properties that $r_{xy}(j) = r_{yx}(-j)$; $r_{12}(j) = x_1(n) \otimes x_2(-n)$ and $r_{21}(j) = x_2(n) \otimes x_1(-n)$. It is evident from examples

Example 3.27

Prove that the auto-correlation of the signal given in tabular form is an even sequence.

n	0	1	2	3
$x(n)$	1	2	1	1
$y(n)$	1	2	1	1

Solution 3.27

$$r_{xx}(j) = x(n) \otimes x(-n)$$

$$x(n) = \left\{ \underset{\uparrow}{1}\ 2\ 1\ 1 \right\}\ x(-n) = \left\{ 1\ 1\ 2\ \underset{\uparrow}{1} \right\}$$

$$\begin{aligned} y(-3) &= 1,\ y(-2) = 3 \\ y(-1) &= 5,\ y(0) = 7 \\ y(1) &= 5,\ y(2) = 3,\ y(3) = 1 \end{aligned}$$

$$r_{yx}(j) = \left\{ 1, 3,\ 5,\ \underset{\uparrow}{7},\ 5,\ 3,\ 1 \right\}$$

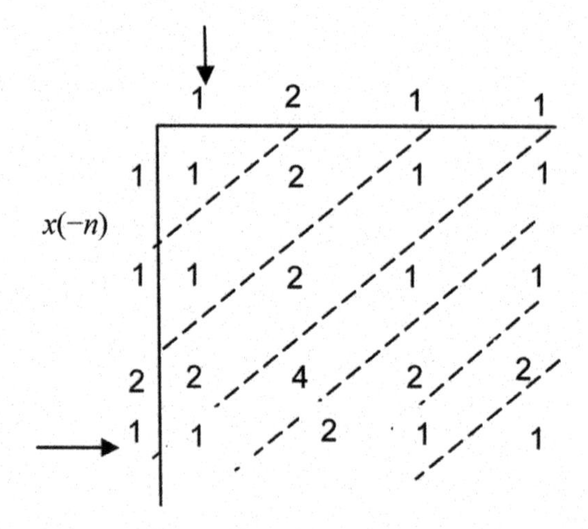

Example 3.28

Find the cross-correlation r_{yx} of the signal given in tabular form using convolution property.

n	0	1	2	3
$x(n)$	4	3	1	6
$y(n)$	5	2	3	0

Solution 3.28

For $r_{yx}(j) = x(-n) \otimes y(n)$

$$x(n) = \left\{\underset{\uparrow}{4}\ 3\ 1\ 6\right\} \quad y(n) = \left\{\underset{\uparrow}{5}\ 2\ 3\right\}$$

$$x(-n) = \left\{6\ 1\ 3\ \underset{\uparrow}{4}\right\} \quad y(n) = \left\{\underset{\uparrow}{5}\ 2\ 3\right\}$$

$$\begin{aligned} y(-3) &= 30,\ y(-2) = 17 \\ y(-1) &= 35,\ y(0) = 29 \\ y(1) &= 17,\ y(2) = 12 \end{aligned}$$

$$r_{yx}(j) = \left\{30, 17, 35, \underset{\uparrow}{29}, 17, 12\right\}$$

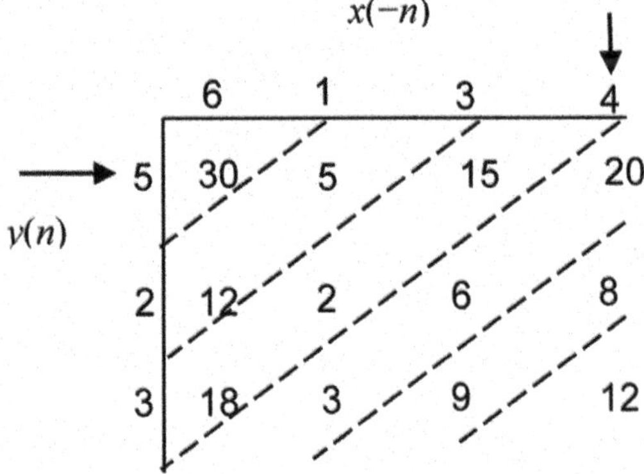

3.13 Cyclic Correlation

Care has to be exercised when cyclic cross-correlating two unequal length sequences when they are periodic. This is because the result of the correlation will be cyclic with the period of the shorter sequence. This result does not represent the full periodicity of the longer sequence and is, therefore, incorrect.

Example 3.29

Compute cyclic correlation of two sequences $a = \{4, 3, 1, 6\}$ and $b = \{5, 2, 3\}$ to obtain $r_{ab}(j)$.

Solution 3.29

The sequence b is placed below sequence a, and b is shifted left, this may be demonstrated by cyclic cross-correlating the sequences $a = \{4, 3, 1, 6\}$ and $b = \{5, 2, 3\}$ to obtain $r_{ab}(j)$. The sequence b is placed below sequence a, and b is shifted left by one lag on each of the subsequent rows, with the value of the cross-correlation appearing in the final column on the right.

	Sequence			Lag	$r_{ab}(j)$
4	3	1	6		
5	2	3	5	1	59
2	3	5	2	2	34
3	5	2	3	3	47
5	2	3	5	4	59 $r_{ab}(j)$ repeats

The result shows that $r_{ab}(j)$ is cyclic, repeating every third lag, that is $r_{ab}(j)$ has the same period as that of the shorter sequence, b.

Thus, the required circular-correlation of a and b is $r_{ab}(j) = \{59, 34, 47\}$.

This procedure is known as cyclic correlation. To obtain the correct value in which each value in a is multiplied by each value in b, all the elements in b have to be shifted in turn below each value.

Example 3.30

Compute cyclic correlation of two sequences $a = \{1, 2, 3, 4\}$ and $b = \{4, 3, 2, 1\}$ to obtain $r_{ab}(j)$.

Solution 3.30

The sequence b is placed below sequence a, and b is shifted left. This may be demonstrated by cyclic cross-correlating the sequences $a = \{1, 2, 3, 4\}$ and $b = \{4, 3, 2, 1\}$ to obtain $r_{ab}(j)$. The sequence b is placed below sequence a, and b is shifted left by one lag on each of the subsequent rows, with the value of the cross-correlation appearing in the final column on the right.

	Sequence			Lag	$r_{ab}(j)$
1	2	3	4		
4	3	2	1	0	20
3	2	1	4	1	11
2	1	4	3	2	26
1	4	3	2	3	25
4	3	2	1	4	20 $r_{ab}(j)$ repeats

The result shows that $r_{ab}(j)$ is cyclic, repeating every third lag, that is $r_{ab}(j)$ has the same period as that of the shorter sequence, b.

Thus, the required circular-correlation of a and b is $r_{ab}(j) = \{20, 11, 26, 25\}$.

3.14 Further Applications of Correlation

It can be shown that $\Im[r_{11}(\tau)] = G_E(f)$, where $G_E(f)$ is the energy spectral density of the waveform, that is the energy spectral density and the auto-correlation function constitute a Fourier Transform pair.

The distribution of energy in the signal is called as spectral density of the waveform. It can further be shown that $r_{11}(0) = E$; where E is the total energy of the waveform.

Example 3.31

Obtain a relationship between the zero-lag correlation functions of two different waveforms and their total energy content.

Solution 3.31

Let the waveforms be $v_1(n)$ and $v_2(n)$, and let their summation be

$$V(n) = u_1(n) + u_2(n).$$

The zero-lag autocorrelation function of $V(n)$ is

$$r_{vv}(0) = E_V = \frac{1}{N}\sum_{n=0}^{N-1} V^2(n) = \frac{1}{N}\sum_{n=0}^{N-1} [v_1(n) + v_2(n)]^2$$

where E_V is the energy of the waveform $V(n)$.

3.15 Problems and Solutions

Problem 3.1

Consider the interconnection of LTI system as shown in the figure.

(a) Express the overall impulse response in terms of $h_1(n)$, $h_2(n)$, $h_3(n)$, and $h_4(n)$.

(b) Determine $h(n)$ when

$$h_1(n) = \{\tfrac{1}{2}, \tfrac{1}{4}, \tfrac{1}{2}\}$$
$$h_2(n) = h_3(n) = (n + 1)u(n)$$
$$h_4(n) = \delta(n - 2)$$

(c) Determine the response of the system in part (b) if

$$x(n) = \delta(n+2) + 3\delta(n-1) - 4\delta(n-3)$$

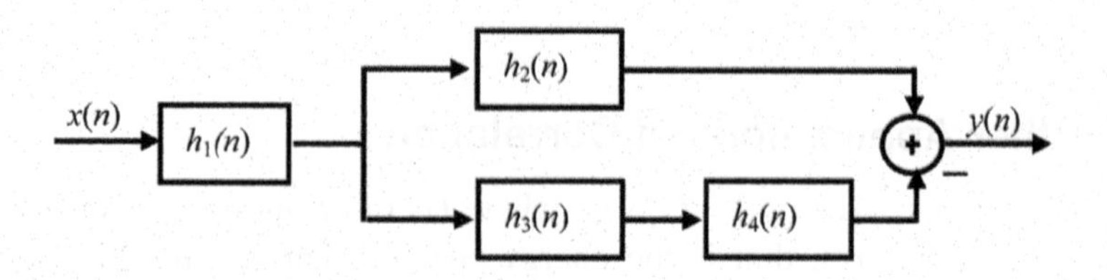

Solution 3.1

(a) $h(n) = h_1(n) \otimes [h_2(n) - \{h_3(n) \otimes h_4(n)\}]$

$h_3(n) \otimes h_4(n) = (n+1)u(n) \otimes \delta(n-2)$

$= \delta(n) + (n-1)u(n)$

$h_2(n) - [h_3(n) \otimes h_4(n)] = [(n+1)u(n)] - [\delta(n) + (n-1)$

$u(n)] = 2\,u(n) - \delta(n)$

(b) $h_1(n) = \frac{1}{2}\delta(n) + \frac{1}{4}\delta(n-1) + \frac{1}{2}\delta(n-2)$

$h(n) = h_1(n)\,[h_2(n) - \{h_3(n)\,h_4(n)\}]$

$\frac{1}{2}\delta(n) + \frac{1}{4}\delta(n-1) + \frac{1}{2}\delta(n-2)]\,[2u(n) - \delta(n)]$

Hence, $h(n) = \frac{1}{2}\delta(n) + \frac{5}{4}\delta(n-1) + 2\delta(n-2) + \frac{5}{2}u(n-3)$

(c) $x(n) = \{1, 0, \underset{\uparrow}{0}, 3, 0, -4\}$

$$h(n) = \{\tfrac{1}{2}, \tfrac{5}{4}, 2, \tfrac{5}{2}\}$$

$$y(n) = \{\tfrac{1}{2}, \tfrac{5}{4}, \underset{\uparrow}{2}, 4, \tfrac{15}{4}, 4, \tfrac{5}{2}, -8, -10\}$$

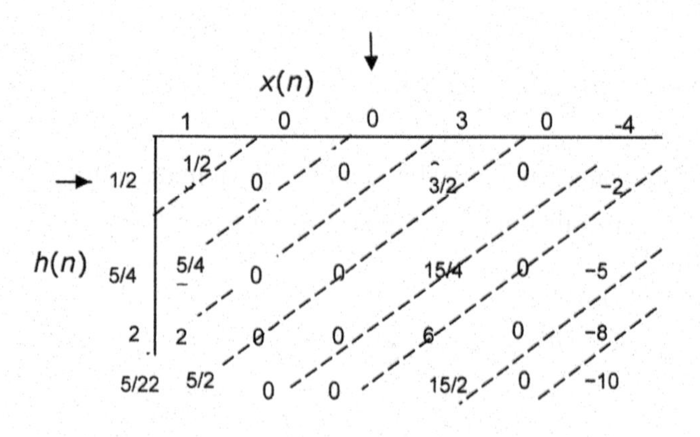

Problem 3.2

(a) If $y(n) = x(n) \otimes h(n)$, show that, $\sum y = \sum x \sum h$

Here $\otimes$ is taken as convolution

(b) Compute the convolution $y(n) = x(n) \otimes h(n)$ of the following signals and check the correctness of the results by using the test in (a).

$$\begin{array}{lll} (i) & x(n) = \{1,\ 2,\ 4\}, & h(n) = \{1,\ 1,\ 1,\ 1,\ 1\} \\ (ii) & x(n) = \{1,\ 2,\ -1\}, & h(n) = x(n) \\ (iii) & x(n) = \{1,\ 2,\ 3,\ 4,\ 5\}, & h(n) = \{1\} \end{array}$$

Solution 3.2

(a)

$$y(n) = \sum h(k)\, x(n-k)$$

$$\sum y(n) = \sum \sum h(k)\, x(n-k) = \sum h(k) \sum_{n=-\infty}^{\infty} x(n-k)$$

$$= \left(\sum h(k)\right)\left(\sum x(n)\right)$$

(b) Using Overlap and add method for easy solution

$$y(n) = x(n) \otimes h(n)$$

$$(i) \quad x(n) = \{1,\ 2,\ 4\},\ h(n) = \{1,\ 1,\ 1,\ 1,\ 1\}$$

$$x(n) = \left\{\underset{\uparrow}{1}\ 2\ 4\right\}\ x(-n) = \{\underset{\uparrow}{1}\ 1\ 1\ 1\ 1\}$$

$$\begin{aligned} 1\ y(0) &= 1,\ y(1) = 3 \\ y(2) &= 7,\ y(3) = 7 \\ y(4) &= 7,\ y(6) = 6,\ y(7) = 4 \end{aligned}$$

$$r_{yx}(j) = \left\{\underset{\uparrow}{1}, 3,\ 7,\ 7,\ 7,\ 6,\ 4\right\}$$

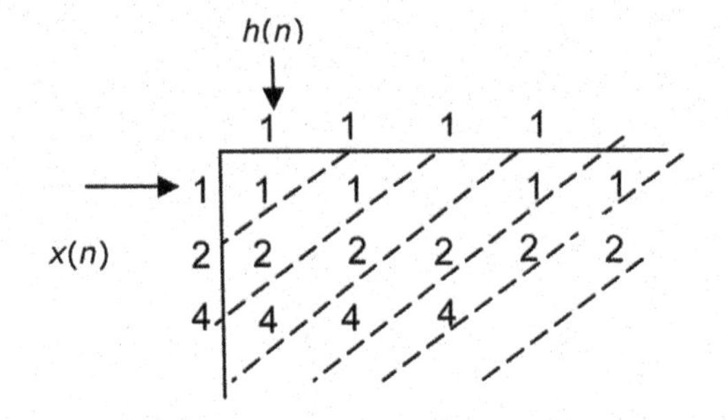

$$\sum y(k) = \left(\sum_k h(k)\right)\left(\sum_n x(n)\right)$$

$$\sum y(n) = 35 \quad \sum x(n) = 7 \quad \sum_k h(k) = 5$$

$$\sum y(k) = (\sum h(k))(\sum x(n))$$
$$\sum y(n) = 4 \quad \sum x(n) = 2 \quad \sum h(k) = 2$$

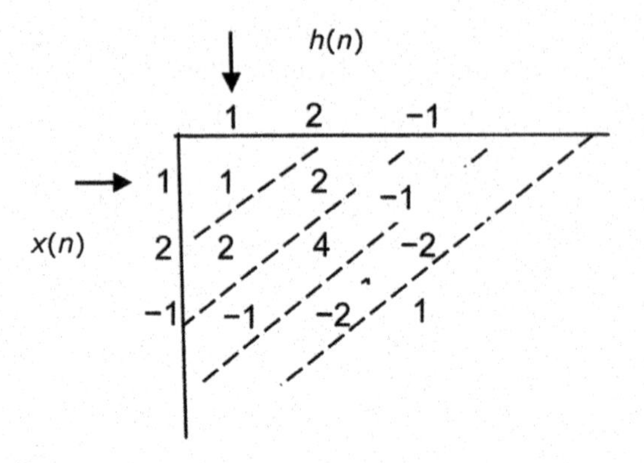

(i) $x(n) = \{1, 2, 3, 4, 5\}$, $h(n) = \{1\}$

$y(0) = 1$, $y(1) = 2$
$y(2) = 3$, $y(3) = 4$
$y(4) = 5$

$r_{yx}(j) = \{\underset{\uparrow}{1}, 2, 3, 4, 5\}$

$$\sum y(k) = (\sum h(k))(\sum x(n))$$
$$\sum y(n) = 15 \quad \sum x(n) = 15 \quad \sum h(n) = 1$$

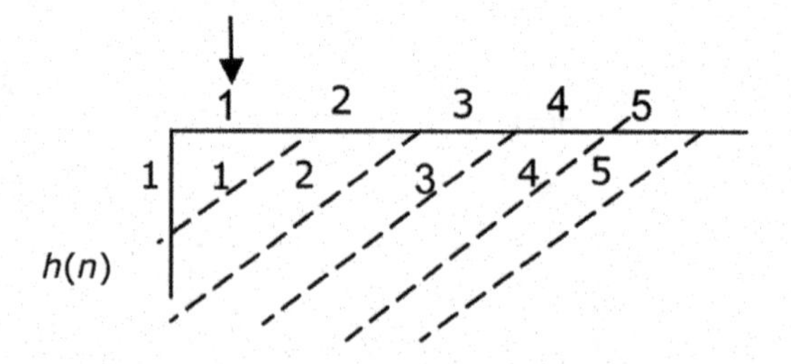

Problem 3.3

Compute the convolution $y(n) = x(n) \otimes h(n)$ of the following signals and check the correctness of the results.

$$x(n) = \left(\frac{1}{2}\right)^n u(n) \quad h(n) = \left(\frac{1}{5}\right)^n u(n)$$

Solution 3.3

The convolution of function of $x(n)$ with $h(n)$ is

$$y(n) = x(n) \otimes h(n) = \sum_{k=-\infty}^{\infty} x(k)h(n-k)$$

Because both sequences are finite in length,

$x(n)$ is changed into $x(k)$ $x(k) = \left(\frac{1}{2}\right)^k u(k)$

$h(n)$ is changed into $h(k)$ $x(k) = \left(\frac{1}{5}\right)^{n-k} u(n-k)$

$$= \sum_{k=-\infty}^{\infty} \left(\frac{1}{2}\right)^k u(k) \left(\frac{1}{5}\right)^{n-k} u(n-k)$$

Due to the step $u(k)$ in the first function, the lower limit on the sum may be changed to $k = 0$, and the upper limit may be changed to $k = n$. $n > 0$ Thus the convolution sum becomes as follows:

$$y(n) = \left(\frac{1}{5}\right)^n \sum_{k=0}^{n} \left(\frac{1}{2}\right)^k \left(\frac{1}{5}\right)^{-k} n \geq 0; \ y(n) = \left(\frac{1}{5}\right)^n \sum_{k=0}^{n} \left(\frac{5}{2}\right)^k n \geq 0$$

Using the series given in Table 3.1, we have

$$y(n) = \left(\frac{1}{5}\right)^n \left\{\frac{1 - \left(\frac{5}{2}\right)^{n+1}}{1 - \frac{5}{2}}\right\} = \left(\frac{1}{5}\right)^n \left\{\frac{1 - \frac{5}{2}\left(\frac{5}{2}\right)^n}{\frac{3}{2}}\right\}$$

$$y(n) = \left[\left(\tfrac{1}{5}\right)^n \left[\frac{2}{3}\left(1 - \frac{5}{2}\left(\tfrac{5}{2}\right)^n\right)\right]\right] u(n)$$

Problem 3.4

Compute the convolutions $x(n) \otimes h(n)$ for the pairs of signals shown in Figure.

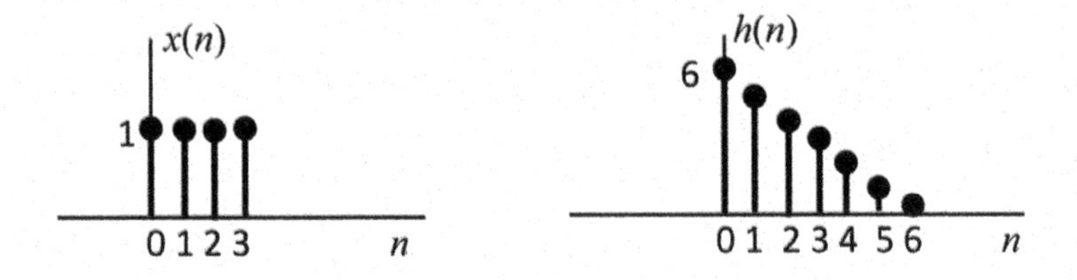

Solution 3.4

$$x(n) = \{\underset{\uparrow}{1}, 1, 1, 1\} \quad h(n) = \{\underset{\uparrow}{6}, 5, 4, 3, 2, 1\}$$

$$y(n) = \sum_{k=0}^{n} x(k)h(n-k)$$

$$\begin{aligned}
y(0) &= x(0)\,h(0) = 6\\
y(1) &= x(0)\,h(1)\ +\ x(1)\,h(0) = 11\\
y(2) &= x(0)\,h(2)\ +\ x(1)\,h(1)\ +\ x(2)\,h(0) = 15\\
y(3) &= x(0)\,h(3)\ +\ x(1)\,h(2)\ +\ x(2)\,h(1)\ +\ x(3)\,h(0) = 18\\
y(4) &= x(0)\,h(4)\ +\ x(1)\,h(3)\ +\ x(2)\,h(2)\ +\ x(3)\,h(1) = 14\\
y(5) &= x(0)\,h(5)\ +\ x(1)\,h(4)\ +\ x(2)\,h(3)\ +\ x(3)\,h(2) = 10\\
y(6) &= x(1)\,h(5)\ +\ x(2)\,h(4)\ +\ x(3)\,h(2) = 6\\
y(7) &= x(2)\,h(5)\ +\ x(3)\,h(4) = 3\\
y(8) &= x(3)\,h(5) = 1
\end{aligned}$$

$$\begin{aligned}
y(-2) &= 0,\ y(-1) = 1/3\\
y(0) &= 1,\ y(1) = 2\\
y(2) &= 1\\
y(n) &= \ \{0, 1/3,\ 1,\ 2,\ 5/3, 1\}
\end{aligned}$$

$$y(n) = \{\underset{\uparrow}{6},\ 11,\ 15,\ 18,\ 14,\ 10,\ 6,\ 3,\ 1\}$$

Problem 3.5

Determine and sketch the convolution $y(n)$ of the signals by overlap and add method

$$x(n) = \begin{cases} \frac{1}{3}n, & 0 \le n \le 3\\ 0, & \text{elsewhere} \end{cases}$$

$$h(n) = \begin{cases} 1, & -2 \le n \le 0\\ 0, & \text{elsewhere} \end{cases}$$

Solution 3.5

$$(a)\quad x(n) = \{\underset{\uparrow}{0}, \tfrac{1}{3}, \tfrac{2}{3}, 1\}$$

$$h(n) = \{1, 1, \underset{\uparrow}{1}\}$$

$$y(n) = x(n) * h(n) = \{0, \tfrac{1}{3}, \underset{\uparrow}{1}, 2, \tfrac{5}{3}, 1\}$$

$$\begin{aligned}
y(-2)\ &=\ 0,\ y(-1) = 1/3\\
y(0)\ &=\ 1,\ y(1) = 2\\
y(2)\ &= 1\\
y(n) &= \left\{0, 1/3,\ 1,\ \underset{\uparrow}{2},\ 5/3, 1\right\}
\end{aligned}$$

$$\sum y(k) = \left(\sum_k h(k)\right)\left(\sum_n x(n)\right)$$
$$\sum y(n) = 6 \quad \sum x(n) = 2 \quad \sum h(n) = 3$$

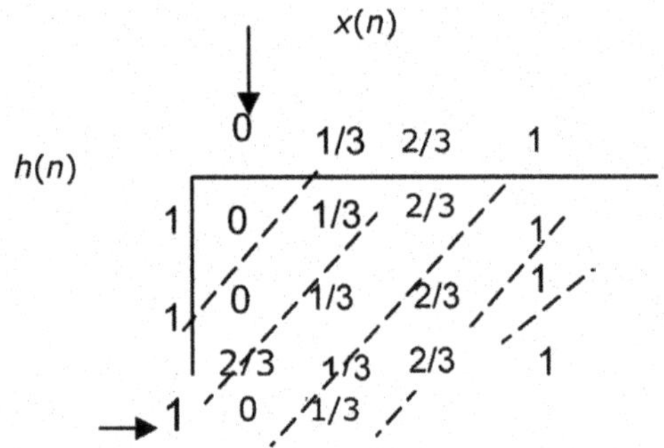

Problem 3.6
Compute the convolution *y*(*n*) of the signals

$$x(n) = \begin{cases} \alpha^n, & -3 \le n \le 5 \\ 0, & \text{elsewhere} \end{cases}$$
$$h(n) = \begin{cases} 1, & 0 \le n \le 4 \\ 0, & \text{elsewhere} \end{cases}$$

Solution 3.6
The limits are from –3 to 5 for *x*(*n*)

$$x(n) = \{\alpha^{-3}, \alpha^{-2}, \alpha^{-1}, \underset{\uparrow}{1}, \alpha, \ldots.\ \alpha^5\}.$$

The output limits are from 0 to 4 for *x*(*n*)

$$y(n) = \sum_{k=0}^{4} h(k)x(n-k)$$
$$h(n) = \{\underset{\uparrow}{1}, 1, 1, 1, 1\}$$
$$y(n) = \sum_{k=0}^{4} x(n-k), \quad -3 \le n \le 9$$
$$= 0, \quad \text{otherwise}$$

Therefore

$$y(-3) = \alpha^{-3},\ y(-2) = x(-3) + x(-2) = \alpha^{-3} + \alpha^{-2}$$
$$y(-1) = \alpha^{-3} + \alpha^{-2} + \alpha^{-1}, \qquad y(0) = \alpha^{-3} + \alpha^{-2} + \alpha^{-1} + 1$$

$$y(1) = \alpha^{-3} + \alpha^{-2} + \alpha^{-1}, \quad y(2) = \alpha^{-2} + \alpha^{-1} + 1 + \alpha + \alpha^2$$
$$y(3) = \alpha^{-1} + 1 + \alpha + \alpha^{-2} + \alpha^3, \quad y(4) = 1 + \alpha + \alpha^2 + \alpha^3 + \alpha^4$$
$$y(5) = \alpha + \alpha^2 + \alpha^3 + \alpha^4 + \alpha^5, \quad y(6) = \alpha^2 + \alpha^3 + \alpha^4 + \alpha^5$$
$$y(7) = \alpha^3 + \alpha^4 + \alpha^5, \quad y(8) = \alpha^4 + \alpha^5, \quad y(9) = \alpha^5$$

Problem 3.7

Consider the following three operations.

1. Multiply the integer numbers: 131 and 122.
2. Compute the convolution of signals: $\{1, 3, 1\} \otimes \{1, 2, 2\}$.
3. Multiply the polynomials: $1 + 3z + z^2$ and $1 + 2z + 2z^2$.
4. Repeat part (a) for the numbers 1.31 and 12.2.
5. Comment on your results.

Solution 3.7

(a) $131 \times 122 = 15982$

(b) $\{\underset{\uparrow}{1}, 3, 1\} \times \{\underset{\uparrow}{1}, 2, 2\} = \{1, 5, 9, 8, 2\}$

(c) $(1 + 3z + z^2)(1 + 2z + 2z^2) = 1 + 5z + 9z^2 + 8z^3 + 2z^4$

(d) $1.31\ x\ 12.2 = 15.982$

These are different ways to perform convolution

Problem 3.8

Compute the convolution $y(n) = x(n) \otimes h(n)$ of the following pairs of signals.

(a) $x(n) = a^n u(n),\ h(n) = b^n u(n)$ when $a \neq b$ and when $a = b$

(b) $x(n) = \begin{cases} 1, & n = -2,\ 1,\ 0 \\ 2, & n = -1 \\ 0, & \text{elsewhere} \end{cases}$

$h(n) = \delta(n) - \delta(n-1) + \delta(n-4) + \delta(n-5)$

(c) $x(n) = u(n+1) - u(n-4) - d(n-5)$

$h(n) = [u(n+2) - u(n-3)]\,(3 - |n|)$

(d) $x(n) = u(n) - u(n-5)$

$h(n) = u(n-2) - u(n-8) + u(n-11) - u(n-17)$

Solution 3.8

(a) $y(n) = \sum\limits_{k=0}^{n} a^k u(k) b^{n-k} u(n-k) = b^n \sum\limits_{k=0}^{n} (ab^{-1})^k$

$$y(n) = \begin{cases} \frac{b^{n+1}-a^{n+1}}{b-a}u(n), & a \neq b \\ b^n(n+1)u(n), & a = b \end{cases}$$

(b) $x(n) = \{1, 2, \underset{\uparrow}{1}, 1\}, h(n) = \{\underset{\uparrow}{1}, -1, 0, 0, 1, 1\},$

$y(n) = \{1, 1, -\underset{\uparrow}{1}, 0, 0, 3, 3, 2, 1\}$

(c) $x(n) = \{1, \underset{\uparrow}{1}, 1, 1, 1, 0, -1\}, \quad h(n) = \{1, 2, \underset{\uparrow}{3}, 2, 1\}$

$y(n) = \{1, 3, 6, \underset{\uparrow}{8}, 9, 8, 5, 1, -2, -2, -1\}$

(d) $x(n) = \{\underset{\uparrow}{1}, 1, 1, 1, 1\}, \quad h(n) = \{\underset{\uparrow}{0}, 0, 1, 1, 1, 1, 1, 1\}$

$$h(n) = h''(n) + h'(n-9)$$
$$y(n) = y'(n) + y'(n-9)$$

where, $y^1(n) = \{\underset{\uparrow}{0}, 0, 1, 2, 3, 4, 5, 5, 4, 3, 2, 1\}$

Problem 3.9

The first non-zero value of a finite-length sequence $x(n)$ occurs at index $n = -6$ and has a value $x(-6) = 3$, and the last non-zero value occurs at index $n = 24$ and has a value $x(24) = -4$. What is the index of the first nonzero value in the convolution? What is its value?

$$y(n) = x(n) \otimes x(n)$$

What about the last non-zero value?

Solution 3.9

Because we are convolving two finite-length sequences, the index of the first non-zero value in the convolution is equal to the sum of the indices of the first non-zero values of the two sequences that are being convolved. In the case, the index is $n = -12$, and the value is

$$y(-12) = x^2(-6) = 9.$$

Similarly, the index of the last non-zero value is at $n = 48$ and the value is

$$y(48) = x^2(24) = 16$$

Problem 3.10

The convolution of two finite-length sequences will be finite in length. Is it true that the convolution of a finite-length sequence with an infinite-length sequence will be infinite in length?

Solution 3.10

It is not necessarily true that the convolution of an infinite-length sequence with a finite-length sequence will be infinite in length. It may be either. Clearly, if $x(n) = \delta(n)$ and $h(n) = (0.5)^n u(n)$, the convolution will be an infinite-length sequence. However, it is possible for the finite-length sequence to remove the infinite-length tail of an infinite-length sequence. For example, note that

$$(0.5)^n u(n) - (0.5)^n u(n-1) = \delta(n).$$

Therefore, the convolution of $x(n) = \delta(n) - \frac{1}{2}\delta(n-1)$ with $h(n) = (0.5)^n u(n)$ will be finite in length:

$$\left[\delta(n) - \frac{1}{2}\delta(n-1)\right] \otimes (0.5)^n u(n) = (0.5)^n u(n) - \frac{1}{2}(0.5)^{n-1} u(n-1) = \delta(n)$$

Problem 3.11

Find the convolution of the two finite-length sequences:

$$x(n) = 0.5n[u(n) - u(n-6)]$$

$$h(n) = 2\sin\left(\frac{n\pi}{2}\right)[u(n+3) - u(n-4)]$$

Solution 3.11

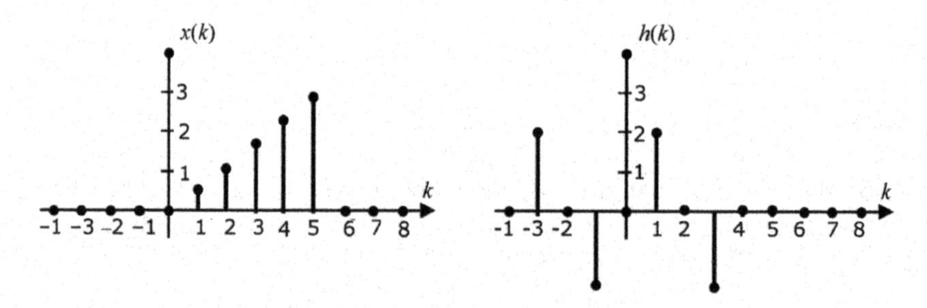

Because $h(n)$ is equal to zero outside the interval [–3, 3], and $x(n)$ is zero outside the interval [1, 5], the convolution $y(n) = x(n) \times h(n)$ is zero outside the interval [–2, 8].

Converting the given function expression in sequence form

$$x(k) = [\,0,\ 0.5,\ 1,\ 1.5,\ 2,\ 2.5] \qquad h(k) = [\,2,\ 0, -2,\ 0,\ 2,\ 0,\ -2]$$

$$y(n) = \ [0,\ 1,\ 2,\ 2,\ 2,\ 3,\ -2,\ -3,\ 2,\ 2,\ -4,\ -5]$$

Problem 3.12

A linear shift-invariant system has a unit sample response

$$h(n) = u(-n-1)$$

Find the output if the input is

$$x(n) = -n3^n u(-n)$$

Shown in figures are the sequences $x(n)$ and $h(n)$.

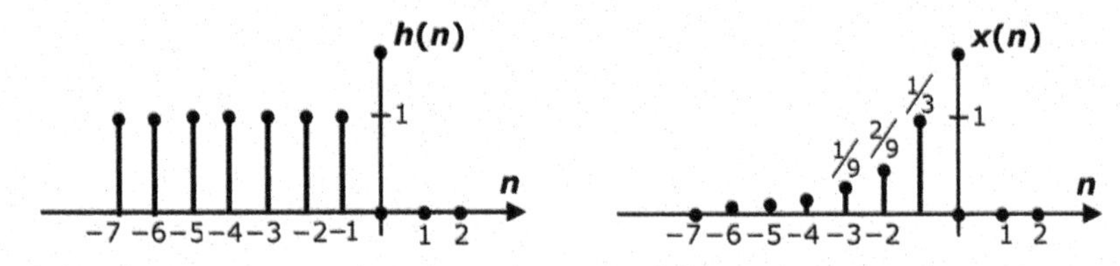

Solution 3.12

Because $x(n)$ is zero for $n > -1$, and $h(n)$ is equal to zero for $n > -1$, the convolution will be equal to zero for $n > 2$. Evaluating the convolution sum directly, converting $h(n)$ into $x(n)$ into $x(k)$ and $h(n)$ into $h(n-k)$

$x(n) = -n3^n u(-n)$, $x(k) = -k3^k u(-k)$ and

$$h(n) = u(-n-1), h(n-k) = u(-(n-k)-1).$$

Substituting in the formula the value of $x(k)$ and $h(n-k)$

$$y(n) = \sum_{k=-\infty}^{\infty} x(k)h(n-k) \qquad y(n) = \sum_{k=-\infty}^{\infty} -k3^k u(-k)u(-(n-k)-1)$$

$$y(n) = \sum_{k=-\infty}^{\infty} x(k)h(n-k) \qquad y(n) = \sum_{k=-\infty}^{\infty} -k3^k u(-k)u(-n+k-1)$$

Because $u(-k) = 0$ for $k > 0$ and $u(-n+k-1) = 0$ for $k < n+1$, the convolution sum becomes

$$y(n) = \sum_{k=-\infty}^{\infty} -k3^k u(-k)u(-n+k-1).$$

Once the expression of $x(k)$ and $h(k)$ is substituted the limits are changed from $-\infty$ to $+\infty$

$$y(n) = \sum_{k=0}^{n+1} -k3^k \quad n \le -2 \;\; y(n) = \sum_{k=0}^{n+1} m(3)^{-m} \;\; y(n) = \sum_{m=0}^{-n-1} m\left(\frac{1}{3}\right)3^m.$$

With the change of variables $m = -k$, and using the series formulas from Table 3.1, we have

$$y(n) = \sum_{m=0}^{-n-1} m\left(\frac{1}{3}\right)^m = \frac{(-n-1)\left(\frac{1}{3}\right)^{-n+1} + n\left(\frac{1}{3}\right)^{-n} + \frac{1}{3}}{\left(1-\frac{1}{3}\right)^2}$$

$$y(n) = \sum_{m=0}^{-n-1} m\left(\frac{1}{3}\right)^m = \frac{9}{4}\left\{(-n-1)\left(\frac{1}{3}\right)^{-n+1} + n\left(\frac{1}{3}\right)^{-n} + \frac{1}{3}\right\}$$

$$y(n) = \frac{9}{4}\left\{(-n-1)\left(\frac{1}{3}\right)^{-n+1} + n\left(\frac{1}{3}\right)^{-n} + \frac{1}{3}\right\}$$

$$y(n) = \frac{9}{4}\left\{\left(\frac{1}{3}\right)\cdot\left(\frac{1}{3}\right)^{-n}(-n-1) + n\left(\frac{1}{3}\right)^{-n} + \frac{1}{3}\right\}$$

$$y(n) = \frac{9}{4}\left\{-\left(\frac{1}{3}\right)\left(\frac{1}{3}\right)^{-n}.\, n - \left(\frac{1}{3}\right)\left(\frac{1}{3}\right)^{-n} + n\left(\frac{1}{3}\right)^{-n} + \frac{1}{3}\right\}$$

$$y(n) = \left\{-\left(\frac{3}{4}\right)\left(\frac{1}{3}\right)^{-n}.\, n - \left(\frac{3}{4}\right)\left(\frac{1}{3}\right)^{-n}\right\} - \left(\frac{9}{4}\right)n\left(\frac{1}{3}\right)^{-n} + \frac{3}{4}$$

$$y(n) = \frac{3}{4} - 3n\left(\frac{1}{3}\right)^{-n} - \frac{3}{4}\left(\frac{1}{3}\right)^{-n}.$$

Let us check this answer for a few values of n using graphical convolution. Time-reversing $x(k)$, we see that $h(k)$ and $x(-k)$ do not overlap for any k and, thus, $y(0) = 0$. In fact, it is not until we shift $x(-k)$ to the left by two that there is any overlap. With $x(-2-k)$ and $h(k)$ overlapping at one point, and the product being equal to $\frac{1}{3}$, it follows that $y(-2) = \frac{1}{3}$.

Evaluating the expression above for $y(n)$ above at index $n = -2$, we obtain the same result. For $n = -3$, the sequences $x(-3-k)$ and $h(k)$ overlap at two points, and the sum of the products gives $y(-3) = \frac{1}{3} + \frac{2}{9} = \frac{5}{9}$, which, again, is the same as the expression above.

Problem 3.13

Prove the commutative property of convolution

$$x(n) \otimes h(n) = h(n) \otimes x(n)$$

Solution 3.13

Proving the commutative property is straightforward and only involves a simple manipulation of the convolution sum. With the convolution of $x(n)$ with $h(n)$ given by

$$x(n) \otimes h(n) = \sum_{k=-\infty}^{\infty} x(k)h(n-k)$$

with the substitution $l = n - k$, we have $k = n - l$, we have

$$x(n) \otimes h(n) = \sum_{l=-\infty}^{\infty} x(n-l)h(l) = h(n) \otimes x(n)$$

and the commutative property is established.

Problem 3.14

Prove the distributive property of convolution

$$h(n) \otimes [x_1(n) + x_2(n)] = h(n) \otimes x_1(n) + h(n) \otimes x_2(n)$$

Solution 3.14

To prove the distributive property, we have

$$x(n) \otimes [x_1(n) + x_2(n)] = \sum_{k=-\infty}^{\infty} h(k)[x_1(n-k) + x_2(n-k)]$$

Therefore,

$$\begin{aligned} h(n) \otimes [x_1(n) + x_2(n)] &= \sum_{k=-\infty}^{\infty} h(k)x_1(n-k) + \sum_{k=-\infty}^{\infty} h(k)x_2(n-k) \\ &= h(n) \otimes x_1(n) + h(n) \otimes x_2(n) \end{aligned}$$

the property is established.

Problem 3.15

Let

$$h(n) = 3\left(\frac{1}{2}\right)^n u(n) - 2\left(\frac{1}{3}\right)^{n-1} u(n)$$

be the unit sample response of a linear shift-invariant system. If the input to this system is a unit step,

$$x(n) = \begin{cases} 1 & n \geq 0 \\ 0 & \text{else} \end{cases}$$

Find $\lim_{n\to\infty} y(n)$ where $y(n) = h(n) \otimes x(n)$

Solution 3.15

Converting $h(n)$ into $x(n)$ into $x(k)$ and $h(n)$ into $h(n-k)$

$x(n) = u(n)$, $x(k) = u(k)$ and

$$h(n) = 3\left(\frac{1}{2}\right)^{n} u(n) - 2\left(\frac{1}{3}\right)^{n-1} u(n), \quad h(n-k) = 3\left(\frac{1}{2}\right)^{n-k}$$

$$u(n-k) - 2\left(\frac{1}{3}\right)^{n-k-1} u(n-k) \quad h(n-k) = u(-(n-k)-1)$$

Substituting in the formula the value of $x(k)$ and $h(n-k)$

$$y(n) = h(n) \otimes x(n) = \sum_{k=-\infty}^{\infty} h(k)x(n-k)$$

$$y(n) = \sum_{k=-\infty}^{\infty} x(k)h(n-k)$$

$$y(n) = \sum_{k=-\infty}^{\infty} 3\left(\frac{1}{2}\right)^{n-k} u(n-k) - 2\left(\frac{1}{3}\right)^{n-k-1} u(n-k)u(k)$$

Because $u(-k) = 0$ for $k < 0$, the convolution sum becomes and evaluating the sum, we have

$$y(n) = 3\sum_{k=0}^{\infty}\left(\frac{1}{2}\right)^{n} - 2\sum_{k=0}^{\infty}\left(\frac{1}{3}\right)^{n-1} = 3\sum_{k=0}^{\infty}\left(\frac{1}{2}\right)^{n} - 2\sum_{k=0}^{\infty}\left(\frac{1}{3}\right)^{n-1}$$

$$3\sum_{k=0}^{\infty}\left(\frac{1}{2}\right)^{n} - 2\left(\frac{1}{3}\right)^{-1}\sum_{k=0}^{\infty}\left(\frac{1}{3}\right)^{n} = 3\sum_{k=0}^{\infty}\left(\frac{1}{2}\right)^{n} - 6\sum_{k=0}^{\infty}\left(\frac{1}{3}\right)^{n}$$

$$= 3\frac{1}{1-\frac{1}{2}} - 6.\frac{1}{1-\frac{1}{3}} = \frac{3}{1-1/2} - \frac{6}{1-1/3} = 3(2-3) = -3$$

Problem 3.16

Perform the convolution

$$y(n) = x(n) \otimes h(n)$$

where

$$h(n) = \left(\frac{1}{2}\right)^{n} u(n)$$

and $x(n) = \left(\frac{1}{3}\right)^{n}[u(n) - u(n-101)]$

Solution 3.16

With $y(n) = x(n) \otimes h(n) = \sum_{k=-\infty}^{\infty} x(k)h(n-k)$

We begin by substituting $x(n)$ and $h(n)$ into the convolution sum

$$x(n) = \left(\frac{1}{3}\right)^n [u(n) - u(n-101)]$$

$$y(n) = \sum_{k=-\infty}^{\infty} \left(\frac{1}{3}\right)^k [u(k) - u(k-101)] \left(\frac{1}{2}\right)^{n-k} u(n-k)$$

or $y(n) = \sum_{k=0}^{100} \left(\frac{1}{3}\right)^k \left(\frac{1}{2}\right)^{n-k} u(n-k)$

To evaluate this sum, which depends on n, we consider three cases. First, for $n < 0$, the sum is equal to zero because $u(n-k) = 0$ for $0 \leq k \leq 100$. Therefore,

$$y(n) = 0 \quad n < 0$$

Second, note that for $0 \leq n \leq 100$, the step $u(n-k)$ is only equal to 1 for $k \leq n$. Therefore,

$$y(n) = \sum_{k=0}^{n} \left(\frac{1}{3}\right)^k \left(\frac{1}{2}\right)^{n-k} = \left(\frac{1}{2}\right)^n \sum_{k=0}^{n} \left(\frac{2}{3}\right)^k$$

$$= \left(\frac{1}{2}\right)^n \frac{1 - \left(\frac{2}{3}\right)^{n+1}}{1 - \frac{2}{3}} = 3\left(\frac{1}{2}\right)^n \left[1 - \left(\frac{2}{3}\right)^{n+1}\right]$$

Finally, for $n \geq 100$, note that $u(n-k)$ is equal to 1 for all k in the range $0 \leq k \leq 100$. Therefore,

$$y(n) = \sum_{k=0}^{100} \left(\frac{1}{3}\right)^k \left(\frac{1}{2}\right)^{n-k} = \left(\frac{1}{2}\right)^n \sum_{k=0}^{100} \left(\frac{2}{3}\right)^k$$

$$= \left(\frac{1}{2}\right)^n \frac{1 - \left(\frac{2}{3}\right)^{101}}{1 - \frac{2}{3}} = 3\left(\frac{1}{2}\right)^n \left[1 - \left(\frac{2}{3}\right)^{101}\right]$$

In summary, we have

$$y(n) = \begin{cases} 0 & n < 0 \\ 3\left(\frac{1}{2}\right)^n \left[1 - \left(\frac{2}{3}\right)^{n+1}\right] & 0 \leq n \leq 100 \\ 3\left(\frac{1}{2}\right)^n \left[1 - \left(\frac{2}{3}\right)^{101}\right] & n \geq 100 \end{cases}$$

Problem 3.17

Let $h(n)$ be a truncated exponential

$$h(n) = \begin{cases} \alpha^n & 0 \le n \le 10 \\ 0 & \text{else} \end{cases}$$

and $x(n)$ a discrete pulse of the form

$$x(n) = \begin{cases} 1 & 0 \le n \le 5 \\ 0 & \text{else} \end{cases}.$$

Find the convolution $y(n) = h(n) \otimes x(n)$.

Solution 3.17

Convert impulse response from $h(n)$ to $h(k)$

$$h(k) = \begin{cases} \alpha^k & 0 \le k \le 10 \\ 0 & \text{else} \end{cases}$$

$$x(n-k) = \begin{cases} 1 & 0 \le n \le 5 \\ 0 & \text{else} \end{cases}.$$

To find the convolution of these two finite-length sequence, we need to evaluate the sum

$$y(n) = h(n) \times x(n) = \sum_{k=-\infty}^{\infty} h(k)x(n-k).$$

To evaluate this sum, it will be useful to make a plot of $h(k)$ and $x(n-k)$ as a function of k as shown in the figure.

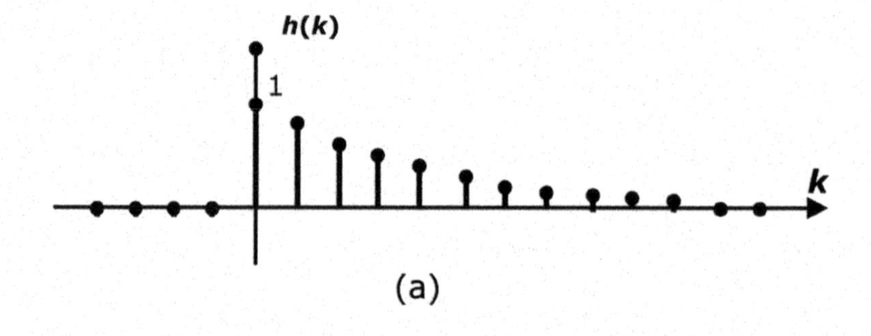

(a)

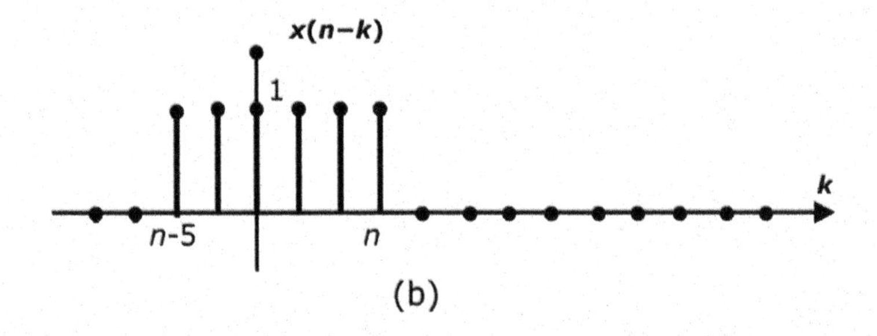

(b)

Note that the amount of overlap between $h(k)$ and $x(n-k)$ depends on the value of n. For example, if $n < 0$, there is no overlap, whereas for $0 \leq n \leq 5$, the two sequences overlap for $0 \leq k \leq n$. Therefore, in the following, we consider five separate cases.

Case 1: $n < 0$. When $n < 0$, there is no overlap between $h(k)$ and $x(n-k)$. Therefore, the product $h(k)\, x(n-k) = 0$ for all k, and $y(n) = 0$.

Case 2: $0 \leq n \leq 5$. For this case, the product $h(k)\, x(n-k)$ is non-zero only for k in the range $0 \leq k \leq n$. Therefore using summation Table of 3.1, we have

$$y(n) = \sum_{k=0}^{n} \alpha^k = \frac{1-\alpha^{n+1}}{1-\alpha}$$

Case 3: $6 \leq n \leq 10$. For $6 \leq n \leq 10$, all of the non-zero values of $x(n-k)$ are within the limits of the sum, and

$$y(n) = \sum_{k=n-5}^{n} \alpha^k = \sum_{k=0}^{5} \alpha^{k+(n-5)}$$

$$y(n) = \alpha^{n-5} \sum_{k=0}^{5} \alpha^k = \alpha^{n-5} \frac{1-\alpha^6}{1-\alpha}$$

Case 4: $11 \leq n \leq 15$. When n is in the range $11 \leq n \leq 15$, the sequences $h(k)$ and $x(n-k)$ overlap for $n-5 \leq k \leq 10$. Therefore,

$$y(n) = \sum_{k=n-5}^{10} \alpha^k = \sum_{k=0}^{15-n} \alpha^{k+(n-5)}$$

$$y(n) = \alpha^{n-5} \sum_{k=0}^{15-n} \alpha^k = \alpha^{n-5} \frac{1-\alpha^{16-n}}{1-\alpha}$$

Case 5: $n > 5$. Finally, for $n > 15$, there is again no overlap between $h(k)$ and $x(n-k)$, and the product $h(k)\,x(n-k)$ is equal to zero for all k. Therefore, $y(n) = 0$ for $n > 15$. In summary, for the convolution we have

$$y(n) = \begin{cases} 0 & n < 0 \\ \frac{1-\alpha^{n+1}}{1-\alpha} & 0 \le n \le 5 \\ \alpha^{n-5}\frac{1-\alpha^6}{1-\alpha} & 6 \le n \le 10 \\ \alpha^{n-5}\frac{1-\alpha^{16-n}}{1-\alpha} & 11 \le n \le 15 \\ 0 & n > 15 \end{cases}$$

Problem 3.18

If the response of a linear shift-invariant system of a unit step is given in the problem (i.e., the step response) is

$$s(n) = n\left(\frac{1}{2}\right)^n u(n)$$

find the unit sample response, $h(n)$.

Solution 3.18

In this problem, we begin by noting that

$$\delta(n) = u(n) - u(n-1)$$

Therefore, the unit sample response, $h(n)$ is related to the step response, $s(n)$, as follows:

$$h(n) = s(n) - s(n-1)$$

Thus, given $s(n)$, we have

$$h(n) = s(n) - s(n-1)$$

$$h(n) = n\left(\frac{1}{2}\right)^n u(n) - (n-1)\left(\frac{1}{2}\right)^{n-1} u(n-1)$$

$$h(n) = \left[n\left(\frac{1}{2}\right)^n - 2(n-1)\left(\frac{1}{2}\right)^n\right]u(n-1)$$

$$h(n) = \left[n\left(\frac{1}{2}\right)^n - 2(n-1)\left(\frac{1}{2}\right)^n\right]u(n-1) = \left[n\left(\frac{1}{2}\right)^n - 2n\left(\frac{1}{2}\right)^n + 2\left(\frac{1}{2}\right)^n\right]u(n-1) \quad h(n) = (2-n)\left(\frac{1}{2}\right)^n u(n-1)$$

Problem 3.19

Consider a system with impulse response

$$h(n) = \begin{cases} (\frac{1}{2})^n, & 0 \leq n \leq 4 \\ 0, & \text{elsewhere} \end{cases}.$$

Determine the input $x(n)$ for $0 \leq n \leq 8$ that will generate the output sequence

$$y(n) = \{\underset{\uparrow}{1}, 2,\ 2.5,\ 3,\ 3,\ 3,\ 2,\ 1,\ 0, \ldots\}$$

Solution 3.19

$$h(n) = \{\underset{\uparrow}{1}, \tfrac{1}{2}, \tfrac{1}{4}, \tfrac{1}{8}, \tfrac{1}{16}\}$$

$$y(n) = \{\underset{\uparrow}{1}, 2, 2.5, 3, 3, 3, 2, 1, 0\}$$

$$\begin{aligned} x(0)\,h(0) &= y(0) \Rightarrow x(0) = 1 \\ \tfrac{1}{2}x(0) + x(1) &= y(1) \Rightarrow x(1) = \tfrac{3}{2} \\ \tfrac{1}{4}x(0) + \tfrac{1}{2}x(1) + x(2) &= y(2) \Rightarrow x(2) = \tfrac{3}{2} \end{aligned}$$

By continuing this process, we obtain

$$x(n) = \{1, \tfrac{3}{2}, \tfrac{3}{2}, \tfrac{7}{4}, \tfrac{3}{2}, \ldots\}$$

Problem 3.20

$$x(n) = \{\underset{\uparrow}{1},\ 2,\ 4\} \qquad h(n) = \{\underset{\uparrow}{1},\ 1,\ 1,\ 1,\ 1\}$$

Find the convolution of the two sequences given above, using analytical method

Solution 3.20

$$y(n) = \sum_{k=0}^{6} x(k)h(n-k) \text{ or } y(0) = \sum_{k=0}^{6} x(k)\,h(0-k)$$

$$y(0) = x(0)\,h(-0) + x(1)\,h(-1) + x(2)\,h(-2) + x(3)\,h(-3)$$

$$+ x(4)\,h(-4) + x(5)\,h(-5) + x(6)\,h(-6)$$

the value of $h(-1)$ to $h(-6)$ are not written because the sequence values for these bits are not given (are zeros). Therefore, the expression for y(0) reduces to

$$y(0) = x(0)\,h(0) = (1)\,(1) = 1$$

$$y(1) = \sum_{k=0}^{6} x(k)\, h(1-k)$$

$$y(1) = x(0)\,h(1) + x(1)\,h(0) + x(2)\,h(-1) + x(3)\,h(-2) + x(4)\,h(-3) + x(5)\,h(-4) + x(6)\,h(-5)$$

the value of $h(-1)$ to $h(-5)$ are not written because the sequence values for these bits are not given (are zeros). Therefore, the expression for $y(1)$ reduces to

$$y(1) = x(0)\,h(1) + x(1)\,x(0) = (1)\,(1) + (2)\,(1) = 3$$

$$y(2) = \sum_{k=0}^{6} x(k)\, h(2-k)$$

$$y(2) = x(0)\,h(2) + x(1)\,h(1) + x(2)\,h(0) + x(3)\,h(-1) + x(4)\,h(-2) + x(5)\,h(-3) + x(6)\,h(-4)$$

the value of $h(-1)$ to $h(-4)$ are not written because the sequence values for these bits are not given (are zeros). Therefore, the expression for $y(2)$ reduces to

$$y(2) = x(0)\,h(2) + x(1)\,h(1) + x(2)\,h(0) = (1)\,(1) + (2)\,(1) + (4) = 7$$

$$y(3) = \sum_{k=0}^{6} x(k)\, h(3-k)$$

$$y(3) = x(0)\,h(3) + x(1)\,h(2) + x(2)\,h(1) + x(3)\,h(0) + x(4)\,h(-1) + x(5)\,h(-2) + x(6)\,h(-3)$$

the value of $h(-1)$ to $h(-3)$ and from $x(3)$ to $x(6)$ are not written because the sequence values for these bits are not given (are zeros). Therefore the expression for $y(3)$ reduces to

$$y(3) = x(0)\,h(3) + x(1)\,h(2) + x(2)\,h(1) + x(3)\,h(0) = (1)\,(1) + (2)\,(1) + (4)\,(1) = 7$$

$$y(4) = \sum_{k=0}^{6} x(k)\, h(4-k)$$

$$y(4) = x(0)\,h(4) + x(1)\,h(3) + x(2)\,h(2) + x(3)\,h(1) + x(4)\,h(0) + x(5)\,h(-1) + x(6)\,h(-2)$$

the value of $h(-1)$ to $h(-2)$ and from $x(3)$ to $x(6)$ are not written because the sequence values for these bits are not given (are zeros). Therefore the expression for $y(4)$ reduces to

$$y(4) = x(0)\,h(4) + x(1)\,h(3) + x(2)\,h(2) = (1)\,(1) + (2)\,(1) + (4)\,(1) = 7$$

$$y(5) = \sum_{k=0}^{6} x(k)\,h(5-k)$$

$$y(5) = x(0)\,h(5) + x(1)\,h(4) + x(2)\,h(3) + x(3)\,h(2) + x(4)\,h(1) + x(5)\,h(0) + x(6)\,h(-1)$$

The value of $h(-1)$, $h(5)$ to $h(6)$ and from $x(3)$ to $x(6)$ are not written because the sequence values for these bits are not given (are zeros). Therefore the expression for $y(5)$ reduces to

$$y(5) = x(0)\,h(5) + x(1)\,h(4) + x(2)\,h(3) = (2)\,(1) + (4)\,(1) = 6$$

$$y(6) = \sum_{k=0}^{6} x(k)\,h(6-k)$$

$$y(6) = x(0)\,h(6) + x(1)\,h(5) + x(2)\,h(4) + x(3)\,h(3) + x(4)\,h(2) + x(5)\,h(1) + x(6)\,h(0)$$

the value of $h(5)$ to $h(6)$ and from $x(3)$ to $x(6)$ are not written because the sequence values for these bits are not given (are zeros). Therefore the expression for $y(6)$ reduces to $y(6) = x(2)h(4) = (4)(1) = 4$

Verification

The correctness of the convolution can be checked and verified by using the formula.

$$\sum y(n) = \left(\sum h(k)\right)\left(\sum x(n)\right)$$

$$\sum y(n) = 35 \quad \sum h(k) = 5 \quad \sum x(k) = 7$$

Problem 3.21

Find the linear convolution of the two sequences given below, using matrix method

$$x(n) = \{\underset{\uparrow}{1},\ 2,\ 4\} \qquad h(n) = \{\underset{\uparrow}{1},\ 1,\ 1,\ 1,\ 1\}$$

Solution 3.21

Max size of the bit in this example is 5, i.e., called N, $N-1$ zeros padding has to be done in either bit for linear convolution. In this example $N-1=5-1=4$ zero padding is required.

$$\begin{bmatrix} & & & x(n) & & & \\ 1 & 0 & 0 & 0 & 0 & 4 & 2 \\ 2 & 1 & 0 & 0 & 0 & 0 & 4 \\ 4 & 2 & 1 & 0 & 0 & 0 & 0 \\ 0 & 4 & 2 & 1 & 0 & 0 & 0 \\ 0 & 0 & 4 & 2 & 1 & 0 & 0 \\ 0 & 0 & 0 & 4 & 2 & 1 & 0 \\ 0 & 0 & 0 & 0 & 4 & 2 & 1 \end{bmatrix} \begin{bmatrix} h(n) \\ 1 \\ 1 \\ 1 \\ 1 \\ 1 \\ 0 \\ 0 \end{bmatrix} = \begin{bmatrix} y(n) \\ 1 \\ 3 \\ 7 \\ 7 \\ 7 \\ 6 \\ 4 \end{bmatrix}$$

Problem 3.22

Find the circular convolution of the two sequences given below, using matrix method

$$x(n) = \{\underset{\uparrow}{1},\ 2,\ 4\} \qquad h(n) = \{\underset{\uparrow}{1},\ 1,\ 1,\ 1,\ 1\}$$

Solution 3.22

Max size of the bit in this problem is 5, therefore only two zeros are required to be padded up in $x(n)$.

$$\begin{bmatrix} & & x(n) & & \\ 1 & 0 & 0 & 4 & 2 \\ 2 & 1 & 0 & 0 & 4 \\ 4 & 2 & 1 & 0 & 0 \\ 0 & 4 & 2 & 1 & 0 \\ 0 & 0 & 4 & 2 & 1 \end{bmatrix} \begin{bmatrix} h(n) \\ 1 \\ 1 \\ 1 \\ 1 \\ 1 \end{bmatrix} = \begin{bmatrix} y(n) \\ 7 \\ 7 \\ 7 \\ 7 \\ 7 \end{bmatrix}$$

Problem 3.23

Determine the overall impulse response of the system of the Figure, where

$$h_1(n) = 2\delta(n-2) - 3\delta(n+1)$$

$$
\begin{aligned}
h_2(n) &= \delta(n-1) + 2\delta(n+2) \\
h_3(n) &= 5\delta(n-5) + 7\delta(n-3) + 2\delta(n-1) - \delta(n) \\
&\quad + 3\delta(n+1)
\end{aligned}
$$

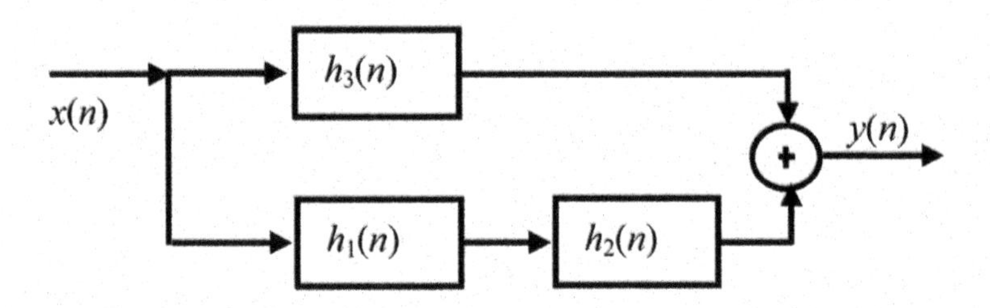

Solution 3.23

$h(n) = [h_1(n) \otimes h_2(n)] + h_3(n)$

now $[h_1(n) \otimes h_2(n)] = [2\delta(n-2) - 3\delta\ (n+1)] \otimes [\delta(n-1) + 2\delta(n+2)]$

$[h_1(n) \otimes h_2(n)] = 2\delta(n-2) \otimes \delta(n-1)] - 3\delta(n+1) \otimes \delta(n-1)$
$\quad + 2\delta(n-2) \otimes 2\delta(n+2) - 3\delta\ (n+1) \otimes 2\delta(n+2)$

$[h_1(n) \otimes h_2(n)] = 2\delta(n-3) - 3\delta(n) + 4\delta(n) - 6\delta(n+3)$

$[h_1(n) \otimes h_2(n)] = 2\delta(n-3) + \delta(n) - 6\delta(n+3)$

$h(n) = [2\delta\ (n-3) + \delta(n) - 6\delta(n+3)] + [5\delta(n-5) + 7\delta(n-3) + 2\delta(n-1)$
$\quad - \delta(n) + 3\delta(n+1)]$

$h(n) = 5\delta\ (n-5) + 9\delta(n-3) + 2\delta(n-1) + 3\delta(n+1) - 6\delta(n+3)$

Problem 3.24

Prove that the convolution operation is commutative and distributive.

Solution 3.24

$$y(n) = x(n) \otimes h(n) = \sum_{k=-\infty}^{\infty} x(n-k)\, h(k)$$

Substituting k by $n - m$ in the above expression we get

$$y(n) = \sum_{m=-\infty}^{\infty} x(m)\, h(n-m) = h(n) \otimes x(n)$$

Hence the convolution operation is commutative.

$$y(n) = x(n) \otimes [h_1(n) + h_2(n)] = \sum_{k=-\infty}^{\infty} x(n-k)\, [h_1(k) + h_2(k)]$$

$$= \sum_{k=-\infty}^{\infty} x(n-k)\,h_1(k) + \sum_{k=-\infty}^{\infty} x(n-k)\,h_1(k) = x_1(n)h(n) + x(n) \otimes h_2(n)$$

Hence the convolution operation is distributive.

Problem 3.25

Find the cross-correlation coefficients $\rho_{12}(0)$, $\rho_{34}(0)$ and $\rho_{12}(2)$, $\rho_{34}(2)$ of the signal given in tabular form.

n	0	1	2	3	4	5	6
$x_1(n)$	0	–4	3	3	3	2	0
$x_2(n)$	1	1	1	1	1	–1	0
$x_3(n)$	0	–8	6	6	6	4	0
$x_4(n)$	4	4	4	4	4	–4	0

Solution 3.25

Cross-correlations $\rho_{12}(0)$ and $\rho_{34}(0)$ which is found as follows

$$\begin{aligned} r_{12}(0) &= 1/7\ (0 \times 1 + (-4) \times 1 + \ 3 \times 1 + 3 \times 1 + 3 \times 1 + 2 \\ &\quad \times (-1) + 0 \times 0) \\ r_{12}(0) &= \ 0.4286 \\ r_{34}(0) &= 1/7\ (0 \times 4 + (-8) \times 4 + 6 \times 4 + 6 \times 4 + 6 \times 4 + 4 \\ &\quad \times (-4) + 0 \times 0) \\ r_{34}(0) &= \ 3.4286 \end{aligned}$$

$$\rho_{12}(j) = \frac{r_{12(j)\text{true}}}{\frac{1}{N}\left[\sum_{n=0}^{N-1} x_1^2(n) \sum_{n=0}^{N-1} x_2^2(n)\right]^{1/2}}$$

The normalizing factor for $r_{12}(j)$ in the above illustration is calculated as below

$$\frac{1}{N}\left[\sum_{n=0}^{N-1} x_1^2(n) \sum_{n=0}^{N-1} x_2^2(n)\right]^{1/2} = \frac{1}{7}(47 \times 6)^{1/2} = 2.399$$

and for $r_{34}(j)$ it is

$$\frac{1}{N}\left[\sum_{n=0}^{N-1} x_3^2(n) \sum_{n=0}^{N-1} x_4^2(n)\right]^{1/2} = \frac{1}{7}(188 \times 96)^{1/2} = 19.1918$$

Therefore

$$\rho_{12}(0) = \frac{r_{12(0)}}{2.3999} = \frac{0.4286}{2.3999} = 0.1786$$

$$\rho_{34}(0) = \frac{\rho_{34(0)}}{19.1918} = \frac{3.4286}{19.1918} = 0.1786$$

Cross-correlations $\rho_{12}(2)$ and $\rho_{34}(2)$ which is found as

$$r_{12}(2) = 1/7\ (0 \times 1 + (-4) \times 1 + \ 3 \times 1 + 3 \times (-1) + 3 \times 0 + 2 \times 0)$$
$$r_{12}(2) = -0.5714$$

$$r_{34}(2) = 1/7\ (0 \times 4 + (-8) \times 4 + 6 \times 4 + 6 \times (-4) + 6 \times 0 + 4 \times 0)$$
$$r_{34}(0) = -4.5714$$

$$r_{12}(2)_{\text{true}} = r_{12}(2) + \frac{2}{7} r_{12}(0)$$

$$r_{12}(2)_{\text{true}} = -0.5714 + \frac{2(0.4286)}{7} = -0.4489$$

$$r_{34}(2)_{\text{true}} = r_{34}(2) + \frac{2}{7} r_{34}(0)$$

$$r_{12}(2)_{\text{true}} = -4.5714 + \frac{2(3.4286)}{7} = -3.5918$$

Therefore

$$\rho_{12(2)} = \frac{r_{12}(2)_{\text{true}}}{2.399} = \frac{-0.4489}{2.399} = -0.187$$

$$\rho_{34(2)} = \frac{r_{34}(2)_{\text{true}}}{19.1918} = \frac{-3.5918}{19.1918} = -0.187$$

Now $\rho_{12}(2) = \rho_{34}(2)$ which again demonstrates that this normalization process indeed allows a comparison of cross-correlations independently of the absolute data values.

Problem 3.26

Find the correlation $r_{yx}(j)$ of the samples of sequences using convolution property $r_{yx}(j) = x(-n) \otimes y(n)$ given in Figure.

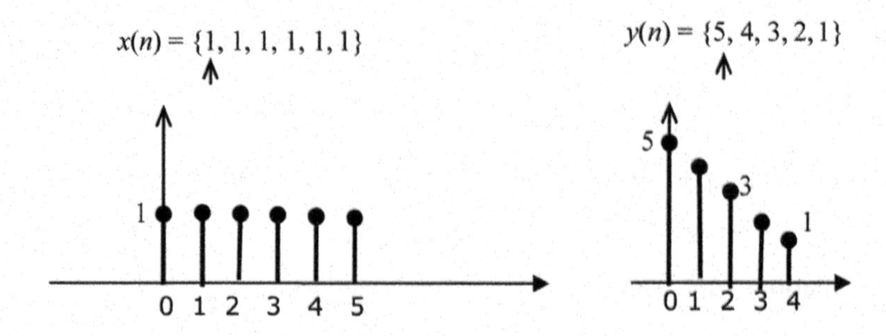

Solution 3.26

$r_{yx}(j) = x(-n) \otimes y(n)$

$y(-5) = 5$

$y(-4) = 9,\ y(-3) = 12$

$y(-2) = 14,\ y(-1) = 15$

$y(0) = 15,\ y(1) = 10$

$y(2) = 6,\ y(3) = 3$

$y(4) = 1$

$r_{yx} = \{5, 9, 12, 14, 15, \underset{\uparrow}{15}, 10, 6, 3, 1\}$

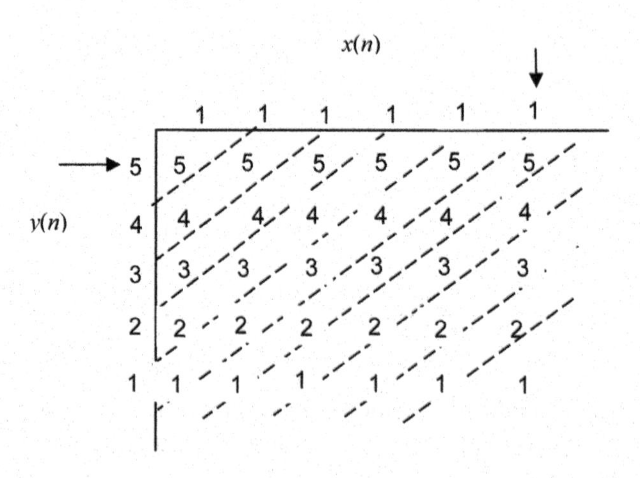

Problem 3.27

Compute the sketch the convolution $y(n)$ and correlation $r_{xy}(j)$, $r_{yx}(j)$ sequences for the following pair of signals.

$$x_1(n) = \{\underset{\uparrow}{1}, 2, 3\}; \quad x_2(n) = \{\underset{\uparrow}{4}, 5, 6\}.$$

Solution 3.27
Convolution: Overlap and Add Method

$y(n) = x_1(n) \otimes x_2(n)$
$y(0) = 4, y(1) = 13$
$y(2) = 28, y(3) = 27$
$y(4) = 18$
$y(n) = \{\underset{\uparrow}{4}, 13, 28, 27, 18\}$

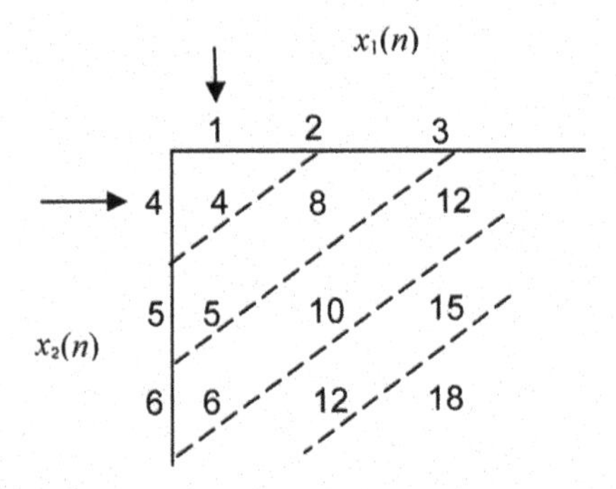

Correlation: Using Convolution Property Method

$r_{12}(j) = x_1(n) \otimes x_2(-n)$
$y(-2) = 6$
$y(-1) = 17$
$y(0) = 32$
$y(1) = 23$
$y(2) = 12$

$$r_{12}(j) = \{6, 17, \underset{\uparrow}{32}, 23, 12\}$$

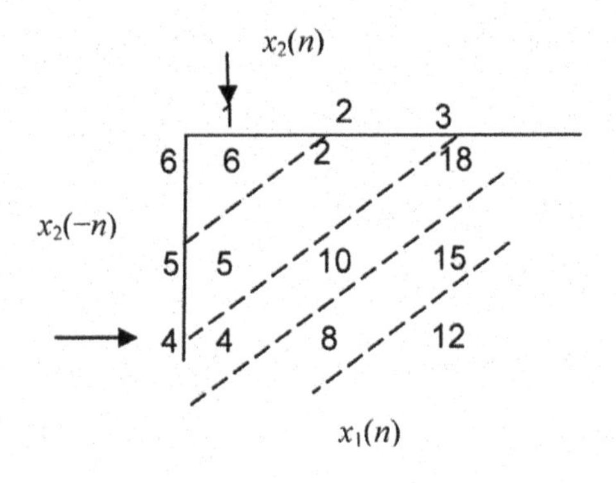

Correlation: Using Convolution Property Method

$r_{21}(j) = x_1(-n) \otimes x_2(n)$

$r_{21}(j) = \{12, 23, \underset{\uparrow}{32}, 17, 6\}$

The cross-correlation satisfies the $r_{12}(j) = r_{21}(-j)$ relationship.

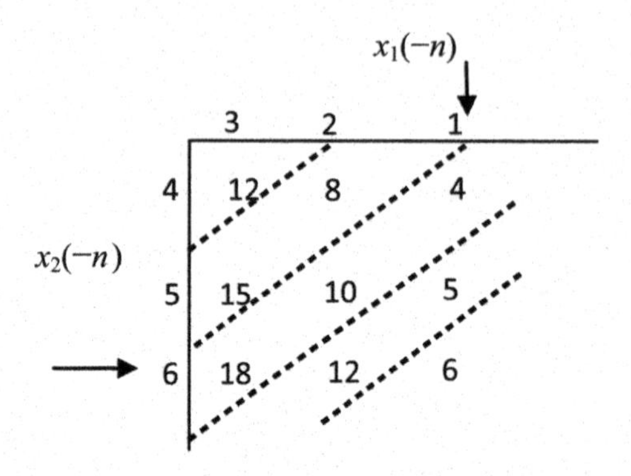

Correlation $r_{12}(j)$: Using Tabular Method

n	0	1	2
$x_1(n)$	1	2	3
$x_2(n)$	4	5	6

$$r_{12}(j) = \sum_{n=-\infty}^{\infty} x_1(n)\, x_2(n-j) \quad j = 0, \pm 1, \pm 2, \ldots$$

Cross correlation using tabular form is carried out here, here the sequence lengths are 3 and 3 and the number of lags necessary is 3 + 3 – 1 = 5. (x_1 is left un-shifted, for x_2 if $j < 0$, i.e., negative, left shift is required; $j > 0$, i.e., positive, right shift is required)

This is now demonstrated for the given sequences x_1 and x_2 to be correlated.

		Sequence				Lag	$r_{12}(j)$
1	2	3	0	0	j		
4	5	6	0	0	0	0 no shifting	32
5	6	0	0	4	–1	I unit left shifting	17
6	0	0	4	5	–2	2 units left shifting	6
0	0	4	5	6	2	2 units right shifting	12
0	4	5	6	0	1	1 unit right shifting	23
4	5	6	0	0		6 repetition starts	32

Thus, the required linear cross-correlation of x and y is
$r_{12}(j) = \{6, 17, \underset{\uparrow}{32}, 23, 12\}$

Correlation $r_{yx}(j)$: Using Tabular Method

Cross correlation using tabular form is carried out here, here the sequence lengths are 3 and 3 and the number of lags necessary is 3 + 3 – 1 = 5.

n	0	1	2
$x_1(n)$	1	2	3
$x_2(n)$	4	5	6

$$r_{21}(j) = \sum_{n=-\infty}^{\infty} x_1(n-j)\, x_2(n) \quad j = 0, \pm1, \pm2, \ldots$$

(x_2 is left unshifted, for x_1 $j < 0$, negative, left shift; for $j > 0$, positive, right shift)

This is now demonstrated for the given sequences x_1 and x_2.

Sequence						Lag	$r_{12}(j)$
4	5	6	0	0	j		
1	2	3	0	0	0	0 no shifting	32
2	3	0	0	1	–1	I unit left shifting	23
3	0	0	1	2	–2	2 units left shifting	12
0	0	1	2	3	2	2 units right shifting	6
0	1	2	3	0	1	1 unit right shifting	17
1	2	3	0	0		Repetition starts	32

Thus, the required linear cross-correlation of x_1 and x_2 is
$r_{21}(j) = \{12, 23, 32, 17, \underset{\uparrow}{6}\}$ $r_{21}(-j) = \{6, 17, 32, 23, \underset{\uparrow}{12}\}$
It proves the correlation property that $r_{12}(j) = r_{21}(-j)$

Correlation $r_{12}(j)$: Analytical Method

n	0	1	2
$x_1(n)$	1	2	3
$x_2(n)$	4	5	6

$$r_{12}(j) = \sum_{n=-\infty}^{\infty} x_1(n)\, x_2(n-j) \quad j = 0, \pm1, \pm2, \ldots$$

$$r_{12}(j) = \sum_{n=-2}^{2} x_1(n)\, x_2(n-j) \quad j = 0, \pm 1, \pm 2, \ldots$$

$$r_{12}(j) = x_1(-2)\, x_2(-2-j) + x_1(-1)\, x_2(-1-j) + x_1(0)\, x_2(0-j) + x_1(1)\, x_2(1-j) + x_1(2)\, x_2(2-j)$$

Note: Here the range of n has been selected as $n = -2$ to $n = 2$.

Here the sequence lengths are 3 and 3 and the number of lags necessary is $3 + 3 - 1 = 5$ (that is sample size after correlation)

$x_1(-2)$ and $x_1(-1)$ values are not given in the problem, therefore it reduces to following

$$r_{12}(j) = x_1(0)\, x_2(0-j) + x_1(1)\, x_2(1-j) + x_1(2)\, x_2(2-j)$$

Now substituting the value of j as 0, –1, +1, –2, +2 in the above expression

$$\begin{aligned} r_{12}(0) &= x_1(0)\, x_2(0) + x_1(1)\, x_2(1) + x_1(2)\, x_2(2) \\ &= (1)(4) + (2)(5) + (3)(6) = 32 \end{aligned}$$

$$\begin{aligned} r_{12}(1) &= x_1(0)\, x_2(-1) + x_1(1)\, x_2(0) + x_1(2)\, x_2(1) \\ &= (1)(0) + (2)(4) + (3)(5) = 23 \\ r_{12}(2) &= x_1(0)\, x_2(-2) + x_1(1)\, x_2(-1) + x_1(2)\, x_2(0) \\ &= (1)(0) + (2)(0) + (3)(4) = 12 \\ r_{12}(-2) &= x_1(0)\, x_2(2) + x_1(1)\, x_2(3) + x_1(2)\, x_2(4) \\ &= (1)(6) + (2)(0) + (3)(0) = 6 \\ r_{12}(-1) &= x_1(0)\, x_2(1) + x_1(1)\, x_2(2) + x_1(2)\, x_2(3) \\ &= (1)(5) + (2)(6) + (3)(0) = 17 \end{aligned}$$

$$r_{12}(j) = \left\{ 6,\ 17,\ 32,\ \underset{\uparrow}{23},\ 12 \right\}$$

4

Z-Transform

This chapter covers: Introduction to Z-transforms, different methods of finding the inverse z transforms, properties of Z-transform, different methods of solution of difference equations, and Problems and Solutions.

4.1 Introduction

Laplace and Fourier transforms are used in signal processing applications. These transformations are defined in continuous- and discrete-time domain. In signal processing applications, we use the discrete version of Fourier and Laplace transformation. These are called the discrete-time Fourier transformation, discrete Fourier transformation and discrete Laplace transformation which is more popularly called the Z-transformation.

Z-transform is a representation of discrete-time signals in the frequency domain or the conversion between discrete-time and frequency domain. The spectrum of a signal is obtained by decomposing it into its constituent frequency components using a discrete transform. Conversion between time and frequency domains is necessary in many DSP applications. For example, it allows for a more efficient implementation of DSP algorithms, such as those for digital filtering, convolution, and correlation.

4.2 Z-Transform

The Z-transform of a number sequence $x(nT)$ or $x(n)$ is defined as the power series in z^{-n} with coefficient equal to the values $x(n)$, where z is a complex variable and $z = \sigma + j\omega$.

The Z-transform of a discrete time signal for a non-causal signal $x(n)$ may be expressed as

$$Z[\{x(n)\}] = X(z) = \sum_{n=-\infty}^{\infty} x(n)z^{-n}. \tag{4.1}$$

The above expression is generally known as two sided Z-transform. The sequence $\{x(n)\}$ is generated from a time function $x(t)$ by sampling every T seconds, $x(n)$ is understood to be $x(nT)$; that is, T is dropped for convenience.

If the discrete time signal is causal signal $x(n) = 0$ for $n < 0$, then the Z-transform is called as one sided Z-transform and is expressed as

$$Z[\{x(n)\}] = X(z) = \sum_{n=0}^{\infty} x(n)z^{-n}. \tag{4.2}$$

Infact, generally we assume that $x(n)$ is a causal discrete-time signal unless it is stated. This means that generally we analyze causal signal. On the other hand, if $x(n)$ is a non-causal discrete-time signal only, i.e., $x(n) = 0$ for $n \geq 0$, otherwise $x(n)$ exists then its z-transform is expressed as

$$Z[\{x(n)\}] = X(z) = \sum_{n=-\infty}^{-1} x(n)z^{-n}. \tag{4.3}$$

The Z-transform of a sequence $x(n)$ is defined as

$$X(z) = \sum_{n=-\infty}^{\infty} x(n)z^{-n}. \tag{4.4}$$

Let us express the complex variable z in polar form as $z = re^{j\omega}$

$$X(z)|_{z=re^{j\omega}} = \sum_{n=-\infty}^{\infty} [x(n)r^{-n}]e^{-j\omega n}. \tag{4.5}$$

From the relationship in Equation (4.5) we note that $X(z)$ can be interpreted as the discrete-time Fourier transform of the signal sequence $x(n)r^{-n}$. Now if $r = 1$ then $|z| = 1$, then Equation (4.5) reduces to discrete transform. Hence the expression in Equation (4.5) will converge if $[x(n)r^{-n}]$ is absolutely sumable.

Mathematically,

$$\sum_{n=-\infty}^{\infty} \left|x(n)r^{-n}\right| \leq \infty. \tag{4.6}$$

Hence for $x(n)$ to be finite, the magnitude of z-transform, $X(z)$ must also be finite.

The region of convergence (ROC) of $X(z)$ is the set of all values of z for which $X(z)$ attains a finite value. Thus any time we cite a z-transform we should also indicate its ROC. Concepts of ROC are illustrated by some simple examples.

Example 4.1

Determine the z-transform of the following finite-duration signals.

$$\begin{aligned}
&(a)\ x_1(n) = \{1, 2, 5, 7, 0, 1\}\\
&(b)\ x_2(n) = \{1, 2, \underset{\uparrow}{5}, 7, 0, 1\}\\
&(c)\ x_3(n) = \{0, 0, 1, 2, 5, 7, 0, 1\}\\
&(d)\ x_4(n) = \{2, 4, \underset{\uparrow}{5}, 7, 0, 1\}\\
&(e)\ x_5(n) = \delta(n)\\
&(f)\ x_6(n) = \delta(n-k), k > 0\\
&(g)\ x_7(n) = \delta(n+k), k > 0.
\end{aligned}$$

Solution 4.1

$$\begin{aligned}
&(a)\ X_1(z) = 1 + 2z^{-1} + 5z^{-2} + 7z^{-3} + z^{-5},\\
&\qquad \text{ROC: entire } z\text{-plane except } z = 0\\
&(b)\ X_2(z) = z^2 + 2z + 5 + 7z^{-1} + z^{-3},\\
&\qquad \text{ROC: entire } z\text{-plane except } z = 0 \text{ and } z = \infty\\
&(c)\ X_3(z) = z^{-2} + 2z^{-3} + 5z^{-4} + 7z^{-5} + z^{-7},\\
&\qquad \text{ROC: entire } z\text{-plane except } z = 0\\
&(d)\ X_4(z) = 2z^2 + 4z + 5 + 7z^{-1} + z^{-3},\\
&\qquad \text{ROC: entire } z\text{-plane except } z = 0 \text{ and } z = \infty\\
&(e)\ X_5(z) = 1[\text{i.e., } \delta(n) \overset{z}{\longleftrightarrow} 1],\ \ \text{ROC: entire } z\text{-plane}\\
&(f)\ X_6(z) = z^{-k}[\text{i.e., } \delta(n-k) \overset{z}{\longleftrightarrow} z^{-k}], k > 0,\\
&\qquad \text{ROC: entire } z\text{-plane except } z = 0\\
&(g)\ X_7(z) = z^{k}[\text{i.e., } \delta(n+k) \overset{z}{\longleftrightarrow} z^{k}], k > 0,\\
&\qquad \text{ROC: entire } z\text{-plane except } z = \infty
\end{aligned}$$

From these examples it is easily seen that the ROC of a finite-duration signal is the entire z-plane, except possibly the points $z = 0$ and or $z = \infty$.

Example 4.2

Find the z-transform of the discrete time unit impulse $\delta(n)$.

Solution 4.2

We know that the Z-transform is expressed as

$$Z[\{x(n)\}] = X(z) = \sum_{n=0}^{\infty} x(n)z^{-n}.$$

We know that the unit impulse sequence $\delta(n)$ is a causal. The signal $x(n)$ consists of a number at $n = 0$, otherwise it is zero. The z-transform of $x(n)$ is the infinite power series

$X(z) = 1+(0)z^{-1}+(0)z^{-2}+\ldots(0)^n z^{-n}+\ldots$. Therefore the transform $Z[\delta(n)] = 1$.

Example 4.3

Find the z-transform of the discrete time unit step signal $u(n)$.

Solution 4.3

We know that the Z-transform is expressed as

$$Z[\{x(n)\}] = X(z) = \sum_{n=0}^{\infty} x(n)z^{-n}.$$

We know that the unit step sequence $u(n)$ is a causal. The z-transform of $x(n)$ is the infinite power series

$$X(z) = 1 + (1)z^{-1} + (1)z^{-2} + \ldots (1)^n z^{-n} + \ldots$$

This is an infinite geometric series. We recall that to sum an infinite series we use the following relation of Geometric series.

$$1 + A + A^2 + A^3 + A^4 + A^5 + A^6 = \frac{1}{1-A} \quad \text{if } |A| < 1$$

Therefore

$$X(z) = \frac{1}{1-z^{-1}} \quad \text{ROC } |z| > 1$$

Example 4.4

Determine the z-transform of the signal

$$x(n) = (\tfrac{1}{2})^n u(n)$$

Solution 4.4

The z-transform of $x(n)$ is the infinite power series

$$X(z) = 1 + \tfrac{1}{2}z^{-1} + \left(\tfrac{1}{2}\right)^2 z^{-2} + \ldots \left(\tfrac{1}{2}\right)^n z^{-n} + \ldots$$
$$= \sum_{n=0}^{\infty} \left(\tfrac{1}{2}\right)^n z^{-n} = \sum_{n=0}^{\infty} \left(\tfrac{1}{2}z^{-1}\right)^n.$$

Consequently, for $|\frac{1}{2}z^{-1}| < 1$, or equivalently, for $|z| > \frac{1}{2}$, $X(z)$ converges to

$$X(z) = \frac{1}{1 - \frac{1}{2}z^{-1}} \quad \text{ROC } |z| > \frac{1}{2}.$$

We see that in this case, the z-transform provides a compact alternative representation of the signal $x(n)$.

Example 4.5

Determine the Z-transform of the signal

$$x(n) = \alpha^n u(n) = \begin{cases} \alpha^n, & n \geq 0 \\ 0, & n < 0 \end{cases}$$

Solution 4.5

From the definition (4.1) we have

$$X(z) = \sum_{n=0}^{\infty} \alpha^{\,n} z^{-n} = \sum_{n=0}^{\infty} (\alpha\, z^{-1})^n$$

If $|\alpha z^{-1}| < 1$ or equivalently, $|z| > |\alpha|$, this power series converges to $1/(1 - \alpha z^{-1})$. Thus, we have the z-transform pair

$$x(n) = \alpha^n u(n) \overset{z}{\longleftrightarrow} X(z) = \frac{1}{1 - \alpha z^{-1}}, \quad \text{ROC}: |z| > |\alpha|. \tag{1}$$

The ROC is the exterior of a circle having radius $|\alpha|$. Figure shows a graph of the signal $x(n)$ and its corresponding ROC. If we set $\alpha = 1$ in Equation (4.1), we obtain the z-transform of the unit step signal.

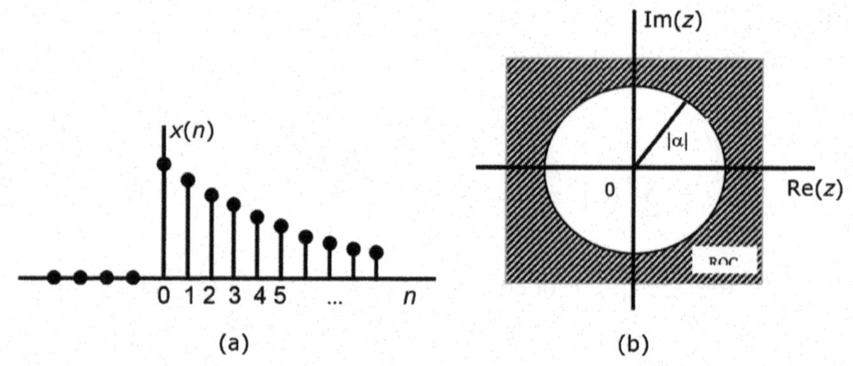

Figure 4.1 (a) The exponential signal $x(n) = \alpha^n u(n)$ and (b) the ROC of its z-transform.

Example 4.6

An expression of discrete time $x(n) = e^{-anT}$ is given, find $X(z)$

Solution 4.6

$$X(z) = z[e^{anT}] = \sum_{n=0}^{\infty} x(n) z^{-n}$$

$$= 1 + e^{-aT} z^{-1} + e^{-2aT} z^{-2} + e^{-3aT} z^{-3} + \ldots$$

$$= 1 + \left(e^{-aT} z^{-1}\right) + \left(e^{-aT} z^{-1}\right)^2 + \ldots$$

$$X(z) = \frac{1}{1 - e^{-aT} z^{-1}} = \frac{z}{z - e^{-aT}}$$

Example 4.7

Find the z-transform of the function $x(n)$, where $x(nT) = nT$

Solution 4.7

$$z[\{x(nT)\}] = \sum_{n=0}^{\infty} x(nT) z^{-n} \quad z[\{x(nT)\}] = \sum_{n=0}^{\infty} nT z^{-n}$$

$$X(z) = 0 + Tz^{-1} + 2Tz^{-2} + 3Tz^{-3} + \ldots$$

$$X(z) = T[z^{-1} + 2z^{-2} + 3z^{-3} + \ldots.]$$

$$= T \frac{z}{(z-1)^2}$$

Table 4.1 of z-transform is given here for ready reference for the students.

Table 4.1 Z-transform

S. No.	Signal $x(n)$	Z-transformation $X(z)$	ROC
1	$\delta(n)$	1	All z
2	$u(n)$	$\frac{1}{1-z^{-1}}$	$\lvert z\rvert > 1$
3	$nu(n)$	$\frac{z^{-1}}{(1-z^{-1})^2}$	$\lvert z\rvert > 1$
4	$a^n u(n)$	$\frac{1}{1-az^{-1}}$	$\lvert z\rvert > \lvert a\rvert$
5	$-a^n u(-n-1)$	$\frac{1}{1-az^{-1}}$	$\lvert z\rvert < \lvert a\rvert$
6	$-na^n u(-n-1)$	$\frac{az^{-1}}{(1-az^{-1})^2}$	$\lvert z\rvert < \lvert a\rvert$
7	$(\cos\omega_0 n)u(n)$	$\frac{1-z^{-1}\cos\omega_0}{1-2z^{-1}\cos\omega_0+z^{-2}}$	$\lvert z\rvert > \lvert 1\rvert$
8	$(\sin\omega_0 n)u(n)$	$\frac{z^{-1}\sin\omega_0}{1-2z^{-1}\cos\omega_0+z^{-2}}$	$\lvert z\rvert > \lvert 1\rvert$
9	$(a^n\sin\omega_0 n)u(n)$	$\frac{az^{-1}\sin\omega_0}{1-2az^{-1}\cos\omega_0+a^2z^{-2}}$	$\lvert z\rvert > \lvert a\rvert$
10	$(a^n\cos\omega_0 n)u(n)$	$\frac{1-az^{-1}\cos\omega_0}{1-2az^{-1}\cos\omega_0+a^2z^{-2}}$	$\lvert z\rvert > \lvert a\rvert$

Before proceeding to Z-transform properties it is better to have knowledge of inverse Z-transform.

4.3 Inverse *Z*-Transform

When a z-transfer function is known we often require finding out its inverse in the discrete-time form. There are different methods of finding inverse Z-transform; we discuss here the most commonly method used in finding z-transform.

4.3.1 Using the Property of *Z*-Transform

Example 4.8

Find the inverse z-transform of the function $X(z) = \log(1 + az^{-1})$, where a is a scalar quantity.

Solution 4.8

Given function is

$$X(z) = \log(1 + az^{-1}) \tag{1}$$

Differentiating both sides of (1)

$$\frac{\mathrm{d}X(z)}{\mathrm{d}z} = \frac{-az^2}{1 + az^{-1}} \tag{2}$$

Multiplying both sides by $-z$ to make right hand side the derivative formula

$$-z\frac{\mathrm{d}X(z)}{\mathrm{d}z} = \frac{az^{-1}}{1+az^{-1}}. \tag{3}$$

We know that inverse z-transform of left hand side is $nx(n)$, and inverse z-transform of right-hand side is

$$Z^{-1}\left(-z\frac{\mathrm{d}X(z)}{\mathrm{d}z}\right) = nx(n) \tag{4}$$

$$Z^{-1}\left(\frac{az^{-1}}{1+az^{-1}}\right) = a(-a)^{n-1}u(n-1). \tag{5}$$

Equating inverse z-transform of Equations (4) and (5)

$$n\,x(n) = a(-a)^{n-1}u(n-1).$$

Therefore

$$\begin{aligned} x(n) &= \frac{1}{n}(-1)(-a)(-a)^{n-1}u(n-1) = \frac{1}{n}(-1)(-a)^{n}u(n-1) \\ &= \frac{1}{n}(-1)^{n+1}(a)^{n}u(n-1) = \frac{1}{n}(-1)^{n+1}(a)^{n} \quad n \geq 1 \end{aligned}$$

4.3.2 Using the Long Division

Problem 4.9

Find the inverse Z-transform of each given $X(z)$ by long method

$$X(z) = \frac{z(z+1)}{(z-1)(z-0.8)} = \frac{z^2+z}{z^2-1.8z+0.8}$$

Solution 4.9

$$\begin{array}{r|l} & 1 + 2.8z^{-1} + 4.24z^{-2} + 5.39z^{-3} \\ z^2 - 1.8z + 0.8 & z^2 + z \\ & \underline{z^2 - 1.8z + 0.8} \\ & 2.8z - 0.8 \\ & \underline{2.8z - 5.04 + 2.24z^{-1}} \\ & 4.24 - 2.24z^{-1} \\ & 4.24 \end{array}$$

$$x(0) = 1, \quad x(1) = 2.8, \quad x(2) = 4.24, \quad x(3) = 5.392$$

4.3.3 Using Residue Method

The partial fraction is done by any known method but one of the residue method is widely in use of finding inverse z transform $x(n)$, by summing the residues at all pole.

$$\text{Residue} = \frac{1}{(m-1)|} \lim_{z\to\alpha}\left[\frac{d^{m-1}}{dz^{m-1}}\{(z-\alpha)^m X(z)\}\right] \quad (4.7)$$

The function $X(z)$ may be expanded into partial fractions in the same manner as used with the Laplace transforms. The Z-transform tables may be used to find the inverse z-transform.
Note: We generally expand the function $X(z)/Z$ into partial fractions, and then multiply z to obtain the expansion in the proper form.

Problem 4.10
Find the inverse Z-transform of each given $X(z)$ by partial expansion.

$$X(z) = \frac{z(z+1)}{(z-1)(z-0.8)} = \frac{z^2+z}{z^2-1.8z+0.8}$$

Solution 4.10

$$\frac{X(z)}{z} = \frac{z+1}{(z-1)(z-0.8)} \Rightarrow \frac{10}{z-1} + \frac{-9}{z-0.8}$$

$$Z^{-1}\{X(z)\} = 10Z^{-1}\left\{\frac{z}{z-1}\right\} - 9Z^{-1}\left\{\frac{z}{z-0.8}\right\}$$

$$X(k) = 10 - 9(0.8)^k$$

4.3.3.1 When the poles are real and non-repeated

The inverse transform for functions that have real and non-repeated poles have been considered here. Residue method is applied here.

Example 4.11
$X(z)$ is given by the following transfer function, find $x(n)$

$$X(z) = \frac{4z^2 - \frac{7}{4}z + \frac{1}{4}}{z^2 - \frac{3}{4}z + \frac{1}{8}}$$

Solution 4.11
Because the given transfer function is not rational that is its numerator value is bigger than the denominator value, so first making the given non-rational

transfer function into rational transfer function by dividing its numerator with denominator. We obtain.

$$X(z) = \frac{4z^2 - \frac{7}{4}z + \frac{1}{4}}{z^2 - \frac{3}{4}z + \frac{1}{8}} = \frac{4z^2 - \frac{7}{4}z + \frac{1}{4}}{\left(z - \frac{1}{2}\right)\left(z - \frac{1}{4}\right)} = 2 + \frac{z\left(2z - \frac{1}{4}\right)}{\left(z - \frac{1}{2}\right)\left(z - \frac{1}{4}\right)}$$

We expand $\frac{X_1(z)}{z}$ into partial fractions, with the result,

$$X(z) = 2 + \frac{z\left(2z - \frac{1}{4}\right)}{\left(z - \frac{1}{2}\right)\left(z - \frac{1}{4}\right)}$$

$$X_1(z) = \frac{z\left(2z - \frac{1}{4}\right)}{\left(z - \frac{1}{2}\right)\left(z - \frac{1}{4}\right)} \quad \frac{X_1(z)}{z} = \frac{\left(2z - \frac{1}{4}\right)}{\left(z - \frac{1}{2}\right)\left(z - \frac{1}{4}\right)} = \frac{X_1(z)}{z}$$

$$= \frac{A}{\left(z - \frac{1}{2}\right)} + \frac{B}{\left(z - \frac{1}{4}\right)}$$

$$A = \left(z - \frac{1}{2}\right)\frac{\left(2z - \frac{1}{4}\right)}{\left(z - \frac{1}{2}\right)\left(z - \frac{1}{4}\right)} \quad z = 1/2 = 3$$

$$B = \left(z - \frac{1}{4}\right)\frac{\left(2z - \frac{1}{4}\right)}{\left(z - \frac{1}{2}\right)\left(z - \frac{1}{4}\right)} \quad z = 1/4 = -1$$

$$\frac{X_1(z)}{z} = \frac{3}{\left(z - \frac{1}{2}\right)} + \frac{-1}{\left(z - \frac{1}{4}\right)}$$

$$X_1(z) = \frac{3z}{\left(z - \frac{1}{2}\right)} + \frac{-z}{\left(z - \frac{1}{4}\right)} \quad X(z) = 2 + X_1(z)$$

$$X(z) = 2 + \frac{z\left(2z - \frac{1}{4}\right)}{\left(z - \frac{1}{2}\right)\left(z - \frac{1}{4}\right)} \quad x(n) = 2\delta(n) + 3\left(\frac{1}{2}\right)^n - \left(\frac{1}{4}\right)^n \quad n \geq 0$$

4.3.3.2 When the poles are real and repeated

The inverse transform for functions that have real and repeated poles have been considered here, in which case residue method is used.

Example 4.12

$X(z)$ is given by the following transfer function, find $x(n)$

$$X(z) = \frac{1}{z^2 - 1.8z + 0.8}$$

Solution 4.12

$$X(z) = \frac{1}{z - 1.8z + 0.8} \qquad \frac{X(z)}{z} = \frac{1}{z(z-1)(z-0.8)}$$

$$\frac{X(z)}{z} = \frac{A}{z} + \frac{B}{z-1} + \frac{C}{z-0.8}$$

$$A = z\frac{1}{z(z-1)(z-0.8)}/_{z} = 0 = 1.25$$

$$X(z) = \frac{1.25}{z} + \frac{5}{z-1} + \frac{-7.813}{z-0.8}$$

$$x(n) = \quad z^{-1}\{X(z)\} = \frac{1.25}{z} + 5\frac{z}{z-1} - 7.813\frac{z}{z-0.8}$$

$$x(n) = Z^{-1}\{X(Z)\} = 1.25\delta(n) + \ 5\ u(n) \ - \ 7.813\ (0.8)^n u(n).$$

4.3.3.3 When the poles are complex

So far the inverse transform only for functions that have real poles have been considered. The same partial-fraction procedure applies for complex poles; however, the resulting inverse transform contains complex functions

$$y[n] = Ae^{\sigma n}\cos(\omega\, n + \theta) = \frac{Ae^{\sigma n}}{2}[e^{j^{\sigma n}} e^{j\theta} + e^{-j^{\sigma n}} e^{-j\theta}]$$

$$= \frac{A}{2}[e^{(\sigma+j\omega)n} e^{j\theta} + e^{(\sigma+j\omega)n} e^{-j\theta}], \tag{4.8}$$

where Σ and Ω are real. The z-transform of this function is given by,

$$Y(z) = \frac{A}{2}\left[\frac{e^{j\theta z}}{z - e^{\sigma+j\omega}} + \frac{e^{-j\theta}z}{z - e^{\sigma-j\omega}}\right] \tag{4.9}$$

$$Y(z) = \left[\frac{(A\ e^{j\theta}/2)z}{z - e^{\sigma+j\omega}} + \frac{(Ae^{-j\theta}/2)z}{z - e^{\sigma-j\omega}}\right] = \frac{k_1}{z-p_1} + \frac{k_1^*}{z-p_1^*}, \tag{4.10}$$

where the asterisk indicates the complex conjugate.

$$p_1 = e^{\sigma}\ e^{j\omega} = e^{\sigma}\ \angle\omega, \quad \sigma = \ln|p_1|, \quad \omega = \arg\ p_1 \tag{4.11}$$

$$k_1 = \frac{Ae^{j\theta}}{2} = \frac{A}{2}\angle\theta \Rightarrow A = 2|k_1|; \quad \theta = \arg k_1. \tag{4.12}$$

Hence, we calculate σ and ω from the poles, and A and θ from the partial-fraction expansion. We can then express the inverse transform as the sinusoid of Equation (4.10).

$$y[n] = Ae^{\omega n}\cos(\omega n + \theta). \tag{4.13}$$

Example 4.13

Find the inverse z-transform of the following transfer function

$$Y(z) = \frac{-3.894z}{z^2 + 0.6065}$$

Solution 4.13

$$Y(z) = \frac{-3.894z}{z^2 + 0.6065} = \frac{-3.894z}{(z - j0.7788)(z + j0.7788)} = \frac{k_1 z}{z - j0.7788} + \frac{k_1^* z}{z + j0.7788}.$$

Dividing both sides by z, we calculate k_1:

$$k_1 = (z - j0.7788)\left[\frac{-3.894}{(z - j0.7788)(z + j0.7788)}\right]_{z=j0.7788}$$

$$= \left.\frac{-3.894}{z + j0.7788}\right|_{z=j0.7788} = \frac{-3.894}{2(j0.7788)} = 2.5\angle 90^\circ$$

$$k_1 = 2.5\angle 90^\circ$$

and

$$p_1 = j0.7788$$

$$p_1 = e^{\sigma} e^{j\omega} = e^{\sigma}\angle\omega$$

$$\sigma = \ln|p_1| \quad \omega = \arg\, p_1,$$

$$\sigma = \ln|p_1| = \ln(0.7788) = -0.250; \quad \omega = \arg\, p_1 = \pi/2$$

$$A = 2|k_1| = 2(2.5) = 5; \quad \theta = \arg k_1 = 90^\circ$$

Hence,

$$y[n] = Ae^{\sigma n}\cos(\omega n + \theta) = 5e^{-0.25n}\cos\left(\frac{\pi}{2}n + 90^\circ\right)$$

4.4 Theorems and Properties of Z-Transform

Following Table 4.2 describe the important properties of z-transform

Table 4.2 Properties of Z-transform

Property	Sequence	Z-transform
Multiplication	$a\,x\,(n)$	$a\,X\,(z)$
Linearity	$a\,x\,(n) + b\,y(n)$	$a\,X\,(z) + b\,Y(z)$
Time shifting	$x(n-k)$	$z^{-k}X(z)$
Scaling	$a^n u(n)$	$X(a^{-1}z)$
Time reversal	$x(-n)$	$X(z^{-1})$
Differentiation	$nx(n)$	$-z.\frac{dX(z)}{dz}$
Convolution	$x(n) \times y(n)$	$X(z)Y(z)$
Correlation	$r_{xy} = x(n)$ and $y(n)$	$X(z)Y(z^{-1})$

4.4.1 Multiplication Property

The z-transform of a number sequence multiplied by constant is equal to the constant multiplied by the z-transform of the number sequence; that is.

$$z[a\{x(n)\}] = az[\{x(n)\}] = aX(z) \tag{4.14}$$

$$Z[a\{x(n)\}] = \sum_{n=0}^{\infty} ax(n)z^{-n} = a\sum_{n=0}^{\infty} x(n)z^{-n} = aX(z)$$

Example 4.14

Determine the z-transform and the ROC of the signal

$$x(n) = 10(2^n)u(n)$$

Solution 4.14

If we define the signals

$$x_1(n) = 2^n u(n)$$

then $x(n)$ can be written as after its z-transform

$$X(z) = 10X_1(z).$$

We recall that

$$\alpha^n u(n) \xleftrightarrow{z} \frac{1}{1-\alpha z^{-1}} \quad \text{ROC}: |z| > |\alpha| \tag{1}$$

By setting $\alpha = 2$ in (1), we obtain

$$x_1(n) = 2^n u(n) \xleftrightarrow{z} X_1(z) = \frac{1}{1-2z^{-1}} \quad \text{ROC}: |z| > 2$$

The ROC of $X(z)$ is $|z| > 2$. Thus the transform $X(z)$ is

$$x(z) = \frac{10}{1 - 2z^{-1}} \quad \text{ROC}: |z| > 2$$

Example 4.15

Determine the z-transform and the ROC of the signal

$$x(n) = 10\left(\frac{1}{2}\right)^n u(n)$$

Solution 4.15

If we define the signals

$$x_1(n) = \left(\frac{1}{2}\right)^n u(n)$$

then $x(n)$ can be written as after its z-transform

$$x_1(n) = \left(\frac{1}{2}\right)^n u(n) \overset{z}{\longleftrightarrow} X_1(z) = \frac{1}{1 - \frac{1}{2}z^{-1}} \quad \text{ROC}: |z| > 2$$

Thus the transform $X(z)$ is

$$x(z) = \frac{10}{1 - \frac{1}{2}z^{-1}} \quad \text{ROC}: |z| > \frac{1}{2}$$

4.4.2 Linearity Property

The z-transform of the sum of number sequences is equal to the sum of the z-transform of the number sequence; that is

$$Z[\{x_1(n) + x_2(n)\}] = X_1(z) + X_2(z)$$

$$Z[a\{x_1(n) + x_2(n)\}] = a\sum_{\text{n}=0}^{\infty}[x_1(n) + x_2(n)]z^{-n}$$

$$X(z) = \sum_{n=0}^{\infty} x_1(n)z^{-\text{n}} + \sum_{n=0}^{\infty} x_2(n)z^{-n} = X_1(z) + X_2(z). \quad (4.15)$$

Example 4.16

Determine the z-transform and the ROC of the signal

$$x(n) = [3(2^n) - 4(3^n)]u(n)$$

Solution 4.16

If we define the signals

$$x_1(n) = 2^n u(n), x_2(n) = 3^n u(n)$$

then $x(n)$ can be written as

$$x(n) = 3x_1(n) - 4x_2(n)$$

Its z-transform is

$$X(z) = 3X_1(z) - 4X_2(z)$$

We recall that

$$\alpha^n u(n) \overset{z}{\longleftrightarrow} \frac{1}{1-\alpha z^{-1}} \quad \text{ROC:} |z| > |\alpha|.$$

By setting $\alpha = 2$ and $\alpha = 3$, we obtain

$$x_1(n) = 2^n u(n) \overset{z}{\longleftrightarrow} X_1(z) = \frac{1}{1-2z^{-1}} \quad \text{ROC:} |z| > 2$$

$$x_2(n) = 3^n u(n) \overset{z}{\longleftrightarrow} X_2(z) = \frac{1}{1-3z^{-1}} \quad \text{ROC:} |z| > 3.$$

The intersection of the ROC of $X_1(z)$ and $X_2(z)$ is $|z| > 3$. Thus the transform $X(z)$ is

$$x(z) = \frac{3}{1-2z^{-1}} - \frac{4}{1-3z^{-1}} \quad \text{ROC:} |z| > 3$$

4.4.3 Time Shifting Property

This property states that

$$\begin{aligned} &\text{If } x(n) \overset{Z}{\longrightarrow} X(z) \quad \text{then} \\ &x(n-n_0) \overset{Z}{\longrightarrow} z^{-n_0} X(z) \end{aligned} \tag{4.16}$$

Example 4.17

Determine the z-transform of

$$\begin{array}{ll} \text{(a) } x_1(n) = \delta(n) & \text{(d) } x_4(n) = u(n) \\ \text{(b) } x_2(n) = \delta(n-k) & \text{(e) } x_5(n) = u(n-k) \\ \text{(c) } x_3(n) = \delta(n+k) & \text{(f) } x_6(n) = u(n+k) \end{array}$$

Solution 4.17

(a) $x_1(n) = \delta(n)$

$X_1(z) = 1$

(b) $x_2(n) = \delta(n-k)$

$X_2(z) = z^{-k}\,(1)$

$X_2(z) = z^{-k}$

(c) $x_3(n) = \delta(n+k)$

$X_3(z) = z^{+k}\,(1)$

$X_3(z) = z^{k}$

(d) $x_4(n) = u(n)$

$$x_4(z) = \frac{1}{1-z^{-1}} = \frac{z}{z-1}$$

(e) $x_5(n) = u(n-k)$

$$X_5(z) = z^{-k}\frac{1}{1-z^{-1}}$$

$$X_5(z) = z^{-k}\left(\frac{z}{z-1}\right)$$

(f) $x_6(n) = u(n+k)$

$$X_6(z) = z^{k}\frac{1}{1-z^{-1}}$$

$$X_6(z) = z^{k}\left(\frac{z}{z-1}\right)$$

Example 4.18

A discrete time signal is expressed as $x(n) = \delta(n+1)+2\delta(n)+5\delta(n-3)-2\delta(n-4)$. Determine its z-transform.

Solution 4.18

According to linear property $X(z) = a_1X_1(z) + a_2X_2(z) + a_3X_3(z) + a_4X_4(z)$

$$\text{or } X(z) = a_1Z\{x_1(n)\,] + a_2Z\{x_2(n)\} + a_3Z\{x_3(n)\} + a_4Z\{x_4(n)\}$$

$$X(z) = Z\{d(n+1)\} + 2Z\{d(n)\} + 5\,Z\{d(n-3)\} - 2\,Z\{d(n-4)\}.$$

$$X(z) = z + 2(1) + 5\left(z^{-3}\right) - 2\left(z^{-4}\right)$$

$$X(z) = z + 2 + 5\,z^{-3} - 2\,z^{-4}$$

4.4.4 Scaling Property

$$x(n) \xrightarrow{Z} X(z) \quad a^n\,x(n) = X(a^{-1}z) \tag{4.17}$$

Example 4.19

Determine Z-transform of

$$x(n) = (-1)^n u(n)$$

Solution 4.19

In the given problem

$x(n) = (-1)^n u(n)$, here we take $x(n) = u(n)$ and

$$X(z) = \frac{1}{1-z^{-1}}$$

For scaling the function by we place $z = a^{-1}z$

$X(z) = \frac{1}{1-\frac{1}{a^{-1}z}}$. Now place $a = -1 X(z) = \frac{1}{1-\frac{1}{(-1)^{-1}z}}$

$$Z(-1)^n\, u(n) = X(z) = \frac{1}{1+z^{-1}}.$$

Example 4.20

A discrete time signal is expressed as $x(n) = 2^n u(n - 2)$. Determine its Z-transform

Solution 4.20

In the given problem

$$x(n) = 2^n u(n-2),$$

Now $Z\{u\ (n)\} = \frac{1}{1-z^{-1}}$.

Therefore, using time shifting property we have

$$Z\{u(n-2)\} = z^{-2}\frac{1}{1-z^{-1}}.$$

Now to find $Z\{2^n u(n-2)\}$ we shall use the scaling property, which states that

$$x(n) \xrightarrow{Z} X(z) \quad a^n\, x(n) = X(a^{-1}z)$$

$$Z\{u(n-2)\} = z^{-2}\frac{1}{1-z^{-1}}, \quad \text{then}$$

$$Z\{2^n u(n-2)\} = \frac{(2^{-1}\,z)^{-2}}{1-(2^{-1}\,z)^{-1}} = \frac{(2\,z^{-1})^2}{1-2\,z^{-1}} = \frac{4\,z^{-2}}{1-2\,z^{-1}}$$

4.4.5 Time Reversal Property

$$x(n) \xrightarrow{Z} X(z) \text{ and } x(-n) \xrightarrow{Z} X(z^{-1}) \tag{4.18}$$

Example 4.21

Determine Z-Transform of $x(n) = \left(\frac{1}{2}\right)^n u(-n)$

Solution 4.21

$$x(-n) \xrightarrow{Z} X(z^{-1})$$

$$\text{we take} \quad x(n) = \left(\frac{1}{2}\right)^n u(n) \quad \text{or} \quad X(z) = \frac{1}{1-\left(\frac{1}{2}\right)z^{-1}}$$

$$x(n) = \left(\frac{1}{2}\right)^n u(-n) \quad \text{or} \quad X(z) = \frac{1}{1-\left(\left(\frac{1}{2}\right)z^{-1}\right)^{-1}} = \frac{1}{1-2z}$$

Example 4.22

Determine Z-transform of $x_1(n) = u(-n)$

Solution 4.22

$$x_1(n) = u(-n)$$

$$\text{we take} \quad x(n) = u(n) \quad \text{or} \quad X(z) = \frac{1}{1-z^{-1}}$$

$$x(n) = u(-n) \quad \text{or} \quad X(z) = \frac{1}{1-z}$$

4.4.6 Differentiation Property

$$x(n) \xrightarrow{Z} X(z)$$

$$nx(n) \xrightarrow{Z} -z\frac{d[X(z)]}{dz} \tag{4.19}$$

Example 4.23

Determine Z-transform of $x(n) = nu(n)$

Solution 4.23

$x(n) = nu(n)$, using the property of derivative

$$nx(n) = -z\frac{d[X(z)]}{dz}u(n) \quad \text{here } x(n) = u(n)$$

$$X(z) = \frac{1}{1-z^{-1}} == \frac{z}{z-1}$$

Substituting $X(z)$ and taking z-transform, using the property of derivative, we have

$$X(z) = -z\,\frac{\mathrm{d}}{\mathrm{d}z}[Z\{u(n)\}].$$

Again using the differentiation property for the factor in bracket, we get.

$$X(z) = -z\frac{\mathrm{d}}{\mathrm{d}z}\left(\frac{z}{(z-1)}\right) = -z\frac{(z-1)(1) - z\cdot 1}{(z-1)^2} = -z\frac{(-1)}{(z-1)^2}.$$

$$\mathrm{Z}[nu(n)] = \frac{z}{(z-1)^2}$$

4.4.7 Convolution Property

If $x_1(n) \xrightarrow{Z} X_1(z)$ and $x_2(n) \xrightarrow{Z} X_2(z)$
Then

$$x(n) = x_1(n) \times x_2(n) \xrightarrow{Z} X(z) = X_1(z)X_2(z) \tag{4.20}$$

Example 4.24
Compute the convolution of two signals.

$$x_1(n) = \{1, \; -2, \; 1\}$$

$$x_2(n) = \begin{cases} 1, & 0 \le n \le 5 \\ 0, & \text{elsewhere} \end{cases}$$

Solution 4.24

$$x_1(n) = \{\underset{\uparrow}{1}, -2, 1\}$$

$$x_2(n) = \{\underset{\uparrow}{1}\, 1, 1, 1, 1, 1\}$$

$$X_1(z) = 1 - 2z^{-1} + z^{-2}; \quad X_2(z) = 1 + z^{-1} + z^{-2} + z^{-3} + z^{-4} + z^{-5}.$$

Using the property of convolution in z domain, we carry out the multiplication.

$$X(z) = X_1(z)X_2(z)$$
$$X(z) = (1 - 2z^{-1} + z^{-2})(1 + z^{-1} + z^{-2} + z^{-3} + z^{-4} + z^{-5})$$
$$X(z) = 1 - z^{-1} - z^{-6} + z^{-7}$$

Hence

$$x(n) = \{\underset{\uparrow}{1}, -1, 0, 0, 0, 0, -1, 1\}.$$

Example 4.25

A discrete time signals are given by the expression

$$x(n) = \left(\frac{1}{2}\right)^n u(n) \qquad h(n) = \left(\frac{1}{4}\right)^n u(n)$$

Using the Z-transform, find the convolution $y(n)$.

Solution 4.25

$$y(n) = x(n) \otimes h(n)$$
$$x(n) = \left(\frac{1}{2}\right)^n u(n)$$
$$Z\{u(n)\} = \frac{z}{z - \left(\frac{1}{2}\right)}$$
$$h(n) = \left(\frac{1}{4}\right)^n u(n)$$
$$Z\{h(n)\} = \frac{z}{z - \left(\frac{1}{4}\right)}$$

We know that convolution in time domain is equal to multiplication in frequency domain

$$y(n) = \left(\frac{1}{2}\right)^n u(n) \otimes \left(\frac{1}{4}\right)^n u(n)$$
$$Z[y(n)] = Z\left[\left(\frac{1}{2}\right)^n u(n)\right] \quad Z\left[\left(\frac{1}{4}\right)^n u(n)\right]$$
$$y(z) = \frac{z}{\left(z - \frac{1}{2}\right)} \frac{z}{\left(z - \frac{1}{4}\right)} \quad \frac{Y(z)}{z} = \frac{z}{\left(z - \frac{1}{2}\right)\left(z - \frac{1}{4}\right)}$$

$$\frac{Y(z)}{z} = \frac{A}{\left(z-\frac{1}{2}\right)} + \frac{B}{\left(z-\frac{1}{4}\right)}$$

$$A = \left(z-\frac{1}{2}\right)\frac{z}{\left(z-\frac{1}{2}\right)\left(z-\frac{1}{4}\right)} \quad z = 1/2 = \frac{\frac{1}{2}}{\frac{1}{4}} = 2$$

$$B = \left(z-\frac{1}{4}\right)\frac{z}{\left(z-\frac{1}{2}\right)\left(z-\frac{1}{4}\right)} \quad z = 1/4 = \frac{\frac{1}{4}}{-\frac{1}{4}} = -1$$

$$\frac{Y(z)}{z} = \frac{2}{\left(z-\frac{1}{2}\right)} + \frac{-1}{\left(z-\frac{1}{4}\right)} = \frac{2z}{\left(z-\frac{1}{2}\right)} + \frac{-z}{\left(z-\frac{1}{4}\right)}$$

$$y(n) = \left\{2\left(\frac{1}{2}\right)^n - \left(\frac{1}{4}\right)^n\right\}u(n).$$

Example 4.26

Compute the convolution of two signals $y(n)$ using Z-transform approach.

$$x(n) = \alpha^n u(n)$$
$$h(n) = \delta(n) - \alpha\delta(n-1)$$

Solution 4.26

$$x(n) = \alpha^n u(n) \qquad X(z) = \frac{1}{1-\alpha z^{-1}}$$

$$h(n) = \delta(n) - \alpha\delta(n-1) \qquad H(z) = 1 - \alpha z^{-1}$$

$$Y(z) = X(z)H(z) = \frac{1}{1-\alpha z^{-1}}(1-\alpha z^{-1}) = 1, \quad \text{which is due to}$$

pole zero cancellation.

Taking inverse Z-transform of $Y(z)$, $y(n) = \delta(n)$.

4.4.8 Correlation Property

Correlation property states that

$$\begin{aligned} \text{If } & x_1(n) \leftrightarrow X_1(z) \\ \text{and } & x_2(n) \leftrightarrow X_2(z) \\ & r_{x1\,x2}(n) = X_1(z)\, X_2(z^{-1}) \end{aligned} \tag{4.21}$$

Example 4.27

Compute the cross-correlation sequence $r_{x1\,x2}\,(j)$ of the sequences

$$X_1(n) = \{1,\ 2,\ 3,\ 4\}$$
$$X_2(n) = \{4,\ 3,\ 2,\ 1\}.$$

Solution 4.27

The cross-correlation sequence can be obtained using its correlation property of the Z-transform.

Therefore, for the given $x_1(n)$ and $x_2(n)$, we have

$$X_1(z) = 1 + 2z^{-1} + 3z^{-2} + 4z^{-3};$$
$$X_2(z) = 4 + 3z^{-1} + 2z^{-2} + z^{-3}$$
$$\text{Thus} \quad X_2(z^{-1}) = 4 + 3z + 2z^2 + z^3.$$

Using the property of convolution in z domain, we carry out the multiplication.

$$R_{x1\,x2}(z) = X_1(z)X_2(z^{-1})$$
$$= (1 + 2z^{-1} + 3z^{-2} + 4z^{-3})(4 + 3z + 2z^2 + z^3)$$
$$R_{x1\,x2}(z) = X_1(z)X_2(z^{-1})$$
$$= (z^3 + 4z^2 + 10z + 20 + 22z^{-1} + 24z^{-2} + 12z^{-3}).$$

4.4.9 Initial Value Theorem

Initial value theorem states that if $x(n)$ is causal discrete time signal with Z-transform $X(z)$, then initial value may be determined by using the following expression.

$$x(0) = \lim_{n \to 0} x(n) = \lim_{|z| \to \infty} X(z). \tag{4.22}$$

4.4.10 Final Value Theorem

Final value theorem states that for a discrete time signal $x(n)$, if $X(z)$ and the poles of $X(z)$ are all inside the unit circle, then the final value of discrete time signal, $x(\infty)$ may be determined using the following expression.

$$x(\infty) = \lim_{n \to \infty} x(n) = \lim_{|z| \to 1} [(1 - z^{-1})X(z)] \tag{4.23}$$

Example 4.28

Given the Z-transform of any signal $X(z) = 2+3z^{-1}+4z^{-2}$, determine the initial and final values of the corresponding discrete time signals.

Solution 4.28

The given expression is $X(z) = 2 + 3z^{-1} + 4z^{-2}$.

We know that the initial value is given as $x(0) = \lim_{n\to\infty} x(n) = \lim_{|z|\to\infty} X(z)$

$$x(0) = \lim_{n\to\infty} x(n) = \lim_{|z|\to\infty} (2 + 3z^{-1} + 4z^{-2}) = 2 + 0 + 0 = 2$$

Also the final value is given as $x(\infty) = \lim_{n\to\infty} x(n) = \lim_{|z|\to 1} [(1 - z^{-1})X(z)]$

$$x(\infty) = \lim_{n\to\infty} x(n) = \lim_{|z|\to 1} [(1 - z^{-1})(2 + 3z^{-1} + 4z^{-2})]$$

$$x(\infty) = \lim_{|z|\to 1} [(2 + z^{-1} + z^{-2} - 4z^{-3})] = 2 + 1 + 1 - 4 = 0$$

4.4.11 Time Delay Property (One-Sided z-Transform)

This property states that

$$\text{If } x(n) \xrightarrow{Z} X(z) \quad \text{then}$$

$$x(n-k) \xrightarrow{Z} z^{-k}\left[X(z) + \sum_{n=1}^{k} x(-n)z^{-n}\right]_{k>0} \tag{4.24}$$

4.4.12 Time Advance Property

This property states that

$$\text{If } x(n) \xrightarrow{Z} X(z) \quad \text{then}$$

$$x(n+k) \xrightarrow{Z} z^{k}\left[X(z) - \sum_{n=0}^{k-1} x(n)z^{-n}\right] \tag{4.25}$$

Example 4.29

Determine the response of the following system:

$x(n+2) - 3x(n+1) + 2x(n) = u(n)$. Assume all the initial conditions are zero.

Solution 4.29

Given system is $x(n+2) - 3x(n+1) + 2x(n) = u(n)$. Taking Z-transform of both the sides of the above equation, we obtain

$$\mathrm{X}(z)\,[z^2 - 3z + 2] = \frac{z}{z-1} \cdot X(z) = \frac{z}{z-1} \cdot \frac{1}{(z-1)(z-2)}$$

$$= \frac{Z}{(z-1)^2(z-2)} \qquad \frac{X(z)}{z} = \frac{A_{12}}{(z-1)^2} + \frac{A_{11}}{(z-1)} + \frac{B}{(z-2)}$$

$$A_{12} = (z-1)^2 \frac{1}{(z-1)^2(z-2)} \qquad A_{12} = \left.\frac{1}{z-2}\right|_{z=1} = -1$$

$$A_{11} = \left.\frac{1}{1!}\frac{d}{dz}\frac{1}{z-2}\right|_{z=1} = -\left.\frac{1}{(z-2)^2}\right|_{z=1} = -1$$

$$B = (z-2)\left.\frac{1}{(z-1)^2(z-2)}\right|_{z=2} = \left.\frac{1}{(z-1)^2}\right|_{z=2} = 1$$

$$\frac{X(z)}{z} = \frac{A_{12}}{(z-1)^2} + \frac{A_{11}}{(z-1)} + \frac{B}{(z-2)} = \frac{-1}{(z-1)^2} + \frac{-1}{(z-1)} + \frac{1}{(z-2)}$$

$$X(z) = \frac{-z}{(z-1)^2} + \frac{-z}{(z-1)} + \frac{z}{(z-2)}$$

$$x(n) = -nu(n) - u(n) + (2)^n u(n)$$

4.5 Problems and Solutions

Problem 4.1

Determine the Z-transform for the pair $x_1(n)$ and $x_2(n)$ expressions.

$$x_1(n) = (\tfrac{1}{4})^n u(n-1); \quad x_2(n) = [1 + (\tfrac{1}{2})^n]u(n)$$

Solution 4.1

Z-transform of the two signals

$$\begin{aligned} x_1(n) &= (\tfrac{1}{4})^n u(n-1) \\ &= (\tfrac{1}{4})^{n-1+1} u(n-1) \end{aligned}$$

$$= \tfrac{1}{4}(\tfrac{1}{4})^{n-1}u(n-1)$$
$$= u(n) + (\tfrac{1}{2})^n u(n)$$
$$x_2(n) = [1 + (\tfrac{1}{2})^n]u(n)$$

Using the scaling and time shifting property using the scaling property

$$X_1(z) = \tfrac{1}{4}\left(\tfrac{z^{-1}}{1-(\frac{1}{4})z^{-1}}\right) \qquad X_2(z) = \tfrac{1}{1-z^{-1}} + \tfrac{1}{1-(\frac{1}{2})z^{-1}}$$
$$X_1(z) = \tfrac{1}{4}\left(\tfrac{\frac{1}{z}}{1-\frac{1}{4z}}\right) \qquad X_2(z) = \tfrac{z}{z-1} + \tfrac{2z}{2z-1}$$
$$X_1(z) = \tfrac{1}{4}\left(\tfrac{\frac{1}{z}}{\frac{4z-1}{4z}}\right) \qquad X_2(z) = \tfrac{(2z-1)z+(z-1)(2z)}{(z-1)(2z-1)}$$
$$X_1(z) = \tfrac{1}{4}\left(\tfrac{4}{4z-1}\right) \qquad X_2(z) = \tfrac{2z^2-z+2z^2-2z}{(z-1)(2z-1)}$$
$$X_1(z) = \tfrac{1}{4z-1} \qquad X_2(z) = \tfrac{4z^2-3z}{(z-1)(2z-1)}$$

Problem 4.2

Determine for the Impulse response for the causal system using z-transform approach.

$y(n)-\frac{3}{4}y(n-1)+\frac{1}{8}y(n-2) = x(n)$, where $x(n)$ is the impulse response.

Solution 4.2

$$y(n) - \frac{3}{4}y(n-1) + \frac{1}{8}y(n-2) = x(n)$$
$$Y(z) - \frac{3}{4}Y(z)z^{-1} + \frac{1}{8}Y(z)z^{-2} = X(z) \quad \text{or} \quad Y(z)\left(1 - \frac{3}{4}z^{-1} + \frac{1}{8}z^{-2}\right)$$
$$= X(z), \quad \text{where } X(z) = 1$$
$$\frac{Y(z)}{X(z)} = \frac{1}{1-\frac{3}{4}z^{-1}+\frac{1}{8}z^{-2}} \quad \text{or} \quad H(z) = \frac{1}{1-\frac{3}{4}z^{-1}+\frac{1}{8}z^{-2}}$$
$$H(z) = \frac{z^2}{z^2 - \frac{3}{4}z + \frac{1}{8}} \quad H(z) = \frac{z^2}{(z-\frac{1}{2})(z-\frac{1}{4})}$$
$$\frac{H(z)}{z} = \frac{z}{(z-\frac{1}{2})(z-\frac{1}{4})} \quad \frac{H(z)}{z} = \frac{A}{z-\frac{1}{2}} + \frac{B}{z-\frac{1}{4}}$$
$$A = (z-\tfrac{1}{2})\left(\frac{z}{(z-\frac{1}{2})(z-\frac{1}{4})}\right)_{z=\frac{1}{2}} \quad A = \frac{\frac{1}{z}}{\frac{1}{2}-\frac{1}{4}} = \frac{1}{2} \times \frac{4}{1} = 2$$

$$B = (z - \tfrac{1}{4})\left[\frac{z}{(z-\frac{1}{2})(z-\frac{1}{4})}\right]_{z=\frac{1}{4}} \qquad B = \frac{\frac{1}{4}}{\frac{1}{4}-\frac{1}{2}} = -1$$

$$\frac{H(z)}{z} = \frac{2}{z-\frac{1}{2}} - \frac{1}{z-\frac{1}{4}}$$

$$H(z) = \frac{2\,z}{z-\frac{1}{2}} - \frac{1\,z}{z-\frac{1}{4}}$$

$$H(z) = 2\frac{1}{1-(\frac{1}{2})z^{-1}} - \frac{1}{1-(\frac{1}{4})z^{-1}}$$

$$h(n) = 2\left(\frac{1}{2}\right)^n u(n) - \left(\frac{1}{4}\right)^n u(n)$$

Problem 4.3

Determine for the step response for the causal system using z-transform approach.

$y(n) - \frac{3}{4}y(n-1) + \frac{1}{8}y(n-2) = x(n)$, where the input $x(n) = u(n)$

Solution 4.3

$$X(z) = \frac{1}{1-z^{-1}}$$

$$Y(z) = H(z)\,X(z)$$

$$Y(z) = \frac{1}{1-\frac{3}{4}z^{-1}+\frac{1}{8}z^{-2}} \times \frac{1}{1-z^{-1}} = \frac{1}{(1-\frac{3}{4}z^{-1}+\frac{1}{8}z^{-2})(1-z^{-1})}$$

$$Y(z) = \frac{z^3}{(z^2-\frac{3}{4}z+\frac{1}{8}) \times (z-1)}$$

$$\frac{Y(z)}{z} = \frac{z^2}{(z-\frac{1}{2})(z-\frac{1}{4})(z-1)} = \frac{-2}{(z-\frac{1}{2})} + \frac{\frac{1}{3}}{(z-\frac{1}{4})} + \frac{\frac{8}{3}}{(z-1)}$$

$$Y(z) = \frac{-2z}{z-\frac{1}{2}} + \frac{\frac{1}{3}(z)}{z-\frac{1}{4}} + \frac{\frac{8}{3}(z)}{(z-1)}$$

$$= -2\left[\frac{1}{1-(\frac{1}{2})z^{-1}}\right] + \frac{1}{3}\left[\frac{1}{1-(\frac{1}{4})z^{-1}}\right] + \frac{8}{3}\left(\frac{1}{1-z^{-1}}\right)$$

$$y(n) = -2\left(\frac{1}{2}\right)^n u(n) + \frac{1}{3}\left(\frac{1}{4}\right)^n u(n) + \frac{8}{3}u(n)$$

$$y(n) = \left[-2\left(\frac{1}{2}\right)^n + \frac{1}{3}\left(\frac{1}{4}\right)^n + \frac{8}{3}\right]u(n)$$

Problem 4.4

Find the inverse z-transform of the following transfer function of the following function $Y(z)$, using residue theorem.

$$Y(z) = \frac{1 + z^{-1}}{1 - z^{-1} + 0.5z^{-2}}$$

Solution 4.4

$$Y(z) = \frac{z^2 + z}{z^2 - z + 0.5} = \frac{z(z+1)}{z^2 - z + 0.5}$$

Dividing both sides by z

$$\frac{y(z)}{z} = \frac{z+1}{z^2 - z + 0.5}$$

$$\frac{y(z)}{z} = \frac{z+1}{(z - \frac{1}{2} + j\frac{1}{2})(z - \frac{1}{2} - j\frac{1}{2})} = \frac{k_1}{z - \frac{1}{2} + j\frac{1}{2}} + \frac{k_1^*}{z - \frac{1}{2} - j\frac{1}{2}}$$

$$p_1 = p_2^* = \frac{1}{\sqrt{2}} e^{j\pi/4} \quad k_1 = \sqrt{2}e^{-j45}$$

$$\sigma = \ln|p_1| = \ln\left(\tfrac{1}{2} + j\tfrac{1}{2}\right); \quad \omega = \arg p_1 = \pi/4$$

$$A = 2|k_1| = \sqrt{2}e^{-j45} \quad \theta = \arg k_1 = 45°.$$

Hence $y[n] = Ae^{\sigma n}\cos(\omega n + \theta) = 1.414 e^{\left(\frac{1}{\sqrt{2}}\right)^n} \cos\left(\frac{\pi}{4} + 45°\right)$

Problem 4.5

Find the inverse z-transform of the following transfer function $Y(z)$.

$$Y(z) = \frac{1}{(1 + z^{-1})(1 - z^{-1})^2}$$

Solution 4.5

$$Y(z) = \frac{1}{4}\frac{1}{1 + z^{-1'}} + \frac{3}{4}\frac{1}{1 - z^{-1}} + \frac{1}{2}\frac{z^{-1}}{(1 - z^{-1})^2}$$

$$y[n] = \frac{1}{4}(-1)^n u(n) + \frac{3}{4}u(n) + \frac{1}{2}n\,u(n)$$

$$y[n] = \left(\frac{1}{4}(-1)^n + \frac{3}{4} + \frac{1}{2}n\right) \quad u(n)$$

Problem 4.6

A discrete time signal is expressed as $x(n) = \cos\omega n$ for $n \geq 0$. Determine its z-transform.

Solution 4.6

$$\cos\ \omega n = \frac{1}{2}[e^{j\omega n} + e^{-j\omega n}]$$

The z-transform is expressed as

$$Z[\{x(n)\}] = X(z) = \sum_{n=0}^{\infty} x(n)z^{-n}$$

$$Z[e^{j\omega\, n}] == \frac{1}{1 - e^{j\omega}\, z^{-1}} \quad \text{for } |z| > 1 \quad [e^{j\omega} = 1]$$

In the same way for $n \geq 0$

$$Z[e^{-j\omega n}] == \frac{1}{1 - e^{-j\omega} z^{-1}} \quad \text{for } |z| > 1 \quad [e^{-j\omega} = 1]$$

$$X(z) = Z\left[\frac{1}{2}\{e^{j\omega n} + e^{-j\omega n}\}\right]$$

$$X(z) = \frac{1}{2}Z\{e^{j\omega n}\} + \frac{1}{2}Z\{e^{-j\omega n}\}$$

$$X(z) = \frac{1}{2}\left[\frac{1}{1 - e^{j\omega} z^{-1}}\right] + \frac{1}{2}\left[\frac{1}{1 - e^{-j\omega} z^{-1}}\right]$$

$$X(z) = \frac{1}{2}\left[\frac{2 - (e^{j\omega} + e^{-j\omega})z^{-1}}{(1 - e^{j\omega} z^{-1})(1 - e^{-j\omega} z^{-1})}\right]$$

$$X(z) = \frac{1 - \ \cos\omega\ z^{-1}}{1 - 2z^{-1}\cos\omega + z^{-2}} = \frac{z^2 - z\cos\omega}{z^2 - 2z\ \cos\omega + 1} \quad \text{for } |z| > 1$$

Problem 4.7

A discrete time signal is expressed as $x(n) = \sin\omega n$ for $n \geq 0$. Determine its z-transform.

Solution 4.7

$$\sin\ \omega n = \frac{1}{2j}[e^{j\omega n} - e^{-j\omega n}]$$

The z-transform is expressed as $Z\left[\{x(n)\}\right] = X(z) = \sum_{n=0}^{\infty} x(n)z^{-n}$

$$Z[e^{j\omega n}] == \frac{1}{1 - e^{j\omega} z^{-1}} \quad \text{for } |z| > e^{j\omega n}$$

$$\text{or} \quad Z[e^{j\omega n}] == \frac{1}{1 - e^{j\omega} z^{-1}} \quad \text{for } |z| > 1[e^{j\omega} = 1].$$

In the same way for $n \geq 0$

$$Z[e^{-j\omega n}] = \frac{1}{1 - e^{-j\omega} z^{-1}} \quad \text{for } |z| > e^{-j\omega n}$$

or

$$Z\left[e^{-j\omega n}\right] == \frac{1}{1 - e^{-j\omega} z^{-1}} \quad \text{for } |z| > 1 \quad [e^{-j\omega} = 1]$$

$$X(z) = Z\left[\frac{1}{2j}\{e^{j\omega n} - e^{-j\omega n}\}\right] \quad X(z) = \frac{1}{2j}Z\{e^{j\omega n}\} - \frac{1}{2j}Z\{e^{-j\omega n}\}$$

$$X(z) = \frac{1}{2j}\left\{\frac{1}{1 - e^{j\omega} z^{-1}}\right\} - \frac{1}{2j}\left\{\frac{1}{1 - e^{-j\omega} z^{-1}}\right\}$$

$$X(z) = \frac{z^{-1}(e^{j\omega} + e^{-j\omega})/2j}{(1 - e^{j\omega} z^{-1})(1 - e^{-j\omega} z^{-1})}$$

$$X(z) = \frac{z^{-1} \sin \omega}{1 - 2z^{-1} \cos \omega + z^{-2}} = \frac{z \sin \omega}{z^2 - 2z \cos \omega + 1} \quad \text{for } |z| > 1$$

Problem 4.8
By applying the time shifting property, determine the $x(n)$ of the signal.

$$X(z) = \frac{z^{-1}}{1 - 3z^{-1}}$$

Solution 4.8

$$X(z) = \frac{z^{-1}}{1 - 3z^{-1}} = z^{-1}X_1(z) \qquad X_1(z) = \frac{1}{1 - 3z^{-1}} x_1(n) = (3)^n \, u(n)$$

Because there is delay z^{-1}, therefore delaying $x_1(n)$ by one unit.

$$x(n) = (3)^{n-1} \, u(n-1)$$

Problem 4.9

The Z-transform of a particular discrete time signal $x(n)$ is expressed

$$X(z) = \frac{1 + \frac{1}{2}z^{-1}}{1 - \frac{1}{2}z^{-1}}$$

Determine $x(n)$ using time shifting property.

Solution 4.9

Splitting the expression in two fractions for convenience

$$X(z) = \frac{1 + \frac{1}{2}z^{-1}}{1 - \frac{1}{2}z^{-1}} = \frac{1}{1 - \frac{1}{2}z^{-1}} + \frac{\frac{1}{2}z^{-1}}{1 - \frac{1}{2}z^{-1}}$$

$$x(n) = Z^{-1}\left[\frac{1}{1 - \frac{1}{2}z^{-1}} + \frac{\frac{1}{2}z^{-1}}{1 - \frac{1}{2}z^{-1}}\right]$$

$$x(n) = Z^{-1}\{X(z)\} = \left[\frac{1}{1 - \frac{1}{2}z^{-1}} + \frac{\frac{1}{2}z^{-1}}{1 - \frac{1}{2}z^{-1}}\right]$$

$$x(n) = Z^{-1}\{X_1(z) + X_2(z)\}$$

$$X_1(z) = \frac{1}{1 - \frac{1}{2}z^{-1}}; \quad X_2(z) = \frac{\frac{1}{2}z^{-1}}{1 - \frac{1}{2}z^{-1}}$$

$$X_1(z) = \frac{1}{1 - \frac{1}{2}z^{-1}} = Z\left[\left(\frac{1}{2}\right)^n u(n)\right]$$

$$X_2(z) = \frac{\frac{1}{2}z^{-1}}{1 - \frac{1}{2}z^{-1}} = \frac{1}{2}Z^{-1}[X_1(z)]$$

Now using time shift property

$$\text{If } x(n) \xrightarrow{Z} X(z) \quad \text{then}$$

$$x(n - n_0) \xrightarrow{Z} z^{-n0}X(z)$$

$$X_2(z) = \frac{1}{2}Z^{-1}[X_1(z)]$$

$$X_2(z) = \frac{1}{2}Z\left[\left(\frac{1}{2}\right)^{n-1} u(n-1)\right]$$

Substituting the value of $X_1(z)$ and $X_2(z)$

$$x(n) = \left(\frac{1}{2}\right)^n u(n) + \frac{1}{2}\left(\frac{1}{2}\right)^{n-1} [u(n-1)]$$

$$x(n) = \left(\frac{1}{2}\right)^n \{u(n) + u(n-1)\}$$

$$\text{or} \quad x(n) = \left(\frac{1}{2}\right)^n \{\delta(n) + 2u(n-1)\}$$

Problem 4.10

Given the Z-transform of $x(n)$ as

$$X(z) = \frac{z}{z-4}.$$

Use Z-transform properties to determine $y(n) = x(n-2)$,

Solution 4.10

$$X(z) = \frac{z}{z-4}. \text{ Using time shifting property}$$

$$Y(z) = Z[x(n-2)] = z^{-2}\left(\frac{z}{z-4}\right) = \left(\frac{z^{-1}}{z-4}\right) = \frac{1}{z(z-4)}$$

We expand $\frac{Y(z)}{z}$ into partial fractions to get $y(n)$

$$\frac{Y(z)}{z} = \frac{1}{z^2(z-4)} = \left(\frac{A_{12}}{z^2}\right) + \left(\frac{A_{11}}{z}\right) + \frac{B}{(z-4)}$$

$$A_{12} = z^2 \; z = 0 \quad z^2 \frac{1}{z^2(z-4)} = -\frac{1}{4}$$

$$A_{11} = z = 0 \quad \frac{\mathrm{d}}{\mathrm{d}z}\left(\frac{1}{(z-4)}\right) = \left(-\frac{1}{(z-4)^2}\right) = -\frac{1}{16}$$

$$B = z = 4 \quad (z-4)\frac{1}{z^2(z-4)} = \frac{1}{z^2} = \frac{1}{16}$$

$$\frac{Y(z)}{z} = -\frac{1}{4}\frac{1}{z^2} - \frac{1}{16}\left(\frac{1}{z}\right) + \frac{1}{16}\frac{1}{(z-4)}$$

$$Y(z) = -\frac{1}{4}\frac{1}{z} - \frac{1}{16} + \frac{1}{16}$$

$$\frac{z}{(z-4)}y(n) = -\frac{1}{4}\delta(n-1) + \frac{1}{16}\delta(n) - \frac{1}{16}4^n \quad n \geq 0$$

Problem 4.11
Using scaling property, determine z-transform of $a^n \cos \omega n$ for $n \geq 0$.

Solution 4.11

$$Z\{\cos\ \omega n\} = \frac{1-\ \cos\omega z^{-1}}{1-2z^{-1}\cos\omega + z^{-2}} = \frac{z^2 - z\cos\omega}{z^2 - 2z\ \cos\omega + 1}$$

Using scaling property

$$Z\{a^n\ \cos\ \omega n\} = \frac{1 - a\ z^{-1}\ \cos\omega}{1 - 2a\ z^{-1}\cos\omega + a^2 z^{-2}} = \frac{z^2 - az\ \cos\omega}{z^2 - 2az\ \cos\omega + a^2}$$

Problem 4.12
Determine Z-transform of $x(n) = \left(\frac{1}{2}\right)^n u(-n)$

Solution 4.12

$$X(z) = \sum_{n=0}^{\infty}\left(\frac{1}{2}\right)^{n} u(-n)\ z^{-n}$$

$$X(z) = \sum_{n=0}^{\infty}\left(\frac{1}{2}\right)^{=n} z^n = \sum_{n=0}^{\infty}(2z)^{\ n}$$

If $2Z < 1$, the sum converges and $X(z) = \frac{1}{1-2z}$
$2Z < 1$ or $|z| < \frac{1}{2}$.

Problem 4.13
Determine the Z-transform of the signal

$$x(n) = -\alpha^n u(-n-1) = \begin{cases} 0, & n \geq 0 \\ -\alpha^n, & n \leq -1 \end{cases}$$

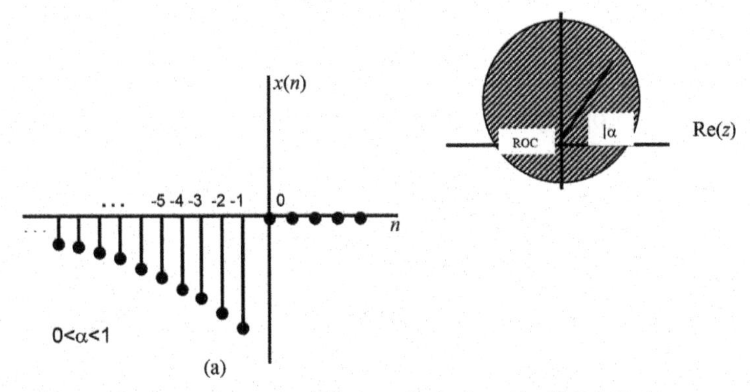

Figure 4.2 (a) Anticausal signal $x(n) = -\alpha^n u(-n-1)$, (b) ROC of its z-transform

Solution 4.13

$$X(z) = \sum_{n=-\infty}^{-1} (-\alpha^n) z^{-n} = -\sum_{L=1}^{\infty} (\alpha^{-1} z)^L$$

where $L = -n$. Using the formula

$$A + A^2 + A^3 + \ldots = A(1 + A + A^2 + \ldots) = \frac{A}{1-A}$$

When $|A| < 1$ gives

$$X(z) = -\frac{\alpha^{-1} z}{1 - \alpha^{-1} z} = \frac{1}{1 - \alpha\, z^{-1}}$$

Provided that $|\alpha^{-1} z| < 1$ or equivalently, $|z| < |\alpha|$. Thus

$$x(n) = -\alpha^n u(-n-1) \stackrel{z}{\longleftrightarrow} X(z) = \frac{1}{1 - \alpha(z)^{-1}}, \quad \text{ROC} : |z| < |\alpha|$$

The ROC is now the interior of a circle having radius $|\alpha|$.

Problem 4.14

Given the Z-Transform of $x(n)$ as

$$X(z) = \frac{z}{z^2 + 4}.$$

Use Z-transform properties to determine $Y(z)$, i.e., $y(n) = x(-n)$,

Solution 4.14

$$x(-n) \xrightarrow{Z} X(z^{-1})$$

Using time reversal property

$$Y(z) = \left(\frac{z^{-1}}{z^{-2}+4}\right) = \left(\frac{z}{1+4z^2}\right)$$

Problem 4.15

A discrete time signal is expressed as $x(n) = n^2 u(n)$. Determine its Z-transform.

Solution 4.15

Given that

$$x(n) = n^2 u(n)$$

$$nx(n) = -zX(z) = Z\{x(n)\} = Z\{n^2\, u(n)\} = Z[n\{nu(n)\}].$$

Using the property of derivative, we have

$$X(z) = -z\,\frac{\mathrm{d}}{\mathrm{d}z}[Z\{nu(n)\}].$$

Again using the differentiation property for the factor in bracket, we get.

$$X(z) = -z\frac{\mathrm{d}}{\mathrm{d}z}\frac{z}{(z-1)^2}$$

$$X(z) = -z\frac{\mathrm{d}}{\mathrm{d}z}\frac{z}{(z-1)^2} = -z\frac{(z-1)^2 - 2z^2 + 2z}{(z-1)^4}$$

$$= -z\frac{(z^2 - 2z + 1) - 2z^2 + 2z}{(z-1)^4}$$

$$X(z) = -z\frac{(-z^2+1)}{(z-1)^4} = z\frac{(z^2-1)}{(z-1)^4}$$

Problem 4.16

Given the Z-transform of a signal $x(n)$ is given as

$$X(z) = \frac{z}{z^2+4}.$$

Use Z-transform properties to determine $Y(z)$.

Solution 4.16

$$x(n) \xrightarrow{Z} X(z) = \frac{z}{z^2+4} \quad nx(n) \xrightarrow{Z} -z\frac{\mathrm{d}[X(z)]}{\mathrm{d}z}$$

Using differentiation property

$$Y(z) = -z\frac{\mathrm{d}}{\mathrm{d}z}\left[\frac{z}{z^2+4}\right] = -z\left(\frac{z^2+4-z(2z)}{(z^2+4)^2}\right)$$

$$Y(z) = -z\left(\frac{-z^2+4}{(z^2+4)^2}\right) = z\left(\frac{-z^2+4}{(z^2+4)^2}\right)$$

Problem 4.17

Determine the Z-transform of the signal

$$y(n) = n[\left(\frac{1}{2}\right)^n u(n) \otimes \left(\frac{1}{2}\right)^n u(n)]$$

Solution 4.17

$$\text{Let } x(n) = \left[\left(\frac{1}{2}\right)^n u(n) \otimes \left(\frac{1}{2}\right)^n u(n)\right]$$

$$Z\{x(n)\} = \frac{z^2}{\left[z-\left(\frac{1}{2}\right)\right]^2} y(n) = nx(n)$$

$$Y(z) = -z\,\frac{\mathrm{d}}{\mathrm{d}z}\frac{z^2}{\left[z-\frac{1}{2}\right]^2} = -z\,\frac{\mathrm{d}}{\mathrm{d}z}\frac{z^2}{\left[z-\frac{1}{2}\right]^2}$$

$$= -z\left[\frac{\left(z-\frac{1}{2}\right)2z - z^2\left(2\left(z-\frac{1}{2}\right)\right)}{\left(z-\frac{1}{2}\right)^4}\right] = \frac{8z^2(2z-1)}{(2z-1)^4}$$

Problem 4.18

A discrete time signal is given by the expression

$$x(n) = n\left(-\frac{1}{2}\right)^n u(n) \otimes \left(\frac{1}{4}\right)^{-n} u(-n)$$

Find the Z-transform.

Solution 4.18

$$x(n) = x_1(n) \otimes x_2(n)$$
$$x_1(n) = n\left(-\frac{1}{2}\right)^n u(n)$$
$$Z\{u(n)\} = \frac{1}{1 - z^{-1}}$$

Using the scaling property we have

$$Z\left[\left(-\frac{1}{2}\right)^n u(n)\right] = \frac{z}{z + \frac{1}{2}} \quad \text{ROC } |z| > \frac{1}{2}$$

Now using Z-domain differentiation property

$$Z\left[\left(n - \frac{1}{2}\right)^n u(n)\right] = -z\frac{\mathrm{d}}{\mathrm{d}z}\left[\frac{z}{z + \frac{1}{2}}\right] = \frac{-\frac{1}{2}z}{\left(z + \frac{1}{2}\right)^2}$$
$$x_2(n) = \left(\frac{1}{4}\right)^{-n} u(-n) \quad Z\{u(n)\} = \frac{1}{1 - z^{-1}}$$

Using the scaling property

$$Z\left[\left(\frac{1}{4}\right)^{-n} u(n)\right] = Z[(4)^n\, u(n)]$$
$$= \frac{1}{[1 - (4)z^{-1}]} = \frac{z}{z - 4} \quad \text{ROC } |z| > \frac{1}{4}$$

Using time reversal property

$$Z\left[\left(\frac{1}{4}^{-n}\right) u(-n)\right] = \frac{z^{-1}}{z^{-1} - 4} \quad \text{ROC } \left|\frac{1}{z}\right| > \frac{1}{4}$$
$$Z\left[\left(\frac{1}{4}^{-n}\right) u(-n)\right] = \frac{1}{1 - 4z} \quad \text{ROC } |z| < \frac{1}{4}.$$

We know that convolution in time domain is equal to multiplication in frequency domain

$$x(n) = n\left(-\frac{1}{2}\right)^n u(n) \otimes \left(\frac{1}{4}\right)^{-n} u(-n) \quad Z[x(n)] = Z[n\left(-\frac{1}{2}\right)^n$$

$$u(n)]\ Z[\left(\frac{1}{4}\right)^{-n} u(-n)]$$

$$X(z) = \frac{-\frac{1}{2}z}{\left(z+\frac{1}{2}\right)^2} \frac{\frac{1}{4}}{\left(z-\frac{1}{4}\right)} = -\frac{\frac{1}{8}z}{\left(z+\frac{1}{2}\right)^2\left(z-\frac{1}{4}\right)} \quad \text{ROC}\frac{1}{2} < |z| < \frac{1}{4}$$

Problem 4.19

Compute the $Y(z)$ using convolution $y(n) = x(n) \otimes h(n)$ of the following signals:

$$x(n) = \left(\frac{1}{2}\right)^n u(n),\ h(n) = \left(\frac{1}{2}\right)^n u(n)$$

Solution 4.19

$$y(n) = [(\tfrac{1}{2})^n\ u(n) \otimes (\tfrac{1}{2})^n\ u(n)]$$

$$Z\{y(n)\} = Y(z) = \frac{1}{(1-(\frac{1}{2})z^{-1})^2}.$$

Problem 4.20

Determine the Z-Transform of the following signal.

Solution 4.20

$$X(z) = \sum_{n=-\infty}^{\infty} \left(\frac{1}{2}\right)^n [u(n) - u(n-10)]z^{-n}$$

$$X(z) = \sum_{n=-\infty}^{\infty} \left(\frac{1}{2z}\right)^n - \sum_{n=-10}^{\infty} \left(\frac{1}{2z}\right)^n = \sum_{n=0}^{9} \left(\frac{1}{2z}\right)^n$$

$$X(z) = \frac{1-\left(\frac{1}{2z}\right)^{10}}{1-\left(\frac{1}{2z}\right)} = \frac{(2z)^{10}-1}{2z-1}\frac{1}{(2z)^9}$$

$$X(z) = \frac{1-\left(\frac{1}{2z}\right)^{10}}{1-\left(\frac{1}{2z}\right)} = \frac{z^{10}-\left[\frac{1}{2}\right]^{10}}{z-\left[\frac{1}{2}\right]}\frac{1}{(z)^9}$$

Problem 4.21

Solve the following difference equation for $x(k)$ using the Z-transform method for $k \leq 4$.

$$x(k) - 4x(k-1) + 3x(k-2) = e(k)$$

$$\text{where } e(k) = \begin{cases} 2, & k = 0, 1 \\ 0, & k \geq 2 \end{cases} \qquad x(-2) = x(-1) = 0.$$

Solution 4.21

Taking transform of the difference equation

$$\left[1 - 4z^{-1} + 3z^{-2}\right] X(z) = E(z) = 2\left(1 + z^{-1}\right)$$

$$X(z) = \frac{2(1 + z^{-1})}{1 - 4z^{-1} + 3z^{-2}} = \frac{2z(z + 1)}{z^2 - 4z + 3}$$

$$\frac{X(z)}{z} = \frac{2(z + 1)}{(z - 1)(z - 3)} = \frac{-2}{z - 1} + \frac{4}{z - 3}$$

$$X(z) = -\frac{2z}{z - 1} + \frac{4z}{z - 3}$$

$$x(k) = \left[-2 + 4(3)^k\right] u(k).$$

Problem 4.22

Given the difference equation:

$$y(k + 2) - 0.75y(k + 1) + 0.125y(k) = e(k),$$

where $e(k) = 1$, $k \geq 0$, and $y(0) = y(1) = 0$.

Solution 4.22

$y(k + 2) - 0.75y(k + 1) + 0.125y(k) = e(k)$, converting the given difference equation into Z transform.

$$Y(z)\left\{z^2 - 0.75z + 0.125\right\} = E(z) \quad \text{where } E(z) = \frac{z}{z - 1} \text{ (given)}$$

$Y(z) = \frac{E(z)}{(z-0.25)(z-0.5)}$. Substituting the value of $E(z)$

$$y(z) = \frac{z}{(z - 1)(z - \frac{1}{4})(z - 1/2)}$$

$$\frac{y(z)}{z} = \frac{1}{(z - 1)(z - 1/4)(z - 1/2)} = \frac{8/3}{z - 1} + \frac{16/3}{z - 1/4} + \frac{-8}{z - 1/2}$$

$$y(k) = 8/3 + \tfrac{16}{3}\left(\tfrac{1}{4}\right)^k - 8(1/2)^k$$

Problem 4.23

Find the inverse Z-transform of each given $E(z)$ by two methods and compare the value of $e(k)$ for $k = 0, 1, 2,$ and 3 obtained by two method.

$$E(z) = \frac{z}{(z-1)(z-0.8)}$$

Solution 4.23

$$E(z) = \frac{z}{(z-1)(z-0.8)}$$

(i)

$$\begin{array}{r|l} & z^{-1} + 1.8z^{-2} + 2.44z^{-3} \\ z^2 - 1.8z + 0.8 & z \\ & z - 1.8 + 0.8z^{-1} \\ \hline & 1.8 - 0.8z^{-1} \\ & 1.8... \end{array}$$

$$e(0) = 0, \quad e(1) = 1, \quad e(2) = 1.8, \quad e(3) = 2.44$$

(ii)

$$\frac{E(z)}{z} = \frac{1}{(z-1)(z-0.8)} = \frac{5}{z-1} + \frac{-5}{z-0.8}$$

$$E(z) = \frac{5z}{z-1} - \frac{5z}{z-0.8}$$

$$z^{-1}\{E(z)\} = 5z^{-1}\left\{\frac{z}{z-1}\right\} - 5z^{-1}\left[\frac{z}{z-0.8}\right]$$

$$e(k) = 5 - 5(0.8)^k, e(k) = 5\left[1 - (0.8)^k\right]$$

$$\begin{aligned} k &= 0, \quad e(0) = 5\left[1 - (0.8)^0\right] = 0 \\ k &= 1, \quad e(1) = 5\left[1 - (0.8)^1\right] = 1 \\ k &= 2, \quad e(2) = 5\left[1 - (0.8)^2\right] = 1.8 \\ k &= 3, \quad e(3) = 5\left[1 - (0.8)^3\right] = 2.44 \end{aligned}$$

Problem 4.24

Find the inverse Z-transform of each given $E(z)$ by two methods and compare the value of $e(k)$ for $k = 0, 1, 2,$ and 3 obtained by two method.

$$E(z) = \frac{1}{(z-1)(z-0.8)} = \frac{1}{z^2 - 1.8z + 0.8}$$

Solution 4.24

(i)

$$\begin{array}{r|l} & z^{-2} + 1.8z^{-3} + 2.44z^{-4} \\ \hline z^2 - 1.8z + 0.8 & 1 \\ & 1 - 1.8z^{-1} + 0.8z^{-2} \\ \hline & 1.8z^{-1} - 0.8z^{-2} \\ & 1.8z^{-1} - 3.24z^{-2} + 1.44z^{3} \\ \hline & 2.44z^{-2} - 1.44z^{-3} \end{array}$$

$$e(0) = 0,\ e(1) = 0,\ e(2) = 1,\ e(3) = 1.8,\ e(4) = 2.44.$$

(ii)

$$\frac{E(z)}{z} = \frac{1}{z(z-1)(z-0.8)} = \frac{A}{z} + \frac{B}{z-1} + \frac{C}{z-0.8}$$

$$A = \lim_{z\to 0} z\frac{1}{z(z-1)(z-0.8)} = \frac{1}{(-1)(-0.8)} = 1.25$$

$$B = \lim_{z\to 1}(z-1)\frac{1}{z(z-1)(z-0.8)} = \frac{1}{1(0.2)} = 5$$

$$C = \lim_{z\to 0.8}(z-0.8)\frac{1}{z(z-1)(z-0.8)} = \frac{1}{(0.8)} = -6.25$$

$$\frac{E(z)}{z} = \frac{1.25}{z} + \frac{5}{z-1} - \frac{6.25}{z-0.8}$$

$$E(z) = 1.25 + 5\frac{z}{z-1} - 6.25\frac{z}{z-0.8}$$

$$\begin{aligned} e(k) &= 1.255d(k) + 5\,u(k) - 6.25\,(0.8)^k u(k) \\ k = 0,\quad e(0) &= 1.25 + 5 - 6.25 = 0 \\ k = 1,\quad e(1) &= 0 + 5 - 6.25\,(0.8) = 0 \\ k = 2,\quad e(2) &= 0 + 5 - 6.25(0.8)^2 = 1.00 \end{aligned}$$

Problem 4.25

Find the inverse Z-transform $x(n)$ of given $X(z)$.

$$X(z) = \frac{3}{z-2}$$

Solution 4.25

$X(z)$ may be written as $X(z) = 3z^{-1}\frac{z}{z-2}$

$Z\{2^n u(n)\} = \frac{z}{z-2}$, using time shifting property

$$Z\{2^{n-1}u(n-1)\} = z^{-1}\left[\frac{z}{z-2}\right] = \frac{1}{z-2}$$

Thus, we conclude that $x(n) = 3\{2^{n-1}u(n-1)\}$.

Problem 4.26

Find the inverse z-transform of the following transfer function

$$X(z) = \frac{z}{z^2 + 1\,z + 0.5}$$

Solution 4.26

$$X(z) = \frac{z}{z^2 + 1\,z + 0.25 + 0.25} = \frac{z}{(z+0.5)^2 + (0.5)^2}$$

$$= \frac{k_1 z}{z + 0.5 - j0.5} + \frac{k_1^* z}{z + 0.5 + j0.5}$$

Dividing both sides by z, we calculate k_1:

$$k_1 = (z + 0.5 - j0.5)\left[\frac{1}{(z + 0.5 - j0.5)(z + 0.5 + j0.5\,)}\right]_{z=-0.5+j0.5}$$

$$= \frac{1}{z + 0.5 + j0.5}\Bigg|_{z=-0.5+j0.5} = \frac{1}{2(j0.5)} = 1\angle{-90^\circ}$$

with $p_1 = -0.5 + j0.5$,

$\Sigma = \ln|p_1| = \ln(0.707) = -0.34; \quad \Omega = \arg p_1 = 3p/4$

$A = 2|k_1| = 2(1) = 2;$

$\theta = \ \arg\ k_1 = -90^\circ$

Hence,

$y[n] = Ae^{\Sigma n}\cos(\Omega n + \theta) = 2e^{-0.34n}\cos\left(\frac{3\pi}{4} - 90^\circ\right)$

Problem 4.27

Determine the convolution of two signals for the convolution of $x(n)$ and $h(n)$ where

$$x(n) = \left(\frac{1}{4}\right)^{n-2} u(n) \quad h(n) = \left(\frac{1}{3}\right)^{n} u(n-3)$$

Solution 4.27

$$x(n) = \left(\frac{1}{4}\right)^{n-2} u(n) \quad x(n) = \left(\frac{1}{4}\right)^{-2}\left(\frac{1}{4}\right)^{n} u(n) \quad x(n) = (16)\left(\frac{1}{4}\right)^{n} u(n)$$

$$X(z) = (16)Z\left(\frac{1}{4}\right)^{n} u(n) = 16\frac{z}{z-\frac{1}{4}}$$

$$h(n) = \left(\frac{1}{3}\right)^{n} u(n-3) = \left(\frac{1}{3}\right)^{3}\left(\frac{1}{3}\right)^{n-3} u(n-3) = \left(\frac{1}{27}\right)z^{-3}\mathrm{Z}\left(\frac{1}{3}\right)^{n} u(n)$$

$$H(z) = \left(\frac{1}{27}\right)z^{-3}\frac{z}{z-\frac{1}{3}} = \left(\frac{1}{27}\right)\frac{z^{-2}}{z-\frac{1}{3}}$$

$$Y(z) = 16\frac{z}{z-\frac{1}{4}}\left(\frac{1}{27}\right)\frac{z^{-2}}{z-\frac{1}{3}} = \left(\frac{16}{27}\right)\frac{z^{-1}}{\left(z-\frac{1}{3}\right)\left(z-\frac{1}{4}\right)}$$

$$Y(z) = 16\frac{z}{z-\frac{1}{4}}\left(\frac{1}{27}\right)\frac{z^{-2}}{z-\frac{1}{3}} = \left(\frac{16}{27}\right)\frac{z^{-1}}{\left(z-\frac{1}{3}\right)\left(z-\frac{1}{4}\right)}$$

$$= \left(\frac{16}{27}\right)\frac{1}{z\left(z-\frac{1}{3}\right)\left(z-\frac{1}{4}\right)}$$

$$\frac{Y(z)}{z} = \frac{16}{27}\frac{1}{z\left(z-\frac{1}{3}\right)\left(z-\frac{1}{4}\right)} = \frac{A_{12}}{z^2} + \frac{A_{11}}{z} + \frac{B}{z-\frac{1}{3}} + \frac{C}{z-\frac{1}{4}}$$

$$A_{12} = \lim_{z\to 0} z^2\frac{1}{z^2\left(z-\frac{1}{3}\right)\left(z-\frac{1}{4}\right)} = \frac{1}{\left(-\frac{1}{3}\right)\left(-\frac{1}{4}\right)} = 12$$

$$A_{11} = \lim_{z\to 0}\frac{\mathrm{d}}{\mathrm{d}z}\left(z^2\frac{1}{z^2\left(z-\frac{1}{3}\right)\left(z-\frac{1}{4}\right)}\right)$$

$$= -\frac{\mathrm{d}}{\mathrm{d}z}\frac{1}{\left(z^2-\frac{7}{12}z+\frac{1}{12}\right)} = -84$$

$$B = \lim_{z\to\frac{1}{3}}\left(z-\frac{1}{3}\right)\frac{1}{z^2\left(z-\frac{1}{3}\right)\left(z-\frac{1}{4}\right)} = \frac{1}{\left(\frac{1}{3}\right)^2\left(\frac{1}{3}-\frac{1}{4}\right)}$$

$$= \frac{1}{\frac{1}{9}\left(\frac{1}{12}\right)} = 108$$

$$C = \lim_{z\to\frac{1}{4}}\left(z-\frac{1}{4}\right)\frac{1}{z^2\left(z-\frac{1}{3}\right)\left(z-\frac{1}{4}\right)} = \frac{1}{\frac{1}{16}\left(\frac{1}{4}-\frac{1}{3}\right)}$$

$$= \frac{1}{\frac{1}{16}\left(-\frac{1}{12}\right)} = -192$$

$$\frac{Y(z)}{z} = \frac{16}{27}\left(\frac{12}{z^2} - \frac{84}{z} + \frac{108}{z-\frac{1}{3}} - \frac{192}{z-\frac{1}{4}}\right)$$

$$= \left(\frac{64/9}{z^2} + \frac{1344/27}{z} + \frac{64}{z-\frac{1}{3}} - \frac{1024/9}{z-\frac{1}{4}}\right)$$

$$Y(z) = 64/9z^{-1} + 1344/27 + 64\frac{z}{z-\frac{1}{3}} - 1024\Big/9\frac{z}{z-\frac{1}{4}}$$

$$y(n) = 64\left(\frac{1}{3}\right)^n - \frac{1024}{9}\left(\frac{1}{4}\right)^n = 64\left(\frac{1}{3}\right)^n\left[1 - \frac{16}{9}\left(\frac{3}{4}\right)^n\right] \quad \text{for } n \geq 3.$$

Problem 4.28

Determine the convolution of two signals for the convolution of $x(n)$ and $h(n)$ where

Using Z-transform

$$x(n) = \left(\frac{1}{2}\right)^n u(n) \quad h(n) = n\,u(n)$$

Solution 4.28

$$X(z) = \frac{z}{z-\frac{1}{2}} \quad H(z) = \frac{z}{(z-1)^2}$$

$$Y(z) = \frac{z}{z-\frac{1}{2}}\frac{z}{(z-1)^2} \quad \frac{Y(z)}{z} = \frac{z}{(z-1)^2\left(z-\frac{1}{2}\right)}$$

$$\frac{Y(z)}{z} = \frac{1}{(z-1)^2\left(z-\frac{1}{2}\right)} = \frac{A_{12}}{(z-1)^2} + \frac{A_{11}}{z-1} + \frac{B}{z-\frac{1}{2}}$$

$$A_{12} = \lim_{z\to 1}(z-1)^2\frac{1}{(z-1)^2\left(z-\frac{1}{2}\right)} = \frac{1}{\left(1-\frac{1}{2}\right)} = 2$$

$$A_{11} = \lim_{z\to 1}\frac{\mathrm{d}}{\mathrm{d}z}\left((z-1)^2\frac{1}{(z-1)^2\left(z-\frac{1}{2}\right)}\right) = \frac{\mathrm{d}}{\mathrm{d}z}\frac{1}{\left(z-\frac{1}{2}\right)}$$

$$= -\frac{1}{\left(z-\frac{1}{2}\right)^2} = -\frac{1}{\left(z-\frac{1}{2}\right)^2} = -\frac{1}{\left(\frac{1}{2}\right)^2} = -4$$

$$B = \lim_{z \to \frac{1}{2}} \left(z - \frac{1}{2} \right) \frac{1}{(z-1)^2 (z - \frac{1}{2})} = \frac{1}{\left(\frac{1}{2} - 1\right)^2} = \frac{1}{\left(\frac{1}{4}\right)} = 4$$

$$\frac{Y(z)}{z} = \left(2\frac{1}{(z-1)^2} - \frac{4}{(z-1)} + \frac{4}{z - \frac{1}{2}} \right)$$

$$Y(z) = \left(2\frac{z}{(z-1)^2} - 4\frac{z}{(z-1)} + 4\frac{z}{z - \frac{1}{2}} \right)$$

$$y(n) = 2\,nu(n) - 4u(n) + 4\left(\frac{1}{2}\right)^n u(n)$$

5

Solution of Difference Equation

This chapter covers: Constant Coefficient Difference Equation, Different methods of solution of difference equations, Problems and solutions.

5.1 Constant-Coefficient Difference Equation

It should be noted the difference equation describes the discrete time systems performance, while differential equation describes the performance of continuous time system. Infinite duration impulse response (IIR) requires an infinite number of memory locations, multiplication, and additions. So it is impossible to implement IIR system by applying convolution. IIR systems are more conveniently described by difference equations.

A recursive system is described with an input–output equation

$$y(n) = \lambda y(n-1) + x(n), \tag{5.1}$$

where λ is constant and can have any value.

The system described by the first-order difference equation in Equation (5.1) is the simplest possible recursive system in the general class of recursive systems described by linear constant coefficient difference equations. The general formula for such equation is

$$y(n) = -\sum_{k=1}^{N} a_k y(n-k) + \sum_{k=0}^{M} b_k x(n-k) \tag{5.2}$$

or equivalently

$$\sum_{k=0}^{N} a_k\, y(n-k) = \sum_{k=0}^{M} b_k\, x(n-k), \quad a_0 = 1 \tag{5.3}$$

the integer N is called the order of the difference equation or the order of the system. The negative sign on the right-hand side of the Equation (5.2) is introduced as a matter of convenience to allow us to express the difference equation in Equation (5.3) without any negative sign.

Given a linear constant-coefficient difference equation as the input–output relationship describing the linear time invariant system, the object is to determine the explicit expression for the output $y(n)$.

Our goal is to determine the output $y(n)$, $n \geq 0$, of the system for the specific input $x(n)$, $n \geq 0$, and a set of initial conditions.

5.2 Solution of Difference Equation

Three techniques are used for solving linear time-invariant difference equations.

(i) Sequential procedure: This technique is used in the digital computer solution of difference equations.
(ii) Using Z-transform.
(iii) Classical technique: It consists of finding the complementary and the particular solution as in the case of solution of differential equation.

5.2.1 Using Sequential Procedure

Example 5.1

We solve for $m(k)$ for the equation $y(k) = x(k) - x(k-1) - y(k-1)$, $k \geq 0$.

$$\text{Where} = x(k) = \begin{cases} 1, & k \text{ even} \\ 0, & k \text{ odd.} \end{cases}$$

and both $x(-1)$ and $y(-1)$ are zero.

Solution 5.1

$$\begin{aligned}
k &= 0, \quad y(0) = x(0) - x(-1) - x(-1) = 1 \\
k &= 1, \quad y(1) = x(1) - x(0) - y(0) = 0 - 1 - 1 = -2 \\
k &= 2, \quad y(2) = x(2) - x(1) - y(1) = 1 - 0 - (-2) = 3 \\
k &= 3, \quad y(3) = x(3) - x(2) - y(2) = 0 - 1 - 3 = -4 \\
k &= 4, \quad y(4) = x(4) - x(3) - y(3) = 1 - 0 - (-4) = 5.
\end{aligned}$$

It is not necessary that all the results can be summed up in closed form.

Example 5.2

Solve the difference equation $y(k) - 4y(k-1) + 3y(k-2) = x(k)$, using the sequential technique. Find $x(k)$ for $k \leq 5$.

$$\text{where} \quad x(k) = \begin{cases} 2, & k = 0, 1 \\ 0, & k \geq 2 \end{cases} \qquad x(-2) = x(-1) = 0.$$

Solution 5.2

$$y(k) - y(k-1) - y(k-2) = x(k)$$

$$\begin{aligned} k &= 0, & y(0) &= 0 - 0 + 2 = 2 \\ k &= 1, & y(1) &= 4 \times 2 - 0 + 2 = 10 \\ k &= 2, & y(2) &= 4 \times 10 - 6 + 0 = 34 \\ k &= 3, & y(3) &= 4 \times 34 - 30 + 0 = 106\ . \end{aligned}$$

Generally it is written as $y(k) = -2 + 4(3)^k$.

5.2.2 Using *Z*-Transform

The second technique for solving the difference equation is that of using Z-transform. Consider the nth order difference equation, where it is assumed that the input sequence $\{x(k)\}$ is known.

$$y(k) + b_1 y(k-1) + \ldots + b_n y(k-n) = a_0 x(k) + a_1 x(k-1) + \ldots + a_n x(k-n).$$

Using the real translation theorem, one can find z-transform of this equation.

$$Y(z) + b_1 z^{-1} Y(z) + \ldots + b_n z^{-n} Y(z) = a_0 X(z) + a_1 z^{-1} X(z) + \ldots + a_n z^{-n} X(z)$$

$$Y(z) = \frac{a_0 + a_1 z^{-1} + \ldots + a_n z^{-n}}{1 + b_1 z^{-1} + \ldots + b_n z^{-n}} X(z)$$

For a given $X(z)$, we find $y(k)$ by taking the inverse Z-transform of the above equation.

5.2.3 Classical Technique of Difference Equation

The classical technique for difference equation solution is also called the direct solution method assumes that the total solution is the sum of two parts, $y_h(n)$

is the homogeneous or complementary solution and $y_p(n)$ is the particular solution $y(n) = y_h(n) + y_p(n)$. There are different cases for complementary function and particular solution.

5.2.4 The Homogeneous Solution

For finding the homogeneous solution, the input, i.e., the forcing function $x(n)$ is assumed to be zero. It is also called zero input response. Mathematician call it complementary solution, engineers cal it transient solution or natural response.

The procedure for solving a linear constant coefficient difference equation directly is very similar to the procedure for solving a linear constant coefficient differential equation. Basically we assume that the solution is in the form of an exponential, that is,

$$y_h(n) = \lambda^n \tag{5.4}$$

We form up the auxiliary equation and substituting instead of y the values $y_h(n) = \lambda^n$, in the auxiliary equation where the subscript h on $y(n)$ is used to denote the solution to the homogeneous difference equation.

This auxiliary polynomial equation is called the characteristic polynomial of the system. In general it has N roots, which is denoted as $\lambda_1, \lambda_2, \ldots \lambda_N$. The roots may be real non- repeated and repeated or complex valued.

5.2.4.1 When the auxiliary polynomial roots are real and distinct

For the moment, let us assume that the roots are distinct; that is, there are no multiple roots. The most general solution to the homogeneous difference equation is, therefore,

$$y_h(n) = c_1\lambda_1^n + \ldots + c_N\lambda_N^n, \tag{5.5}$$

where $c_1, c_2, \ldots, c_N$ are weighting coefficients. These coefficients are determined from the initial conditions specified for the system. Since the input $x(n) = 0$, Equation (5.5) may be used to obtain the zero input response of the system. The following example illustrates the procedure when the auxiliary equation has the real and distinct roots.

Example 5.32

Determine the zero-input response of the system described by the homogeneous second-order difference equation, with initial condition $y(-1) = 5$ and $y(-2) = 0$.

$$y(n) - 3y(n-1) - 4y(n-2) = 0.$$

Solution 5.32

First determine the solution to the homogeneous equation. We assume the solution to be exponential. The assumed solution is obtained by setting $x(n) = 0$ is $y_{\rm h}(n) = \lambda^n$.

When we substitute this solution in (1), we obtain [with $x(n) = 0$]

$$\lambda^n - 3\lambda^{n-1} - 4\lambda^{n-2} = 0.$$

Dividing both sides by λ^{n-2}

$$\lambda^2 - 3\lambda - 4 = 0.$$

Therefore, the two distinct roots are $\lambda = -1, 4$ and we write the general form of the solution as

$$y_{\rm h}(n) = C_1\lambda_1^n + C_2\lambda_2^n \quad \text{Substituting the value of two } \lambda.$$

$$y_{\rm h}(n) = C_1(-1)^n + C_2(4)^n.$$

Because the initial conditions for this problem are known we can find the value of c_1 and c_2.

$$y_{\rm h}(-1) = C_1(-1)^{-1} + C_2(4)^{-1} = 5 \quad \text{or} \quad -C_1 + C_2(\frac{1}{4}) = 5$$

$$y_h(-2) = C_1(-1)^{-2} + C_2(4)^{-2} = 0 \quad or \quad = C_1 + C_2(\frac{1}{16}) = 0$$

Solving the simultaneous equation to find the value of c_1 and c_2

$$C_2 = \frac{1}{16}, \quad C_1 = -1 \quad y_{\rm h}(n) = -(-1)^n + 16(4)^n = (-1)^{n+1} + (4)^{n+2}$$

5.2.4.2 When the characteristics polynomial roots are real and repeated

For the moment, let us assume that the roots are repeated; that is, there are multiple roots. The most general solution to the homogeneous difference equation is

$$y_{\rm h}(n) = C_1\lambda_1^n + C_2 n\lambda_2^n + \ldots \tag{5.6}$$

where $C_1, C_2, \ldots, C_N$ are weighting coefficients. These coefficients are determined from the initial conditions specified for the system. The following example illustrates the procedure when the characteristic polynomial have the repeated roots.

Example 5.33

Find the homogeneous solution of the second-order difference equation

$$y(n) - 1.8y(n-1) + 0.81y(n-2) = 0, \quad n \geq 0.$$

Solution 5.33

With $y(n) = \lambda^n$ substituted into the homogeneous counterpart, the following is obtained

$$\lambda^2 - 1.8\lambda + 0.81 = 0,$$

which results in the repeated roots. It means that while writing the repeated root expression, it should be multiplied by a factor n.

$$\lambda_1 = \lambda_2 = 0.9.$$

Thus, as in the case of differential equations, we consider the complementary solution to be

$$y_c(n) = c_1(0.9)^n + c_2 n(0.9)^n,$$

where c_1 and c_2 can be evaluated if two initial conditions are specified.

5.2.5 The Particular Solution of Difference Equation

We assume for selecting the $y_p(n)$ a form that depends on the form of the input $x(n)$. The following examples illustrate the procedure and show also the rule for choosing the particular solution.

5.2.6 Rules for Choosing Particular Solutions

As is the case with the solution of differential equations, there are a set of rules, one must use to form appropriate particular solutions while solving difference equations, as summarized in Table 5.1.

5.2.6.1 When the forcing function is having term different from the value of the roots of the auxiliary equation

For the moment, let us assume that the roots are distinct; that is, there are no multiple roots. The most general solution to the homogeneous difference equation is

$$y_h(n) = C_1\lambda_1^n + C_2\lambda_2^n + \ldots\ldots + C_N\lambda_N^n, \tag{5.7}$$

Table 5.1 Rules for choosing particular solutions

Terms in Forcing Function	Choice of Particular Solution
1. Constant	Select c as constant name it other than it is if used in auxiliary equation.
2. $1 + 3^{-n}$	If a term is present in the homogeneous solution then modify the particular solution by multiplying it by n.
3. $\left.\begin{matrix} b_3 \cos(n\omega n) \\ b_4 \sin(n\omega n) \end{matrix}\right\}$ b_3 and b_4 are constants	$c_1 \sin(n\omega) + c_2 \cos(n\omega)$

If a term in any of the particular solutions in this column is a part of the complementary solution, it is necessary to modify the corresponding choice by multiplying it by n before using it. If such a term appears r times in the complementary solution, the corresponding choice must be multiplied by n^r.

where $C_1, C_2, \ldots, C_N$ are weighting coefficients. These coefficients are determined from the initial conditions specified for the system. The complementary function is calculated in same manner as discussed earlier in case of homogenous equation. In the following case the value of λ is different than the forcing function.

Example 5.34

Find the general solution of the second-order difference equation

$$y(n) - \frac{5}{6}y(n-1) + \frac{1}{6}y(n-2) = 5^{-n}, \quad n \geq 0 \tag{1}$$

with initial conditions $y(-2) = 25$ and $y(-1) = 6$.

Solution 5.34

The general (or closed-form) solution $y(n)$ of (1) is obtained in three steps that are similar to those used for solving second-order differential equations. They are as follows:

The complementary solution $y_c(n)$ in terms of two arbitrary constants c_1 and c_2.

Obtain the particular solution $y_p(n)$, and write

$$y(n) = y_c(n) + y_p(n). \tag{2}$$

Steps are elaborated as follows:

We assume that the complementary solution $y_c(n)$ has the form

$$y_c(n) = c_1\lambda_1^n + c_2\lambda_2^n \tag{3}$$

where the λ_1 and λ_2 are real constants.

Next substitute $y(n) = \lambda^n$ in the homogeneous equation to get

$$\lambda^n - \frac{5}{6}\lambda^{n-1} + \frac{1}{6}\lambda^{n-2} = 0. \tag{4}$$

Dividing both sides of (4) by λ^{n-2}, we obtain or $\left(\lambda - \frac{1}{2}\right)\left(\lambda - \frac{1}{3}\right) = 0$ which yields the characteristic roots

$$\lambda_1 = \frac{1}{2} \text{ and } \lambda_2 = \frac{1}{3}.$$

Thus, the complementary solution is

$$y_c(n) = c_1 2^{-n} + c_2 3^{-n},$$

where c_1 and c_2 are arbitrary constants.

In this case the forcing function is different than the roots of the auxiliary equation. Then particular solution $y_p(n)$ is assumed to be

$$y_p(n) = c_3 5^{-n}$$

since the forcing function is 5^{-n}; substitution of $y_p(n) = c_3 5^{-n}$ in (1) leads to

$$c_3[5^{-n} - \left(\frac{5}{6}\right)5^{-(n-1)} + \left(\frac{1}{6}\right)5^{-(n-2)}] = 5^{-n}.$$

Dividing both sides of this equation by 5^{-n}, we obtain

$$c_3[1 - \left(\frac{5}{6}\right)5 + \left(\frac{1}{6}\right)5^2] = 1$$

which implies that $c_3 = 1$. Thus

$$\begin{aligned} y(n) &= y_c(n) + y_p(n) \\ &= c_1 2^{-n} + c_2 3^{-n} + 5^{-n}. \end{aligned} \tag{5}$$

Since the initial conditions are

$$y(-2) = 25 \text{ and } y(-1) = 6.$$

Using the above given initial conditions (5) yields the simultaneous equations

$$\begin{aligned} 4c_1 + 9c_2 &= 0 \\ \text{and} \quad 2c_1 + 3c_2 &= 1 \end{aligned} \tag{6}$$

Solving (6) for c_1 and c_2, we obtain

$$c_1 = \frac{3}{2} \text{ and } c_2 = -\frac{2}{3}. \tag{7}$$

Thus the desired general solution is given by

$$y(n) = \frac{3}{2}(2^{-n}) - \frac{2}{3}(3^{-n}) + 5^{-n}, \quad n \geq 0 \tag{8}$$

$y(n)$ can be interpreted as the output of a DT system when it is subjected to the exponential input (forcing function) 5^{-n}, which is the right-hand side of the given difference equation in (1).

5.2.6.2 When the forcing function is having same term as in the roots of the auxiliary equation

Example 5.35

Solve the second-order difference equation

$$y(n) - \frac{3}{2}y(n-1) + \frac{1}{2}y(n-2) = 1 + 3^{-n}, \quad n \geq 0 \tag{1}$$

with the initial conditions $y(-2) = 0$ and $y(-1) = 2$.

Solution 5.35

The solution consists of three steps.

Assume the complementary solution as $y_c(n) = c_1\lambda_1{}^n + c_2\lambda_2{}^n$. Substituting $y(n) = \lambda^n$ in the homogeneous counterpart of (1)we obtain the characteristic equation

$$\lambda^2 - \frac{3}{2}\lambda + \frac{1}{2} = 0$$

the roots of which are $\lambda_1 = \frac{1}{2}$ and $\lambda_2 = 1$.

Thus

$$y_c(n) = c_1 2^{-n} + c_2 1^n = c_1 2^{-n} + c_2. \tag{2}$$

To choose an appropriate particular solution, we refer to Table 5.1. From the given forcing function and lines 1 and 3 of Table 5.1, it follows that a choice for the particular solution is $c_3 + c_4 3^{-n}$.

However, we observe that this choice for the particular solution and $y_c(n)$ in (2) have common terms, each of which is a constant; that is, c_3 and c_2, respectively. Thus in accordance with the footnote of Table 5.1, we modify the choice $c_3 + c_4 3^{-n}$ to obtain.

$$y_p(n) = c_3 n + c_4 3^{-n}. \tag{3}$$

Next, substitution of $y_p(n)$ in (3) into (1) leads to

$$c_3 n + c_4 3^{-n} - \frac{3}{2}c_3 n + \frac{3}{2}c_3 - \frac{9}{2}c_4 3^{-n} + \frac{1}{2}c_3 n$$

$$-c_3 + \frac{9}{2}c_4 3^{-n} = 3^{-n} + 1 \tag{4}$$

equating the coefficient of 3^{-n} and constant, from (4) it follows that

$$\tfrac{1}{2}c_3 = 1$$
$$\text{and} \quad c_4\left[1 - \tfrac{9}{2} + \tfrac{9}{2}\right]3^{-n} = 3^{-n}$$

which results in

$$c_3 = 2; \ c_4 = 1.$$

Thus, combining (2) and (3), we get

$$y(n) = c_1 2^{-n} + c_2 + 2n + 3^{-n}. \tag{5}$$

To evaluate c_1 and c_2 in (5), the given initial conditions are used; that is, $y(-2) = 0$ and $y(-1) = 2$. This leads to the simultaneous equations.

$$\begin{aligned} 4c_1 + c_2 &= -5 \\ 2c_1 + c_2 &= 1 \end{aligned}$$

Solving, we obtain $c_1 = -3$ and $c_2 = 7$, which yields the desired solution as

$$y(n) = (-3)2^{-n} + 7 + 2n + 3^{-n}, \quad n \geq 0$$

Example 5.36

Find the general solution of the first-order difference equation.

$$y(n) - 0.9y(n-1) = 0.5 + (0.9)^{n-1}, n \geq 0 \tag{1}$$

with $y(-1) = 5$.

Solution 5.36

Substituting $y(n) = \lambda^n$ in the homogeneous equation

$$y(n) - 0.9\,y(n-1) = 0$$

we obtain

$$y_c(n) = c_1(0.9)^n \tag{2}$$

since it is a first-order difference equation.

From the forcing function in (1), the complementary solution in (2), and lines 1 and 3 of Table 5.1, it follows that

$$y_p(n) = c_2 n(0.9)^n + c_3$$

Substitution of $y(n) = y_p(n)$ in (1) results in

$$c_3 + c_2 n(0.9)^n - 0.9c_2(n-1)(0.9)^{n-1} - 0.9c_3 = 0.5 + (0.9)^{n-1}$$

equating the coefficient of constant 0.5 and $(0.9)^{n-1}$ leads to

$$0.1c_3 = 0.5$$

and

$$(0.9)^n c_2 = (0.9)^{n-1}.$$

Thus we have

$$c_3 = 5 \text{ and } c_2 = \frac{10}{9}$$

which implies that

$$y_p(n) = \frac{10}{9} n\,(0.9)^n + 5. \tag{3}$$

Combining (2) and (3), the following equation is obtained

$$y(n) = c_1(0.9)^n + \frac{10}{9} n\,(0.9)^n + 5 \tag{4}$$

From (4) and the initial condition $y(-1) = 5$, it follows that $c_1 = \frac{10}{9}$. Hence the desired solution can be written as

$$y(n) = (n+1)(0.9)^{n-1} + 5, \quad n \geq 0.$$

Example 5.37

Find the general solution of the second-order difference equation

$$y(n) - 1.8y(n-1) + 0.81y(n-2) = (0.9)^n, \quad n \geq 0. \tag{1}$$

Leave the answer in terms of unknown constants, which one can evaluate if the initial conditions are given.

Solution 5.37

The complementary solution is given by

$$y_c(n) = c_1(0.9)^n + c_2 n(0.9)^n \tag{2}$$

Since the forcing function is $(0.9)^n$, line 3 of Table 5.1 implies that a choice for the particular solution is $c_3(0.9)^n$.

However, since this choice and the preceding $y_c(n)$ have a term in common, we must modify our choice according to the footnote of Table 5.1 to obtain $c_3 n(0.9)^n$.

But this choice again has a term in common with $y_c(n)$.

Thus we refer to the footnote of Table 5.1 once again to obtain.

$$y_p(n) = c_3 n^2(0.9)^n \tag{3}$$

which has no more terms in common with $y_c(n)$.

Hence $y_p(n)$ in (3) is the appropriate choice for the particular solution for the difference equation in (1) when the forcing function is $(0.9)^n$.

5.2.6.3 When the forcing function is having sinusoidal forcing function

When the forcing function is having a term of sinusoidal forcing function then in selection of the particular solution we do as following.

$$y_p(n) = c_1 \sin\left(\frac{n\pi}{2}\right) + c_2 \cos\left(\frac{n\pi}{2}\right)$$

Example 5.38

Find the particular solution for the first-order difference equation.

$$y(n) - 0.5y(n-1) = \sin\left(\frac{n\pi}{2}\right), \quad n \geq 0 \tag{1}$$

Solution 5.38

Since the forcing function is sinusoidal; we refer to line 3 of Table 5.1 and choose a particular solution of the form

$$y_p(n) = c_1 \sin\left(\frac{n\pi}{2}\right) + c_2 \cos\left(\frac{n\pi}{2}\right) \tag{2}$$

Substitution of $y(n) = y_{\mathrm{p}}(n)$ in (1) leads to

$$c_1 \sin\left(\frac{n\pi}{2}\right) + c_2 \cos\left(\frac{n\pi}{2}\right) - 0.5c_1 \sin\left[\frac{(n-1)\pi}{2}\right]$$
$$-0.5c_2 \cos\left[\frac{(n-1)\pi}{2}\right] = \sin\left(\frac{n\pi}{2}\right) \quad (3)$$

using the following identities:

$$\sin\left[\frac{(n-1)\pi}{2}\right] = \sin\left(\frac{n\pi}{2} - \frac{\pi}{2}\right) = -\cos\left(\frac{n\pi}{2}\right) \quad (4)$$

$$\cos\left[\frac{(n-1)\pi}{2}\right] = \cos\left(\frac{n\pi}{2} - \frac{\pi}{2}\right) = \sin\left(\frac{n\pi}{2}\right)$$

Substituting (2) in (1), we obtain

$$(c_1 - 0.5c_2) \sin\left(\frac{n\pi}{2}\right) + (0.5c_1 + c_2) \cos\left(\frac{n\pi}{2}\right) = \sin\left(\frac{n\pi}{2}\right)$$

which yields the simultaneous equations.

$$\begin{aligned} c_1 - 0.5c_2 &= 1 \\ 0.5c_1 + c_2 &= 0. \end{aligned} \quad (5)$$

The solution of (5) yields $c_1 = \frac{4}{5}$ and $c_2 = -\frac{2}{5}$. Hence the desired result is given by (2) to be

$$y_{\mathrm{p}}(n) = \frac{4}{5} \sin\left(\frac{n\pi}{2}\right) - \frac{2}{5} \cos\left(\frac{n\pi}{2}\right) \quad n \geq 0.$$

5.3 Problems and Solutions

K and *n* has been used in the difference equation representation.

Problem 5.1

Solve the following difference equation for $x(k)$ using the Z-transform method for $k \leq 5$.

$$y(k) - 4y(k-1) + 3y(k-2) = x(k)$$

$$\text{where} \quad x(k) = \begin{cases} 2, & k = 0, 1 \\ 0, & k \geq 2 \end{cases} \qquad y(-2) = y(-1) = 0.$$

Solution 5.1

Taking transform of the above difference equation

$[1-4z^{-1}+3z^{-2}]\,Y(z) = X(z)$. Where $X(z) = 2(1+z^{-1})$ from the given problem

$Y(z) = \frac{2(1+z^{-1})}{1-4z^{-1}+3z^{-2}}$.

Multiply numerator and denominator by z^2 $Y(z) = \frac{2z(z+1)}{z^2-4z+3}$

$$\frac{Y(z)}{z} = \frac{2(z+1)}{(z-1)(z-3)} = \frac{-2}{z-1} + \frac{4}{z-3}$$

$$Y(z) = -\frac{2z}{z-1} + \frac{4z}{z-3}$$

$$y(n) = [-2+4(3)^n]\,u(n).$$

Problem 5.2

Given the difference equation:

$y(n+2) - 0.75y(n+1) + 0.125y(n) = x(n)$

Where $x(n) = 1,\ n \geq 0,$ and $y(0) = y(1) = 0.$

Solution 5.2

$y(n+2) - 0.75y(n+1) + 0.125y(n) = x(n)$, converting the given difference equation into Z transform.

$$Y(z)\{z^2 - 0.75z + 0.125\} = E(z) \quad \text{where } X(z) = \frac{z}{z-1}(\text{given})$$

$Y(z) = \frac{X(z)}{(z-0.25)(z-0.5)}$ Substituting the value of $X(z)$

$$y(z) = \frac{z}{(z-1)(z-\frac{1}{4})(z-1/2)}$$

$$\frac{y(z)}{z} = \frac{1}{(z-1)(z-1/4)(z-1/2)} = \frac{8/3}{z-1} + \frac{16/3}{z-1/4} + \frac{-8}{z-1/2}$$

$$y(n) = \left[\frac{8}{3} + \frac{16}{3}\left(\frac{1}{4}\right)^n - 8\left(\frac{1}{2}\right)^n\right]$$

Problem 5.3

Find the inverse Z-transform of each given $E(z)$ by two methods and compare the value of $e(k)$ for $k = 0$, 1, 2, and 3 obtained by two method.

$$E(z) = \frac{z}{(z-1)(z-0.8)}$$

Solution 5.3

$$E(z) = \frac{z}{(z-1)(z-0.8)}$$

(i)

$$\begin{array}{r|l} & z^{-1} + 1.8z^{-2} + 2.44z^{-3} \\ \hline z^2 - 1.8z + 0.8 & z \\ & z - 1.8 + 0.8z^{-1} \\ \hline & 1.8 - 0.8z^{-1} \\ & 1.8 \end{array}$$

$$e(0) = 0, \quad e(1) = 1, \quad e(2) = 1.8, \text{ and } e(3) = 2.44$$

(ii) $\frac{E(z)}{z} = \frac{1}{(z-1)(z-0.8)} = \frac{5}{z-1} + \frac{-5}{z-0.8}$

$E(z) = \frac{5z}{z-1} - \frac{5z}{z-0.8}$,

$z^{-1}\{E(z)\} = 5z^{-1}\left\{\frac{z}{z-1}\right\} - 5z^{-1}\left[\frac{z}{z-0.8}\right]$

$e(k) = 5 - 5(0.8)^k, e(k) = 5\left[1 - (0.8)^k\right]$

$$k = 0, \quad e(0) = 5\left[1 - (0.8)^0\right] = 0$$

$$k = 1, \quad e(1) = 5\left[1 - (0.8)^1\right] = 1$$

$$k = 2, \quad e(2) = 5\left[1 - (0.8)^2\right] = 1.8$$

$$k = 3, \quad e(3) = 5\left[1 - (0.8)^3\right] = 2.44.$$

Problem 5.4

Find the inverse Z-transform of each given $E(z)$ by two methods and compare the value of $e(k)$ for $k = 0$, 1, 2, and 3 obtained by two method.

$$E(z) = \frac{1}{(z-1)(z-0.8)} = \frac{1}{z^2 - 1.8z + 0.8}$$

Solution 5.4

(i)

$$
\begin{array}{r|l}
 & z^{-2} + 1.8z^{-3} + 2.44z^{-4} \\
z^2 - 1.8z + 0.8 & 1 \\
 & 1 - 1.8z^{-1} + 0.8z^{-2} \\ \hline
 & 1.8z^{-1} - 0.8z^{-2} \\
 & 1.8z^{-1} - 3.24z^{-2} + 1.44z^{3} \\ \hline
 & 2.44z^{-2} - 1.44z^{-3}
\end{array}
$$

$$e(0) = 0, \quad e(1) = 1, \quad e(2) = 1.8, \text{ and } e(3) = 2.45$$

(ii) $\frac{E(z)}{z} = \frac{1}{z(z-1)(z-0.8)} = \frac{A}{z} + \frac{B}{z-1} + \frac{C}{z-0.8}$

$$A = \lim_{z \to 0} z\frac{1}{z(z-1)(z-0.8)} = \frac{1}{(-1)(-0.8)} = 1.25$$

$$B = \lim_{z \to 1} (z-1)\frac{1}{z(z-1)(z-0.8)} = \frac{1}{1(0.2)} = 5$$

$$C = \lim_{z \to 0.8} (z-0.8)\frac{1}{z(z-1)(z-0.8)} = \frac{1}{(0.8)} = -6.25$$

$$\frac{E(z)}{z} = \frac{1.25}{z} + \frac{5}{z-1} - \frac{6.25}{z-0.8}$$

$$E(z) = 1.25 + 5\frac{z}{z-1} - 6.25\frac{z}{z-0.8}$$

$$
\begin{aligned}
e(k) &= 1.255\delta(k) + 5u(k) - 6.25\,(0.8)^k u(k) \\
k = 0, &\quad e(0) = 1.25 + 5 - 6.25 = 0 \\
k = 1, &\quad e(1) = 0 + 5 - 6.25\,(0.8) = 0 \\
k = 2, &\quad e(2) = 0 + 5 - 6.25(0.8)^2 = 1.00
\end{aligned}
$$

Problem 5.5

Find the inverse Z-transform $x(n)$ of given $X(z)$.

$$X(z) = \frac{3}{z - 2}$$

Solution 5.5

$X(z)$ may be written as $X(z) = 3z^{-1}\frac{z}{z-2}$

$Z\{2^n u(n)\} = \frac{z}{z-2}$, using time shifting property

$$Z\{2^{n-1}u(n-1)\} = z^{-1}[\frac{z}{z-2}] = \frac{1}{z-2}$$

Thus we conclude that $x(n) = 3\{2^{n-1}u(n-1)\}$.

Problem 5.6

Find the inverse z-transform of the following transfer function

$$X(z) = \frac{z}{z^2 + 1\,z + 0.5}$$

Solution 5.6

$$X(z) = \frac{z}{z^2 + 1\,z + 0.25 + 0.25} = \frac{z}{(z+0.5)^2 + (0.5)^2}$$

$$= \frac{k_1 z}{z + 0.5 - j0.5} + \frac{k_1^* z}{z + 0.5 + j0.5}$$

Dividing both sides by z, we calculate k_1:

$$k_1 = (z + 0.5 - j0.5\,)\left[\frac{1}{(z + 0.5 - j0.5\,)(z + 0.5 + j0.5\,)}\right]_{z=-0.5\,+\,j0.5}$$

$$= \frac{1}{z + 0.5 + j0.5}\Bigg|_{z=-0.5+j0.5} = \frac{1}{2(j0.5)} = 1\angle -90^{\circ}$$

with $p_1 = -0.5 + \; j0.5$,

$$\sum = \; ln|p_1| \;=\; ln(0.707) \;= -0.34; \;\; \Omega = \; \arg p_1 = 3p/4$$

$$A = 2|k_1| = 2(1) \;=\; 2; \quad \theta = \arg k_1 = -90^{\circ}$$

Hence,

$$y[n] = Ae^{\Sigma n}\cos(\Omega n + \theta) = 2e^{-0.34n}\cos\left(\frac{3\pi}{4} - 90^{\circ}\right)$$

Problem 5.7

Determine the convolution of two signals for the convolution of $x(n)$ and $h(n)$ where

$$x(n) = \left(\frac{1}{4}\right)^{n-2} u(n) \quad h(n) = \left(\frac{1}{3}\right)^{n} u(n-3)$$

Solution 5.7

$$x(n) = \left(\frac{1}{4}\right)^{n-2} u(n) \quad x(n) = \left(\frac{1}{4}\right)^{-2}\left(\frac{1}{4}\right)^{n} u(n)$$

$$x(n) = (16)\left(\frac{1}{4}\right)^{n} u(n)$$

$$X(z) = (16)Z\left(\frac{1}{4}\right)^n u(n) = 16\frac{z}{z-\frac{1}{4}}$$

$$h(n) = \left(\frac{1}{3}\right)^n u(n-3) = \left(\frac{1}{3}\right)^3 \left(\frac{1}{3}\right)^{n-3} u(n-3)$$

$$= \left(\frac{1}{27}\right) z^{-3} Z\left(\frac{1}{3}\right)^n u(n)$$

$$H(z) = \left(\frac{1}{27}\right) z^{-3} \frac{z}{z-\frac{1}{3}} = \left(\frac{1}{27}\right) \frac{z^{-2}}{z-\frac{1}{3}}$$

$$Y(z) = 16\frac{z}{z-\frac{1}{4}}\left(\frac{1}{27}\right)\frac{z^{-2}}{z-\frac{1}{3}} = \left(\frac{16}{27}\right)\frac{z^{-1}}{(z-\frac{1}{3})(z-\frac{1}{4})}$$

$$Y(z) = 16\frac{z}{z-\frac{1}{4}}\left(\frac{1}{27}\right)\frac{z^{-2}}{z-\frac{1}{3}} = \left(\frac{16}{27}\right)\frac{z^{-1}}{(z-\frac{1}{3})(z-\frac{1}{4})}$$

$$= \left(\frac{16}{27}\right)\frac{1}{z(z-\frac{1}{3})(z-\frac{1}{4})}$$

$$\frac{Y(z)}{z} = \frac{16}{27}\frac{1}{z(z-\frac{1}{3})(z-\frac{1}{4})} = \frac{A_{12}}{z^2} + \frac{A_{11}}{z} + \frac{B}{z-\frac{1}{3}} + \frac{C}{z-\frac{1}{4}}$$

$$A_{12} = \lim_{z \to 0} z^2 \frac{1}{z^2(z-\frac{1}{3})(z-\frac{1}{4})} = \frac{1}{(-\frac{1}{3})(-\frac{1}{4})} = 12$$

$$A_{11} = \lim_{z \to 0} \frac{d}{dz}\left(z^2 \frac{1}{z^2(z-\frac{1}{3})(z-\frac{1}{4})}\right)$$

$$= -\frac{d}{dz}\frac{1}{(z^2 - \frac{7}{12}z + \frac{1}{12})} = -84$$

$$B = \lim_{z \to \frac{1}{3}} \left(z - \frac{1}{3}\right)\frac{1}{z^2(z-\frac{1}{3})(z-\frac{1}{4})}$$

$$= \frac{1}{(\frac{1}{3})^2(\frac{1}{3}-\frac{1}{4})} = \frac{1}{\frac{1}{9}(\frac{1}{12})} = 108$$

$$C = \mathop{Lim}_{z \to \frac{1}{4}} (z - \frac{1}{4}).\frac{1}{z^2(z-\frac{1}{3})(z-\frac{1}{4})} = \frac{1}{\frac{1}{16}(\frac{1}{4}-\frac{1}{3})}$$

$$= \frac{1}{\frac{1}{16}(-\frac{1}{12})} = -192$$

$$\frac{Y(z)}{z} = \frac{16}{27}\left(\frac{12}{z^2} - \frac{84}{z} + \frac{108}{z-\frac{1}{3}} - \frac{192}{z-\frac{1}{4}}\right)$$

$$= \left(\frac{64/9}{z^2} + \frac{1344/27}{z} + \frac{64}{z-\frac{1}{3}} - \frac{1024/9}{z-\frac{1}{4}}\right)$$

$$Y(z) = 64/9z^{-1} + 1344/27 + 64\frac{z}{z-\frac{1}{3}} - 1024\Big/9\frac{z}{z-\frac{1}{4}}$$

$$y(n) = 64\left(\frac{1}{3}\right)^n - \frac{1024}{9}\left(\frac{1}{4}\right)^n$$

$$= 64\left(\frac{1}{3}\right)^n\left[1 - \frac{16}{9}\left(\frac{3}{4}\right)^n\right] \text{for } n \geq 3$$

Problem 5.8

Determine the convolution of two signals for the convolution of $x(n)$ and $h(n)$ where

Using Z-transform

$$x(n) = \left(\frac{1}{2}\right)^n u(n) \quad h(n) = n\, u(n)$$

Solution 5.8

$$X(z) = \frac{z}{z-\frac{1}{2}}; H(z) = \frac{z}{(z-1)^2}$$

$$Y(z) = \frac{z}{z-\frac{1}{2}}\frac{z}{(z-1)^2} \quad \frac{Y(z)}{z} = \frac{z}{(z-1)^2(z-\frac{1}{2})}.$$

$$\frac{Y(z)}{z} = \frac{1}{(z-1)^2(z-\frac{1}{2})} = \frac{A_{12}}{(z-1)^2} + \frac{A_{11}}{z-1} + \frac{B}{z-\frac{1}{2}}$$

$$A_{12} = \lim_{z \to 1} (z-1)^2\frac{1}{(z-1)^2(z-\frac{1}{2})} = \frac{1}{(1-\frac{1}{2})} = 2$$

$$A_{11} = \lim_{z \to 1} \frac{d}{dz}\left((z-1)^2\frac{1}{(z-1)^2(z-\frac{1}{2})}\right) = \frac{d}{dz}\frac{1}{z-\frac{1}{2}}$$

$$= -\frac{1}{\left(z-\frac{1}{2}\right)^2} = -\frac{1}{\left(1-\frac{1}{2}\right)^2} = -\frac{1}{\left(\frac{1}{2}\right)^2} = -4$$

$$B = \lim_{z \to \frac{1}{2}} (z - \frac{1}{2}).\frac{1}{(z-1)^2(z-\frac{1}{2})} = \frac{1}{(\frac{1}{2}-1)^2} = \frac{1}{(\frac{1}{4})} = 4$$

$$\frac{Y(z)}{z} = \left(2\frac{1}{(z-1)^2} - \frac{4}{(z-1)} + \frac{4}{z-\frac{1}{2}}\right)$$

$$Y(z) = \left(2\frac{z}{(z-1)^2} - 4\frac{z}{(z-1)} + 4\frac{z}{z-\frac{1}{2}}\right)$$

$$y(n) = 2\,nu(n) - 4u(n) + 4\left(\frac{1}{2}\right)^n u(n)$$

Example 5.9

Determine the impulse response of the system described by the difference equation

$y(n) - 0.6y(n-1) + 0.08y(n-2) = x(n)$, the initial condition are $y(0) = 1$.

Solution 5.9

$y(n) - 0.6y(n-1) + 0.08y(n-2) = x(n)$, where $x(n) = \delta(n)$

Homogenous Solution:

Keeping the forcing function $x(n) = 0$ for finding the zero input response.
The characteristic equation is $\lambda^2 - 0.6\lambda + 0.08 = 0$; $\lambda = 0.2, 0.4$
Hence, $y_h(n) = c_1(\frac{1}{5})^n + c_2(\frac{2}{5})^n$ with $x(n) = \delta(n)$,

Particular Solution:

Because the initial conditions of this problem is not given we find the initial condition from given problem and the input $x(n)$.

$y(n) - 0.6y(n-1) + 0.08y(n-2) = \delta(n)$, substituting the value of $n = 1$

$$y(1) - 0.6y(0) = 0,\ y(1) = 0.6$$

$$y_h(n) = c_1(\tfrac{1}{5})^n + c_2(\tfrac{2}{5})^n n = 0, \quad y(0) = 1 = c_1 1 + c_2 1 \ \ n = 1,$$

$$y(1) = 0.6 = c_1(\tfrac{1}{5}) + c_2(\tfrac{2}{5}) \Rightarrow c_1 = -1, \quad c_2 = 2$$

Therefore,

$$h(n) = [-(\tfrac{1}{5})^n + 2(\tfrac{2}{5})^n]u(n)$$

Problem 5.10

Solve the following second-order linear difference equations with constant coefficients.

$2y(n-2) - 3y(n-1) + y(n) = 3, n \geq 0,$ with $y(-2) = -3$ and $y(-1) = -2$.

Hint: $y_c(n) = c_1 + c_2 n,$ and $y_p(n) = c_3 n^2$ $(why?)$.

Solution 5.10

Homogeneous Solution:

Let the assumed $y_h(n) = \lambda^n$ by substitution into original equation, we get

$$2\lambda^{n-2} - 3\lambda^{n-1} + \lambda^n = 0 \quad \text{dividing whole equation } \lambda^{n-2}$$
$$2 - 3\lambda + \lambda^2 = 0, \text{ rearranging in descending format}$$
$$\lambda^2 - 3\lambda + 2 = 0 \quad \lambda = 2,\ 1 \tag{1}$$
$$y_h(n) = c_1(2)^n + c_2(1)^n$$
$$y_h(n) = c_1 2^n + c_2$$

Particular Solution:

The assumed $y_p = c_3$, but because in the selection of forcing function the constant C_2 is also present in the homogeneous solution therefore the selection of forcing function is to be modified form $y_p = nc_3$. Substituting y_p in the main Equation (1)

$[2c_3(n-2)] - 3[c_3(n-1)] + nc_3 = 3$

$2c_3 n - 4c_3 - 3c_3 n + 3c_3 + nc_3 = 3$

Equating the coefficients of the constant we get $c_3 = -3$.

The complete solution

$$y(n) = y_p(n) + y_c(n)$$
$$y(n) = c_1\, 2^n + c_2 - 3n \tag{2}$$

Now for finding c_1, c_2 has to be calculated using condition $y(-2) = -3$, $y(-1) = -2$

$$c_1 + 4\, c_2 = -44/3,$$
$$c_1 + 2\, c_2 = -14/3,$$
$$c_1 = 16/3,\ c_2 = -5 \quad \text{Substituting in (2)}$$

$$y(n) = -13 + 16\, 2^n - 3n$$

Problem 5.11

Solve the following second-order linear difference equations with constant coefficients.

$y(n-2) - 2y(n-1) + y(n) = 1,\ n \geq 0,$ with $y(-1) = -0.5$ and $y(-2) = 0.$

$$\text{Hint: } y_c(n) = c_1 + c_2 n, \quad \text{and} \quad y_p(n) = c_3 n^2$$

Solution 5.11

Homogeneous Solution:

Let

$$\begin{aligned}
&y_h(n) = \lambda^n \text{ by substituting into original equation} \\
&\lambda^{n-2} - 2\lambda^{n-1} + \lambda^n = 0 \\
&\lambda^2 - 2\lambda + 1 = 0 \\
&(\lambda - 1)^2 = 0 \\
&\lambda = 1,\ 1.
\end{aligned}$$

Since roots turned out to be same. It falls in the case two when the roots of the auxiliary equation are real and repeated. So

$$y_c(n) = c_1 1^n + n c_2 1^n \qquad y_c(n) = c_1 + n c_2$$

Particular Solution:

The assumed $y_p = c_3$, but because in the selection of forcing function the constant c_2 is also present in the homogeneous solution therefore the selection of forcing function is to be modified form $y_p = n^2 c_3$. Substituting y_p in the main Equation (1)

$$y_p = n^2 c_3 \tag{3}$$

$$y(n-2) - 2y(n-1) + y(n) = 1$$

Put (3) in original equation.

$$\left[c_3(n-2)^2\right] - 2\left[c_3(n-1)^2\right] + c_3 n^2 = 1,\ c_3 = 1/2$$

The complete solution

$y(n) = c_1 + n c_2 + \frac{n^2}{2}$

using initial conditions $y(-1) = -0.5$ and $y(-2) = 0$

$$c_1 - c_2 = -1,$$
$$c_1 - 2c_2 = -2$$
$$c_1 = 0, \ c_2 = 1$$
$$y(n) = n + 0.5\,(n^2)$$

Problem 5.12

Solve the following second-order linear difference equations with constant coefficients

$$y(n) - 0.8y(n-1) = (0.8)^n, \ n \geq 0, \quad \text{with}\, y(0) = 6.$$

The answer may be left in terms of two arbitrary constants

$$y_{\mathrm{p}}(n) = c_3 n^2 + c_4 n + c_5.$$

Solution 5.12

$y(n) - 0.8y(n-1) = 0.8^n, \quad n \geq 0 \; y(0) = 6$
the y_{c} part can be assumed as $y = l^n$
by substitution $\lambda^n - 0.8a^{n-1} = 0$
$\lambda = 0.8$
$y_{\mathrm{c}} = c_1 0.8^n$
for y_{p} part (0.8^n) is same so assumed solution is $y_{\mathrm{p}} = nc_2 0.8^n$
$c_2 = 8$
$y(n) = y_{\mathrm{c}} + y_{\mathrm{p}}$
but $y(n) = c_1 0.8^n + (n0.8^n x8)$
$= c_1 0.8^n + 8n0.8^n$
Using $y(0) = 6c_1 = 6$
$y(n) = 6(0.8)^n + 8n(0.8)^n$

Problem 5.13

Solve the following second-order linear difference equations with constant coefficients

$$y(n) - y(n-1) = 1 + (0.5)^n, \quad n \geq 0, \quad \text{with} \quad y(0) = 1.$$

The answer may be left in terms of two arbitrary constants

$$y_{\mathrm{p}}(n) = c_3 n^2 + c_4 n + c_5.$$

Solution 5.13

$y(n) - y(n-1) = 1 + 0.5^n$ for y_c part we have equation

$$\lambda^n - \lambda^{n-1} = 0$$
$$\lambda - 1 = 0;\ \lambda = 1$$
$$y_c = c_1(1)^n = c_1$$

for $y_p = n\,c_2 + c_3(0.5)^n$, we use the modified form of for selection y_p. by substitution it into original equation

$$(nc_2 + c_3(0.5)^n) - \left(c_2(n-1) + c_3(0.5)^{n-1}\right) = 1 + 0.5^n c_3 = -1$$

and $c_2 = 1$ (by comparing coefficients)

Total solution $y(n) = c_1 + nc_2 + c_3 0.5^n$

$$y = c_1 + n - 0.5^n$$

using $y(0) = 1$

$c_1 = 2$ and $y(n) = 2 + n - 0.5^n$

Problem 5.14

Solve the following second-order linear difference equations with constant coefficients

$$5y(n-2) + 5y(n) = 1,\ n \geq 0, \quad \text{with} \quad y(-2) = 2 \text{ and } y(-1) = 1.$$

The answer may be left in terms of two arbitrary constants

$$y_p(n) = c_3 n^2 + c_4 n + c_5.$$

Solution 5.14

$$-5y(n-2) + 5y(n) = 1 \quad n \geq 0 \quad y(-2) = 2\ y(-1) = 1$$

for $y_c = -5y(n-2) + 5y(n) = 0$ put $y(n) = a^n$

$-5\lambda^{n-2} + 5\lambda^n = 0$

$5\lambda^2 - 5 = 0 \quad a = \pm 1$

$y_c(n) = c_1(1)^n + c_2(-1)^n = c_1 + c_2(-1)^n$

let $y_p = nc_3$ (to distinguish from c_1) by substitution in main equation

$-5((n-2)c_3) + 5(nc_3) = 1\ c_3 = 1/10 = 0.1$

$$\begin{aligned} y(n) &= y_p + y_c \\ &= c_1 + c_2(-1)^n + 0.1n \end{aligned}$$

using $y(-2) = 2$

$y(-1) = 1$

we get

$c_1 + c_2 = 2.2$

$c_1 - c_2 = 1.1$

$c_1 = 1.65, \ c_2 = 0.55$

$y(n) = 1.65 + 0.55(-1)^n + n(0.1)$

Problem 5.15

Find the general solution of the second-order difference equation

$$y(n) - 1.8y(n-1) + 0.81y(n-2) = 2^{-n}, \quad n \geq 0. \tag{1}$$

Leave the answer in terms of unknown constants, which one can evaluate if the initial conditions are given.

Solution 5.15

With $y(n) = \lambda^n$ substituted into the homogeneous counterpart, the following is obtained

$$\lambda^2 - 1.8\lambda + 0.81 = 0$$

which results in the repeated roots

$$\lambda_1 = \lambda_2 = 0.9.$$

Thus, as in the case of differential equations, we consider the complementary solution to be

$$y_c(n) = c_1(0.9)^n + c_2 n(0.9)^n. \tag{2}$$

From the given forcing function in (1), $y_c(n)$ in (2), and line 3 of Table 5.1, it is clear that

$$y_p(n) = c_3 2^{-n} \tag{3}$$

Substitution of (3) in (1) leads to

$$c_3[1 - (1.8)(2) + (0.81)(4)]2^{-n} = 2^{-n}$$

which yields $c_3 = \frac{1}{0.64} = 1.5625.$

Thus the desired solution is given by (2) and (3) to be

$$y(n) = c_1(0.9)^n + c_2 n(0.9)^n + (1.5625)\, 2^{-n}$$

where c_1 and c_2 can be evaluated if two initial conditions are specified.

Problem 5.16

Determine the zero-input response of the system described by the second-order difference equation

$$3y(n-1) - 4y(n-2) = x(n)$$

Solution 5.16

We have $y(n-1) + \frac{4}{3}y(n-2) = 0$ with $x(n) = 0$,

$$\begin{aligned} y(-1) &= -\tfrac{4}{3}y(-2) \\ y(0) &= \left(-\tfrac{4}{3}\right)^2 y(-2) \\ y(1) &= \left(-\tfrac{4}{3}\right)^3 y(-2) \\ y(k) &= \left(-\tfrac{4}{3}\right)^{k+2} y(-2) \leftarrow \text{Zero-input response} \end{aligned}$$

Problem 5.17

Determine the particular solution of the difference equation

$$y(n) - y(n-1) + y(n-2) = x(n),\ n \geq 0$$

when the forcing function is $x(n) = 2^n$

Solution 5.17

Homogenous Solution:
Consider the homogeneous equation:

$$y(n) - \frac{5}{6}y(n-1) + \frac{1}{6}y(n-2) = 0$$

The characteristic equation is
$\lambda^2 - \frac{5}{6}\lambda + \frac{1}{6} = 0$, $\lambda = 1/2, 1/3$.

Hence $y_h(n) = C_1(\frac{1}{2})^n + C_2(\frac{1}{3})^n$

The particular solution to $x(n) = 2^n u(n)$. $Y_P(n) = C_3 2^n u(n)$, Substitute this into the difference equation.

Then, we obtain $C_3(2^n) - C_3\frac{5}{6}(2^{n-1})u(n-1) + C_3\frac{1}{6}2^{n-2}u(n-2)$
$= 2^n u(n)$
For $n = 2$, $4C_3 - \frac{5}{3}C_3 + \frac{C_3}{6} = 4 \Rightarrow C_3 = \frac{8}{5}$
Therefore, the total solution is
$y(n) = y_p(n) + y_h(n) = \frac{8}{5}(2)^n u(n) + c_1(\frac{1}{2})^n u(n) + c_2(\frac{1}{3})^n u(n)$
To determine c_1 and c_2, assume that $y(-2) = y(-1) = 0$.
Then, $y(0) = 1$ and $y(1) = \frac{5}{6}y(0) + 2 = 17/6$
Then, $\frac{8}{5} + c_1 + c_2 = 1 \Rightarrow \quad c_1 + c_2 = -\frac{3}{5}$
$\frac{16}{5} + \frac{1}{2}c_1 + \frac{1}{3}c_2 = \frac{17}{6} \Rightarrow 3 \quad c_1 + 2c_2 = -\frac{11}{5}$
and, therefore, $c_1 = -1$, $c_2 = \frac{2}{5}$
The total solution is

$$y(n) = \left[\tfrac{8}{5}(2)^n - (\tfrac{1}{2})^n + \tfrac{2}{5}(\tfrac{1}{3})^n\right] u(n)$$

Problem 5.18

Determine the response $y(n)$, $n \geq 0$, of the system described by the second-order difference equation

$$y(n) - 3y(n-1) - 4y(n-2) = x(n) \text{ to the input } x(n) = 4^n u(n).$$

Solution 5.18

Homogenous Solution:

The characteristic equation is $\lambda^2 - 3\lambda - 4 = 0$

Hence, $\lambda = 4, -1$, and $y_h(n) = c_1 4^n + c_2(-1)^n$

Particular Solution: We assume a particular solution of the modified form $y_p(n) = c_3 n 4^n u(n)$.

Then

$c_3 n\, 4^n u(n) - 3c_3(n-1)\, 4^{n-1}u(n-1) - 4c_3(n-2)\, 4^{n-2}u(n-2)$
$= 4^n u(n) + 2(4)^{n-1}u(n-1)$
For $n = 2$, $c_3(32 - 12) = 42 + 8 = 24 \Rightarrow c_3 = \frac{6}{5}$
The total solution is $y(n) = y_p(n) + y_h(n)$
$= \left[\frac{6}{5}n4^n + c_1 4^n + c_2(-1)^n\right]u(n)$

Problem 5.19

Determine the response $y(n)$, $n \geq 0$, of the system described by the second-order difference equation

$$y(n) - 3y(n-1) - 4y(n-2) = x(n) + 2x(n-1) \text{ to the input}$$
$$x(n) = 4^n u(n).$$

Solution 5.19

The characteristic equation is $\lambda^2 - 3\lambda - 4 = 0$

Hence, $\lambda = 4, -1$, and $y_h(n) = c_1 4^n + c_2(-1)^n$

we assume a particular solution of the form $y_p(n) = kn4^n u(n)$. Then

$kn\, 4^n u(n) - 3k(n-1)\, 4^{n-1} u(n-1) - 4k(n-2)\, 4^{n-2} u(n-2)$

$= 4^n u(n) + 2(4)^{n-1} u(n-1)$

$For\ n = 2,\ k(32 - 12) = 42 + 8 = 24 \Rightarrow k = \frac{6}{5}$

The total solution is

$y(n) = y_p(n) + y_h(n) = [\frac{6}{5} n4^n + c_1 4^n + c_2(-1)^n] u(n)$

To solve for c_1 and c_2, we assume that $y(-1) = y(-2) = 0$.

Then, $y(0) = 1$ and $y(1) = 3y(0) + 4 + 2 = 9$

Hence, $c_1 + c_2 = 1$ and $\frac{24}{5} + 4c_1 - c_2 = 9$

$$4c_1 - c_2 = \frac{21}{5}$$

Therefore, $c_1 = \frac{26}{25}$ and $c_2 = -\frac{1}{25}$

The total solution is

$$y(n) = [\tfrac{6}{5} n\, 4^n + \tfrac{26}{25} 4^n - \tfrac{1}{25}(-1)^n] u(n)$$

Problem 5.20

Determine the impulse response of the following causal system:

$$y(n) - 3y(n-1) - 4y(n-2) = x(n) + 2x(n-1)$$

Solution 5.20

Homogenous Solution:

Auxiliary equation is $\lambda^2 - 3\lambda - 4 = 0$

The characteristic values are $\lambda = 4, -1$.

Hence

$$y_\lambda(n) = c_1 4^n + c_2(-1)^n$$

When $x(n) = \delta(n)$, we find that $y(0) = 1$ and

$$y(1) - 3y(0) = 2 \text{ or } y(1) = 5.$$

Hence

$$c_1 + c_2 = 1 \text{ and } 4c_1 - c_2 = 5$$

This yields, $c_1 = 6/5$ and $c_2 = -1/5$

Therefore, $h(n) = [\frac{6}{5} 4^n - \frac{1}{5}(-1)^n] u(n)$

Problem 5.21

Determine the impulse response and the unit step response of the systems described by the difference equation

$$y(n) = 0.6y(n - 1) - 0.08y(n - 2) + x(n)$$

Solution 5.21

Homogenous Solution:

$$y(n) - 0.6y(n-1) + 0.08y(n-2) = x(n)$$

The characteristic equation is $\lambda^2 - 0.6\lambda + 0.08 = 0$;

$$\lambda = 0.2,\ 0.5.$$

Hence, $y_\lambda(n) = c_1(\frac{1}{5})^n + c_2(\frac{2}{5})^n$ with $x(n) = \delta(n)$, the initial condition are $y(0) = 1$,

$$y(1) - 0.6y(0) = 0 \Rightarrow y(1) = 0.6$$

Hence,

$$c_1 + c_2 = 1 \text{ and } \frac{1}{5}c_1 + \frac{2}{5}c_2 = 0.6 \Rightarrow c_1 = -1, \quad c_2 = 2$$

Therefore, $h(n) = [-(\frac{1}{5})^n + 2(\frac{2}{5})^n]u(n)$

The step response is

$$\begin{aligned} s(n) &= \sum_{k=0}^{n} h(n-k), \quad n \geq 0 \\ &= \sum_{k=0}^{n} [2(\tfrac{2}{5})^{n-k} - (\tfrac{1}{5})^{n-k}] \\ &= 2(\frac{2}{5})^n[(\tfrac{5}{2})^{n+1} - 1]\, u(n) - (\frac{1}{5})^n[(5)^{n+1} - 1]\, u(n) \end{aligned}$$

Problem 5.22

Determine the impulse response and the unit step response of the systems described by the difference equation

$$y(n) = 0.7y(n - 1) - 0.1y(n - 2) + 2x(n) - x(n - 2)$$

Solution 5.22

$$y(n) - 0.7y(n-1) + 0.1y(n-2) = 2x(n) - x(n-2)$$

The characteristic equation is

$$\lambda^2 - 0.7\lambda + 0.1 = 0; \ \lambda = \frac{1}{2}, \frac{1}{5}$$

Hence, $y_{\lambda}(n) = c_1(\frac{1}{2})^n + c_2(\frac{1}{5})^n$ with $x(n) = \delta(n)$, we have $y(0) = 2$,

$$y(1) - 0.7y(0) = 0 \Rightarrow y(1) = 1.4$$

Hence, $c_1 + c_2 = 2$ and $\frac{1}{2}c_1 + \frac{1}{5}c_2 = 1.4 = \frac{7}{5} \Rightarrow c_1 + \frac{2}{5}c_2 = \frac{14}{5}$
These equations yield $c_1 = 10/3$, $c_2 = -4/3$

$$s(n) = \sum_{k=0}^{n} h(n-k) = \tfrac{10}{3}\sum_{k=0}^{n}(\tfrac{1}{2})^{n-k} - \tfrac{4}{3}\sum_{h=0}^{n}(\tfrac{1}{5})^{n-h}$$

$$= \tfrac{10}{3}(\tfrac{1}{2})^n \sum_{k=0}^{n} 2^h - \tfrac{4}{3}(\tfrac{1}{5})^n \sum_{k=0}^{n} 5^k$$

$$= \tfrac{10}{3}(\tfrac{1}{2})^n(2^{n+1} - 1)u(n) - \tfrac{1}{3}(\tfrac{1}{5})^n(5^{n+1} - 1)u(n).$$

Problem 5.23

Consider the system in Figure with $h(n) = a^n u(n)$, $-1 < a < 1$. Determine the response $y(n)$ of the system to the excitation $x(n) = u(n + 5) - u(n - 10)$

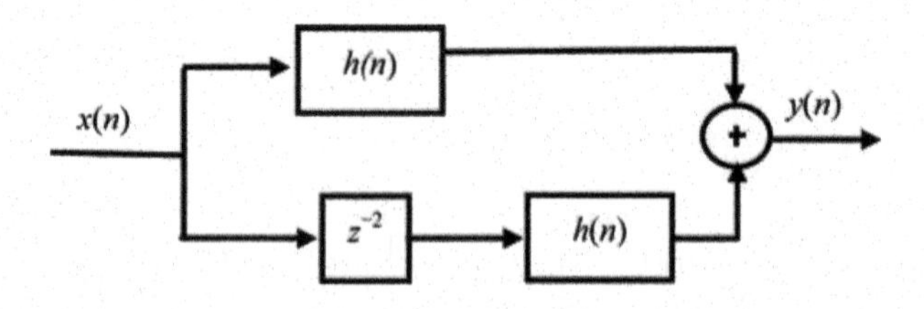

Solution 5.23

First, we determine $s(n) = u(n) \otimes h(n)$

$$s(n) = \sum_{k=0}^{\infty} u(k)h(n-k) = \sum_{k=0}^{n} h(n-k) = \sum_{k=0}^{\infty} a^{n-k} = \frac{a^{n+1} - 1}{a - 1}, n \geq 0.$$

For $x(n) = u(n + 5) - u(n - 10)$, we have the response

$$s(n+5) - s(n-10) = \frac{a^{n+6} - 1}{a - 1}u(n+5) - \frac{a^{n-9} - 1}{a - 1}u(n-10)$$

From figure, $y(n) = x(n) \otimes h(n) + x(n) \otimes h(n - 2)$

$$y(n) = \frac{a^{n+6} - 1}{a - 1}u(n+5) - \frac{a^{n-9} - 1}{a - 1}u(n-10) - \frac{a^{n+4} - 1}{a - 1}u(n+3) + \frac{a^{n-11} - 1}{a - 1}u(n-12).$$

Problem 5.24

Determine the response of the system with impulse response

$$h(n) = a^n u(n)$$

to the input signal

$$x(n) = u(n) - u(n - 10)$$

(Hint: The solution can be obtained easily and quickly by applying the linearity and time-invariance properties).

Solution 5.24

$h(n) = a^n u(n)$. The response to $u(n)$ is

$$y_1(n) = \sum_{k=0}^{\infty} u(k)h(n-k) = \sum_{k=0}^{n} a^{n-k} = a^n \sum_{k=0}^{n} a^{-k}$$
$$= \frac{1 - a^{n+1}}{1 - a} y(n) \quad (4)$$

Then, $y(n) = y_1(n) - y_1(n - 10)$

$$= \tfrac{1}{1-a}[(1 - a^{n+1})u(n) - (1 - a^{n+9})u(n-10)]$$

Problem 5.25

Determine the response of the (relaxed) system characterized by the impulse response

$$h(n) = (\frac{1}{2})^n u(n)$$

to the input signal

$$x(n) = \begin{cases} 1, & n \le n < 10 \\ 0, & \text{otherwise} \end{cases}$$

Solution 5.25

With $\lambda = \frac{1}{2}$. Then,

$$y(n) = 2[1 - (\tfrac{1}{2})^{n+1}]u(n) - 2[1 - (\tfrac{1}{2})^{n-9}]u(n-10)$$

Problem 5.26

Determine the response of the (relaxed) system characterized by the impulse response

$$h(n) = (\frac{1}{2})^n u(n)$$

to the input signals

(a) $x(n) = 2^n u(n)$
(b) $x(n) = u(-n)$

Solution 5.26

$$\text{(a) } y(n) = \sum_{k=-\infty}^{\infty} h(k)x(n-k) = \sum_{k=0}^{n} (\tfrac{1}{2})^k 2^{n-k} = 2^n \sum_{k=0}^{n} (\tfrac{1}{4})^k$$

$$= 2^n [1 - (\tfrac{1}{4})^{n+1}](\tfrac{4}{3}) = \tfrac{2}{3}[2^{n+1} - (\tfrac{1}{2})^{n+1}]u(n)$$

$$\text{(b) } y(n) = \sum_{h=-\infty}^{\infty} h(k)x(n-k) = \sum_{k=0}^{\infty} h(k) = \sum_{k=0}^{\infty} (\tfrac{1}{2})^n = 2, n < 0$$

$$y(n) = \textstyle\sum_{h=n}^{\infty} h(k) = \sum_{k=n}^{\infty} (\tfrac{1}{2})^h = \sum_{k=0}^{\infty} (\tfrac{1}{2})^k - \sum_{k=0}^{n-1} (\tfrac{1}{2})^k$$

$$= 2 - \frac{1-(\frac{1}{2})^n}{1/2} = 2(\tfrac{1}{2})^n, \ n \ge 0$$

Problem 5.27

Three systems with impulse responses $h_1(n) = \delta(n) - \delta(n-1)$, $h_2(n) = h(n)$, and $h_3(n) = u(n)$, are connected in cascade.

(a) What is the impulse response, $h_c(n)$, of the overall system?
(b) Does the order of the interconnection affect the overall system?

Solution 5.27

(a)
$$\begin{aligned} h_c(n) &= h_1(n) \otimes h_2(n) \otimes h_3(n) \\ &= [\delta(n) - \delta(n-1)] \otimes u(n) \otimes h(n) \\ &= [u(n) - u(n-1)] \otimes h(n) = \delta(n) \otimes h(n) = h(n) \end{aligned}$$

(b) No.

Problem 5.28

Compute the zero-state response of the system described by the difference equation

$$h(n) + \frac{1}{2}y(n-1) = x(n) + 2x(n-2)$$

to the input $x(n) = \{1, 2, 3, \underset{\uparrow}{4}, 2, 1\}$ by solving the difference equation recursively.

Solution 5.28

$$\begin{aligned} y(n) &= -\tfrac{1}{2}y(n-1) + x(n) + 2x(n-2) \\ y(-2) &= -\tfrac{1}{2}y(-3) + x(-2) + 2x(-4) = 1 \\ y(-1) &= -\tfrac{1}{2}y(-2) + x(-1) + 2x(-3) = \tfrac{3}{2} \\ y(0) &= -\tfrac{1}{2}y(-1) + 2x(-2) + x(0) = \tfrac{17}{4} \\ y(1) &= -\tfrac{1}{2}y(0) + x(1) + 2x(-1) = \tfrac{47}{8}, \text{ etc.} \end{aligned}$$

Problem 5.29

Consider the system described by the difference equation

$$y(n) = ay(n-1) + bx(n)$$

(a) Determine b in terms of a so that

$$\sum_{n=-\infty}^{\infty} h(n) = 1$$

(b) Compute the zero-state step response $s(n)$ of the system and choose b so that $s(\infty) = 1$.

(c) Compare the values of b obtained in parts (a) and (b). What did you notice?

Solution 5. 29

(a) $y(n) = ay(n-1) + bx(n) \Rightarrow h(n) = ba^n u(n)$

$$\sum_{n=0}^{\infty} h(n) = \frac{b}{1-a} = 1 \Rightarrow b = 1 - a$$

(b) $s(n) = \sum_{n=0}^{\infty} h(n-k) = b\left[\frac{1-a^{n+1}}{1-a}\right]u(n)$

$$s(\infty) = \frac{b}{1-a} = 1 \Rightarrow b = 1 - a$$

(c) $b = 1 - a$ in both cases.

Problem 5.30

A discrete-time system is realized by the structure shown in Figure.

(a) Determine the impulse response.

(b) Determine a realized for its inverse system, that is, the system which produces $x(n)$ as an output when $y(n)$ is used as an input.

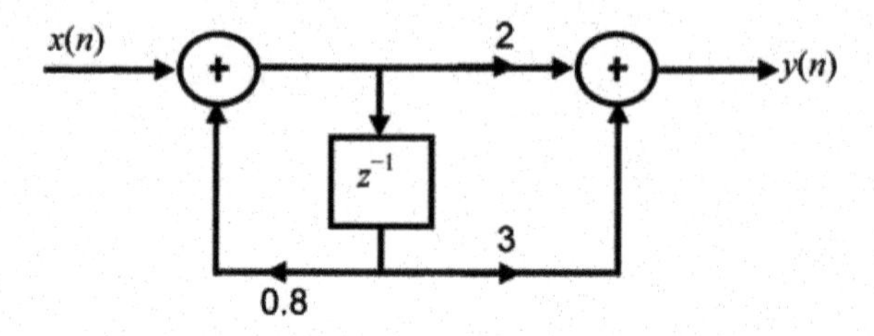

Solution 5.30

(a) $$\begin{aligned} y(n) &= 0.8\,y(n-1) + 2x(n) + 3x(n-1) \\ y(n) - 0.8\,y(n-1) &= 2x(n) + 3x(n-1) \end{aligned}$$

The characteristic equation is $\lambda - 0.8 = 0$, $\lambda = 0.8$

$$y_\lambda(n) = c(0.8)^n.$$

Let us first consider the response of the system $y(n) - 0.8y(n-1) = x(n)$
To $x(n) = \delta(n)$. Since $y(0) = 1$, it follows that $c = 1$.
Then, the impulse response of the original system is

$$\begin{aligned} h(n) &= 2(0.8)^n u(n) + 3(0.8)^{n-1} u(n-1) \\ &= 2\delta(n) + 5.6\,(0.8)^{n-1} u(n-1) \end{aligned}$$

(b) The inverse system is characterized by the difference equation

$$x(n) = -1.5x(n-1) + \frac{1}{2}y(n) - 0.4y(n-1)$$

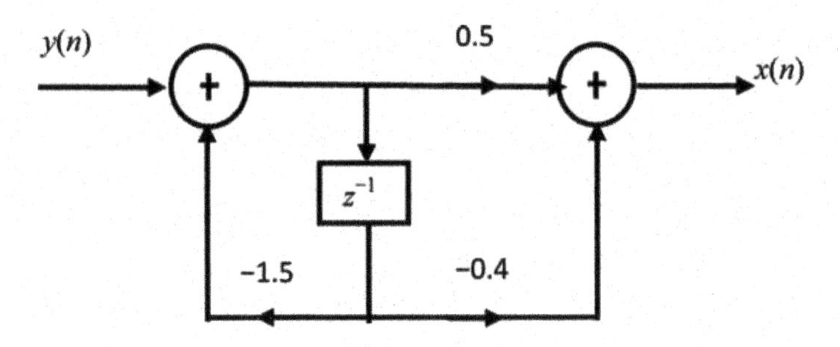

Problem 5.31

Consider the discrete-time system shown in Figure.

(a) Compute the first six values of the impulse response of the system.
(b) Compute the first six values of the zero-state step response of the system.
(c) Determine an analytical expression for the impulse response of the system.

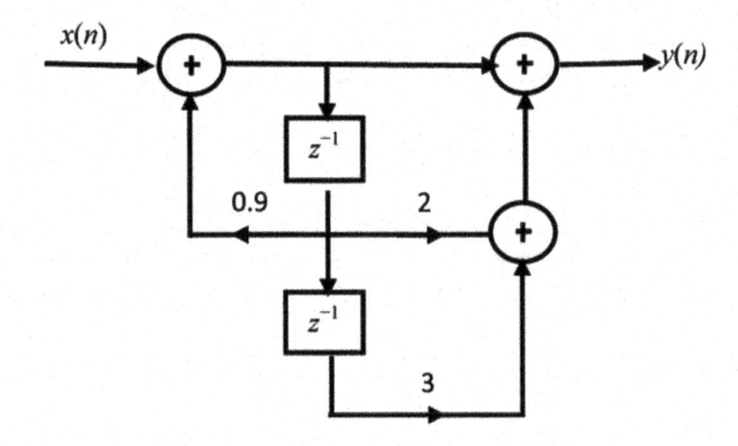

Solution 5.31

$$y(n) = 0.9y(n-1) + x(n) + 2x(n-1) + 3x(n-2)$$

(a) for $x(n) = d(n)$, we have $y(0) = 1$, $y(1) = 2.9$,
$y(2) = 5.61$, $y(3) = 5.049$, $y(4) = 5.544$, $y(5) = 5.090, \ldots$

(b) $s(0) = y(0) = 1, s(1) = y(0) + y(1) = 3.91,$
$s(2) = y(1) + y(1) + y(2) = 9.51$
$s(3) = y(0) + y(1) + y(2) + y(3) = 15.56,$

$$s(4) = \sum_{0}^{4} y(n) = 19.10$$

$$s(5) = \sum_{0}^{5} y(n) = 23.19$$

(c) $h(n) = (0.9)^n u(n) + 2(0.9)^{n-1}u(n-1) + 3(0.9)^{n-2}u(n-2)$
$= \delta(n) + 2.9d(n-1) + 5.61\,(0.9)^{n-2}u(n-2)$

Problem 5.32

Consider the systems shown in Figure. Determine and sketch their impulse responses $h_1(n)$, $h_2(n)$, and $h_3(n)$.

(a) Is it possible to choose the coefficients of these systems in such a way that $h_1(n) = h_2(n) = h_3(n)$

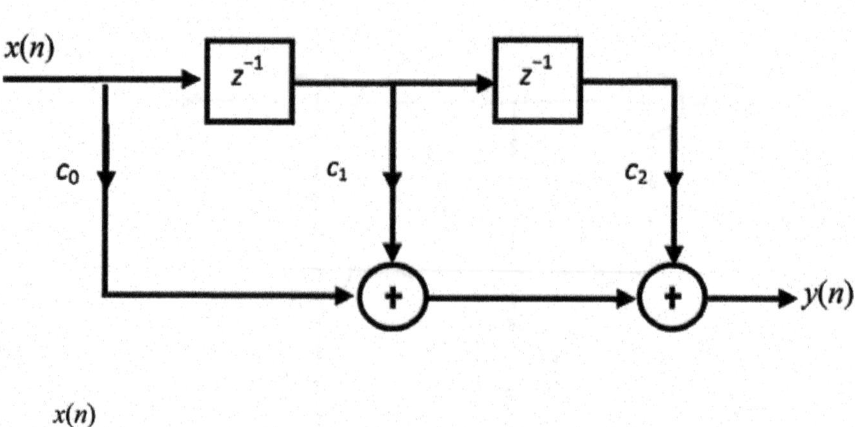

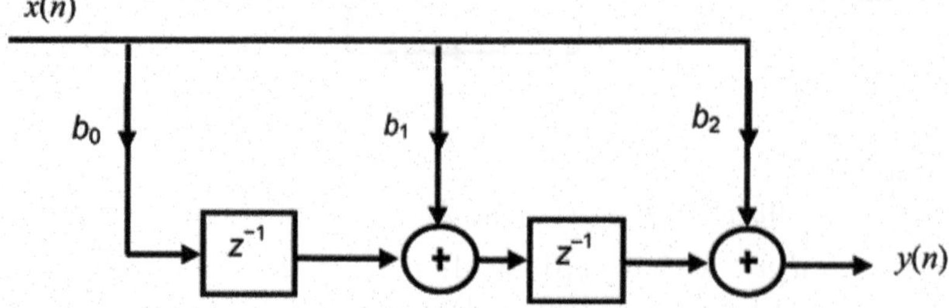

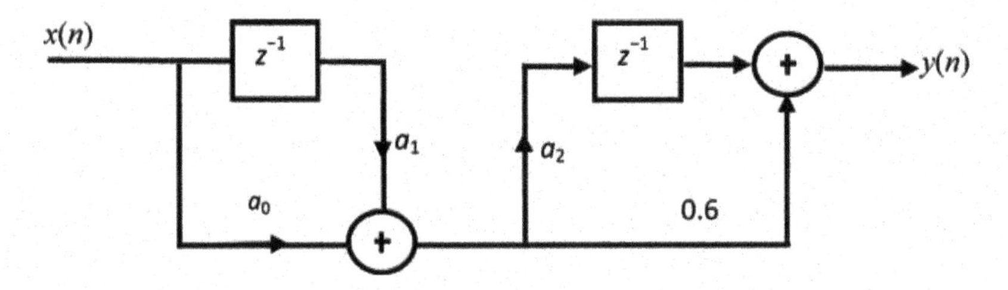

Solution 5.32

(a)
$$\begin{aligned} h_1(n) &= c_o\delta(n) + c_1\delta(n-1) + c_2\delta(n-2) \\ h_2(n) &= b_2\delta(n) + b_1\delta(n-1) + b_0\delta(n-2) \\ h_3(n) &= a_o\delta(n) + (a_1 + a_o\, a_2)\delta(n-1) + a_1 a_2\delta(n-2) \end{aligned}$$

(b) The only question is whether $h_3(n) = h_2(n) = h_1(n)$

Let $a_0 = c_o$, $a_1 + a_2 c_o = c_1$, $a_2\, a_1 = c_2$. Hence,

$$\frac{c_2}{a_2} + a_2 c_0 - c_1 = 0 \Rightarrow c_0 a_2^2 - c_1 a_2 + c_2 = 0$$

For $c_0 \neq 0$, the quadratic has a real solution if and only if $c_1^2 - 4c_o c_2 \geq 0$.

Problem 5.33

Consider the system shown in Figure.

1. Determine its impulse response $h(n)$.
2. Show that $h(n)$ is equal to the convolution of the following signals.

$$\begin{aligned} h_1(n) &= \delta(n) + \delta(n-1) \\ h_2(n) &= \left(\tfrac{1}{2}\right)^n u(n) \end{aligned}$$

Solution 5.33

(a) $y(n) = \frac{1}{2}y(n-1) + x(n) + x(n-1)$

For $y(n) - y(n-1) = \delta(n)$, the solution is $h_1(n) = (\frac{1}{2})^n u(n)$

Hence, $h(n) = (\frac{1}{2})^n u(n) + (\frac{1}{2})^{n-1} u(n-1)$

(b) $h1(n) \otimes [\delta(n) + \delta(n-1)] = (\frac{1}{2})^n u(n) + (\frac{1}{2})^{n-1} u(n-1)$.

Problem 5.34

The zero-state response on a causal LTI system to the input *x*(*n*)

$$x(n) = \{\underset{\uparrow}{1}, 3, 3, 1, \}$$
$$\text{is } y(n) = \{\underset{\uparrow}{1}, 4, 6, 4, 1\}.$$

Determine its impulse response.

Solution 5.34

Obviously, the length of $h(n)$ is 2, i.e., $h(n) = \{h_0, h_1\}$ $h_0 \Rightarrow h_0 = 1$, $h_1 = 1$, $3h_0 + h_1 = 4$

Problem 5.35

Determine the response $y(n)$, $n \geq 0$ of the system described by the second-order difference equation

$$y(n) - 4y(n-1) + 4y(n-2) = x(n) - x(n-1)$$

when the input is $x(n) = (-1)^n u(n)$ and the initial conditions are $y(0) = 1$; $y(1) = 2$.

Solution 5.35

$$y(n) - 4y(n-1) + 4y(n-2) = x(n) - x(n-1)$$

The characteristic equation is $\lambda^2 - 4\lambda + 4 = 0\lambda = 2, 2$. (case of repeated roots)

Hence,

$$y_\lambda(n) = c_1 2^n + c_2 n 2^n$$

The particular solution is $y_p(n) = C_3(-1)^n u(n)$ substituting this solution into the difference equation, we obtain

$$C_3(-1)^n u(n) - 4C_3(-1)^{n-1} u(n-1) + 4C_3(-1)^{n-2} u(n-2)$$
$$= (-1)^n u(n) - (-1)^{n-1} u(n-1).$$

For $n = 2$, $C_3[1 + 4 + 4] = 2 \Rightarrow C_3 = 2/9$. The total solution is $y(n) = [c_1 2^n + c_2 n 2^n + \frac{2}{9}(-1)^n]u(n)$. From the initial conditions, in

$$y(0) = 1,\ y(1) = 2, \text{ we obtain, } c_1 + \frac{2}{9} = 1 \Rightarrow c_1 = 7/9;$$

$$2c_1 + 2c_2 - \frac{2}{9} = 2 \Rightarrow c_2 = 1/3.$$

$$y_\lambda(n) = c_1 2^n + c_2 n 2^n + 2/9(-1)^n u(n)$$
$$y_\lambda(n) = 7/9 2^n + 1/3 n 2^n + 2/9(-1)^n u(n)$$

Practice Problem 5.36

Find the particular solution of the second-order difference equation

$$8y(n) - 6y(n-1) + y(n-2) = 5\sin\left(\frac{n\pi}{2}\right), \quad n \geq 0$$

6

Discrete-Time Fourier Transform Discrete Fourier Transform

This chapter covers: Fourier synthesis, Discrete-Time Fourier transforms, Discrete Fourier transforms, Comparison between DTFT and DFT. Inverse Discrete Fourier Transforms, Fast Fourier transforms (Decimation in Time and Decimation in Frequency algorithms), Problems and solutions.

6.1 Introduction

A transformation normally involves changes in coordinates and domain of operations. Discrete Fourier transformation is a representation of discrete-time signals in the frequency domain or the conversion between time and frequency domain. The spectrum of a signal is obtained by decomposing it into its constituent frequency components using a discrete transform. Conversion between time and frequency domains is necessary in many DSP applications. For example, it allows for a more efficient implementation of DSP algorithms, such as those for digital filtering, convolution and correlation.

6.2 Periodic Function and Fourier Synthesis

Jean Baptiste Joseph Baron de Fourier (1768–1830), a French physicist, discovered that any periodic waveform can be broken down into a combination of sinusoidal waves. All the sinusoidal waves that can be combined together to produce any periodic wave form are collectively called a basis and each individual wave is an element of the basis.

The amplitudes of the sinusoids are called the Fourier coefficients. The relationship between the frequencies of sinusoids is such that they are all harmonics (integer multiples) of a fundamental frequency. The amplitude of the periodic signal as a function of time is its time-domain representation, where as the Fourier coefficients correspond to its frequency-domain representation.

6.2.1 Constructing a Waveform with Sine Waves

In Figures 6.1(a)–(c) given below shows three basic sine waves 2, 4, and 6 Hz with amplitude of 7, 2, and 4, respectively. They add together to form the wave in Figure(d).

We can write the equation for the wave in Figure 6.1(d) as

$$s(t) = 7\sin(2\pi 2t) + 2\sin(2\pi 4t) + 4\sin(2\pi 6t). \tag{6.1}$$

In this case the fundamental frequency is 2 Hz, the frequency of the second harmonic is 4 Hz, and the frequency of the third harmonic is 6 Hz.

If the fundamental frequency is denoted by f_1, then Equation (6.1) can be written as

$$s\,(t) \;=\; 7\sin(2\pi[f_1]t) + 2\sin(2\pi[2f_1]t) + 4\sin(2\pi[3f_1]t). \tag{6.2}$$

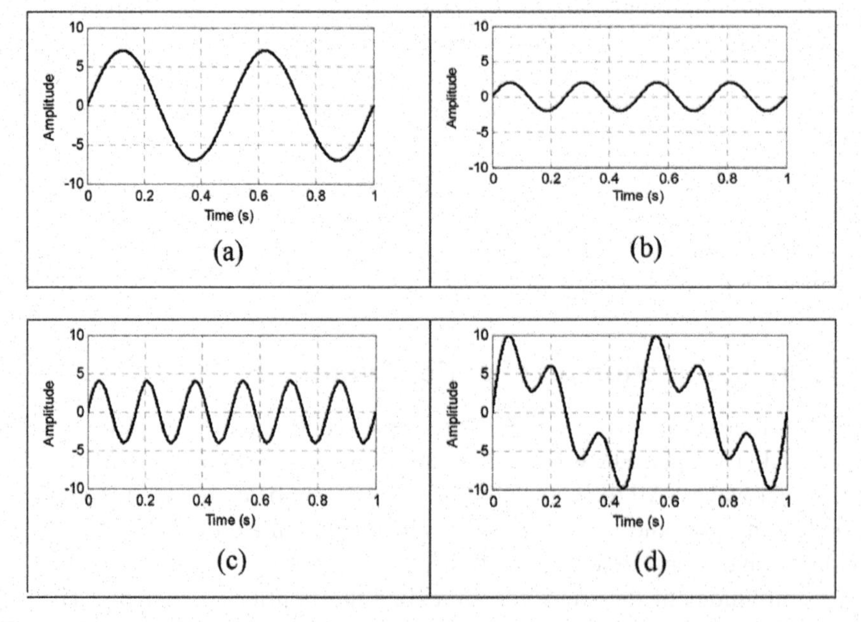

Figure 6.1 (a) Sine wave 1 with f = 2 Hz, amplitude = 7; (b) Sine wave 2 with f = 4 Hz, amplitude = 2; (c) Sine wave 3 with f = 6 Hz, amplitude = 4 and (d) Addition of sine waves 1, 2, 3.

Therefore, $s(t)$ is a combination of a fundamental and the second and the third harmonics. Observe that the frequency of the more complicated periodic wave $s(t)$ is 2 Hz, which is the same as the frequency of the lowest frequency sine wave (the fundamental).

6.2.2 Constructing a Waveform with Cosine Waves

In fact, this waveform Figure 6.2(a) has its maximum value at $t = 0$, so it can be created using a combination of cosine wave. Once again the frequency of the more complicated periodic wave, which is 2 Hz, which is the same as frequency of the lowest frequency cosine wave (the fundamental).

Now look at the Figure 6.2(d). These waveform does not start at $t = 0$; that the value is not equal to zero at $t = 0$. Since sine waves always have a value of zero at $t = 0$, we cannot use combination of sine wave to produce it. In fact, this wave form has its maximum value at $t = 0$, so it can be

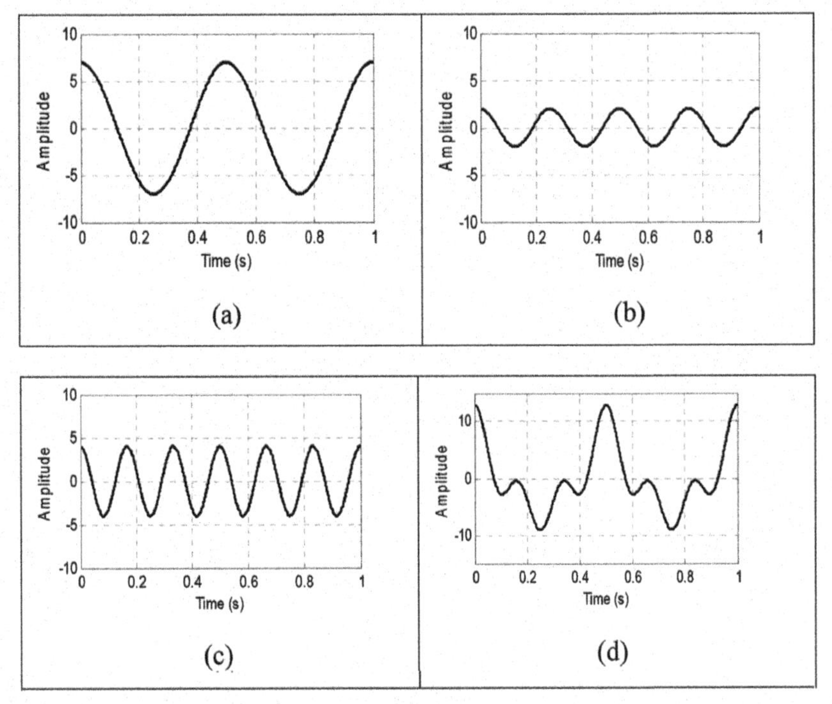

Figure 6.2 (a) Cosine wave 1 with $f = 2$ Hz, amplitude = 7; (b) Cosine wave 2 with $f = 4$ Hz, amplitude = 2; (c) Cosine wave 3 with $f = 6$ Hz, amplitude = 4 and (d) Addition of cosine waves in (a), (b), and (c). The maximum value occurs at $t = 0$.

created using combination of cosine waves. The equation for the wave form of Figure 6.2(d) is

$$c\,(t) = 7\cos(2\pi 2t) + 2\cos(2\pi 4t) + 4\cos(2\pi 6t). \qquad (6.3)$$

Therefore, $s(t)$ is a combination of a fundamental and the second and the third harmonics. In this case the fundamental frequency is 2 Hz, the frequency of the of the second harmonic is 4 Hz, and the frequency of the third harmonic is 6 Hz.

If the fundamental frequency is denoted by f_1, then Equation (6.3) can be written as

$$c(t) = 7\cos(2\pi[f_1]t) + 2\cos(2\pi[2f_1]t) + 4\cos(2\pi[3f_1]t). \qquad (6.4)$$

Observe that the frequency of the more complicated periodic wave $c(t)$ is 2 Hz, which is the same as the frequency of the lowest frequency cosine wave (the fundamental).

6.2.3 Constructing a Waveform with Cosine and Sine Waves

What would a wave form look like that is a combination of three sine waves Figure 6.1(a)–(c) and three cosine wave in Figure 6.2(a)–(c)? It is shown in Figure 6.1(d) and 6.2(d). Observe that the value at $t = 0$ is neither zero nor maximum value of the wave. This brings us a very important point: by adding sines and cosines of appropriate amplitude we construct a periodic wave form starting with any value at $t = 0$. The equation for the wave form for Figure 6.3 is $f(t)$ = cosine wave 1 + cosine wave 2 + cosine wave 3 + sine wave 1 + sine wave 2 + sine wave 3.

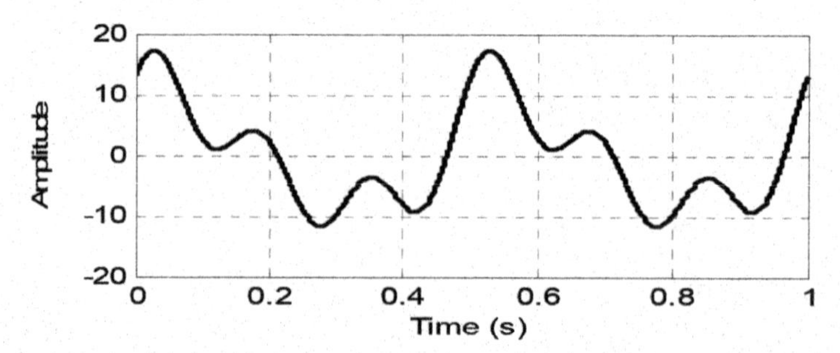

Figure 6.3 Combination of sine waves from 6.1(a)–(c) and the cosine waves from 6.2(a)–(c). The value at $t = 0$ is neither 0 nor maximum, but somewhere in between.

$$f(t) = 7\cos(2\pi[f_1]t) + 2\cos(2\pi[2f_1]t) + 4\cos(2\pi[3f_1]t) \\ + 7\sin(2\pi[f_1]t) + 2\sin(2\pi[2f_1]t) + 4\sin(2\pi[3f_1]t) \quad (6.5)$$

We have kept these examples simple and are only using the fundamental and the second and the third harmonics to produce periodic waves. Periodic signals actually encountered in practice can be combinations of extremely large number of harmonics. Therefore, equation for such harmonics waves can be written as

$$f(t) = a_1\cos(2\pi[f_1]t) + a_2\cos(2\pi[2f_1]t) + \ldots\ldots a_n\cos(2\pi[nf_1]t) \\ + b_1\sin(2\pi[f_1]t) + b_2\sin(2\pi[2f_1]t) + \ldots\ldots b_n\sin(2\pi[nf_1]t), \quad (6.6)$$

where $a_1, a_2, a_3,$ are the amplitude of cosine wave
$b_1, b_2, b_3,$ are the amplitude of sine wave
$f_1, f_2, f_3,$ are the frequencies of the fundamental and harmonics

6.2.4 Constructing a Waveform with Sine, Cosine, and a DC

Finally, this wave form will like the same as of Figure 6.3, except it would be shifted upwards due to addition of a DC value. This leads finally to Equation (6.7)

$$f(t) = a_0 + a_1\cos(2\pi[f_1]t) + a_2\cos(2\pi[2f_1]t) + \ldots\ldots a_n\cos(2\pi[nf_1]t) \\ + b_1\sin(2\pi[f_1]t) + b_2\sin(2\pi[2f_1]t) + \ldots\ldots b_n\sin(2\pi[nf_1]t). \quad (6.7)$$

Equation (6.7) is known as Fourier series. It consists of a series of frequencies used to form any periodic function. The amplitude of sines and cosines are given a special name- they are called the Fourier coefficients. Thus $a_1, a_2, \ldots, a_n, b_1, b_2, \ldots, b_n$ are the Fourier coefficients.

The sines and cosines, those have a fundamental frequency f_1, and all those corresponding harmonics, are said to form the basis for all periodic wave forms. Each of sine and cosine waveform is an element of basis. This means that not only can sines and cosines be combined to construct any complicated periodic wave form, but also any complicated periodic wave form can be broken down into sum of sines and cosines.

The Equation (6.7) can be written as

$$f(t) = a_0 + \sum_{n=1}^{\infty}(a_n \cos n\omega_0 t + b_n \sin n\omega_0 t) \tag{6.8}$$

$$\omega_0 = \frac{2\pi}{T} = 2\pi f_0 \quad a_0 = \frac{1}{T}\int_o^T f(t)dt \tag{6.9}$$

$$a_n = \frac{2}{T}\int_o^T f(t)\cos n\omega_0 t\, dt \qquad b_n = \frac{2}{T}\int_o^T f(t)\sin n\omega_0 t\, dt \tag{6.10}$$

Even symmetry $f(t) = f(-t)$
Odd symmetry $f(t) = -f(-t)$
Half Wave symmetry $f(t) = -f(t - T/2) = -f(t + T/2)$.

Following Table 6.1 shows that which of the coefficients for Fourier series have to be calculated. It is always worthwhile spending few minutes investigating the symmetry of a function for which a Fourier series for the wave form is to be determined.

6.2.5 Gibbs' Phenomenon

Fourier series can be used to construct any complicated periodic wave form. In general the more the number of elements in Fourier series, the better is the construction of corresponding wave form. However, there is limitation in that if the periodic wave form has any discontinuity, that is the vertical edges, then even an infinite number of elements in Fourier series cannot construct that discontinuity exactly. Adding more and more harmonics to the sine waves, one gets a better representation of a square wave. Keep on adding the higher harmonics, the resulting wave looks more and more ideal square wave. By including the higher odd harmonics, the following changes in the resulting wave form as the number of harmonics is increased:

1. The number of oscillation increases, but the amplitudes of the oscillations decrease.
2. The vertical edge gets steeper.
3. Overshoot exist at the vertical edges.
4. The approximation is worst at the vertical edges.

It is observed that the overshoots at the vertical edges do not really go away, irrespective how many harmonics are added. Even if an infinite number of harmonics are added, the overshoot remains and its size settles to about

Table 6.1 Used for selection of the Fourier series coefficients

Even Symmetry	$a_0 = \frac{1}{T}\int_o^T f(t)\mathrm{d}t$ $a_n = \frac{2}{T}\int_o^T f(t)\cos nw_0t\ \mathrm{d}t$	$b_n = 0$
Odd Symmetry	$a_0 = 0,\ a_n = 0$	$b_n = \frac{2}{T}\int_o^T f(t)\sin nw_0t\ \mathrm{d}t$
Half-wave Symmetry	$a_n = \frac{4}{T}\int_o^{T/2} f(t)\cos nw_0t\ \mathrm{d}t; n$ odd $= 0$ n even	$b_n = \frac{4}{T}\int_o^{T/2} f(t)\sin nw_0t\ \mathrm{d}t; n$ odd $= 0$ n even
Half-wave and even symmetry	$a_n = \frac{8}{T}\int_o^{T/4} f(t)\cos nw_0t\ \mathrm{d}t; n$odd $a_n = 0$ n even	$b_n = 0$ for all n
Half-wave and odd symmetry	$a_n = 0$ for all n	$b_n = \frac{8}{T}\int_o^{T/4} f(t)\sin nw_0t\ \mathrm{d}t$ n odd $b_n = 0$ n even

8.95% of the size of the vertical edge. This phenomenon is called the Gibbs phenomenon in honour of Willard Gibbs who described it occurrence in the late 19th century (Figure 6.4).

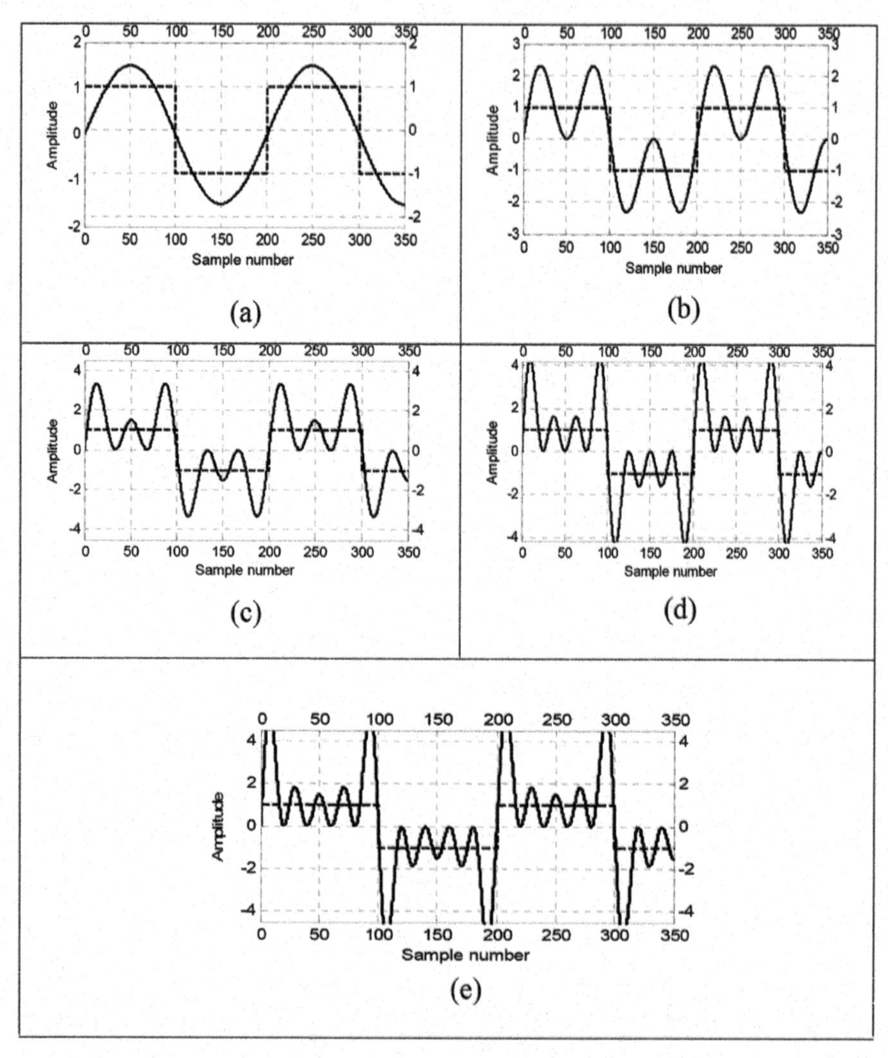

Figure 6.4 (a) Square wave with fundamental harmonic; (b) Square wave with fundamental and third harmonic; (c) Square wave with fundamental, third, fifth harmonic; (d) Square wave with fundamental, third, fifth and seventh harmonic and (e) Square wave with fundamental, third, fifth, seventh and ninth harmonic.

6.3 Introduction to Fourier Transforms

Transformations are used in signal processing systems mostly for converting signals in time domain to signals in frequency domain. By signal in time domain we mean a signal whose amplitude with respect to time. Similarly, by term signal in frequency domain, we mean a signal whose amplitude varies with respect to frequency.

Fourier and inverse Fourier are the mathematical tools that are used to switch from time-domain to frequency-domain and vice versa, in order to view a signal.

This conversion between time and frequency domains is very helpful in enhancing the understanding of signals and systems. This is called Fourier analysis.

There is an entire family of Fourier analysis techniques, where each specific technique is applied to a particular type of signals – continuous or discrete, and periodic or aperiodic in order to determine its frequency content. These distinctions results in four categories of signals:

(i) Continuous-time (CT) and periodic
(ii) CT and aperiodic
(iii) Discrete-time (DT) and periodic
(iv) DT and aperiodic

For a CT periodic signal, the Fourier analysis technique is CT Fourier series. For CT aperiodic signal, it is CT Fourier transform. In DSP, our main interest lies in signals that are discrete in time. The transform techniques used in this case are the DFS or DT Fourier series (DTFS) for DT periodic signals and DT Fourier transform (DTFT) for DT aperiodic signals. The table lists the transform that are used for different types of signals.

Periodic	Continuous	Discrete
	CT Fourier series	DTFSeries
Aperiodic	CTFT	DTFT

Continuous-time or analog systems are modeled by a set of differential equations and the transform used in the analysis of LTI analog systems is the Laplace transform, which is strongly related with CTFT. (Laplace Transform becomes CTFT on imaginary axis in s-plane.) DT systems are modeled by a set of difference equations and the transform used in the analysis of LTI discrete systems is the Z-transform, which is strongly related with DTFT. (z-transform becomes DTFT on unit circle in z-plane.)

As most of the practical signals of interest are aperiodic signals, hence it is DTFT that we are frequently coming across. DTFT is a continuous function and consequently in computer, its discretized or sampled version is computed. This sampled version of DTFT is called Discrete Fourier Transform (DFT).

The DFT is one of the most common and powerful algorithms of digital signal processing. It is used in a wide variety of applications in many branches of science and engineering. For example, it allows for more efficient implementation of DSP algorithms, such as those for digital filtering, convolution and correlation. The plot of DTFT/DFT is called frequency spectrum, i.e., magnitude spectrum and phase spectrum.

6.4 DT Fourier Transform

In the DSP, we are concerned with the discrete version of Fourier transformation, which is defined as

$$X(\omega) = \sum_{n=-\infty}^{\infty} x(n)e^{-j\omega nT} \tag{6.11}$$

Considering that $T = 1$

$$X(\omega) = \sum_{n=-\infty}^{\infty} x(n)e^{-j\omega n}. \tag{6.12}$$

The DTFT general equation can also be developed from the expression of the Z-transform, where $x(n)$ is defined as

$$X(z) = \sum_{n=-\infty}^{\infty} x(n)z^{-n}, \tag{6.13}$$

Where $r_2 < |z| < r_1$ is the region of convergence of $X(z)$. Let us express the complex variable z in polar form as

$$z = re^{jw} \tag{6.14}$$

Where $|z| = r$ and $\angle z = \omega$. For the region of convergence of $X(z)$, we may substitute

$z = re^{j\omega}$ into Equation (6.13). This yield

$$X(z)|_{z=re^{j\omega}} = \sum_{n=-\infty}^{\infty} [x(n)r^{-n}]e^{-j\omega n}. \tag{6.15}$$

From the relationship in Equation (6.15) we note that $X(z)$ can be interpreted as the DTFT of the signal sequence $x(n)r^{-n}$.

The weighting factor r^{-n} is growing with n if $r < 1$ and decaying if $r > 1$. Alternatively, if $X(z)$ converges for $|z| = 1$, then Equation (6.15) for DTFT becomes

$$X(z)|_{z=e^{j\omega}} = \sum_{n=-\infty}^{\infty} x(n)e^{-j\omega n}. \tag{6.16}$$

It means a DTFT) can be expressed as

$$\text{or} \quad X(\omega) = \sum_{n=-\infty}^{\infty} x(n)e^{-j\omega n} \tag{6.17}$$

or DTFT $x(n) = X(\omega)$ and inverse DT Fourier Transform (IDTFT) is expressed as

$$\text{or} \quad x(n) = \frac{1}{2\pi}\int_{-\pi}^{\pi} X(\omega)\, e^{j\omega n}\, d\omega \tag{6.18}$$

Equations (6.17) and (6.18) are called DTFT pairs.

Few points to note at this stage

(i) Like CT Fourier transform (CTFT), the frequency spectrum in DTFT is also continuous.

(ii) There is a major difference between CTFT and DTFT. The frequency spectrum is not periodic in CTFT where as in DTFT; the frequency spectrum is periodic with period 2π.

We also define the IDTFT as

$$h(n) = \frac{1}{2\pi}\int_{-\pi}^{\pi} H(\omega)\, e^{j\omega n}\, d\omega.$$

The angular frequency ω is not discretized. Consider the situation, where ω is discretized by defining the relation

$$\omega = \frac{2\pi k}{N}.$$

After substituting the value of ω, Equation (6.17) can be written as

$$X[k] = \sum_{n=0}^{N-1} x(n)e^{-j\left(\frac{2\pi\, kn}{N}\right)}.$$

This is called the DFT of the sequence $x(n)$.

Example 6.1

Obtain the impulse response of the system described by

$$H(\omega) = 1 \quad \text{for } |\omega| \leq \omega_C$$
$$H(\omega) = 0 \quad \text{for } \omega_C \leq |\omega| \leq \pi$$

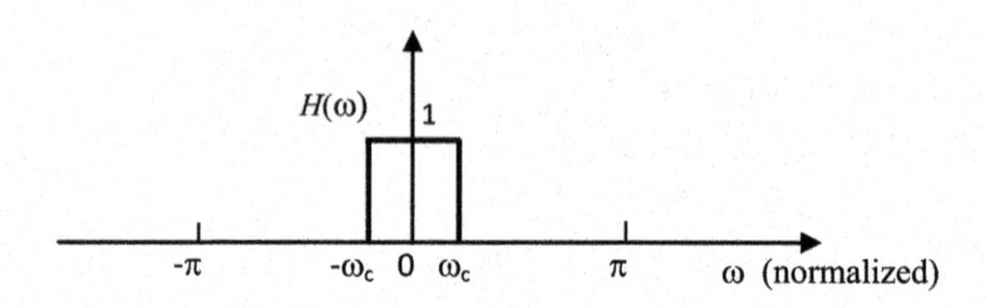

Solution 6.1

DTFT $h(n) = H(\omega)$ and IDTFT is expressed as

$$\text{or} \quad h(n) = \frac{1}{2\pi} \int_{-\pi}^{\pi} H(\omega)\, e^{j\omega n}\, \mathrm{d}\omega = \frac{1}{2\pi} \int_{-\omega_C}^{\omega_C} 1 e^{j\omega n}\, \mathrm{d}\omega$$

$$\text{or} \quad h(n) = \frac{1}{2\pi j n}\, e^{j\omega n}\Big]_{-\omega_C}^{\omega_C} = \frac{e^{j\omega_C n} - e^{-j\omega_C n}}{2\pi j n} = \frac{\sin\, \omega_C n}{\pi n}$$

It means the impulse response can be calculated at each discrete value of n from the above expression. But when we want to calculate $h(0)$ it becomes infinite so the result is not valid.

In this case using L' Hopital's Rule

$$h(0) = \frac{\omega_c\ \cos\ \omega_C\ n}{\pi}$$

$$h(0) = \frac{\omega_c\ \cos\ 0}{\pi} = \frac{\omega_c}{\pi}.$$

In Table 6.2 few properties of DTFT are given in tabular form.

6.5 Properties of the DTFT

The properties of DTFT is same as DFT but here they have been dealt here separately

Table 6.2 Properties of DTFT

Property	Sequence	DTFT
Periodicity	$x(n) = x(n+N)$	$X(\omega) = X[(\omega + 2k\pi)]$
Linearity	$ax(n) + by(n)$	$aX(\omega) + bY(\omega)$
Time Shift	$x(n - n_0)$	$e^{-j\omega n_0} X(\omega)$
Time Reversal	$x(-n)$	$X(e^{-j\omega})$
Frequency Shift	$x(n)$	$e^{j\omega n_0} x(n) \leftrightarrow X[(\omega - \omega_0)]$
Scaling	$y(n) = x(pn)$	$Y(\omega) = X(\frac{\omega}{P})$
Multiplication by n	$n\, x(n)$	$j\frac{d[X(\omega)]}{d\omega}$
Convolution	$x(n) \times y(n)$	$X(\omega).Y(\omega)$
Time Reversal	$x(-n)$	$X[-\omega]$
Multiplication in time domain	$x(n).y(n)$	$\frac{1}{2\pi}[X(\omega) \otimes Y(\omega)]$
Complex Conjugation	$x(n)$ and $x^*(n)$	$X(\omega)$ and $X(-\omega)$

6.5.1 Periodicity

The DT transform is always periodic in ω with period 2π

$$\text{or} \quad X[(\omega + 2k\pi)] = X(\omega) \tag{6.19}$$

Example 6.2

Find the DTFT of the unit step sequence

$$x(n) = u(n)$$

Solution 6.2

$$X(\omega) = \sum_{n=-\infty}^{\infty} x(n)e^{-j\omega n} = \sum_{n=-\infty}^{\infty} 1e^{-j\omega n} X(\omega) = \frac{1}{1 - e^{-j\omega}}.$$

The relation is not convergent for $\omega = 0$. This is because of the fact $x(n)$ is not absolutely summable sequence. However $X(\omega)$ may be evaluated for other values of ω.

Rearranging

$$X(\omega) = \frac{1}{e^{-j\omega/2}.e^{j\omega/2} - e^{-j\omega/2}.e^{-j\omega/2}} = \frac{1}{e^{-j\omega/2}[e^{j\omega/2} - e^{-j\omega/2}]}$$

using Euler's identity we may write

$$X(\omega) = \frac{1}{e^{-j\omega/2}.2j\ \sin\frac{\omega}{2}} = \frac{e^{j\omega/2}}{2j\ \sin\frac{\omega}{2}}, \omega \neq 0$$

6.5.2 Linearity

According to this property, if the DTFT is linear:

$$\begin{aligned} x_1(n) &\leftrightarrow X_1(\omega) \\ x_2(n) &\leftrightarrow X_2(\omega) \end{aligned} \tag{6.20}$$

then according to this property $ax_1(n) + bx_2(n) = aX_1(\omega) + bX_2(\omega)$.

6.5.3 Time Shifting

This property states that if a DT signal is shifted in time domain by n samples, its magnitude spectrum remains unchanged, however the phase spectrum is changed by an amount $-\omega n_0$, mathematically

$$\begin{aligned} x(n) &\leftrightarrow X(\omega) \\ x(n - n_0) &\leftrightarrow e^{-j\omega n_0} X(\omega) \end{aligned} \tag{6.21}$$

where n_0 is an integer.

Example 6.3

Find the DTFT of

$$x(n) = \left(\frac{1}{2}\right)^{n-1} u(n-1)$$

Solution 6.3

$$X(\omega) = \sum_{n=-\infty}^{\infty} x(n)e^{-j\omega n} \quad X(\omega) = \sum_{n=1}^{\infty} \left(\frac{1}{2}\right)^{n-1} u(n-1)e^{-j\omega n}$$

$$X(\omega) = e^{-jw} + \frac{1}{2}e^{-j2w} + \ldots\ldots X(\omega) = \frac{e^{-jw}}{1 - \frac{1}{2}e^{-jw}}$$

Example 6.4

Compute the DTFT of the following signal

$$x(n) = u(n) - u(n-6)$$

Solution 6.4

Using

$$X(\omega) = \sum_{n=-\infty}^{\infty} x(n)\, e^{-j\omega n} \Rightarrow X(\omega) = \sum_{n=-\infty}^{\infty} [u(n) - u(n-6)]\; e^{-j\omega n}$$

$$X(\omega) = \frac{1}{1-e^{-jw}} - \left[\frac{e^{-6jw}}{1-e^{-jw}}\right] \quad X(\omega) = \frac{1-e^{-j6\omega}}{1-e^{-j\omega}}.$$

Using z-transform

$$X(z) = \frac{1}{1-z^{-1}} - \frac{z^{-6}}{1-z^{-1}} \quad X(z) = \frac{1-z^{-6}}{1-z^{-1}}$$

Put $z = e^{jw}$ for $X(w)$ $X(\omega) = \frac{1-e^{-j6\omega}}{1-e^{-j\omega}}$

Example 6.5

Compute the DTFT of the following signal

$$x(n) = \left(\frac{1}{4}\right)^n u(n+4)$$

Solution 6.5

Using

$$X(\omega) = \sum_{n=-\infty}^{\infty} x(n)e^{-j\omega n} = \sum_{n=-\infty}^{\infty} \left(\frac{1}{4}\right)^n u(n+4)e^{-j\omega n}$$

$$X(\omega) = \left(\frac{1}{4}\right)^{=4} \sum_{n=-\infty}^{\infty} \left(\frac{1}{4}\right)^{n+4} u(n+4)e^{-j\omega n}$$

$$= (4)^4 \sum_{n=-\infty}^{\infty} \left(\frac{1}{4}\right)^{n+4} u(n+4)e^{-j\omega n}$$

$$X(\omega) = \frac{256e^{4j\omega}}{1 - \frac{1}{4}e^{-j\omega}}$$

Using Z-transform

$$z[u(n+4)] = \frac{z^4}{1 - z^{-1}}.$$

Using scaling property of z-transform: $a^n x(n) \overset{Z}{\longleftrightarrow} x(a^{-1}z)$

$$X(z) = \frac{[(\frac{1}{4})^{-1}z]^4}{1 - \left[(\frac{1}{4})^{-1} z\right]^{-1}} == \frac{256z^4}{1 - \frac{1}{4}z^{-1}} \quad X(\omega) = \frac{256e^{4j\omega}}{1 - \frac{1}{4}e^{-j\omega}}$$

6.5.4 Frequency Shifting

This property states that multiplication of a sequence *x*(*n*) by the spectrum $e^{-j\omega_0 n}$ is equivalent to a frequency translation $X(\omega)$ by ω_0. Since the spectrum $X(\omega)$ is periodic, the shift ω_0 applies to the spectrum of the signal in every period.

Mathematically

$$\begin{aligned} x(n) &\leftrightarrow X(\omega) \\ e^{j\omega n_0} x(n) &\leftrightarrow X[(\omega - \omega_0)] \end{aligned} \tag{6.22}$$

6.5.5 Scaling

Let the DT sequence can be scaled as *y*(*n*)=*x*(*pn*) for *p* integer. Mathematically

$$\begin{aligned} x(n) &\leftrightarrow X(\omega) \\ y(n) = x(pn) &\leftrightarrow Y(\omega) = X\left(\tfrac{\omega}{P}\right) \end{aligned} \tag{6.23}$$

6.5.6 Multiplication by *n* (Frequency Differentiation)

This property states that

$$x(n) \leftrightarrow X(\omega) \; nx(n) \leftrightarrow j\frac{d[X(\omega)]}{d\omega} \tag{6.24}$$

6.5.7 Time Reversal

This property states that if a DT signal is folded about the origin in time, its magnitude spectrum remains unchanged; however the phase spectrum undergoes a change in sign.

Mathematically

$$\begin{aligned} x(n) &\leftrightarrow X(\omega) \\ x(-n) &\leftrightarrow X[-\omega] \end{aligned} \tag{6.25}$$

Example 6.6

Find the DTFT of the following signal $x(n) = 2^n u(-n)$

Solution 6.6

$$X(\omega) = \sum_{n=-\infty}^{\infty} x(n)\, e^{-j\omega n} = \sum_{n=-\infty}^{\infty} 2^n u(-n) e^{-j\omega n}$$

$$X(\omega) = \sum_{n=-\infty}^{0} \left(\tfrac{2}{e^{j\omega}}\right)^n.\text{ Inverting the limit of the summation sign}$$

$$X(\omega) = \sum_{n=0}^{\infty} \left(\frac{e^{j\omega}}{2}\right)^n = \frac{1}{1-\frac{e^{j\omega}}{2}} = \frac{2}{2-e^{j\omega}}$$

Using Z-transform

$$z\left[u(-n)\right] = \frac{1}{1-z}$$

Using Scaling property: $a^n x(n) \overset{Z}{\longleftrightarrow} X(a^{-1}z)$

$$X(z) = \frac{1}{1-2^{-1}z} X(z) = \frac{2}{2-z}$$

Putting $z = e^{j\omega}$ for $X(\omega)$, $X(\omega) = \frac{2}{2-e^{j\omega}}$

6.5.8 Convolution

This property states that if a DT signal is folded about the origin in time, its magnitude spectrum remains unchanged; however the phase spectrum undergoes a change in sign.

Mathematically

$$\begin{aligned} x(n) &\leftrightarrow X(\omega) \\ Y(n) &\leftrightarrow y(\omega) \\ x(n) \otimes y(n) &\leftrightarrow X(\omega).Y(\omega) \end{aligned} \tag{6.26}$$

Example 6.7

Consider the following frequency response

$$H(\omega) = \frac{1}{j\omega + 3}$$

And with applied input $e^{-4t}u(t)$. Determine the output spectrum $Y(\omega)$

Solution 6.7

$$y(\omega) = H(\omega)\ X(\omega)$$

$$X(\omega) = \Im\{\mathrm{e}^{-4\mathrm{t}}u(t)\} = \frac{1}{j\omega + 4} \quad H(\omega) = \left(\frac{1}{3 + j\omega}\right)$$

$$y(\omega) = \left(\frac{1}{4 + j\omega}\right)\left(\frac{1}{3 + j\omega}\right)$$

Which can be combined as:

$$y(\omega) = \frac{1}{-\omega^2 + 7\omega + 12}$$

6.5.9 Multiplication in Time Domain

This property states that the multiplication of two time domain sequence is equivalent to the convolution of their DTFT.

Mathematically

$$x(n) \leftrightarrow X(\omega) \quad y(n) \leftrightarrow Y(\omega)$$

$$x(n) \cdot y(n) \leftrightarrow \frac{1}{2\pi}[X(\omega) \otimes Y(\omega)] \tag{6.27}$$

6.5.10 Complex Conjugation and Conjugate Symmetry

Mathematically

$$x(n) \leftrightarrow X(\omega) \quad x^*(n) \leftrightarrow (-\omega) \tag{6.28}$$

6.5.11 Parseval's Theorem

Parseval's theorem states that the total energy of a DT signal $x(n)$ may be determined by the knowledge of its DTFT.

Mathematically
If $x(n) \leftrightarrow X(\omega)$
Then according to Parseval's theorem, the energy E of a DT signal $x(n)$ is expressed as

$$E == \frac{1}{2\pi}\int_{-\pi}^{\pi} |X(\omega)|^2 \; d\omega. \tag{6.29}$$

6.5.12 Energy Density Spectrum

The energy in a DT signal $x(n)$ is given as:

$$E = \sum_{n=-\infty}^{\infty} |x(n)|^2 . \tag{6.30}$$

According to Parseval's theorem, this energy may also be expressed in term of DTFT as under:

$$E = \frac{1}{2\pi}\int_{-\pi}^{\pi} |X(\omega)|^2 \; d\omega. \tag{6.31}$$

Example 6.8

Calculate the DTFT of the signal $x(n) = \{1, -1, 1, -1\}$ at the digital frequencies of π and $\pi/2$.

Solution 6.8

$$X[\omega] = \sum_{n=0}^{N-1} x(n)e^{-j\omega n} = \sum_{n=0}^{3} x(n)e^{-j\pi n} = \sum_{n=0}^{3} x(n)$$
$$\{\cos(\pi n) - j\sin(\pi n)\}$$

$$X[\omega] = \sum_{n=0}^{3} x(n)\{\cos(\pi n)\} = (1)(1) + (1)(1) + (-1)(-1)$$
$$+ (1)(1) + (-1)(-1) = 4$$

For $\omega = \pi/2$

$$X[\omega] = \sum_{n=0}^{N-1} x(n)e^{-j\omega n} = \sum_{n=0}^{3} x(n)e^{-j\pi n/2} = \sum_{n=0}^{3} x(n)$$
$$\left\{\cos\left(\frac{\pi}{2}n\right) - j\sin\left(\frac{\pi}{2}n\right)\right\}$$

$$X[\omega]=\sum_{n=0}^{3} x(n)\left\{\cos\left(\frac{\pi}{2}n\right)+j\sin\left(\frac{\pi}{2}n\right)\right\}=(1)(1)+(-1)(j)+(1)(-1)$$
$$+(-1)(-j)=0$$

1. All practical calculations of the DTFT of a signal using
$$X[\omega]=\sum_{n=-\infty}^{\infty} x(n)e^{-j\omega n}=\sum_{n=0}^{N-1} x(n)e^{-j\omega n}$$
$\omega = 2\pi\frac{f_c}{F_s}\omega$ is the digital frequency, ω is the digital frequency, F_s is sampling frequency. f_c is the cutoff frequency, so units of ω is samples per cycle.
2. The frequency resolution of a rectangular windowed signal is given by equation
$\Delta\omega = \frac{4\pi}{N}$, where *N* is the number of samples in the signal or alternatively, the width of rectangular window.
3. To suppress the side lobes in the DTFT we must use a window function that tapers the abrupt edges of the rectangular window. With the Hamming window, the side lobes are attenuated by at least –40 dB relative to the main lobes. Other widow function attenuates the side lobes in a side lobes different degree.
4. The price paid for attenuation is a degradation of the frequency resolution. In case of Hamming window the resolution is $\Delta\omega = \frac{4\pi r}{N}$, where *r* is approximately equal to 2. It turns out that all of window functions, the rectangular window has the best resolution to some desired value using tapering windows is to increase the signal length.

Example 6.9

How many samples in a sinusoidal signal are required to achieve DTFT resolution of 0.01π rad if the signal is windowed with a Hamming window?

Solution 6.9

$$\Delta\omega=\frac{4\pi r}{N}$$
$$N=\frac{4\pi r}{\Delta\omega} \quad N=\frac{4\pi(2)}{0.01\pi}=\frac{8}{0.01}=800 \quad \text{samples.}$$

Example 6.10

For the desired resolution namely, $\Delta\omega = 0.02\pi$, what would be equivalent resolution in Hz if the signal signals were sampled at $F_s = 1000$ Hz?

Solution 6.10

Using the equation

$$\omega = 2\pi\frac{f}{F_s}$$

$$\Delta\omega = 2\pi\frac{\Delta f}{F_s} = 0.02\pi$$

$$\Delta f = 0.02\frac{\pi F_s}{2\pi} = \frac{(0.01)(1000)}{2} = 5 \text{ Hz.}$$

Example 6.11

How many samples in a sinusoidal signal are required to achieve DTFT resolution of 0.02π rad if the signal is windowed with a Hamming window?

Solution 6.11

$$\Delta\omega = \frac{4\pi r}{N}$$

$$N = \frac{4\pi r}{\Delta\omega} \quad N = \frac{4\pi(2)}{0.02\pi} = \frac{8}{0.02} = 400 \quad \text{samples}$$

Example 6.12

For the desired resolution namely, $\Delta\omega = 0.01\pi$, what would be equivalent resolution in Hz if the signal signals were sampled at $F_s = 2000$ Hz?

Solution 6.12

Using the equation

$$\omega = 2\pi\frac{f}{F_s}\Delta\omega = 2\pi\frac{\Delta f}{F_s} = 0.01\pi$$

$$\Delta f = 0.01\frac{\pi F_s}{2\pi} = \frac{(0.01)(2000)}{2} = 10 \text{ Hz}$$

6.6 Why the DFT?

The FT of a DT signal $x(n)$ is called DTFT and is denoted as $X(\omega)$. $X(\omega)$ is a continuous function of frequency ω. Therefore this type representation is not computationally convenient representation of DT signal $x(n)$. Thus taking one step further, we represent a sequence by samples of its continuous spectrum. This type of frequency domain representation of signal is known as DFT.

Let us pause for a moment and re-examine why DFT is the only possible choice that we have when we want to determine the frequency content of a signal using a digital computer. We cannot resort to the Fourier series or the FT because they apply only to CT signals.

Digital computers work with signals that are discrete in time. One moment, “Wait a minute”. What about the DTFT? It also applies to signals that are discrete in time. A very good point indeed! The difficulty in using the DTFT is that it is not practical because it applies to DT aperiodic signal which is made up of an infinite number of sines and cosines. To calculate the amplitude of each of the constituent of an aperiodic signal would thus take an incredibly long time. The only practical option is the DFT.

The primary application of DFT is to approximate the FT of signals. The other applications we consider – convolution, filtering, correlation, and energy spectral density estimation – are all based on the DFT being an approximation of the FT.

DFT is used as discrete-frequency approximation of the DTFT. The DFT is also used to calculate approximations of the FTs of analog signals.

6.6.1 Window

It is important to note that each of the transform assumes that the signal exists for infinite time in the past to infinite time in the future. Of course, we cannot measure a signal for such long time.

In practice, the measurement is made for finite duration that could be typically as small as few milliseconds to large as several hours. The duration over which the signal is measured is known as window because it is as you are looking signal through a small window that restricts your field of vision. You can think of it as using a telescope to view the vastness of the universe – even though you can see tiny part of the universe as you peer through the telescope, it nevertheless extends to infinity in all directions.

6.6.2 Orthogonal Signals

The DFT determine the frequency content of the sampled digital signal by correlating it with samples of sinusoids of different frequencies. Before we delve into the DFT, it is important to emphasize that the sines and cosines that are used by the DFT for correlation are orthogonal as shown in Figure 6.5.

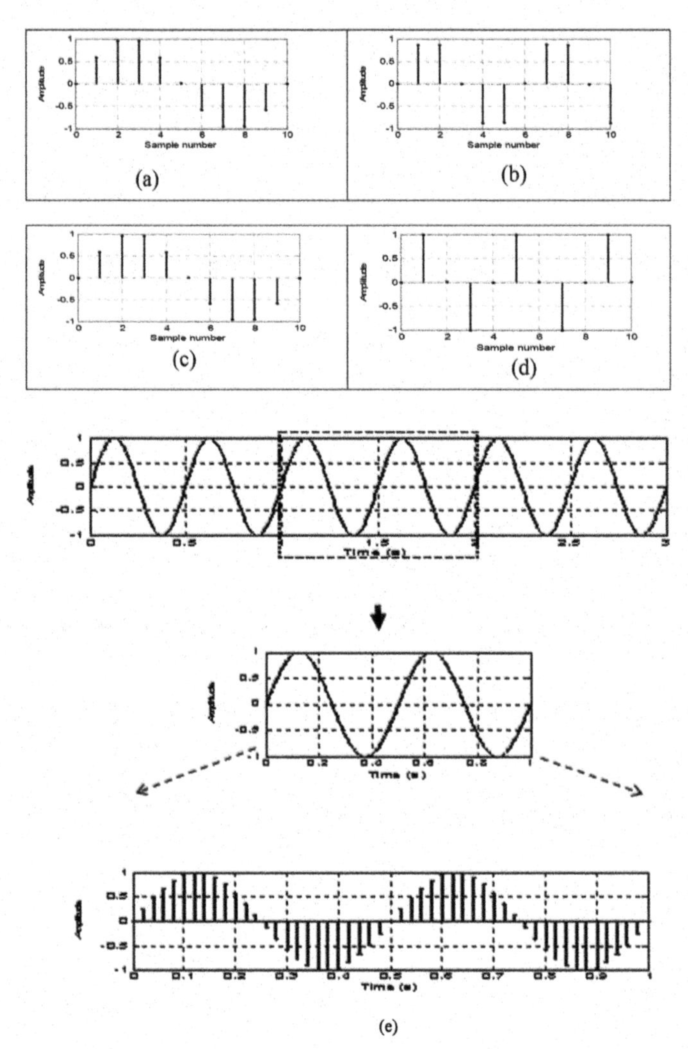

Figure 6.5 (a) Sample of one cycle of a sine wave; (b) Sample of two cycles of a sine wave; (c) Sample of one cycle of a sine wave; (d) Sample of three cycles of a sine wave and (e) Inside DFT.

What does that mean? You ask. When two signals are multiplied together and the result is zero, they are said to be orthogonal.

The sines and cosines used by DFT for correlation have this property-multiply any two of them, and the result will be zero.

6.6.3 Inside the DFT

If the sampled digital signal contains a frequency that exactly matches the frequency of the sinusoids, the correlation value gives the amplitude of the sinusoids. When there is no match, the correlation value is zero.

But how does the DFT select the frequencies of sinusoids with which sample digital signal will correlate? The answer is quite simple and is as follows:

(a) The correlation is done with sines and cosines of frequencies such that an integer number of cycles fit within window. The lowest-frequency sinusoid is that for which exactly one cycle fits with in window. This is fundamental frequency-its period is exactly equal to the time duration of the window.

(b) The other frequencies are those for which exactly 2, 3, 4,..., cycles fit with in the window. These are harmonic frequencies.

Thus, correlation is done with sinusoids having the frequencies; f_1 is the fundamental frequency whose period is exactly equal to the time duration of window. The other frequencies are harmonics of fundamental frequency.

The DFT algorithm used to determine the frequency content of the sampled signal is as follows.

1. Correlate the samples with a sine wave having a frequency such that exactly one cycle fits with in the window. The window fits for one 1-s duration. So the frequency of the sine wave used for correlation is 1 Hz. Interestingly the correlation value turns out to be zero (Figure 6.6).

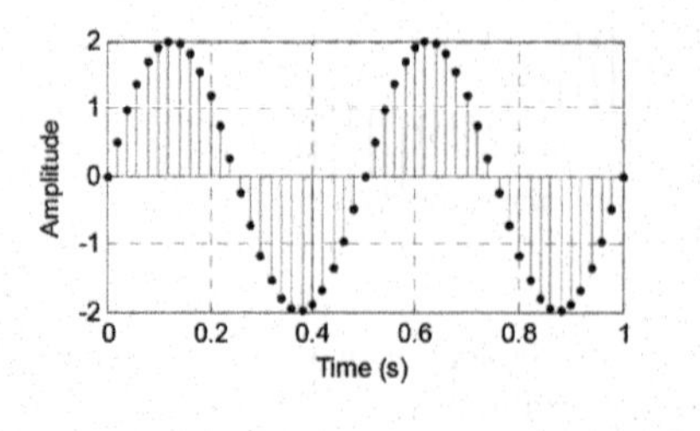

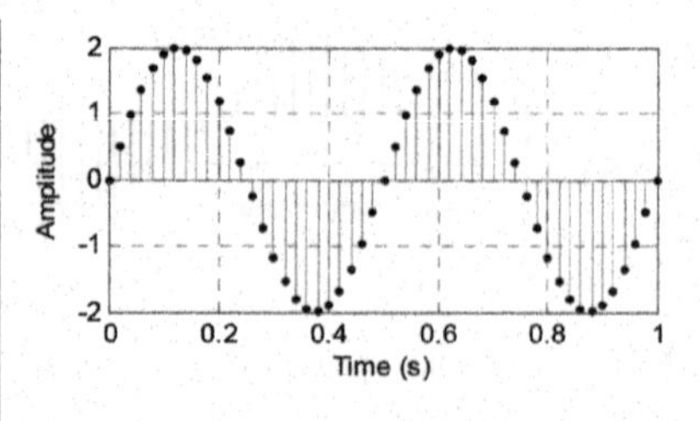

Figure 6.6 Samples of the windowed sine wave and samples of one cycle of a sine wave (f = 1 Hz) used for correlation.

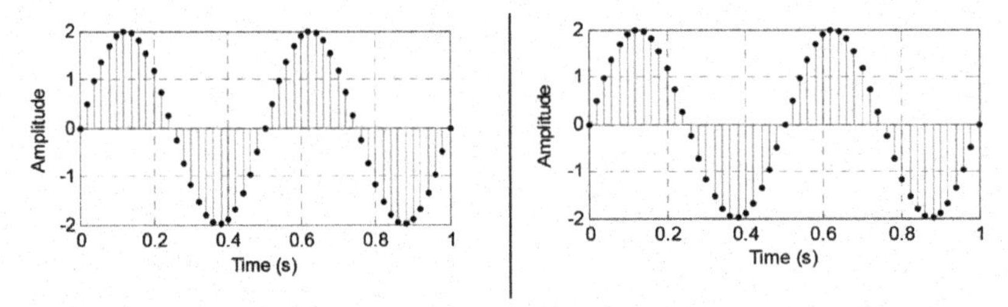

Figure 6.7 Samples of the windowed sine wave and samples of two cycles of a sine wave (f = 2 Hz) used for correlation.

2. Correlate the samples with in a sine wave having a frequency such that exactly two cycles fit with in the window. So the frequency of sine wave used for correlation is 2 Hz, and it exactly matches the frequency of the sampled windowed signal. This time the correlation value is not zero. Can you see why? When the signal of interest contains a sine wave exactly the same frequency sine wave that is used for the correlation, we get a non-zero value (Figure 6.7).
3. Correlate the samples with a sine wave having a frequency, such that exactly three cycles fit with in the window. This time the frequency of the sine wave used for correlation is 3 Hz. The correlation value is again equal to zero. In-fact , the correlation value for all other harmonics will be zero because the positive and negative product will be exactly equal in amplitude and so will cancel out (Figure 6.8).
4. Now correlate with samples of a cosine wave of frequencies 1, 2, 3,. . . . All the correlation will be zero because our signal of interest does not consists of any cosine waves (Figures 6.9–6.11).

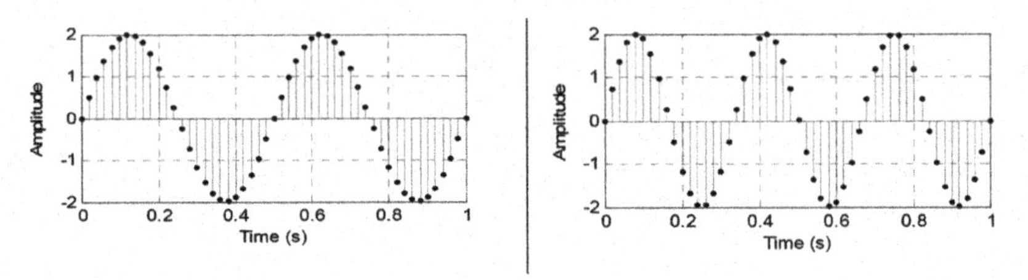

Figure 6.8 Samples of the windowed sine wave and samples of three cycles of a sine wave (f = 3 Hz) used for correlation.

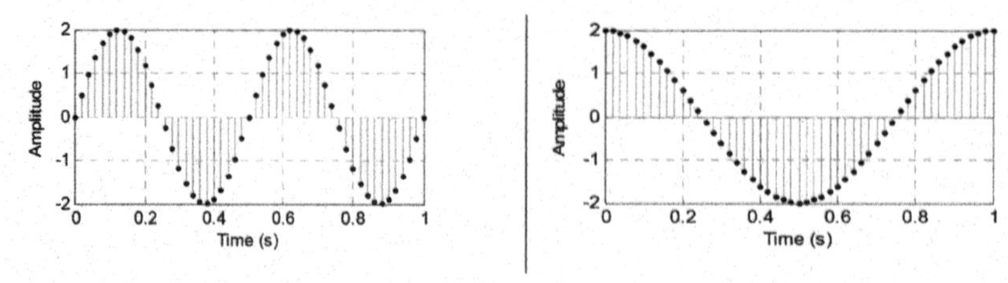

Figure 6.9 Samples of the windowed sine wave and samples of one cycle of a cosine wave ($f = 1$ Hz) used for correlation.

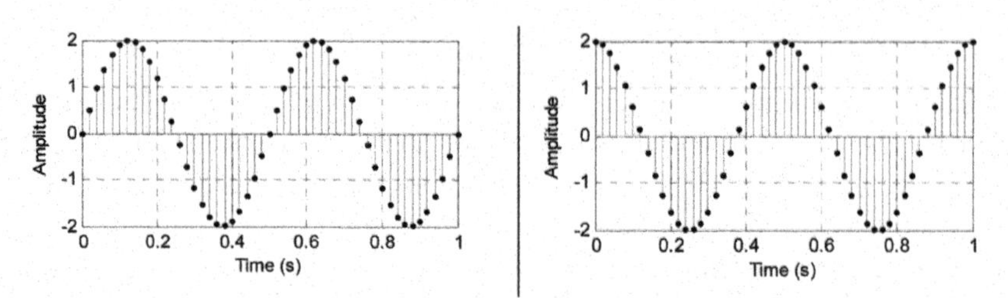

Figure 6.10 Samples of the windowed sine wave and samples of two cycles of a cosine wave ($f = 2$ Hz) used for correlation.

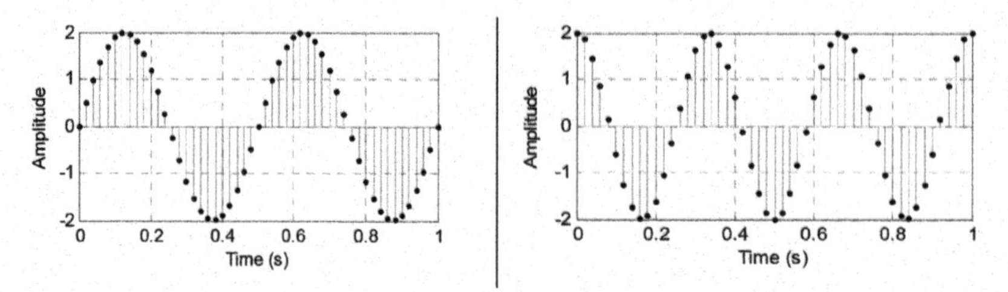

Figure 6.11 Samples of the windowed sine wave and samples of three cycles of a cosine wave ($f = 3$ Hz) used for correlation.

6.6.4 DFT Frequencies and Frequency Resolution

The DFT correlates the input signal with sinusoids of different frequencies. If a sine or cosine wave of these frequencies exists in the input signal, the correlation values is nonzero and, with appropriate scaling, provides the amplitudes of corresponding sine or cosines. On the other hand, if a sine or cosine wave of that frequency does not exist in the input signal, the correlation will be zero.

An easy mathematical formula that provides the frequency of the fundamental sine and cosine used by the DFT for correlation with fundamental frequency is

$$f_1 = \frac{f_s}{N} \tag{6.32}$$

Where f_s is sampling frequency and N is the number of samples of the time domain input signal given to the DFT.

$$\Delta f = f_1 = \frac{f_s}{N} \tag{6.33}$$

Δf is appropriately termed the frequency resolution. If the input time-domain signal has two frequencies that are at least Δf apart, then the DFT is able to separate these two frequencies. Since the frequencies of harmonics are chosen that they are integer multiples of the fundamental (i.e., Δf), the harmonic frequencies are separated by Δf.

If the input time-domain signal contains two frequencies that are less than Δf apart, then the DFT will be unable to distinguish them as separate frequency components. Equation (6.33) indicates that better frequency resolution (smaller Δf) can be obtained by either of the following:

1. Reduce the sampling frequency but keeping N the same,
2. Increase N but keeping F_s the same.

If we have frequencies of 320 and 640 Hz, and the chosen sampling rate of 2560 Hz and the number of samples = 512, the frequency resolution (and also the fundamental frequency used for correlation by DFT) is

$$\Delta f = f_1 = \frac{f_s}{N} = \frac{2560}{512} = 5 \text{ Hz}.$$

The sine wave in the input signal, 320 and 640 Hz were exact integer multiples of Δf; that is, they were exactly equal to the 64th ($64 \times 5 = 320$) and 128th ($128 \times 5 = 640$) harmonics of the fundamental frequency.

Thus DFT will show the two lines in the frequency spectrum corresponding to the correlation of the 64th and 128th harmonics with the input signal, but when using a signal that consists of a sine wave 322.5 Hz sine wave will not perfectly match any of the frequency that the DFT uses for correlation.

6.6.4.1 Spectral leakage due to correlation

Signals that consist of a 322.5 Hz sine wave will not perfectly match any of the frequency that the DFT uses for correlation, therefore, the input signal of frequency 322.5 Hz looks as it is composed of many sine waves, when in fact all it had a single frequency? This phenomenon is called spectral leakage because it is as if the energy in the single frequency of 322.5 Hz has leaked out to all other frequencies.

Sine and cosine of different frequencies are orthogonal. That means that if we correlate an integer number of cycles, we will get zero – no match. But if one of the signals involved in the correlation has a non-integer number of cycles, then the correlation value will be non-zero!

This is what happens in practice – you usually do not know the frequencies present in the signal of interest. So you measure the signal over a certain time interval and perhaps attain 1¼, 3½, or 6.1 cycles or any other non-integer number of cycle. When DFT correlates the non-integer number of cycles with its sines and cosines, the process of correlation return non-zero numbers.

In practice it is very difficult to obtain an integer number of cycles for all sinusoids in the signal of interest during the measurement interval, and hence it will be quite rare to obtain exactly zero correlation values.

6.6.4.2 Spectral leakage due to discontinuities

Spectral leakage occurs because of non-zero correlation values between signal being analyzed and the frequencies of DFT. There is another way to explain spectral leakage. This explanation will prepare us for the topic of windows.

The basic assumption behind the DFT is that the input time domain signal is periodic. Suppose we measure the signal and sampled it for duration of I s. these samples are given to DFT for calculating the frequency contents of the signal.

When using the DFT, samples come from a signal that is periodic this assumption results in the signal which is created by repeating the windowed signal over and over again. The signal is not quite the same as the original time-domain signal. These sudden changes require a large number of frequencies. It is the discontinuities, or sudden changes, that result in large number of frequencies.

These transition results in spreading of energy in the frequency domain referred as spectral leakage. Most practical signals consist of many frequencies, and there is no way in advance what these frequencies are. So we cannot determine the window duration such that we will always obtain an integer number of cycles of all constituent sinusoids. The best method of attack is to reduce the abruptness of the transitions. This is what window accomplishes.

6.7 Discrete Fourier Transform

It may be noted that the DFT is itself a sequence rather than a function of continuous variable and it corresponds to equally spaced frequency samples of DTFT of a signal. Also Fourier series representation of the periodic sequence corresponds to discrete Fourier transform of the finite length sequence. In short, we can say that DFT is used for transforming DT sequence $x(n)$ of finite length into discrete frequency sequence $X[k]$ of finite length. This means that by using DFT, the DT sequence $x(n)$ is transformed into corresponding discrete frequency sequence $X[k]$.

Assume that a wave form has been sampled at regular time intervals T to produce the sample sequence $\{x(nT)\} = x(0), x(T), \ldots..x[(N-1)T]$ of N sample values, where n is the sample number from $n = 0$ to $n = N - 1$.

The data values $x(nT)$ will be real only when representing the values of a time series such as a voltage waveform. The DFT of $x(nT)$ is then defined as the sequence of complex values $\{X[k\omega]\} = X(0), X(\varpi), \ldots..X[(N-1)\omega]$ in the frequency domain, where ω is the first harmonic frequency given by $\omega = 2\pi/NT$.

Thus the $X[k\omega]$ have real and imaginary components in general so that for the kth harmonic

$$X(k) = R(k) + jI(k) \tag{6.34}$$

and

$$|X(k)| = \left[R^2(k) + I^2(k)\right]^{1/2} \tag{6.35}$$

and $X(k)$ has the associated phase angle

$$\varphi(k) = \tan^{-1}\left[I(k)/R(k)\right], \tag{6.36}$$

where $X(k)$ is understood to represent $X(k\omega)$. These equations are therefore analogous to those for the FT.

Note that N real data values (in the time domain) transform to N complex DFT values (in the frequency domain). The DFT values, $X(k)$, are given by

$$X[k] = F_D\left[x(nT)\right] = \sum_{n=0}^{N-1} x(nT)e^{-jk\omega nT}, k = 0, 1, \ldots..N-1, \tag{6.37}$$

where $\omega = 2\pi/NT$ and F_D denotes the discrete Fourier transformation.

$$X[k] = \sum_{n=0}^{N-1} x(nT)e^{-jk2\pi \, nT/NT} \tag{6.38}$$

$$X[k] = \sum_{n=0}^{N-1} x(nT)e^{-jk2\pi \, n/N} \tag{6.39}$$

Example 6.13

Find DFT of the sequence $x(n) = \{1, 0, 0, 1\}$.

Solution 6.13

The DFT of the sequence $\{1, 0, 0, 1\}$ is evaluated, [N = 4], it is required to find the complex values $X(k)$ for k = 0, k = 1, k = 2, and k = 3

$$X[k] = \sum_{n=0}^{N-1} x(nT)e^{-jk2\pi n/N}$$

$$X[k] = \sum_{n=0}^{3} x(nT)e^{-jk2\pi n/4}$$

$$X[k] = x(0) \; + x(1)e^{-jk2\pi 1/4} + x(2)e^{-jk2\pi 2/4} + x(3)e^{-jk2\pi 3/4}. \tag{1}$$

With k = 0, Equation (1) becomes

$$X[0] = x(0) \; + x(1) + x(2) + x(3)$$

$$= 1 + 0 + 0 + 1 = 2$$

$X[0]$ = 2 is entirely real, of magnitude 2 and phase angle $\phi(0) = 0$.

With k = 1, Equation (1) becomes

$$X[1] = x(0) \; + x(1)e^{-j\pi 1/2} + x(2)e^{-j\pi} + x(3)e^{-j\pi 3/2}$$

$$= 1 + 0 + 0 + 1e^{-j\pi 3/2} = 1 + e^{-j3\pi/2}$$

$$= 1 + \cos\left(\frac{3\pi}{2}\right) - j\sin\left(\frac{3\pi}{2}\right) = 1 + j$$

Thus $X[1] = 1 + j$ and is complex with magnitude $\sqrt{2}$ and phase angle $\varphi(\Omega) = \tan^{-1} 1 = 45$.

For $k = 2$, (1) becomes

$$X[2] = x(0) + x(1)e^{-j\pi} + x(2)e^{-j2\pi} + x(3)e^{-j3\pi}$$

$$= 1 + 0 + 0 + 1e^{-j\pi 3} = 1 - 1 = 0$$

Thus $X[2] = 0$, of magnitude zero and phase angle $\phi(2) = 0$.

Finally, for $k = 3$, (1) becomes

$$X[k] = x(0) + x(1)e^{-j3\pi/2} + x(2)e^{-j3\pi} + x(3)e^{-j9\pi/2}$$

$$= 1 + 0 + 0 + e^{-j9\pi/2} = 1 - j$$

Thus $X[3] = 1 - j$, of magnitude $\sqrt{2}$ and phase angle $\varphi(3) = -45°$.

It has, therefore, been shown that the time series $\{1, 0, 0, 1\}$ has the DFT, given by the complex sequence $\{2, 1 + j, 0, 1 - j\}$.

It is common practice to represent the DFT by the plots of $|X(k)|$ versus $k\omega$ and of $\phi(k)$ versus $k\omega$. This may be done in terms of harmonics of ω, or in terms of frequency if ω is known.

To find ω it is necessary to know the value of T, the sampling interval. If it is assumed that the above data sequence had been sampled at 8 kHz then $T = 1/(8 \times 10^3) = 125\ \mu s$.

Then $\omega = 2\pi/NT = 2\pi/(4x125x10^{-6}) = 12.57K$ rad/s.

Hence $2\omega = 25.14$ and $3\omega = 37.71K$ rad/s.

Figure 6.12(a) is a plot of $x(nT)$ versus t.

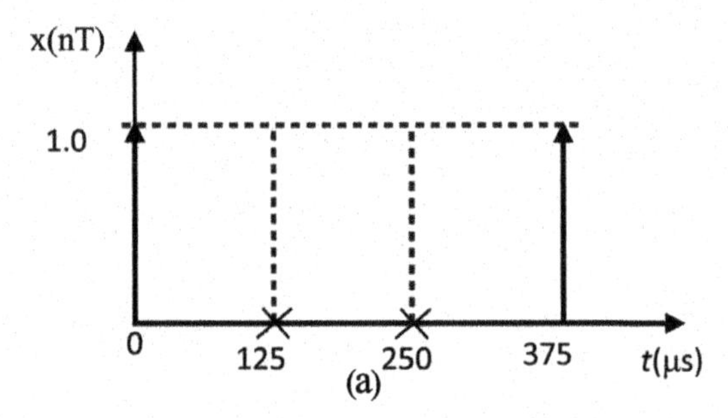

Figure 6.12 (a) $x(nT)$ versus t.

It is noteworthy that the 'amplitude' plot of Figure 6.12(b) is symmetrical about the second harmonic component that is about harmonic number N/2

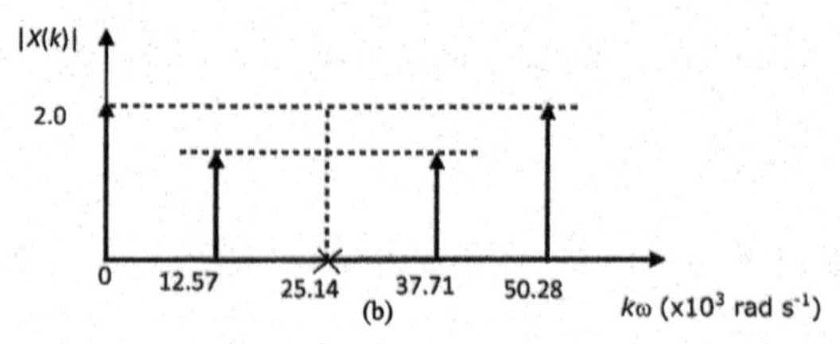

Figure 6.12 (b) $|X(k)|$ versus k.

and that in Figure 6.12(c) the phase angles are an odd function centered round this component. These results are more generally true.

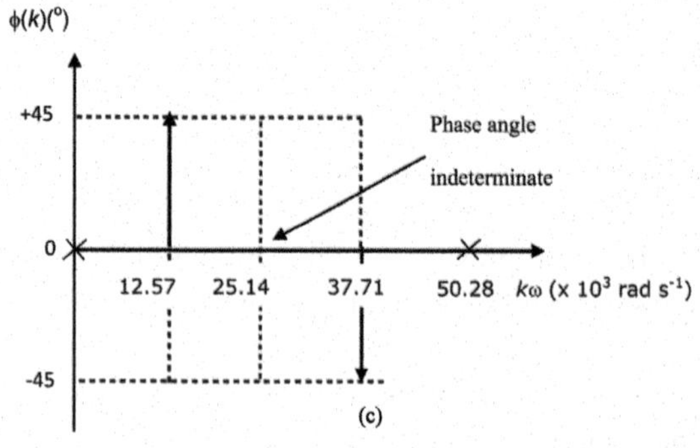

Figure 6.12 (c) $\phi(k)$.

An important property of the DFT may be deduced if the kth component of the DFT, $X[k]$, is compared with the $(k+N)$th component, $X(k+N)$.

Thus

$$X[k] = \sum_{n=0}^{N-1} x(nT)e^{-jk\omega nT}$$

$$= \sum_{n=0}^{N-1} x(nT)e^{-jk2\pi n/N}$$

$$X[k+N] = \sum_{n=0}^{N-1} x(nT)e^{-jk2\pi n/N}e^{-jN2\pi n/N}$$

$$= \sum_{n=0}^{N-1} x(nT)e^{-jk2\pi n/N}e^{-j2\pi n}$$

$$= \sum_{n=0}^{N-1} x(nT)e^{-jk2\pi n/N} = X[k]$$

Since *n* is integer, therefore $e^{-j2\pi n} = 1$.

The fact that $X[k+N] = X[k]$ shows that the DFT is periodic with period *N*. This is the cyclical property of the DFT. The values of the DFT components are repetitive. If $k = 0$, then $k + N = N$ and $X[0] = X[N]$.

In the above example $X[0) = 2$ and therefore $X[4] = 2$ also. This is illustrated in Figure 6.12(b) where the fourth harmonic amplitude is drawn at 50.28 kHz. The symmetry of the amplitude distribution about the second harmonic is obvious.

Finally the values of the FT components, the data {1, 0, 0, 1} may be obtained by multiplying the DFT components by $T = 125\ \mu$s. Therefore

$$f[0] = 250\ \mu\text{VHz}^{-1}, f[12.57\text{ kHz}] = (125 + j125)\mu\text{VHz}^{-1},$$
$$f[25.14\text{ kHz}] = 0\text{ VHz}^{-1}, f[37.71\text{ kHz}] = (125 - j125)\mu\text{VHz}^{-1}$$

6.7.1 Inverse Discrete Fourier Transform

It is also necessary to be able to carry out discrete transformation from the frequency to the time domain. This may be achieved using the IDFT, defined by the formula as in (6.40).

$$x(nT) = F_D^{-1}\ X[k] = \frac{1}{N}\sum_{k=0}^{N-1} X[k]e^{jk\,\omega T}, n = 0, 1, \ldots..N - 1 \quad (6.40)$$

$$x(n) = \frac{1}{N}\sum_{k=0}^{N-1} X[k]e^{jk2\pi\ n/N} \quad (6.41)$$

Where F_D^{-1} denotes the inverse discrete Fourier transformation.

Example 6.14

Find the inverse FT of the sequence $[2, 1 + j, 0, 1 - j]$.

Solution 6.14

It is useful to illustrate the IDFT by using it to derive the time series $\{1, 0, 0, 1\}$ from its DFT components $[2, 1 + j, 0, 1 - j]$

$$x(n) = \frac{1}{N}\sum_{k=0}^{N-1} X[k]e^{jk2\pi n/N}, \quad x(n) = \frac{1}{4}\sum_{k=0}^{3} X[k]e^{+jk2\pi n/4}$$

$$x(n) = \frac{1}{4}[X[0] + X[1]e^{+j\pi n/2} + X[2]e^{+jn\pi} + X[3]e^{+j3n\pi/2}$$

With $n = 0$,

$$x(0) = \frac{1}{4}[X[0] + x[1] + x[2] + x[3]] = \frac{1}{4}[2 + (1+j) + 0 + (1-j)] = 1$$

With $n = 1$

$$\begin{aligned} x(1) &= \frac{1}{4}[X[0] + X[1]e^{+j\pi/2} + X[2]e^{+j\pi} + X[3]e^{+j3\pi/2} \\ &= \frac{1}{4}[2 + (1+j)e^{j\pi/2} + 0 + (1-j)e^{j3\pi/2}] \\ &= \frac{1}{4}[2 + (1+j)j + (1-j)(-j)] \\ x(1) &= \frac{1}{4}(2 + j - 1 - j - 1) = 0 \end{aligned}$$

With $n = 2$,

$$\begin{aligned} x(2) &= \frac{1}{4}[X[0] + X[1]e^{+j\pi} + X[2]e^{2j\pi} + X[3]e^{+j3\pi} \\ &= \frac{1}{4}[2 + (1+j)e^{j\pi} + (1-j)e^{j3\pi}] = \frac{1}{4}[2 - (1+j) - (1-j)] = 0 \end{aligned}$$

Finally, with $n = 3$,

$$\begin{aligned} x(3) &= \frac{1}{4}[X[0] + X[1]e^{+j3\pi/2} + X[2]e^{j3\pi} + X[3]e^{+j9\pi/2} \\ &= \frac{1}{4}[2 + (1+j)e^{j3\pi/2} + (1-j)e^{j9\pi/2}] \\ &= \frac{1}{4}[2 + (1+j)(-j) + (1-j)j] = \frac{1}{4}(2 - j + 1 + j + 1) = 1 \end{aligned}$$

6.7.2 DFT: Matrix Method

The matrix method of evaluating DFT is the easiest method (for numerical calculation with paper and pencil), where a matrix W_N^{nk} is generated called the Twiddle factor matrix, which can be easily remembered.

$$X[k] = \sum_{n=0}^{N-1} x(n)W_N^{kn}, k = 0, 1, 2, \ldots .N-1 \quad X[k] = W_N^{nk} x(n) \quad (6.42)$$

$$W_N^{nk} = e\frac{-jk2n\pi}{N}$$

$$W_N = \begin{bmatrix} 1 & 1 & 1 & 1 \\ 1 & W_N^1 & W_N^2 & W_N^{N-1} \\ 1 & W_N^2 & W_N^4 & W_N^{2(N-1)} \\ 1 & W_N^{N-1} & W_N^{2(N-1)} & W_N^{(N-1)(N-1)} \end{bmatrix}$$

Example 6.15

Find DFT of the sequence $x(n) = \{1, 2, 0, 1\}$ using matrix method.

$$X[k] = W_N^{nk}\, x(n)$$

$$W_N^{nk} = \begin{bmatrix} 1 & 1 & 1 & 1 \\ 1 & W_N^1 & W_N^2 & W_N^{N-1} \\ 1 & W_N^2 & W_N^4 & W_N^{2(N-1)} \\ 1 & W_N^{N-1} & W_N^{2(N-1)} & W_N^{(N-1)(N-1)} \end{bmatrix}$$

Solution 6.15

$$\begin{bmatrix} X[0] \\ X[1] \\ X[2] \\ X[3] \end{bmatrix} = \begin{bmatrix} 1 & 1 & 1 & 1 \\ 1 & W_4^1 & W_4^2 & W_4^{(4-1)} \\ 1 & W_4^2 & W_4^4 & W_4^{2(4-1)} \\ 1 & W_4^{(4-1)} & W_4^{2(4-1)} & W_4^{(4-1)(4-1)} \end{bmatrix} \begin{bmatrix} x(0) \\ x(1) \\ x(2) \\ x(3) \end{bmatrix}$$

$$\begin{bmatrix} X[0] \\ X[1] \\ X[2] \\ X[3] \end{bmatrix} = \begin{bmatrix} 1 & 1 & 1 & 1 \\ 1 & W_4^1 & W_4^2 & W_4^3 \\ 1 & W_4^2 & W_4^4 & W_4^6 \\ 1 & W_4^3 & W_4^6 & W_4^9 \end{bmatrix} \begin{bmatrix} 1 \\ 2 \\ 0 \\ 1 \end{bmatrix}$$

$$\begin{bmatrix} X[0] \\ X[1] \\ X[2] \\ X[3] \end{bmatrix} = \begin{bmatrix} 1 & 1 & 1 & 1 \\ 1 & -j & -1 & j \\ 1 & -1 & 1 & -1 \\ 1 & j & -1 & -j \end{bmatrix} \begin{bmatrix} 1 \\ 2 \\ 0 \\ 1 \end{bmatrix} = \begin{bmatrix} 1+2+1 \\ 1-2j+j \\ 1-2-1 \\ 1+2j-j \end{bmatrix} = \begin{bmatrix} 4 \\ 1-j \\ -2 \\ 1+j \end{bmatrix}$$

$$X[k] = \{4,\ 1-j\ ,-2,\ 1+j\}$$

6.7.3 IDFT: Matrix Method

The matrix method of evaluating IDFT is the easiest method (for numerical calculation with paper and pencil), where a matrix W_N^{-nk} is generated called the Inverse Twiddle factor matrix, which can be easily remembered. The traditional formula is converted into matrix methods as below

$$x(n) = \frac{1}{N}\sum_{k=0}^{N-1} X[k]e^{jk2\pi n/N} \quad x(n) = \frac{1}{N}W_N^{-1}X[k] \tag{6.43}$$

$$W_N^{-1} = \begin{bmatrix} 1 & 1 & 1 & 1 \\ 1 & W_N^{-1} & W_N^{-2} & W_N^{-(N-1)} \\ 1 & W_N^{-2} & W_N^{-4} & W_N^{-2(N-1)} \\ 1 & W_N^{-(N-1)} & W_N^{-2(N-1)} & W_N^{-(N-1)(N-1)} \end{bmatrix} \tag{6.44}$$

Example 6.16

Find IDFT of the sequence $X[k] = \{4, 1-j,\ -2, 1+j\}$ using matrix method.
IDFT can be calculated as

$$x(n) = \frac{1}{N}W_N^{-1}\ X[k]$$

Solution 6.16

$$\begin{bmatrix} x(0) \\ x(1) \\ x(2) \\ x(3) \end{bmatrix} = \frac{1}{N}\begin{bmatrix} 1 & 1 & 1 & 1 \\ 1 & W_4^{-1} & W_4^{-2} & W_4^{-(4-1)} \\ 1 & W_4^{-2} & W_4^{-4} & W_4^{-2(4-1)} \\ 1 & W_4^{-(4-1)} & W_4^{-2(4-1)} & W_4^{-(4-1)(4-1)} \end{bmatrix}\begin{bmatrix} X(0) \\ X(1) \\ X(2) \\ X(3) \end{bmatrix}$$

$$\begin{bmatrix} x(0) \\ x(1) \\ x(2) \\ x(3) \end{bmatrix} = \frac{1}{4}\begin{bmatrix} 1 & 1 & 1 & 1 \\ 1 & W_4^{-1} & W_4^{-2} & W_4^{-3} \\ 1 & W_4^{-2} & W_4^{-4} & W_4^{-6} \\ 1 & W_4^{-3} & W_4^{-6} & W_4^{-9} \end{bmatrix}\begin{bmatrix} 4 \\ 1-j \\ -2 \\ 1+j \end{bmatrix}$$

$$\begin{bmatrix} x(0) \\ x(1) \\ x(2) \\ x(3) \end{bmatrix} = \frac{1}{N} \begin{bmatrix} 1 & 1 & 1 & 1 \\ 1 & W_4^{-1} & W_4^{-2} & W_4^{-3} \\ 1 & W_4^{-2} & W_4^{0} & W_4^{-2} \\ 1 & W_4^{-3} & W_4^{-2} & W_4^{-1} \end{bmatrix} \begin{bmatrix} 4 \\ 1-j \\ -2 \\ 1+j \end{bmatrix}$$

$$\begin{bmatrix} x(0) \\ x(1) \\ x(2) \\ x(3) \end{bmatrix} = \frac{1}{4} \begin{bmatrix} 1 & 1 & 1 & 1 \\ 1 & j & -1 & -j \\ 1 & -1 & 1 & -1 \\ 1 & -j & -1 & j \end{bmatrix} \begin{bmatrix} 4 \\ 1-j \\ -2 \\ 1+j \end{bmatrix}$$

$$x(n) = \{1,\ 2,\ 0,\ 1\}$$

6.8 Properties of the DFT

The properties of DFT are quite useful in the practical techniques for processing signals, which can be used to simplify problems or which lead to useful applications. The data sequences $x(nT)$ are written as $x(n)$. The properties of DFT can be listed as under:

1. Periodicity
2. Linearity
3. Time reversal
4. Circular time shift
5. Circular frequency shift
6. Complex conjugate property
7. Circular convolution
8. Circular correlation
9. Multiplication of two sequences
10. Parseval's theorem

6.8.1 Periodicity

This property states that if a DT signal is periodic then its DFT will also be periodic. Also, if a signal or sequence repeats its wave form after N number of samples then it is called a periodic signal or sequence and N is called the period of the signal.

Mathematically

$$\begin{aligned} x(n+N) &= x(n) \quad \text{for all values of } n \\ x(k+N) &= x(k) \quad \text{for all values of } k \end{aligned} \tag{6.45}$$

6.8.2 Linearity

According to this property, the DFT is linear.

$$\begin{aligned} x_1(n) &\leftrightarrow X_1(k) \\ x_2(n) &\leftrightarrow X_2(k) \end{aligned} \tag{6.46}$$

then according to this property $ax_1(n) + bx_2(n) = aX1(k) + bX_2(k)$

6.8.3 Time Reversal

This property states that

$$\begin{aligned} &\text{If } x(n) \ \leftrightarrow X(k) \\ &x[(-n), (\text{mod } N)] = x(N-n) \leftrightarrow [X(K), (\text{mod } N)] = X(N-k). \end{aligned} \tag{6.47}$$

Thus when the *N* point sequence is time reversed, it is equivalent to reversing the DFT values.

6.8.4 Circular Time Shift

This property states that

$$\begin{aligned} &\text{If } x(n) \leftrightarrow X(k) \\ &x[(n-l), (\text{mod } N)] \leftrightarrow X(K)\, e^{-j2k\pi l/N}. \end{aligned} \tag{6.48}$$

This means the shifting of the sequence by l units in time domain is equivalent to multiplication of $e^{-jk\pi l/N}$ in the frequency domain.

6.8.5 Circular Frequency Shift

This property states that

$$\begin{aligned} &\text{If } x(n) \leftrightarrow X(k) \\ &x(n)\, e^{-j2k\pi l/N} \leftrightarrow X((k-l), (\text{mod } N)) \end{aligned} \tag{6.49}$$

This means when the sequence $x(n)$ is multiplied by the complex exponential sequence $e^{2jk\pi l/N}$, it is equivalent to circular shift of the DFT by l units in the frequency domain.

6.8.6 Circular Convolution

Circular convolution property states that

$$\begin{aligned} x(n) &\leftrightarrow X(k) \quad y(n) \leftrightarrow Y(k) \\ z(n) &= x(n) \otimes y(n) \leftrightarrow X(k).Y(k) \end{aligned} \tag{6.50}$$

where $z(n) = \sum_{m=0}^{N-1} x(m)y(n-m), (\text{mod } N))$

6.8.7 Circular Correlation

This property states that for complex valued sequences $x(n)$ and $y(n)$

$$\begin{aligned} x(n) &\leftrightarrow X(k) \quad y(n) \leftrightarrow Y(k) \\ r_{XY}(j) &= R_{XY}(k) \leftrightarrow X(k)Y^*(k) \end{aligned} \tag{6.51}$$

where $r_{XY}(j) = \sum_{n=0}^{N-1} x(n)y^*(n-j), (\text{mod } N))$

6.8.8 Multiplication of Two Sequences

This property mathematically states

$$\begin{aligned} x(n) &\leftrightarrow X(k) \quad y(n) \leftrightarrow Y(k) \\ x(n).y(n) &\leftrightarrow \tfrac{1}{N}[X(k).Y(k)] \end{aligned} \tag{6.52}$$

6.8.9 Even Functions

If $x(n)$ is an even function $x_e(n)$, that is $x_e(n) = x_e(-n)$ then

$$F_D\left[x_e(n)\right] = X_e(k) = \sum_{n=0}^{N-1} x_e(n)\cos(k\omega nT) \tag{6.53}$$

6.8.10 Odd Functions

If $x(n)$ is an odd function $x_0(n)$, that is $x_0(n) = -x_0(-n)$ then

$$F_D\left[x_0(n)\right] = X_0(k) = -j\sum_{n=0}^{N-1} x_0(n)\sin(k\omega nT) \tag{6.54}$$

6.8.11 Parseval's Theorem

This theorem states that for complex valued sequence $x(n)$ and $y(n)$

$$x(n) \leftrightarrow X(k) \quad y(n) \leftrightarrow Y(k) \quad x(n) \cdot y(n) \leftrightarrow \frac{1}{N}[X(k).Y(k)]$$

$$\sum_{n=0}^{N-1} x(n)y^*(n) = \frac{1}{N} \sum_{n=0}^{N-1} X(K)Y^*(k) \tag{6.55}$$

If $y(n) = x(n)$, then the above equation reduces to

$$\sum_{n=0}^{N-1} |x(n)|^2 = \frac{1}{N} \sum_{k=0}^{N-1} |X(k)|^2. \tag{6.56}$$

This expression relates the energy in the finite duration sequence $x(n)$ to power in frequency component $X(k)$.

6.9 Comparison between DTFT and DFT

We have seen that DFT is a sampled version of DTFT, where the frequency term ω is also sampled. A physical feeling of the similarity and the difference between the two can be obtained by considering an actual numerical example.

Example 6.17

Compute the values of DTFT and DFT, given

$X(n) = \{0, 1, 2, 3\}$

Solution 6.17

The DTFT of this is given by

$$X(\omega) = \sum_{n=-\infty}^{\infty} x(n)e^{-j\omega n} \quad \text{and} \quad X(\omega) = \sum_{n=0}^{3} x(n)e^{-j\omega n}$$

$$X(\omega) = 0 + e^{-j\omega} + 2\,e^{-2j\omega} + 3\,e^{-3j\omega} \tag{1}$$

$$X(\omega) = 0 + (\cos\,\omega - j\,\sin\,\omega) + 2(\cos\,2\omega - j\,\sin\,2\omega) + 3(\cos\,3\omega - j\,\sin\,3\omega) \tag{2}$$

$X(\omega)$ may be separated into real and imaginary part as:

$$X(\omega) = (\cos\ \omega + 2\cos 2\omega + 3\cos 3\omega) - j(\sin\omega + 2\sin\ 2\omega + 3\sin\ 3\omega) \quad (3)$$

$$X(\omega) = a - jb = M\angle\theta \quad (4)$$

First for our ease a and b are computed, the sum $a - jb$ is calculated, the complex summation yield the $M\angle\theta$. In the following table rows 1, 3, 6, 8, and 11 are in bold faced letters against which the corresponding values of DFT and DTFT exactly coincide with each other according to relation that $\omega = \frac{2\pi k}{N}$.

DTFT				DFT				
Trial	ω (rad/s)	M	$\angle\theta$ (deg.)	Trial	K	$X[k]$	Mod $X[K]$	$\angle\theta$ (deg.)
1	**0**	**6**	**0**	**1**	**0**	**6**	**6**	**0**
2	1	4.48	136.63					
3	**$\pi/2$**	**2.82**	**135**	**2**	**1**	**$-2 + j2$**	**2.84**	**135**
4	2	1.84	–51.4					
5	3	1.99	156.5					
6	**π**	**2**	**180**	**3**	**2**	**–2**	**2**	**180**
7	4	1.64	–13.7					
8	**$3\pi/2$**	**2.84**	**–135**	**4**	**3**	**$-2 - j2$**	**2.84**	**–135**
9	5	3.68	–178.6					
10	6	6.86	–37.87					
11	**2π**	**6**	**0**	**5**	**4**	**6**	**6**	**0**

For example, we find that

$$X[k = 0] = X(\omega = 0) \quad (5)$$
$$X[k = 1] = X(\omega = \pi/2) \quad (6)$$
$$X[k = 2] = X(\omega = \pi) \quad (7)$$
$$X[k = 3] = X(\omega = 3\pi/2) \quad (8)$$

The result in Equations (5)–(8) are not surprising because DFT is a special case of DTFT, where the frequency ω is sampled frequency of $(2\pi k/N)$, where $k = 0, 1, 2, 3, \ldots, N - 1$. Hence the DTFT and DFT must be exactly the same at sampling intervals.

$$\omega = \frac{2\pi}{N}$$

In the current example, we have N = 4, therefore, the sampling frequency interval is

$$\omega = \frac{2\pi}{4} = \frac{\pi}{2}$$

this means that the values of DTFT and DFT must be exactly the same $\omega = \frac{\pi}{2}$, $2\frac{\pi}{2}$, $3\frac{\pi}{2}$, and $4\frac{\pi}{2}$ and so on. And by inspecting rows 1, 3, 6, 8, and 11 in the above Table we find that this statement is complete true. The above example tells us very important facts about DTFT and DFT:

- DFT is sampled version of DTFT, i.e., DFT is derived by sampling DTFT. But, we know that DTFT is obtained by using the sampled from of the input signal $x(t)$ So, we find that DFT is obtained by the double sampling of $x(t)$ and it gives lesser number of frequency components that DTFT.
- DFT gives only positive frequency values, whereas DTFT can give both positive and negative frequency values. DTFT and DFT coincide at intervals $\omega = \frac{2\pi k}{N}$, where $k = 0, 1, \ldots, N - 1$.
- To get more accurate values of DFT, number of samples N must be very high. Where N is very high, the required computation time will also be very high.

Example 6.18

Compute the values of DTFT and DFT, given

$$X(n) = \{1,\ 2,\ 4,\ 6\}$$

Solution 6.18

The DTFT of this is given by

$$X(\omega) = \sum_{n=-\infty}^{\infty} x(n)e^{-j\omega n} \tag{1}$$

$$X(\omega) = 1\,e^{-j0} + 2\,e^{-j\omega} + 4\,e^{-2j\omega} + 6\,e^{-3j\omega}$$

$$X(\omega) = 1 + 2(\cos\ \omega - j\ \sin\ \omega) + 4(\cos\ 2\omega - j\ \sin\ 2\omega) + 6(\cos\ 3\omega - j\ \sin\ 3\omega) \tag{2}$$

Equation (2) may be separated into real and imaginary part as:

$$X(\omega) = (1 + 2\cos\ \omega + 4\cos 2\omega + 6\cos 3\omega) - j(2\sin\omega + 4\ \sin\ 2\omega + 6\sin\ 3\omega) \quad (3)$$

$$X(\omega) = a - jb \quad (4)$$

After a and b are computed, the complex sum $a - jb$ is calculated, the complex summation yield the magnitude and phase angle θ. In the following table rows 1, 3, 6, 8, and 11 are in bold faced letters.

Now consider the DFT of the same sequence $x(n)$, the value of $X[k]$ is compared for the rows 1, 3, 6, 8, and 11. We find that corresponding value of DFT and DTFT exactly coincide with each other.

DTFT				DFT				
Trial	ω (rad/s)	$a - jb$	$M\angle\theta$ (deg.)	Trial	K	$X[k]$	$\lvert X[k]\rvert$	$M\angle\theta$ (deg.)
1	**0**	**13**	$13\angle 0°$	**1**	**0**	**13**	**13**	**0**
2	1	$-5 - j6.1$	$8.2\angle{-31.8°}$					
3	**π/2**	**$-3 + j4$**	$5\angle 126°$	**2**	**1**	**$-3 + j4$**	**5**	**126.8**
4	2	$3.3 + j2.8$	$4.3\angle 0°$					
5	3	$-2.6 - j1.6$	$3.06\angle{-147.8°}$					
6	**π**	**-3**	$3\angle{-180°}$	**3**	**2**	**-3**	**3**	**-180**
7	4	$4.1 + 0.7j$	$4.24\angle 10.5°$					
8	**3π/2**	**$-3 - j4$**	$5\angle{-126.8°}$	**4**	**3**	**$-3 - j4$**	**5**	**-126.8**
9	5	$-6.3-j0.19$	$6.34\angle 178°$					
10	6	$10.2 + 7.2j$	$12.5\angle 36.1°$					
11	**2π**	13	**0**	**5**	**0**	**13**	**13**	**0**

For example, we find that

$$X[k = 0] = X(\omega = 0) \quad (5)$$

$$X[k = 1] = X(\omega = \pi/2) \quad (6)$$

$$X[k = 2] = X(\omega = \pi) \quad (7)$$

$$X[k = 3] = X(\omega = 3\pi/2) \quad (8)$$

The result in Equations (5)–(8) are not surprising because DFT is a special case of DTFT, where the frequency ω is sampled frequency of $(2\pi k/N)$, where

$k = 0, 1, 2, 3, \ldots, N - 1$. Hence the DTFT and DFT must be exactly the same at sampling intervals.

$$\omega = \frac{2\pi}{N}$$

In the above example, we have $N = 4$, therefore, the sampling frequency interval is

$$\omega = \frac{2\pi}{4} = \frac{\pi}{2}$$

this means that the values of DTFT and DFT must be exactly the same $\omega = \frac{\pi}{2}$, $2\frac{\pi}{2}$, $3\frac{\pi}{2}$, $4\frac{\pi}{2}$ and so on. And by inspecting rows 1, 3, 6, 8, and 11 in the above table we find that this statement is complete true.

6.10 Fast Fourier Transform

We can define Fast Fourier Transform (FFT) as the DFT but its algorithm is written in such a way that number of addition and multiplication becomes very less and the operation becomes fast, therefore it is called FFT, it is a preferred method of computing the frequency content of a signal. Any number of samples, N, can be given to DFT algorithm; generally a power of 2 is chosen such as $256(2^8)$, $512(2^9)$, for calculation. The fundamental principle behind the FFT algorithms is to break down DFT of N samples into successfully smaller DFTs.

The DFT is given by

$$X[k] = \sum_{n=-\infty}^{\infty} x(n)e^{-j\frac{2\pi kn}{N}} \qquad k = 0, 1, 2, \ldots\ldots, N - 1 \qquad (6.57)$$

When the exponential $e^{-j\frac{2\pi kn}{N}}$ is replaced here with the twiddle factor W_N^{kn}. It is replaced due to mathematics easiness.

The discrete transform pair (FT & inverseFT) is given by

In twiddle factor form.

$$X[k] = \sum_{n=0}^{N-1} x(n)W_N^{kn}, k = 0, 1, 2, \ldots\ldots, N - 1$$

$$x(n) = \frac{1}{N}\sum_{k=0}^{N-1} X[k]W_N^{-kn}, \qquad (6.58)$$

This section has been focused on an algorithm which is used to compute the DFT more efficiently. The collections of efficient algorithms that are generally used to compute the DFT is known as the FFT.

6.10.1 Decomposition-in-Time (DIT) FFT Algorithm

In butterfly diagram the left hand side is time and the right hand side is frequency. In decimation in time, because the decomposition is being done in time, the left hand side which is denoted for time is to be written as in even and odd sequence, while right hand side is frequency, it is written in the same sequence i.e. no order is changed. An efficient algorithm for computing the DFT is developed for cases in which the number of samples to be computed is a power of 2 ($N = 2^m$). Where, m is called the stage of the butterfly diagram.

6.10.1.1 Two-point FFT

For drawing two point butterfly diagram a power of 2 ($N = 2^m$). Where m is called the stage of the butterfly diagram, it means in drawing two point butterfly 2^1. Only one stage is required. We try to generalize the process. The result of the effort is known as the DIT, radix-2 FFT.

$$X[k] = \sum_{n=0}^{N-1} x(n)W_N^{nk} \tag{6.59}$$

For calculating a two-point DFT, it does not need to be decimated in its even and odd components

$$X[k] = \sum_{n=0}^{1} x(n)W_2^{nk} \tag{6.60}$$

$$X[k] = x(0)W_2^{0k} + x(1)W_2^{1k}$$

$$X[0] = x(0)W_2^{0} + x(1)W_2^{(1)(0)}$$

$$X[1] = x(0)W_2^{0} + x(1)W_2^{1}$$

Because $W_2^{0k} = e^{-j0} = 1$ and $W_2^{1k} = e^{-j\pi k} = (-1)^k$

The following can be

$$\begin{aligned} X[0] &= x(0) + x(1) \\ X[1] &= x(0) - x(1) \end{aligned} \tag{6.61}$$

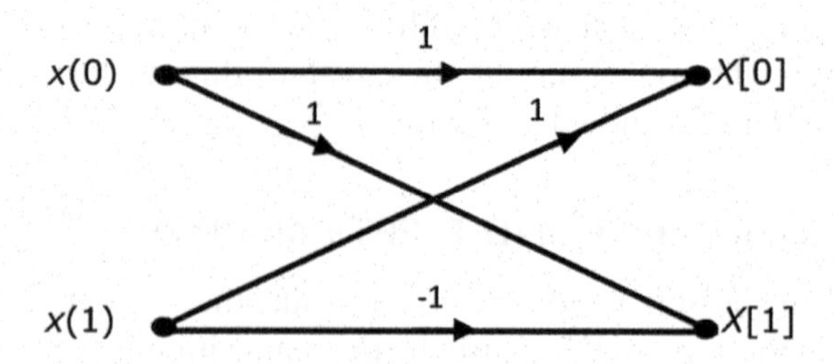

Figure 6.13 Butterfly diagram for a 2-point DFT.

The signal flow graph of Figure 6.13 illustrates the process for computing the two-point DFT. This signal flow graph is known as a butterfly diagram because of its shape.

6.10.1.2 Four-point FFT

The four point DFT is carried out from the generalized formula as

$$X[k] = \sum_{n=0}^{N-1} x(n)W_N^{nk} \tag{6.62}$$

It is decimated into $N/2$ point even and odd components, we use the concept that DFT is periodic.

$$X[k] = \sum_{n=0}^{N/2-1} x(2n)W_N^{2nk} + \sum_{n=0}^{N/2-1} x(2n+1)W_N^{(2n+1)k} \tag{6.63}$$

$$X[k] = \sum_{n=0}^{N/2-1} x(2n)W_{N/2}^{nk} + \quad W_N^k \sum_{n=0}^{N/2-1} x(2n+1)W_{N/2}^{nk} \tag{6.64}$$

$$X[k] = G(k) + W_N^k\ H(k) \tag{6.65}$$

Substituting the value of k = 0–3 in Equation (6.65)

$$X[0] = G[0] + W_4^0\ H[0]$$
$$X[1] = G[1] + W_4^1\ H[1]$$
$$X[2] = G[2] + W_4^2\ H[2] = G[0] + W_4^2\ H[0]$$
$$X[3] = G[3] + W_4^3\ H[3] = G[1] + W_4^3\ H[1]$$

The Equation (6.65) can be written for $N/2$ even components of four-point DFT

$$G[k] = \sum_{n=0}^{N/2-1} x(2n)W_{N/2}^{nk} = \sum_{n=0}^{1} x(2n)W_2^{nk} \tag{6.66}$$

$$G[k] = x(0)W_2^{0k} + x(2)W_2^{1k}$$
$$G[0] = x(0)W_2^0 + x(2)W_2^0$$
$$G[1] = x(0)W_2^0 + x(2)W_2^1$$

The Equation (6.67) can be written for $N/2$ odd components of four-point DFT

$$H[k] = \sum_{n=0}^{N/2-1} x(2n+1)W_{N/2}^{nk} = \sum_{n=0}^{1} x(2n+1)W_2^{nk} \tag{6.67}$$

$$H[k] = x(1)W_2^{0k} + x(3)W_2^{1k}$$
$$H[0] = x(1)W_2^0 + x(3)W_2^0$$
$$H[1] = x(1)W_2^0 + x(3)W_2^1$$

It is to be noted that four-point DFT can be computed by the generation of two two-point DFTs followed by a re-composition of terms as shown in the signal flow graph of Figure 6.14.

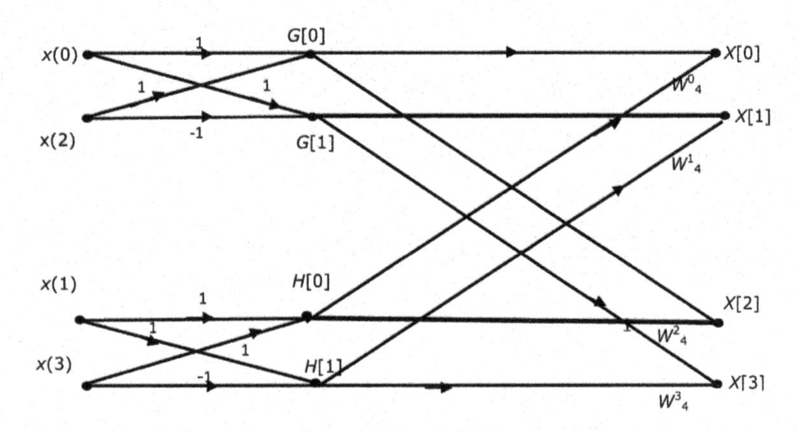

Figure 6.14 Signal flow graph for a four-point DFT.

6.10.1.3 Eight-point FFT

Eight Point DFT needs 2, 2, 2, 2, 4, 4, 8 equations to be developed. The eight point DFT is proceeded as

$$X[k] = \sum_{n=0}^{N-1} x(n) W_N^{nk}.$$

For calculating eight point DFT, it is decimated into $N/2$, i.e., two four-point DFT, the main expression of the DFT is written in even and odd components.

$$X[k] = \sum_{n=0}^{N/2-1} x(2n) W_{N/2}^{nk} + W_N^k \sum_{n=0}^{N/2-1} x(2n+1) W_{N/2}^{nk} \tag{6.68}$$

$$X[k] = G[k] + W_N^k \; H[k] \tag{6.69}$$

where $G[k] = \sum_{n=0}^{N/2-1} x(2n) W_{N/2}^{nk}$.

Even components can be further decimated into $N/4$, i.e., two two-point DFT, To avoid any further confusion in derivation the variable name have been changed, even in g and in odd in *h*.

$$\sum_{n=0}^{N/2-1} x(2n) W_{N/2}^{nk} = \sum_{m=0}^{N/4-1} g(2m) W_{N/2}^{2mk} + \sum_{m=0}^{N/4-1} g(2m+1) W_{N/2}^{(2m+1)k}$$

$$\sum_{n=0}^{N/2-1} x(2n) W_{N/2}^{nk} = \sum_{m=0}^{N/4-1} g(2m) W_{N/4}^{mk} + W_{N/2}^k \sum_{m=0}^{N/4-1} g(2m+1) W_{N/4}^{mk}$$

Even sequence is further decimated into $N/4$, i.e., two four-point DFT even components.

$$G[k] = A[k] + W_{N/2}^k \; B[k] \tag{6.70}$$

Equation (6.70) can be expressed as

$$\begin{aligned} G[k] &= A(k) + W_{N/2}^k \; B(k) \\ G[0] &= A(0) + W_4^0 \; B(0) \\ G[1] &= A(1) + W_4^1 \; B(1) \\ G[2] &= A(2) + W_4^2 \; B(2) \\ G[3] &= A(3) + W_4^3 \; B(3). \end{aligned} \tag{6.71}$$

Equation (6.71) gives the four equation of drawing the butterfly diagram, where $A[k]$ can be expressed as:

$$A[k] = \sum_{m=0}^{N/4-1} g(2m) W_{N/4}^{mk}$$
$$A[k] = \sum_{m=0}^{2-1} g[2m] W_2^{mk} = g(0) W_2^0 + g(2) W_2^k \tag{6.72}$$
$$A[0] = g(0) + g(2)$$
$$A[1] = g(0) + g(2) W_2^1 = g(0) - g(2).$$

Equation (6.71) gives the two equation of drawing the butterfly diagram, where $B[k]$ can be expressed as:

$$B[k] = \sum_{m=0}^{N/4-1} g(2m+1) W_{N/4}^{mk}$$
$$B[k] = \sum_{m=0}^{2-1} g(2m+1) W_2^{mk} = g(1) W_2^0 + g(3) W_2^k$$
$$B[0] = g(1) + g(3)$$
$$B[1] = g(1) + g(3) W_2^1 = g(0) - g(3). \tag{6.73}$$

Odd components can be further decimated into $N/4$, i.e., two two-point DFT

$$H[k] = \sum_{m=0}^{N/4-1} h[2m] W_{N/4}^{mk} + W_{N/2}^k \sum_{M=0}^{N/4-1} h[2m+1] W_{N/4}^{mk}$$
$$H[k] = C[k] + W_{N/2}^k\, D[k] \tag{6.74}$$
$$H[k] = C(k) + W_{N/2}^k\, D(k) \tag{6.75}$$
$$H[0] = C(0) + W_4^0\, D(0) \quad H[1] = C(1) + W_4^1\, D(1)$$
$$H[2] = C(2) + W_4^2\, D(2) \quad H[3] = C(3) + W_4^3\, D(3)$$
$$C[k] = \sum_{m=0}^{N/4-1} h(2m) W_{N/4}^{mk} \quad D[k] = \sum_{m=0}^{N/4-1} h(2m+1) W_{N/4}^{mk}$$

$$C[k] = \sum_{m=0}^{2-1} h(2m)W_2^{mk} = h(0)W_2^0 + h(2)W_2^k \tag{6.76}$$

$$C[0] = h(0) + h(2) \quad C[1] = h(0) + h(2)W_2^1 = h(0) - h(2)$$

$$D[k] = \sum_{m=0}^{2-1} h(2m+1)W_2^{mk} = h(1)W_2^0 + h(3)W_2^k$$

$$D[0] = h(1) + h(3) \quad D[1] = h(1) + h(3)W_2^1 = h(1) - h(3) \tag{6.77}$$

Equation (6.74) can be expressed as

$$H[k] = \sum_{n=0}^{N/2-1} x[2n+1]W_{N/2}^{nk} \tag{6.78}$$

$$H[k] = \sum_{m=0}^{N/4-1} h[2m]W_{N/4}^{mk} + W_{N/2}^{k} \sum_{M=0}^{N/4-1} h[2m+1]W_{N/4}^{mk}. \tag{6.79}$$

It is worth wile to understand that only one value has to be calculated, rest are the calculated from squaring or cubing the equation or by multiplying two calculated values of twiddle factor. The weighting factors for the eight-point DFT are

$$\begin{aligned}
W_8^0 &= 1, W_8^1 = e^{-j(\pi/4)}, W_8^{2k} = e^{-j(\pi/4)2k} = e^{-j(\pi/2)k} = W_4^{1k} \\
W_8^3 &= e^{-j(\pi/4)3} = \left[e^{-j(\pi/4)2}\right] e^{-j(\pi/4)} = W_8^1 W_4^1 W_8^4 = e^{-j(\pi/4)4} \\
&= e^{-j\pi} = W_4^2; W_8^6 = e^{-j(\pi/4)6} = W_4^3 \; W_8^5 = e^{-j(\pi/4)5} \\
&= e^{-j(\pi/4)4}e^{-j(\pi/4)} = W_8^1 W_4^2, \\
W_8^7 &== W_8^1 W_4^3 = e^{-j(\pi/4)7} = e^{-j(\pi/4)}e^{-j(\pi/4)6} = W_8^1 W_4^3.
\end{aligned}$$

The factors in brackets as the four-point DFTs is recognized as $g(n)$ and $h(n)$ respectively. Therefore, it is seen that the eight-point FFT is found by the re-composition of two, four-point FFTs.

Figure 6.15 illustrates the procedure for computing the eight-point FFT.

In general, the N-point, radix-2 FFT is computed by the re-composition of two (N/2)-point FFTs. The generalized procedure is illustrated in Figure 6.15.

The reason for deriving the FFT algorithm is for computational efficiency in calculating the DFT. Table 6.3 shows the computational cost of DFT and

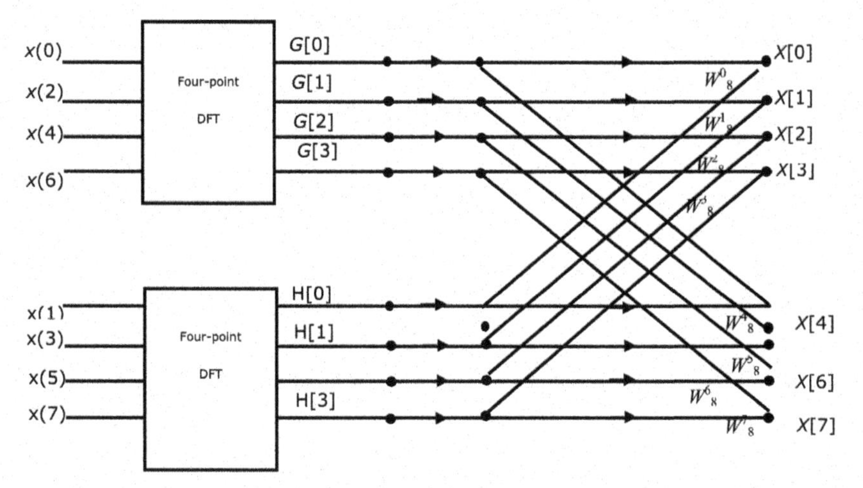

Figure 6.15 Decomposition-in-time fast Fourier transforms.

Table 6.3 Computational cost

	Computational Cost of the DFT and FFT	
	N-point DFT	N-point FFT
Algorithm	Solution of N equations in N unknown	$N/2$ butterflies/stage, For m stages, total butterflies = $N_m/2$
Multiplications per step	N per equation	1 per butterfly
Addition per step	$N - 1$ per equation	2 per butterfly
Total multiplications	N^2	$N_m/2 = (N/2)\log_2 N$
Total addition	$N(N-1)$	$N_m = N\log_2 N$

FFT. Table 6.4 shows only the number of complex multiplications required for both the DFT and the FFT for several values of N.

Table 6.4 DFT and FFT comparison (number of complex multiplications required)

N	Standard DFT	FFT
2	4	1
4	16	4
8	64	12
16	256	32
32	1024	80
64	4,096	192
128	16,384	448
N (a power of 2)	N^2	$\frac{N}{2}\log_2 N$

We see that the increased efficiency of the radix-2 FFT algorithm becomes more significant as the number of points in the DFT becomes larger.

Example 6.19

The DIT method of the FFT will be used to compute the DFT of the discrete sequence $x(n) = [1, 2, 3, 4]$

Solution 6.19

Referring to Figure 6.16 for the four point FFT, we find the following.

$$\begin{aligned}
&G[0] = x(0) + x(2) = 1 + 3 = 4 \\
&G[1] = x(0) - x(2) = 1 - 3 = -2 \\
&H[0] = x(1) + x(3) = 2 + 4 = 6 \\
&H[1] = x(1) - x(3) = 2 - 4 = -2 \\
&X[0] = G[0] + H[0] = 4 + 6 = 10 \quad X[1] = G[1] + H[1]W_4^1 \\
&\qquad = -2 + (-j)(-2) = -2 + 2j \\
&X[2] = G[0] - H[0] = -2 \quad X[3] = G[1] + H[1]W_4^1 = -2 - 2j \\
&X\,[k] = \{10, \ -2 + 2j, \ -2, \ -2 - 2j\}.
\end{aligned}$$

6.11 Decomposition-in-Frequency (DIF) FFT Algorithm

In DIF FFT frequency component has to be decomposed into it even and odd components. The idea behind the DIF FFT algorithm is similar to that of the DIT FFT presented previously. The DIT FFT and the DIF FFT require the same number of complex multiplications to compute. Consider dividing the output sequence $X\,[k]$ into smaller and smaller sub sequences.

6.11.1 Two-point DFT

To develop FFT algorithms, let us again restrict the discussion to N a power of 2 and consider computing separately the even numbered frequency samples and odd number of frequency samples.

$$X\,[k] = \sum_{n=0}^{N-1} x(n)W_N^{kn}, k = 0, 1, 2, \ldots\ldots, \ N \tag{6.80}$$

For calculating a two-point DFT, it does not need to be decimated in its even and odd components of frequencies.

$$X[k] = \sum_{n=0}^{1} x(n) W_2^{nk} \tag{6.81}$$

$$X[k] = x(0) W_2^{0k} + x(1) W_2^{1k}$$

$$X[0] = x(0) + x(1) \quad X[1] = x(0) + x(1) W_2^1$$

6.11.2 Four-point DFT

Since $X[k]$ is and even numbered frequency samples are

$$X\,[2k] = \sum_{n=0}^{N-1} x(n) W_N^{n(2k)} k = 0, 1, \ldots, \left(\frac{N}{2}\right) - 1 \tag{6.82}$$

which can be decomposed as

$$x[2k] = \sum_{n=0}^{N/2-1} x(n) W_N^{2nk} + \sum_{n=N/2}^{N-1} x(n) W_N^{2nk} \tag{6.83}$$

With a substitution of variables in the second summation in (6.80), we obtain

$$X[2k] = \sum_{n=0}^{N/2-1} x(n) W_N^{2nk} + \sum_{n=0}^{N/2-1} x(n + N/2) W_N^{2k(n+N/2)} \tag{6.84}$$

Finally, because of the periodicity of W_N^{2kn}

$$W_N^{2k(n+N/2)} = W_N^{2kn} W_N^{kN} = W_N^{2kN};\ W_N^{kN} = 1 \tag{6.85}$$

Substituting above values in Equation (6.81)

$$X\,[2k] = \sum_{n=0}^{(N/2)-1} [x\,[n] + x[n + N/2]]\ W_N^{2kn}, k = 0, 1, 2, \ldots, \frac{N}{2} - 1$$
$$X\,[2k] = \sum_{n=0}^{(N/2)-1} [x\,[n] + x[n + N/2]]\ W_{N/2}^{kn}, k = 0, 1, 2, \ldots, \frac{N}{2} - 1 \tag{6.86}$$

Equation (6.88) is the N/2 point DFT of the (N/2)-point sequence obtained by adding first half and the last half of the sequence.

Substituting $x(n) + x(n + N/2) = g(n)$

$$X[2k] = \sum_{n=0}^{(N/2)-1} g(n)\, W_{N/2}^{kn}, k = 0, 1, 2, \ldots\ldots, \frac{N}{2} - 1 \tag{6.87}$$

$$X[2k] = g(0)W_2^{0k} + g(1)W_2^{1k}$$

$$X[0] = g(0)W_2^{0k} + g(1) \quad X[2] = g(0)W_2^{0k} + g(1)W_2^1$$

We now consider obtaining the odd numbered frequency points given by

$$X[2k+1] = \sum_{n=0}^{N-1} x[n]\; W_N^{n(2k+1)} k = 0, 1, \ldots, \left(\frac{N}{2}\right) - 1$$

$$X[2k+1] = \sum_{n=0}^{N/2-1} x[n]\; W_N^{n(2k+1)} + \sum_{n=N/2}^{N-1} x[n]\; W_N^{n(2k+1)} \tag{6.88}$$

Alternative form of the (6.88)

$$x[2k+1] = \sum_{n=0}^{N/2-1} x(n)\; W_N^{n(2k+1)} + \sum_{n=0}^{N/2-1} x\left(n + \frac{N}{2}\right)\; W_N^{[n+(N/2)](2k+1)} \tag{6.89}$$

Now we consider the weighting factors of the two sequences

$$W_N^{2kn} = W_{N/2}^{kn} W_N^{(2k+1)n} = W_{N/2}^{kn} W_N^{n} \quad W_N^{Nk} = -1 \tag{6.90}$$

substituting

$$x(n) - x(n + N/2) = h(n) \tag{6.91}$$

$$x[2k+1] = \sum_{n=0}^{(N/2)-1} [x(n) - x(n + N/2)] W_{N/2}^{nk} \; W_N^n \tag{6.92}$$

$$X[2k+1] = \sum_{n=0}^{(N/2)-1} h(n)\; W_{N/2}^{k\,n}\; W_N^n \; k = 0, 1, 2, \ldots\ldots, \frac{N}{2} - 1$$

$$X[2k+1] = h(0)W_2^{0k} W_N^0 + h(1)W_2^{1k} W_N^1$$

$$X[1] = h(0) - h(1)\; W_4^1 \quad X[3] = h(0) - h(1)W_4^1$$

Equations (6.87), (6.88), (6.91) and (6.92) are used for drawing butterfly diagram.

Figure 6.16 shows the butterfly diagram for the DIF FFT algorithm. Figure 6.17 illustrates the DIF FFT process for a four-point DIF FFT.

Example 6.20

The DIF method of the FFT will be used to compute the DFT of the discrete sequence.

$$x(n) = [1, 2, 3, 4]$$

Solution 6.20

Referring to Figure 6.17 for the four-point DIF FFT, we find the following:

$$g[0] = x(0) + x(2) = 1 + 3 = 4$$

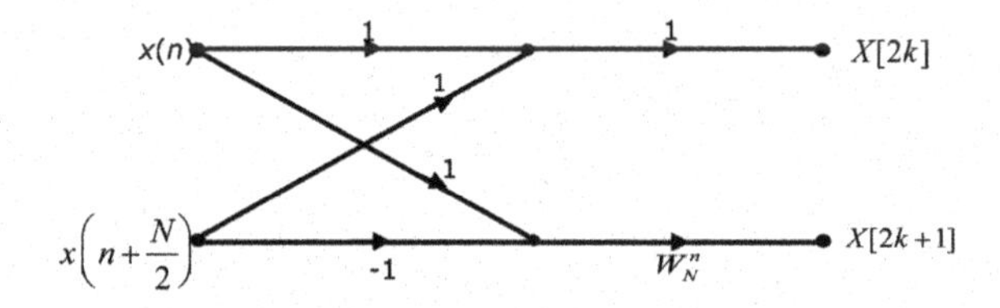

Figure 6.16 A general decomposition-in-frequency FFT.

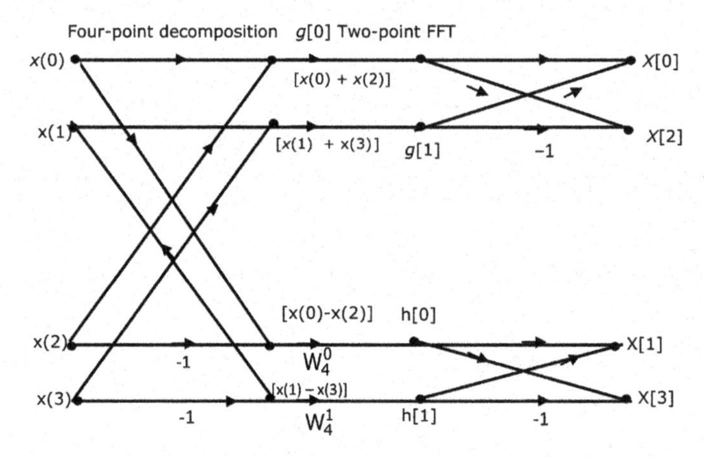

Figure 6.17 A 4-point Decomposition-in-frequency FFT.

$$g[1] = x(1) + x(3) = 2 + 4 = 6$$
$$h[0] = W_4^0[x(0) - x(2)] = 1 - 3 = -2$$
$$h[1] = W_4^1[x(1) - x(3)] = -j[2 - 4] = j2$$

and

$$X[0] = g[0] + g[1] = 4 + 6 = 10$$
$$X[2] = g[0] - g[1] = 4 - 6 = -2$$
$$X[1] = h[0] + h[1] = -2 + j2$$
$$X[3] = h[0] - h[1] = -2 - j2.$$

6.12 Problems and Solutions

Problem 6.1

Compute the DTFT of the following signal

$$x(n) = \begin{cases} 2 - (\frac{1}{2})n & |n| \leq 4 \\ 0 & \text{elsewhere} \end{cases}$$

Solution 6.1

Using

$$\mathrm{X}(\omega) = \sum_{n=-\infty}^{\infty} x(n)e^{-j\omega n} = \sum_{n=0}^{3} x(n)e^{-j\omega n} = \sum_{n=0}^{3}(2 - \frac{1}{2}n)e^{-j\omega n}$$
$$X(w) = 2 + 1.5e^{-jw} + e^{-2jw} + 0.5^{-3jw}$$

Using Z-transform:

$$x(n) = \{\underset{\uparrow}{2},\ 1.5,\ 1,\ 0.5\} \quad X(z) = 2 + 1.5z^{-1} + z^{-2} + 0.5z^{-3}$$

Taking FT by substituting $z = e^{j\omega}$

$$X(w) = 2 + 1.5e^{-jw} + e^{-2jw} + 0.5^{-3jw}$$

Problem 6.2

Compute the DTFT of the following signal

$$x(n) = \{-2, -1, \underset{\uparrow}{0}, 1, 2\}$$

Solution 6.2

Using

$$X(\omega) = \sum_{n=-\infty}^{\infty} x(n)e^{-j\omega n} = \sum_{n=-2}^{2} x(n)e^{-j\omega n}$$
$$= x(-2)e^{2j\omega} + x(-1)e^{j\omega} + x(0) + x(1)e^{-j\omega} + x(2)e^{-2j\omega}$$
$$X(w) = -2e^{-jw(-2)} - 1e^{-jw(-1)} + 1e^{-jw(1)} + 2^{-jw(2)}$$
$$X(w) = -2e^{2jw} - e^{jw} - e^{-jw} + 2e^{-2jw}$$

Using Z-transform:

$$X(z) = -2z^2 - z - z^{-1} + 2z^{-2}$$
$$X(w) = -2(e^{jw})^2 - e^{jw} - (e^{jw})^{-1} + 2(e^{jw})^{-2}$$
$$X(w) = -2e^{2jw} - e^{jw} - e^{-jw} + 2^{-2jw}$$

Problem 6.3

Compute the DTFT of the following signal

$$x(n) = (0.5)^n u(n) + (2)^{-n} u(-n-1)$$

Solution 6.3

Using

$$X(\omega) = \sum_{n=-\infty}^{\infty} (05)^n u(n)e^{-j\omega n} + \sum_{n=-\infty}^{\infty} (2)^{-n} u(-n-1)e^{-j\omega n}$$

$$X(\omega) = \sum_{n=0}^{\infty}(0.5)^n e^{-j\omega n} + \sum_{n=-\infty}^{-1} (2)^{-n} e^{-j\omega n} \;\; X(\omega)$$
$$= \sum_{n=0}^{\infty}(0.5e^{-j\omega})^n + \sum_{n=1}^{\infty}(2^{-1}e^{-j\omega})^n$$

$$X(\omega) = \frac{1}{1-0.5e^{-j\omega}} + \frac{0.5e^{j\omega}}{1-0.5e^{j\omega}} = \frac{1-0.5e^{j\omega}+0.5e^{j\omega}-0.25}{1-2\cos\,\omega+0.25}$$

$$X(\omega) = \frac{0.75}{1.25-2\cos\,\omega}$$

Problem 6.4

Compute the DTFT of the following signal

$$x(n) = \sin\left(\frac{\pi\, n}{2}\right) u(n)$$

Solution 6.4

$$\mathrm{X}(\omega) = x(n)e^{-j\omega n} = \sum_{n=0}^{\infty}\left[\frac{e^{j\pi n/2} - e^{-j\pi n/2}}{2j}\right]$$

$$e^{-j\omega n} = \frac{e^{-j\omega}\,\sin\frac{\pi}{2}}{1+e^{-j2\omega}} = \frac{e^{-j\omega}}{1+e^{-j2\omega}}$$

Problem 6.5

Find the DTFT output $y(n)$ of a causal DT LTI system which is characterized by the difference equation

$$y(n)-\frac{3}{4}y(n-1)+\frac{1}{8}y(n-2) = 2\,x(n) \quad \text{for the input } x(n) = \left[\frac{1}{4}\right]^n u(n)$$

Solution 6.5

$$\text{DTFT}\left[y(n) - \frac{3}{4}y(n-1) + \frac{1}{8}y(n-2)\right] = \text{DTFT } 2\left[\frac{1}{4}\right]^n u(n)$$

$$\text{DTFT }[y(n)] - \frac{3}{4}\text{ DTFT }[y(n-1)] + \frac{1}{8}\text{ DTFT }[y(n-2)]$$

$$= \text{DTFT } 2\left[\frac{1}{4}\right]^n u(n)$$

$$Y(\omega) - \frac{3}{4}e^{-j\omega}Y(\omega) + \frac{1}{8}e^{-j2\omega}Y(\omega) = \frac{2}{1-\frac{1}{4}e^{-j\omega}}$$

$$Y(\omega) = \frac{2}{[1 - \frac{3}{4}e^{-j\omega} + \frac{1}{8}e^{-j2\omega}][1 - \frac{1}{4}e^{-j\omega}]} = \frac{2}{[1 - \frac{1}{2}e^{-jw}][1 - \frac{1}{4}e^{-jw}]^2}$$

$$y(n) = 8\left(\frac{1}{2}\right)^n u(n) - 4\left(\frac{1}{4}\right)^n u(n) - 2(n+1)\left[\frac{1}{4}\right]^n u(n)$$

Problem 6.6

Compute the DTFT of the following signal

$$x(n) = (2)^n u(n)$$

Solution 6.6

The sequence $x(n) = (2)^n u(n)$ is not absolutely summable. Therefore, DTFT does not exist.

Problem 6.7

Find DFT of the sequence $x(n) = \{1,\ 2,\ 0,\ 4\}$ using Matrix method.

Solution 6.7

$$X[k] = \sum_{n=0}^{N-1} x(n)\, W_N^{kn}, k = 0, 1, 2, \ldots\ldots, N-1, \quad W_N^1 = e^{-j2\pi/N}$$

$X(k) = \mathbf{W}_N x(n)$, where $x(n)$ is the input vector of N DFT samples
$W_N = N \times N$ DFT matrix
$X(k) =$ the output vector composed of N DFT samples

$$W_N = \begin{bmatrix} 1 & 1 & 1 & 1 \\ 1 & W_N^1 & W_N^2 & W_N^{N-1} \\ 1 & W_N^2 & W_N^4 & W_N^{2(N-1)} \\ 1 & W_N^{N-1} & W_N^{2(N-1)} & W_N^{(N-1)(N-1)} \end{bmatrix}$$

$$\begin{bmatrix} X[0] \\ X[1] \\ X[2] \\ X[3] \end{bmatrix} = \begin{bmatrix} 1 & 1 & 1 & 1 \\ 1 & W_4^1 & W_4^2 & W_4^{(4-1)} \\ 1 & W_4^2 & W_4^4 & W_4^{2(4-1)} \\ 1 & W_4^{(4-1)} & W_4^{2(4-1)} & W_4^{(4-1)(4-1)} \end{bmatrix} \begin{bmatrix} x(0) \\ x(1) \\ x(2) \\ x(3) \end{bmatrix}$$

$$\begin{bmatrix} X[0] \\ X[1] \\ X[2] \\ X[3] \end{bmatrix} = \begin{bmatrix} 1 & 1 & 1 & 1 \\ 1 & W_4^1 & W_4^2 & W_4^3 \\ 1 & W_4^2 & W_4^4 & W_4^6 \\ 1 & W_4^3 & W_4^6 & W_4^9 \end{bmatrix} \begin{bmatrix} 1 \\ 2 \\ 0 \\ 4 \end{bmatrix}$$

$$\begin{bmatrix} X[0] \\ X[1] \\ X[2] \\ X[3] \end{bmatrix} = \begin{bmatrix} 1 & 1 & 1 & 1 \\ 1 & -j & -1 & j \\ 1 & -1 & 1 & -1 \\ 1 & j & -1 & -j \end{bmatrix} \begin{bmatrix} 1 \\ 2 \\ 0 \\ 4 \end{bmatrix}$$

$$= \begin{bmatrix} 1+2+4 \\ 1-2j+4j \\ 1-2-4 \\ 1+2j-4j \end{bmatrix} = \begin{bmatrix} 7 \\ 1+2j \\ -5 \\ 1-2j \end{bmatrix}$$

$$X[k] = \{7, 1+2j, -5, 1-2j\}$$

Problem 6.8

Find the inverse FT of the sequence $\{7,\ 1+2j,\ -5,\ 1\ -2\,j\}$.

Solution 6.8

It is useful to illustrate the inverse DFT by using it to derive the time series $\{1, 2, 0, 4\}$ from its DFT components $\{7, 1+2j, -5, 1-2j\}$.

$$x(n) = F_D^{-1}[X(k)] = \frac{1}{N}\sum_{k=0}^{N-1} X[k]e^{jk\Omega nT}, n = 0, 1, \ldots\ldots..N-1$$

$$x(n) = \frac{1}{N}\sum_{k=0}^{N-1} X[k]e^{+jk2\pi n/N} \quad x(nT) = x(n) = \frac{1}{N}\sum_{k=0}^{N-1} X[k]$$

With $n = 0$

$$x(0) = \frac{1}{4}\left[X[0] + X[1] + X[2] + X[3]\right]$$
$$= \frac{1}{4}\left[7 + (1+2j) - 5 + (1-2j)\right] = 1$$

With $n = 1$

$$x(n) = \frac{1}{N}\sum_{k=0}^{N-1} X[k]e^{jk2\pi/N} = \frac{1}{4}\sum_{k=0}^{N-1} X[k]e^{jk\pi/2}$$

$$x(1) = \frac{1}{4}[7 + (1 + 2j)e^{j\pi/2} - 5e^{j\pi} + (1 - 2j)e^{j3\pi/2}]$$

$$x(1) = \frac{1}{4}[7 + (1 + 2j)j - 5\,(-1) + (1 - 2j)(-j)]$$

$$= \frac{1}{4}(7 + j - 2 + 5 - j - 2) = 2$$

With $n = 2$,

$$x(nT) = x(n) = \frac{1}{N}\sum_{k=0}^{N-1} X[k]e^{jk\pi}$$

$$x(2) = \frac{1}{4}\left[7 + (1 + 2j)e^{j\pi} - 5\,e^{j2\pi}(1 - j)e^{j3\pi}\right]$$

$$= \frac{1}{4}\left[7 + (1 + 2j)(-1) - 5 + (1 - 2j)(-1)\right] = 0$$

Finally, with $n = 3$,

$$x(n) = x(3) = \frac{1}{N}\sum_{k=0}^{N-1} X(k)e^{jk3\pi/2}$$

$$x(3) = \frac{1}{4}[7 + (1 + 2j)e^{j3\pi/2} - 5\,e^{j3\pi} + (1 - 2j)e^{j9\pi/2}]$$

$$x(3) = \frac{1}{4}\left[7 + (1 + 2j)(-j) - 5(-1) + (1 - 2j)j\right]$$

$$= \frac{1}{4}(7 - j + 2 + 5 + j + 2) = 4$$

Problem 6.9

Find IDFT of the sequence $X(k) = \{7, 1 + 2j, -5, 1 - 2j\}$ using matrix method.

Solution 6.9

$$X[k] = \sum_{n=0}^{N-1} x(n)W_N^{kn}, \quad k = 0, 1, 2, \ldots\ldots\ldots N - 1 \text{ DFT}$$

$$x(n) = \frac{1}{N}\sum_{k=0}^{N-1} X[k]W_N^{-kn}, \quad W_N^1 = e^{-j2\pi/N}\text{IDFT}$$

$x(n) = {}_{W_N^1}X[k]$, where $x(n)$ is the input vector of N DFT samples
$W_N = N \times N$ DFT matrix
$X[k]$ = the output vector composed of N DFT samples

IDFT can be calculated as

$$W_N^{-1} = \begin{bmatrix} 1 & 1 & 1 & 1 \\ 1 & W_N^{-1} & W_N^{-2} & W_N^{-(N-1)} \\ 1 & W_N^{-2} & W_N^{-4} & W_N^{-2(N-1)} \\ 1 & W_N^{-(N-1)} & W_N^{-2(N-1)} & W_N^{-(N-1)(N-1)} \end{bmatrix}$$

$$[W_N^{-1}] = \begin{bmatrix} 1 & 1 & 1 & 1 \\ 1 & W_4^{-1} & W_4^{-2} & W_4^{-(4-1)} \\ 1 & W_4^{-2} & W_4^{-4} & W_4^{-2(4-1)} \\ 1 & W_4^{-(4-1)} & W_4^{-2(4-1)} & W_4^{-(4-1)(4-1)} \end{bmatrix}$$

$$\begin{bmatrix} x(0) \\ x(1) \\ x(2) \\ x(3) \end{bmatrix} = \frac{1}{N}\begin{bmatrix} 1 & 1 & 1 & 1 \\ 1 & W_4^{-1} & W_4^{-2} & W_4^{-(4-1)} \\ 1 & W_4^{-2} & W_4^{-4} & W_4^{-2(4-1)} \\ 1 & W_4^{-(4-1)} & W_4^{-2(4-1)} & W_4^{-(4-1)(4-1)} \end{bmatrix}\begin{bmatrix} X[0] \\ X[1] \\ X[2] \\ X[3] \end{bmatrix}$$

$$\begin{bmatrix} x(0) \\ x(1) \\ x(2) \\ x(3) \end{bmatrix} = \frac{1}{4}\begin{bmatrix} 1 & 1 & 1 & 1 \\ 1 & W_4^{-1} & W_4^{-2} & W_4^{-3} \\ 1 & W_4^{-2} & W_4^{-4} & W_4^{-6} \\ 1 & W_4^{-3} & W_4^{-6} & W_4^{-9} \end{bmatrix}\begin{bmatrix} 7 \\ 1+2j \\ -5 \\ 1-2j \end{bmatrix}$$

$$x(n) = \{1, 2, 0, 4\}$$

Problem 6.10

Find DFT of the sequence $x(n) = \{1,\ 2,\ 0,\ 1\}$ using Matrix method.

Solution 6.10

$X[k] = W_N x(n)$

$$W_N = \begin{bmatrix} 1 & 1 & 1 & 1 \\ 1 & W_N^1 & W_N^2 & W_N^{N-1} \\ 1 & W_N^2 & W_N^4 & W_N^{2(N-1)} \\ 1 & W_N^{N-1} & W_N^{2(N-1)} & W_N^{(N-1)(N-1)} \end{bmatrix}$$

$$\begin{bmatrix} X[0] \\ X[1] \\ X[2] \\ X[3] \end{bmatrix} = \begin{bmatrix} 1 & 1 & 1 & 1 \\ 1 & W_4^1 & W_4^2 & W_4^{(4-1)} \\ 1 & W_4^2 & W_4^4 & W_4^{2(4-1)} \\ 1 & W_4^{(4-1)} & W_4^{2(4-1)} & W_4^{(4-1)(4-1)} \end{bmatrix} \begin{bmatrix} x(0) \\ x(1) \\ x(2) \\ x(3) \end{bmatrix}$$

$$\begin{bmatrix} X[0] \\ X[1] \\ X[2] \\ X[3] \end{bmatrix} = \begin{bmatrix} 1 & 1 & 1 & 1 \\ 1 & W_4^1 & W_4^2 & W_4^3 \\ 1 & W_4^2 & W_4^4 & W_4^6 \\ 1 & W_4^3 & W_4^6 & W_4^9 \end{bmatrix} \begin{bmatrix} 1 \\ 2 \\ 0 \\ 1 \end{bmatrix}$$

$$\begin{bmatrix} X[0] \\ X[1] \\ X[2] \\ X[3] \end{bmatrix} = \begin{bmatrix} 1 & 1 & 1 & 1 \\ 1 & -j & -1 & j \\ 1 & -1 & 1 & -1 \\ 1 & j & -1 & -j \end{bmatrix} \begin{bmatrix} 1 \\ 2 \\ 0 \\ 1 \end{bmatrix} = \begin{bmatrix} 1+2+1 \\ 1-2j+j \\ 1-2-1 \\ 1+2j-j \end{bmatrix} = \begin{bmatrix} 4 \\ 1-j \\ -2 \\ 1+j \end{bmatrix}$$

$$X[k] = \{4, 1-j, -2, 1+j\}$$

Problem 6.11

The DIT method of the FFT will be used to compute the DFT of the following discrete sequence.

$$x(n) = [1, 2, 0, 4]$$

Solution 6.11

Referring to the four point FFT, we find the following.

$$\begin{aligned} G[0] &= x(0) + x(2) = 1 + 0 = 1 \\ G[1] &= x(0) - x(2) = 1 - 0 = 1 \\ H[0] &= x(1) + x(3) = 2 + 4 = 6 \\ H[1] &= x(1) - x(3) = 2 - 4 = -2 \quad \text{and} \\ X[0] &= G[0] + H[0] = 1 + 6 = 7 \end{aligned}$$

$$X[1] = G[1] + H[1]W_4^1 = 1 + (-j)(-2) = 1 + j2$$
$$X[2] = G[0] - H[0] = 1 - 6 = -5X[3] = G[1] - H[1]W_4^1$$
$$= 1 - (2)(j) = 1 - 2j$$

which is in agreement with the previous results found.

Problem 6.12

Compute the values of DTFT and DFT, given

$$X(n) = \{4,\ 3,\ 2,\ 1\}$$

Solution 6.12

The DTFT of this is given by

$$X(\omega) = \sum_{0}^{3} x(n)e^{-j\omega n}$$
$$X(\omega) = 4 + 3e^{-j\omega} + 2\,e^{-2j\omega} + 1\,e^{-3j\omega} \quad (1)$$
$$X(\omega) = 4 + 3(\cos\,\omega - j\,\sin\,\omega) + 2(\cos\,2\omega - j\,\sin\,2\omega)$$
$$+ 1(\cos\,3\omega - j\,\sin\,3\omega) \quad (2)$$

Equation (2) may be separated into real and imaginary part as:

$$X(\omega) = (4 + 3\cos\,\omega + 2\cos 2\omega + \cos 3\omega)$$
$$- j(3\sin\omega + 2\,\sin\,2\omega + \sin\,3\omega) \quad (3)$$
$$X(\omega) = a - jb \quad (4)$$

After a and b are computed separately, the complex sum $a - jb$ is calculated, the complex summation yield the magnitude and phase angle θ. In the following table rows 1, 3, 6, 8, and 11 are in bold faced letters.

Now consider the DFT of the same sequence $x(n)$, the value of $X[k]$ is compared for the rows 1, 3, 6, 8, and 11. We find that corresponding value of DFT and DTFT exactly coincide with each other.

DTFT					DFT			
				$\angle\theta$				$M\angle\theta$
Trial	ω (rad/s)	$a - jb$	M	(deg.)	Trial	K	$X[k]$	(deg.)
1	**0**	**10**	**10**	**0**	**1**	**0**	**10**	**0**
2	1	$0.556 + j1.89$	1.97	136.63				
3	$\pi/2$	$-2 + 2j$	**2.84**	**135**	**2**	**1**	**2.84**	**135**
4	2	$4.9 - j0.127$	4.9	–51.4				
5	3	$7.979 + j0.273$	7.98	1.96				
6	π	**–2**	**2**	**180**	**3**	**2**	**2**	**–180**
7	4	$6.51 - j0.30$	6.51	–2.63				
8	$3\pi/2$	$-2 - 2j$	**2.84**	**–135**	**4**	**3**	**2.84**	**–135**
9	5	$0.711 - j0.85$	1.108	–50°				
10	6	$3.46 - j1.567$	3.8	–24.32				
11	2π	**10**	**10**	**0**	**5**	**4**	**10**	**0**

7

Structure for FIR and IIR Filters

This chapter covers: Realization structure of the FIR and IIR filters filter in different forms, i.e., direct form, Lattice, Frequency sampling and Fast convolution for FIR in brief and Direct form-I and II, Cascade, Parallel, Lattice and Lattice–Ladder network for IIR filters, Problems and solutions.

7.1 Introduction

Realization involves in converting a given transfer function $H(z)$ into a suitable filter structure. Block or flow diagrams are often used to depict filter structures and they show the computational procedure for implementing the digital filter. The structure used depends on whether the filter is an IIR or FIR filter. The method of realization can be recursive or non recursive. Few of the following methods have been discussed keeping in mind the level of the students.

The following are the methods commonly used in realization structures for FIR and IIR filters:

(a) Lattice (FIR)
(b) Frequency Sampling Realization (FIR)
(c) Fast convolution (FIR)
(d) Direct form I (IIR)
(e) Direct form II, Transversal (IIR & FIR)
(f) Cascade (IIR)
(g) Parallel (IIR)
(h) Lattice (All Pole Filters) (IIR)
(i) Lattice Ladder (IIR)

7.2 Structure Form of FIR Filters

7.2.1 Direct Form (Transversal)

FIR filter are related by convolution sum

$$y(n) = \sum_{k=0}^{N-1} x(k)h(n-k) \quad \text{or} \quad Y(z) = H(z)X(z).$$

Sequence of FIR filter with system transfer functions $H(z)$ coefficients polynomial

$$H(z) = A_m(z) \quad m = 0, 1, 2, \ldots M-1, \tag{7.1}$$

where by definition $A_m(z)$ is the polynomial.

$$A_m(z) = 1 + \sum_{k=1}^{m} \alpha_m(k)\, z^{-k}; \quad m \geq 1 \tag{7.2}$$

and $A_0(z) = 1$.

The subscript m on the polynomial $A_m(z)$ and $y(n)$ is the output sequence, we have

$$y(n) = x(n) + \sum_{k=1}^{m} \alpha_m(k)\, x(n-k) \tag{7.3}$$

7.2.2 Lattice Structure

Lattice structure may be used to represent FIR as well as IIR filter. Lattice structures are used extensively in digital speech processing and in the implementation of adaptive filters.

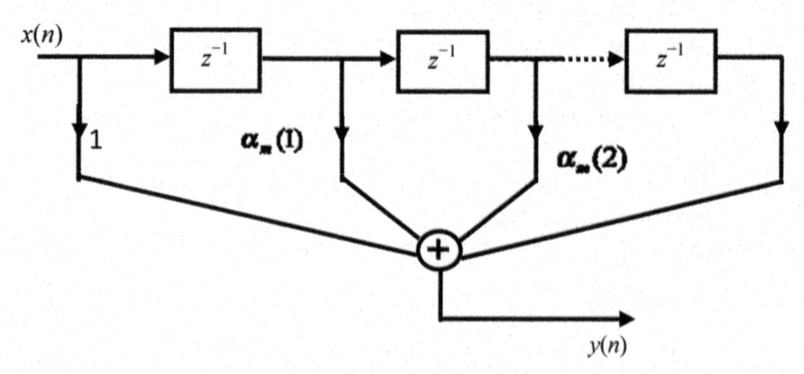

Figure 7.1 The direct form of FIR filter.

FIR filter structures are initially related with the topic of linear predictor where

$$\widehat{x}(n) = -\sum_{k=1}^{m} \alpha_m(k)x(n-k), \tag{7.4}$$

where $y(n) = x(n) - \widehat{x}(n)$ represents the prediction error sequence

$$f_m(n) = \sum_{k=0}^{m} \alpha_m(k)x(n-k)$$

$$f_m(n) = \alpha_m(0)x(n) + \sum_{k=1}^{m} \alpha_m(k)x(n-k)$$

$$y(n) = x(n) + \sum_{k=1}^{m} \alpha_m(k)x(n-k). \tag{7.5}$$

For $m = 1$, a single lattice filter. The output of such filter is

$$y(n) = x(n) + \alpha_1(1)x(n-1) \tag{7.6}$$

The input $x(n)$ and the output $f_1(n)$ and $g_1(n)$ are related by

$$\begin{aligned} f_1(n) &= x(n) + k_1 x(n-1) \\ g_1(n) &= k_1 x(n) + x(n-1) \end{aligned} \tag{7.7}$$

Comparing Equations (7.6) and (7.7)

We get the value of $\alpha_1(1) = k_1$. Here, k_1 is called reflection coefficient of lattice structure as shown in Figure 7.2

For $m = 2$, a two stage lattice filter. From Equation (7.5) output of such filter is

$$y(n) = x(n) + \alpha_2(1)\, x(n-1) + \alpha_2(2)\, x(n-2) \tag{7.8}$$

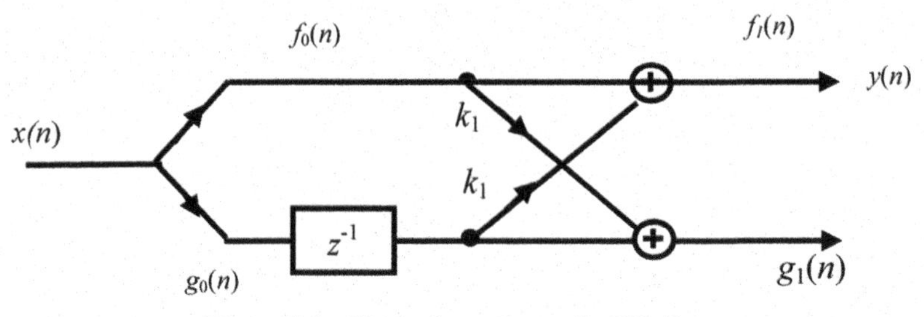

Figure 7.2 The lattice structure for FIR filter.

The input $x(n)$ and the output $f_2(n)$ and $g_2(n)$ are related by

$$\begin{aligned} f_2(n) &= f_1(n) + k_2 g_1(n-1) \\ g_2(n) &= k_2 f_1(n) + g_1(n-1) \end{aligned} \tag{7.9}$$

if we focus our attention on $f_2(n)$ and substitute for $g_1(n - 1)$ from Equation (7.7) into Equation (7.9)

$$\begin{aligned} f_2(n) &= x(n) + k_1 x(n-1) + k_2[k_1 x(n-1) + x(n-2)] \\ f_2(n) &= x(n) + [k_1 + k_1 k_2] x(n-1) + k_2 x(n-2) \end{aligned} \tag{7.10}$$

By comparing Equations (7.8) and (7.10)

$$\alpha_2(1) = k_1(1 + k_2) \text{ and } \alpha_2(2) = k_2$$

$$k_2 = \alpha_2(2) \quad k_1 = \frac{\alpha_2(1)}{1 + \alpha_2(2)}.$$

Now, in a two-stage lattice structure $g_2(n)$ can be expressed in the form

$$\begin{aligned} g_2(n) &= k_2 f_1(n) + g_1(n-1) \\ g_2(n) &= k_2[x(n) + k_1 x(n-1)] + k_1 x(n-1) + x(n-2) \\ g_2(n) &= k_2 x(n) + k_1(1 + k_2)\, x(n-1)] + x(n-2) \\ g_2(n) &= [\alpha_2(2)] x(n) + [\alpha_2(1)]\, x(n-1) + x(n-2) \end{aligned} \tag{7.11}$$

The above Equations (7.10) and (7.11) reveals that the filter coefficients of $g_2(n)$ are $\{\alpha_2(2), \alpha_2(1), 1\}$, where as the coefficients for the filter $f_2(n)$ are $\{1, \alpha_2(1), \alpha_2(2)\}$. Again the two sets of filter coefficients are in reverse order.

The above comparison process for finding the reflection coefficients is not a valid process for higher order filter, therefore one can easily demonstrate by induction, the equivalence between an *m*th order direct form FIR filter and an *m*th order *m* stage lattice structure.

The generalized equations for the lattice structure are described by the following set of recursive equations.

$$\begin{aligned} f_0(n) &= g_0(n) = x(n) \\ f_m(n) &= f_{m-1}(n) + k_m g_{m-1}(n-1) \\ g_m(n) &= k_m f_{m-1}(n) + g_{m-1}(n-1). \end{aligned} \tag{7.12}$$

As a consequence of the equivalence between an FIR filter and a Lattice structure, the one output $f_m(n)$ of a m stage lattice structure.

$$f_m(n) = \sum_{k=0}^{m} \alpha_m(k)\, x(n-k); \quad \alpha_m(0) = 1 \tag{7.13}$$

Equation (7.13) is a convolution sum, it follows that the z-transform relationship is

$$F_m(z) = A_m(z)X(z)$$

$$A_m(z) = \frac{F_m(z)}{X(z)} = \frac{F_m(z)}{F_0(z)}. \quad (7.14)$$

For the further development it follows that the output $g_m(n)$ for an m-stage lattice filter can be expressed by the convolution sum of the form

$$g_m(n) = \sum_{k=0}^{m} \beta_m(k)\, x(n-k) \quad (7.15)$$

$$\text{where} \quad \beta_m(k) = \alpha_m(m-k) \quad k = 0, 1, 2. \ldots \mathrm{m}$$

Equation (7.15) can be written in Z-transform form

$$G_m(z) = B_m(z)\, X(z)$$

$$B_m(z) = \frac{G_m(z)}{X(z)}$$

$$B_m(z) = \sum_{k=0}^{m} \beta_m(k)\, z^{-k} \qquad \beta_m(k) = \alpha_m(m-k)$$

$$B_m(z) = \sum_{k=0}^{m} \alpha_m(m-k)\, z^{-k}$$

$$B_m(z) = \sum_{L=0}^{m} \alpha_m(L)\, z^{L-m} = z^{-m}\, A_m(z^{-1})$$

Hence $B_m(z)$ is called the reciprocal or reverse polynomial of $A_m(z)$ Equation (7.12) can be written in Z-transform form

$$F_0(z) = G_0(z) = X(z)$$

$$F_m(z) = F_{m-1}(z) + k_m\, z^{-1}\, G_{m-1}(z) \quad m = 0,\, 1,\, 2. \ldots,\, M-1$$

$$G_m(z) = k_m\, F_{m-1}(z) + z^{-1}\, G_{m-1}(z) \quad m = 0,\, 1,\, 2. \ldots,\, M-1 \quad (7.16)$$

If we divide each equation by $X(z)$, we obtain the desired result in the form

$$A_0(z) = B_0(z) = 1$$

$$A_m(z) = A_{m-1}(z) + k_m\, z^{-1}\, B_{m-1}(z) \quad m = 0,\, 1,\, 2. \ldots,\, M-1$$

$$B_m(z) = k_m\, A_{m-1}(z) + z^{-1}\, B_{m-1}(z) \quad m = 0,\, 1,\, 2. \ldots,\, M-1. \quad (7.17)$$

The lattice stage is described in the z-domain by the matrix equation as follows and shown in the Figure 7.3.

$$\begin{bmatrix} A_m(z) \\ B_m(z) \end{bmatrix} = \left[\begin{bmatrix} 1 & k_m \\ k_M & 1 \end{bmatrix}\right] \begin{bmatrix} A_{m-1}(z) \\ z^{-1}B_{m-1}(z) \end{bmatrix} \quad (7.18)$$

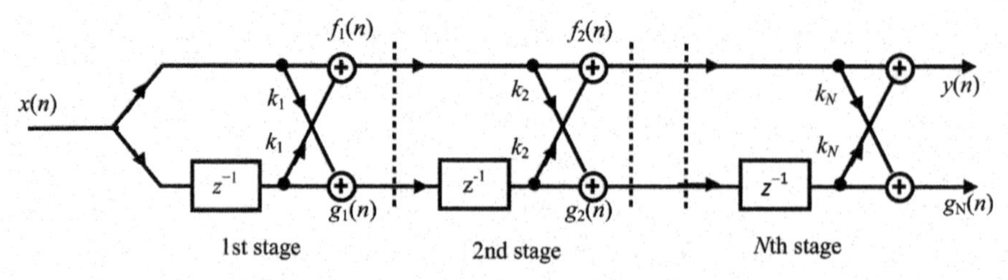

Figure 7.3 An N-stage FIR lattice structure.

7.2.2.1 Direct form filter-to-lattice coefficients

From Equation (7.17)

$$
\begin{aligned}
A_0(z) &= B_0(z) = 1 \\
A_m(z) &= A_{m-1}(z) + k_m\, z^{-1}\, B_{m-1}(z) \quad m = 0, 1, 2. \ldots, M-1 \\
B_m(z) &= k_m\, A_{m-1}(z) + z^{-1}\, B_{m-1}(z) \quad m = 0, 1, 2. \ldots, M-1 \\
A_m(z) &= A_{m-1}(z) + k_m\, z^{-1}\, B_{m-1}(z) \quad m = 0, 1, 2. \ldots, M-1 \\
A_m(z) &= A_{m-1}(z) + k_m\, [B_m(z) - k_m\, A_{m-1}(z)] \\
A_{m-1}(z) &= \tfrac{A_m(z) - k_m\, B_m(z)}{1-k_m^2} \quad m = 0, 1, 2. \ldots, M-1
\end{aligned}
$$

Example 7.1

Determine the lattice coefficients corresponding to the FIR filter with system function

$$H(z) = A_3(z) = 1 + \frac{13}{24}\, z^{-1} + \frac{5}{8}\, z^{-2} + \frac{1}{3}\, z^{-3}$$

Solution 7.1

We note that $k_3 = \alpha_3(3) = 1/3$

$$B_3(z) = \frac{1}{3} + \frac{5}{8}\, z^{-1} + \frac{13}{24}\, z^{-2} + z^{-3}$$

The step-down relationship with $m = 3$ yields

$$A_2(z) = \frac{A_3(z) - k_3\, B_3(z)}{1 - k_3^2}$$

$$A_2(z) = \frac{\left[1 + \frac{13}{24}\, z^{-1} + \frac{5}{8}\, z^{-2} + \frac{1}{3}\, z^{-3}\right] - \frac{1}{3}\left[\frac{1}{3} + +\frac{5}{8}\, z^{-1} + \frac{13}{24}\, z^{-2} + z^{-3}\right]}{1 - \frac{1}{9}}$$

$$A_2(z) = \frac{9}{8}\left[1 - \frac{1}{9} + \left(\frac{13}{24} - \frac{5}{24}\right) z^{-1} + \frac{1}{3} z^{-3}\right]$$

$$-\left(\frac{5}{8} - \frac{13}{72}\right) z^{-2} + \left(\frac{1}{3} - \frac{1}{3}\right) z^{-3}$$

$$A_2(z) = 1 + \frac{3}{8} z^{-1} + \frac{1}{2} z^{-2}$$

$$k_2 = \alpha_2(2) = 1/2$$

$$B_2(z) = \frac{1}{2} + \frac{3}{8} z^{-1} + z^{-2}.$$

By replacing the step-sown recursion, we obtain

$$A_1(z) = \frac{A_2(z) - k_2 B_2(z)}{1 - k_2^2}$$

$$A_1(z) = \frac{\left[1 + \frac{3}{8} z^{-1} + \frac{1}{2} z^{-2}\right] - \frac{1}{2}\left[\frac{1}{2} + + \frac{3}{8} z^{-1} + z^{-2}\right]}{1 - \frac{1}{4}}$$

$$A_1(z) = \frac{4}{3}\left[\left(1 - \frac{1}{4}\right) + \left(\frac{3}{8} - \frac{3}{16}\right) z^{-1} - \left(\frac{1}{2} - \frac{1}{2}\right) z^{-2}\right]$$

$$A_2(z) = 1 + \frac{1}{4} z^{-1}$$

$$k_1 = \alpha_1(1) = 1/4.$$

In the above example, one method is given for calculating the reflection coefficient, which in short is explained below.

From the step down recursive equation, it is easily easy to obtain a formula for recursively computing k_m, beginning with $m = M - 1$ and stepping down to $m = 1$, For $m = M - 1, M - 2, \ldots 1$.

$$\alpha_m(m) = K_m \quad \alpha_m(0) = 1$$

$$\alpha_{m-1}(k) = \frac{\alpha_m(k) - k_m\, \beta_m(k)}{1 - k_m^2} \quad \text{and } \beta_m(k) = \alpha_m(m - k)$$

$$\alpha_{m-1}(k) = \frac{\alpha_m(k) - \alpha_m(m)\, \alpha_m(m-k)}{1 - \alpha_m^2(m)} \quad 1 \le k \le m - 1.$$

Example 7.2

Determine the lattice coefficients corresponding to the FIR filter with system function

$$H(z) = A_3(z) = 1 + \frac{13}{24} z^{-1} + \frac{5}{8} z^{-2} + \frac{1}{3} z^{-3}$$

Solution 7.2

$m = 3$, $\alpha_m(m) = K_m \quad \alpha_3(3) = k_3 = 1/3$

$$\alpha_{m-1}(k) = \frac{\alpha_m(k) - k_m\, \beta_m(k)}{1-k_m^2}$$
$$\alpha_2(k) = \frac{\alpha_3(k) - \alpha_3(3)\, \alpha_3(m-k)}{1-\alpha_3^2(3)} \quad 1 \le k \le 2.$$

It means value of k to be substituted in the above equation, i.e., $k = 1$ and $k = 2$

$k = 1$

$$\alpha_2(1) = \frac{\alpha_3(1) - k_3\, \alpha_3(2)}{1 - (\frac{1}{3})^2} = \frac{9}{24}$$

$k = 2$

$$\alpha_2(2) = \frac{\alpha_3(2) - k_3\, \alpha_3(1)}{1 - (\frac{1}{3})^2} = \frac{1}{2}$$
$$\alpha_2(2) = k_2 = \frac{1}{2}$$

$m = 2$,

$$\alpha_1(k) = \frac{\alpha_2(k) - k_2 \alpha_2(2-k)}{1 - k_2^2}.$$

It means value of k to be substituted in the above equation

$k = 1$

$$\alpha_1(1) = \frac{\alpha_2(1) - k_2\, \alpha_2(1)}{1 - k_2^2} = \frac{1}{4}$$

7.2.2.2 Lattice-to-direct form coefficients

The direct form FIR filter coefficients $\{\alpha_m(k)\}$ can be obtained from the lattice coefficients $\{k_i\}$ by using the following relationship

$$A_0(z) = B_0(z) = 1 \tag{7.19}$$

$$A_m(z) = A_{m-1}(z) + k_m\, z^{-1}\, B_{m-1}(z) \quad m = 0, 1, 2\ldots, M-1 \tag{7.20}$$

$$B_m(z) = z^{-m} A_m(z^{-1}) \quad m = 0, 1, 2\ldots, M-1. \tag{7.21}$$

The solution is obtained recursively, beginning with $m = 1$.

Example 7.3

Given a three-stage Lattice filter with coefficients $k_1 = \frac{1}{4}, k_2 = \frac{1}{2}, k_3 = \frac{1}{3}$. Determine the FIR filter coefficients corresponding to the direct form structure.

Solution 7.3

$$A_0(z) = B_0(z) = 1$$

$$A_m(z) = A_{m-1}(z) + k_m z^{-1} B_{m-1}(z) \quad m = 0, 1, 2 \ldots, M-1$$

for $m = 1$

$$A_1(z) = A_0(z) + k_1 z^{-1} B_0(z) = 1 + \frac{1}{4} z^{-1}.$$

Hence, the coefficients of an FIR filter corresponding to a single stage lattice are:

$$\alpha_1(0) = 1,\ \alpha_1(1) = 1/4 = k_1$$

Since $B_m(z)$ is the reverse polynomial of $A_m(z)$

$$B_1(z) = \frac{1}{4} + z^{-1}.$$

We add second stage $m = 2$

$$\begin{aligned} A_2(z) &= A_1(z) + k_2 z^{-1} B_1(z) \\ &= 1 + \tfrac{1}{4} z^{-1} + \tfrac{1}{2} z^{-1}(\tfrac{1}{4} + z^{-1}) \end{aligned}$$

$$A_2(z) = 1 + \frac{3}{8} z^{-1} + \frac{1}{2} z^{-2}$$

$$\alpha_2(0) = 1, \alpha_2(1) = 3/8, \alpha_2(2) = 1/2$$

Also

$$B_2(z) = \frac{1}{2} + \frac{3}{8} z^{-1} + z^{-2}$$

Finally, we add second stage $m = 3$

$$\begin{aligned} A_3(z) &= A_2(z) + k_3 z^{-1} B_2(z) \\ &= 1 + \tfrac{3}{8} z^{-1} + \tfrac{1}{2} z^{-2} + \tfrac{1}{3} z^{-1}[\tfrac{1}{2} + \tfrac{3}{8} z^{-1} + z^{-2}] \end{aligned}$$

$$A_3(z) = 1 + \tfrac{13}{24} z^{-1} + \tfrac{5}{8} z^{-2} + \tfrac{1}{3} z^{-3}$$

$$\alpha_3(0) = 1, \alpha_3(1) = 13/24, \alpha_3(2) = 5/8, \text{ and } \alpha_3(3) = 1/3$$

A formula for determining the filter coefficients $\{\alpha_m(k)\}$ recursively can be easily derived from polynomial relationships (7.19) through (7.21). From the relationship of (7.21) we have

$$A_m(z) = A_{m-1}(z) + k_m z^{-1} B_{m-1}(z)$$

$$\sum_{k=0}^{m} \alpha_m(k)\, z^{-k} = \sum_{k=0}^{m-1} \alpha_{m-1}(k)\, z^{-k} + k_m \sum_{k=0}^{m-1} \alpha_{m-1}(m-1-k)\, z^{-(k+1)} \quad (1)$$

By equating the coefficients of equal powers of z^{-1} and recalling that $\alpha_m(0) = 1$ for $m = 1, 2, \ldots, M - 1$, we obtain the desired recursive equation for the FIR filter coefficients in the form

$$\alpha_m(m) = K_m \quad \alpha_m(0) = 1$$

$$\alpha_m(k) = \alpha_{m-1}(k) + k_m\, \alpha_{m-1}(m-k) \qquad (2)$$

$$\alpha_m(k) = \alpha_{m-1}(k) + \alpha_m(m)\, \alpha_{m-1}(m-k) \quad 1 \leq k \leq m-1$$

$$m = 1, 2, \ldots, M-1$$

Equation (2) is called as Levinson–Durbin recursive equation.

Example 7.4

Given a 3-stage Lattice filter with coefficients $k_1 = \frac{1}{4}$, $k_2 = \frac{1}{2}$, and $k_3 = \frac{1}{3}$. Determine the FIR filter coefficients corresponding to the direct form structure.

Solution 7.4

By equating the coefficients of equal powers of z^{-1} and recalling that $\alpha_m(0) = 1$ for $m = 1, 2, \ldots, M - 1$, we obtain the desired recursive equation for the FIR filter coefficients in the form

$$\alpha_m(m) = K_m \quad \alpha_{m-1}(0) = 1$$

$$\alpha_m(k) = \alpha_{m-1}(k) + k_m\, \alpha_{m-1}(m-k)$$

$$\alpha_m(k) = \alpha_{m-1}(k) + \alpha_m(m)\alpha_{m-1}(m-k) \quad 1 \leq k \leq m-1$$

$$m = 1, 2, \ldots, M-1$$

$$\alpha_3(3) = K_3 = \frac{1}{3}$$

For $m = 1$

$$\alpha_m(k) = \alpha_{m-1}(k) + k_m\, \alpha_{m-1}(m-k)$$
$$\alpha_1(k) = \alpha_0(k) + \alpha_1(1)\, \alpha_0(1-k) \quad 1 \leq k \leq m-1 \quad m = 1, 2, \ldots, M-1.$$

Hence the coefficients of an FIR filter corresponding to a single stage lattice are

$$\alpha_1(0) = 1,\ \alpha_1(1) = 1/4 = k_1$$

$$\alpha_m(k) = \alpha_{m-1}(k) + k_m\, \alpha_{m-1}(m-k)$$
$$\alpha_1(k) = \alpha_0(k) + \alpha_1(1)\, \alpha_0(1-k) \quad 1 \le k \le 0$$

The limit of *k* is not a valid limit, therefore, omitting this expression, we add second stage *m* = 2, by seeing the limit of *k*, only *k* = 1 is considered.

$$\alpha_m(k) = \alpha_{m-1}(k) + k_m\, \alpha_{m-1}(m-k)$$
$$\alpha_2(k) = \alpha_1(k) + \alpha_2(2)\, \alpha_1(2-k) \quad 1 \le k \le 1$$

$$\alpha_2(1) = \alpha_1(1) + \alpha_2(2)\, \alpha_1(1)$$
$$\alpha_2(1) = \tfrac{1}{4} + \left(\tfrac{1}{2}\right)\left(\tfrac{1}{4}\right) = \tfrac{3}{8}.$$

Finally, we add second stage *m* = 3, by seeing the limit of *k*, *k* = 1, and *k* = 2 is considered.

$$\alpha_m(k) = \alpha_{m-1}(k) + k_m\, \alpha_{m-1}(m-k)$$
$$\alpha_3(k) = \alpha_2(k) + \alpha_3(3)\, \alpha_2(3-k) \quad 1 \le k \le 2$$
$$\alpha_3(1) = \alpha_2(1) + \alpha_3(3)\, \alpha_2(2)$$
$$\alpha_3(1) = \tfrac{3}{8} + \left(\tfrac{1}{3}\right)\left(\tfrac{1}{2}\right) = \tfrac{13}{24}$$
$$\alpha_3(2) = \alpha_2(2) + \alpha_3(3)\, \alpha_2(1)$$
$$\alpha_3(2) = \tfrac{1}{2} + \left(\tfrac{1}{3}\right)\left(\tfrac{3}{8}\right) = \tfrac{5}{8}$$

Therefore, $\alpha_3(0) = 1$, $\alpha_3(1) = 13/24$, $\alpha_3(2) = 5/8$, and $\alpha_3(3) = 1/3$.

7.2.3 Frequency Sampling Form

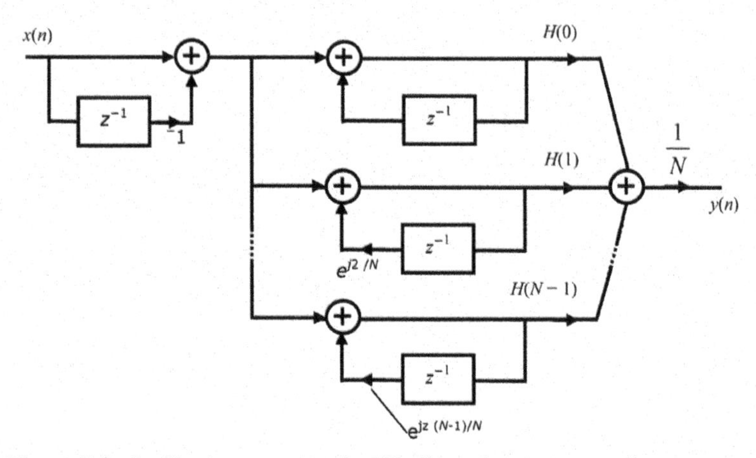

Figure 7.4 Realization structures for FIR filters: frequency sampling structure.

7.2.4 Fast Convolution Form

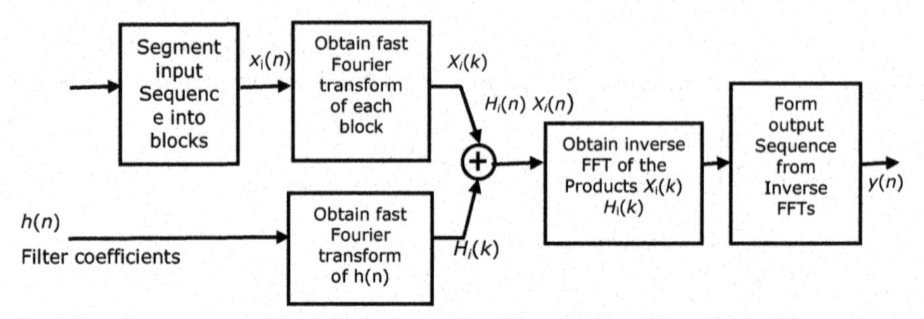

Figure 7.5 Realization structures for FIR filters: fast convolution.

7.3 Realization Form of IIR Filters

7.3.1 Direct Form I

A transfer function in z-transform form is given below

$$H(z) = \frac{y(z)}{x(z)} = \frac{N(z)}{D(z)} = \frac{a_0 + a_1 z^{-1} + \ldots + a_N z^{-N}}{1 + b_1 z^{-1} + \ldots + b_M z^{-M}} = \frac{\sum_{k=0}^{N} a_k z^{-k}}{1 + \sum_{k=1}^{M} b_k z^{-k}} \tag{7.22}$$

Therefore, the above equation can also be written as

$$Y(z) = H(z)X(z) \tag{7.23}$$

We obtain by cross multiplying

$$Y(z)\left[1 + b_1 z^{-1} + \ldots + b_k z^{-k}\right] = \left[a_0 + a_1 z^{-1} + \ldots + a_k z^{-k}\right] X(z) \tag{7.24}$$

Use of the IZT in connection with Equation (7.24) leads to

$$y(n) + b_1 y(n-1) + \ldots + b_k y(n-k) = a_0 x(n) + a_1 x(n-1) + \ldots + a_k x(n-k), \tag{7.25}$$

which can be written in a more compact form as

$$y(n) = \sum_{k=0}^{N} a_k x(n-k) - \sum_{k=1}^{M} b_k\, y(n-k) \tag{7.26}$$

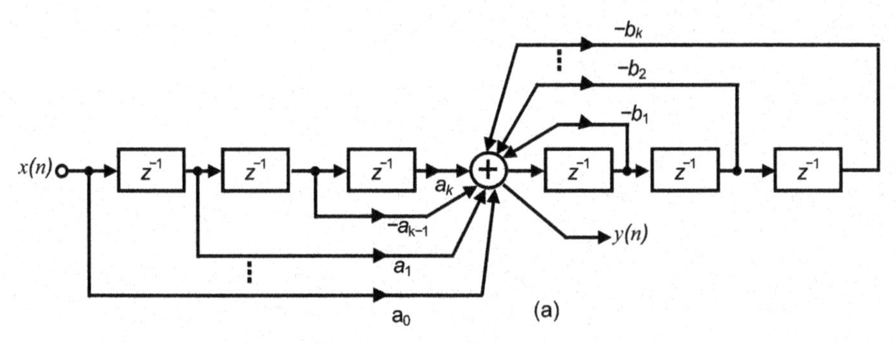

Figure 7.6 Direct form-I realization: k-th order $H(z)$.

Equation (7.26) can be realized as shown in Figure 7.6

From Figure 7.6 it is apparent that $2k$ registers or memory locations are required to store the $x(n-i)$ and $y(n-i)$, $1 \leq i \leq k$.

It is worthwhile verifying that the direct form I realization does indeed represent the fundamental relation $Y(z) = H(z)X(z)$ by tracing through it. To this end, we consider the direct form-I realization for the second-order case, which is shown in Figure 7.7. Examining Figure 7.7 at the summing junction, we obtain

$$a_0X(z) + a_1z^{-1}X(z) + a_2z^{-2}X(z) + (-b_1)\,z^{-1}Y(z) + (-b_2)\,z^{-2}\,Y(z) = Y(z) \tag{7.27}$$

which yields

$$Y(z)\left(1 + b_1z^{-1} + b_2z^{-2}\right) = X(z)\left(a_0 + a_1z^{-1} + a_2z^{-2}\right)$$

$$Y(z) = \left[\frac{a_0 + a_1z^{-1} + a_2z^{-2}}{1 + b_1z^{-1} + b_2z^{-2}}\right]X(z),$$

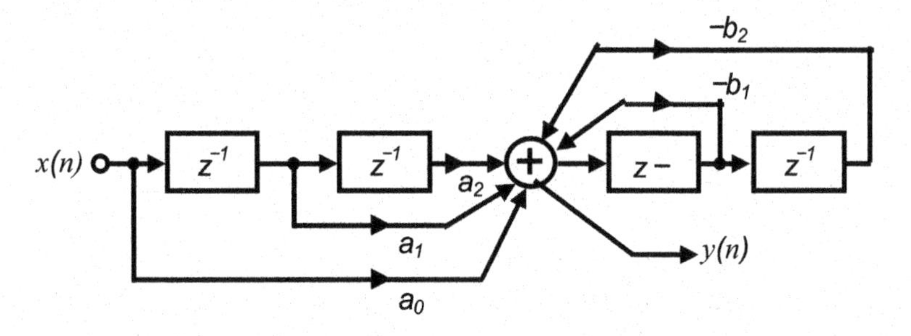

Figure 7.7 Direct form-I realization: second-order $H(z)$.

which is the desired result since the quantity in brackets is the transfer function $H(z)$ in Equation (7.22) for $k = 2$.

7.3.2 Direct Form II

From Equations (7.22) and (7.23) we have

$$Y(z) = N(z)\frac{X(z)}{D(z)}. \tag{7.28}$$

Let us define

$$W(z) = \frac{X(z)}{D(z)} = \frac{X(z)}{1 + b_1 z^{-1} + \ldots + b_M z^{-M}}. \tag{7.29}$$

Again, Equation (7.29) yields

$$X(z) = W(z)\left[1 + b_1 z^{-1} + b_2 z^{-2} + \ \ldots \ + b_k z^{-k}\right],$$

which leads to

$$x(n) = w(n) + b_1 w(n-1) + b_2 w(n-2) + \ldots + b_k w(n-k), \quad n \geq 0. \tag{7.30}$$

Then Equations (7.28) and (7.29) imply that

$$Y(z) = N(z)W(z) = (a_0 + a_1 z^{-1} + \ldots + a_N z^{-N})W(z). \tag{7.31}$$

Next Equation (7.31) can be written as

$$Y(z) = W(z)\left[a_0 + a_1 z^{-1} + a_2 z^{-2} + \ \ldots \ + a_k z^{-k}\right]$$

which yields

$$y(n) = a_0 w(n) + a_1 w(n-1) + a_2 w(n-2) + \ldots + a_k w(n-k), \quad n \geq 0. \tag{7.32}$$

We now combine Equations (7.31) and (7.32) to obtain the direct form realization shown in Figure 7.8(a) for the general case and Figure 7.8(b) for the special case $k = 2$.

We observe that this realization requires k registers to realize the kth order $H(z)$ in Equation (7.28), as opposed to $2k$ registers required for the direct form-I realization. Hence the direct form II realization is also referred to as a canonical form.

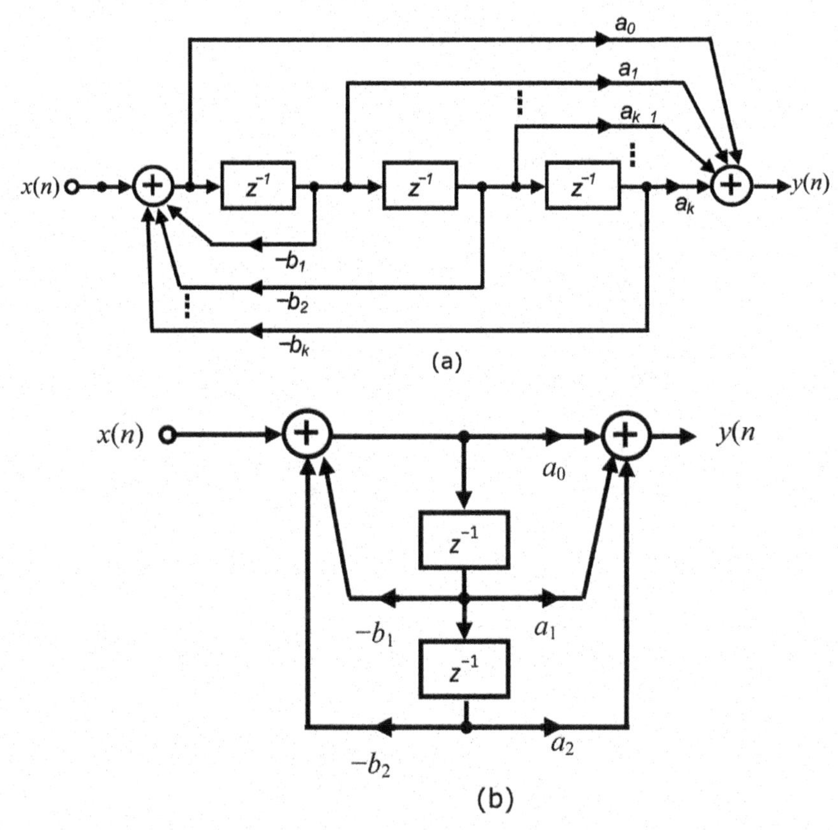

Figure 7.8 Direct form II realization: (a) kth-order $H(z)$; (b) second-order $H(z)$.

7.3.3 Cascade (Series) Form

Using straightforward polynomial factorization techniques, the numerator and denominator polynomials of $H(z)$ in Equation (7.22) can be factored into first- and second-order polynomials. As such, $H(z)$ in Equation (7.22) can be expressed as

$$H(z) = a_0 H_1(z)\, H_2(z)\, \ldots\, H_i(z) \tag{7.33}$$

Where i is a positive integer, and each $H_i(z)$ is a first- or second-order transfer function; that is,

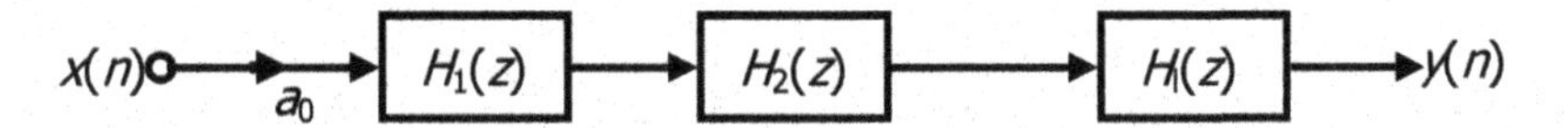

Figure 7.9 Cascade or series realization.

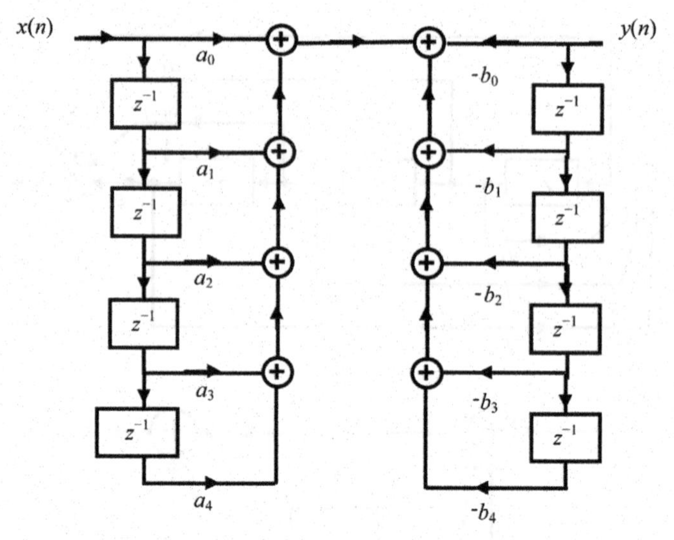

Figure 7.10 Direct form I realization of fourth order IIR filter.

$$H_i(z) = \frac{1 + a_{i1}z^{-1}}{1 + b_{i1}z^{-1}}$$

$$H_i(z) = \frac{1 + a_{i1}z^{-1} + a_{i2}z^{-2}}{1 + b_{i1}z^{-1} + b_{i2}z^{-2}}. \tag{7.34}$$

Substitution of Equation (7.33) in Equation (7.23) leads to a

$$Y(z) = a_0[H_1(z)\, H_2(z)\, \ldots\, H_l(z)]\, X(z)$$

which yields the cascade or series realization shown in Figure 7.8.

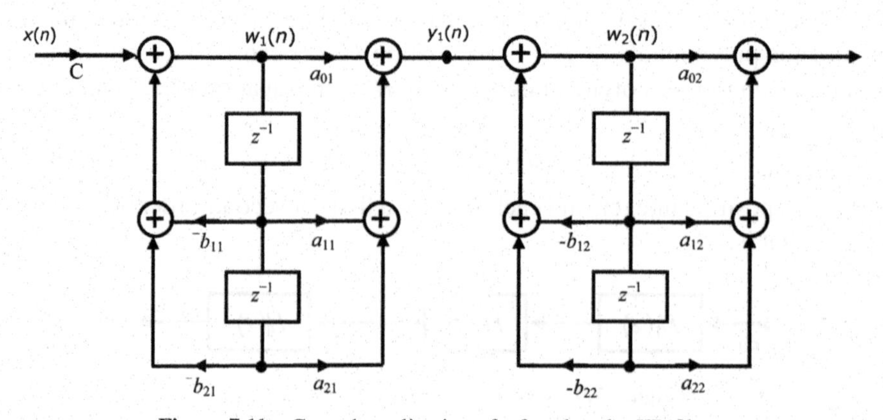

Figure 7.11 Cascade realization of a fourth order IIR filter.

$$H(z) = C\prod_{k=1}^{2}\frac{1 + a_{1k}z^{-1} + b_{2k}z^{-2}}{1 + b_{1k}z^{-1} + b_{2k}z^{-2}}$$

$$\begin{aligned}
w_1(n) &= Cx(n) - b_{11}w_1(n-1) - b_{21}w_1(n-2)\\
y_1(n) &= a_{01}w_1(n) + a_{11}w_1(n-1) + a_{21}w_1(n-2)\\
w_2(n) &= y_1(n) - b_{12}w_2(n-1) - b_{22}w_2(n-2)\\
y(n) &= a_{02}w_2(n) + a_{12}w_2(n-1) + a_{22}w_2(n-2)
\end{aligned}$$

7.3.4 Parallel Form

Here we express $H(z)$ in Equation (7.33) as

$$H(z) = C + H_1(z) + H_2(z) + \ \ldots \ + H_l(z) \tag{7.35}$$

Where C is a constant, r is a positive integer, and $H_i(z)$ is a first- or second-order transfer function; that is,

$$H_i(z) = \frac{a_{i0}}{1 + b_{i1}z^{-1}}$$

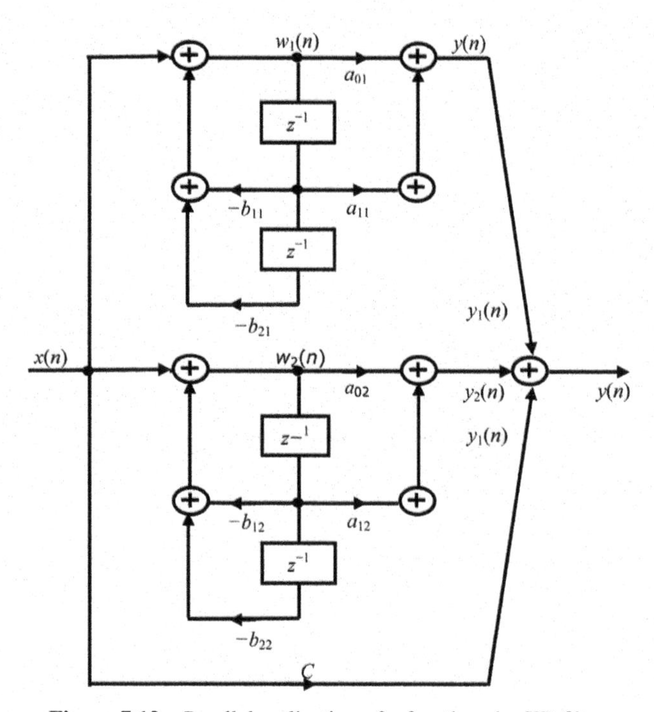

Figure 7.12 Parallel realization of a fourth-order IIR filter.

$$H_i(z) = \frac{a_{i0} + a_{i1}z^{-1}}{1 + b_{i1}z^{-1} + b_{i2}z^{-2}} \tag{7.36}$$

A given $H(z)$ can be expressed as indicated in Equation (7.35) by resorting to a PFE of $H(z)/z$. The desired parallel realization is shown in Figure 7.11, and follows directly from Equation (7.35)

$$w_1(n) = x(n) - b_{11}w_1(n-1) - b_{21}w_1(n-2)$$
$$w_2(n) = x(n) - b_{12}w_2(n-1) + b_{22}w_2(n-2)$$
$$y_1(n) = a_{01}w_1(n) + a_{11}w_1(n-1)$$
$$y_2(n) = a_{02}w_2(n) + a_{12}w_2(n-2)$$
$$y_3(n) = Cx(n)$$

Example 7.5

Given the second-order transfer function

$$H(z) = \frac{0.7(z^2 - 0.36)}{z^2 + 0.1z - 0.72}$$

Obtain the following realization.

(a) The direct form II (or canonical).
(b) Series form in terms of first-order sections.
(c) Parallel form in terms of first-order sections.

Solution 7.5

Direct Form II

(a) We rewrite $H(z)$ as

$$H(z) = \frac{0.7(1 - 0.36z^{-2})}{1 + 0.1z^{-1} - 0.72z^{-2}} = \frac{0.7(a_0 + a_1z^{-1} + a_2z^{-2})}{1 + b_1z^{-1} + b_2z^{-2}}.$$

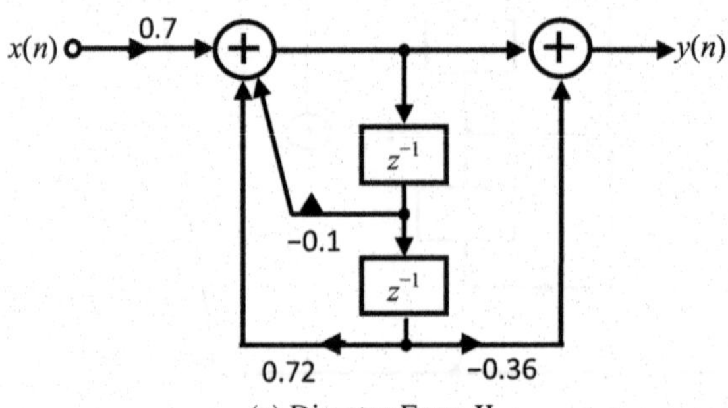

(a) Director Form II

From figure below, it is clear that the canonical form realization is as shown in the following sketch.

Series Form

(b) To obtain the series form realization, we factor the numerator and denominator polynomials of $H(z)$ to obtain

$$H(z) = \frac{0.7(1 + 0.6z^{-1})(1 - 0.6z^{-1})}{(1 + 0.9z^{-1})(1 - 0.8z^{-1})}.$$

There is nothing unique about how one combines the first-order polynomials above to obtain corresponding first-order transfer functions. For example, one choice is as follows:

$$H_1(z) = \frac{1 + 0.6z^{-1}}{1 - 0.8z^{-1}}$$

$$H_2(z) = \frac{1 - 0.6z^{-1}}{1 + 0.9z^{-1}}.$$

The above transfer functions in can be realized in terms of either of the two direct forms (i.e., 1 or 2). To illustrate, we use the direct form 2 realization in Figure given below with $a_2 = b_2 = 0$, since $H_1(z)$ and $H_2(z)$ are first-order transfer functions. This results in the following series form realization.

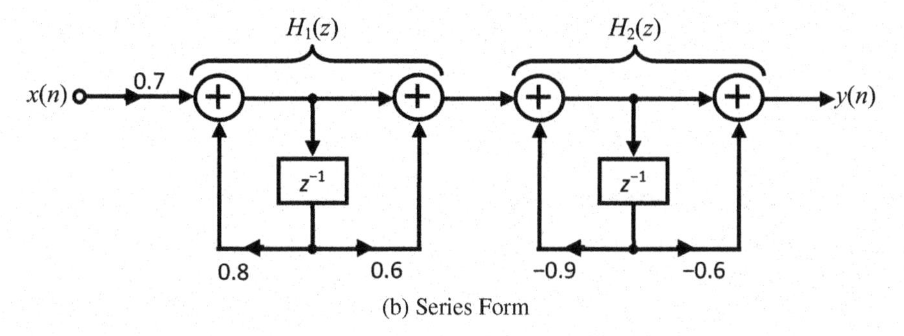

(b) Series Form

Parallel Form

(c) Next we seek a parallel form realization, and hence obtain a PFE of $H(z)/z$ is given as below,

$$H(z) = \frac{H(z)}{z} = \frac{0.7(z + 0.6)(z - 0.6)}{z(z + 0.9)(z - 0.8)}$$

$$= \frac{A}{z} + \frac{B}{z + 0.9} + \frac{C}{z - 0.8}$$

where

$$A = \widehat{H}(z)\Big|_{z=0} = 0.35 \quad B = (z + 0.9)\widehat{H}(z)\Big|_{z=-0.9} = 0.206$$

$$C = (z - 0.8)\widehat{H}(z)|_{z=0.8} = 0.144$$

Thus $H(z) = 0.35 + \frac{0.206z}{z+0.9} + \frac{0.144z}{z-0.8}$ $H(z) = 0.35 + \frac{0.206}{1+0.9z^{-1}} + \frac{0.144}{1-0.8z^{-1}}$

$$= H_1(z) + H_2(z) + H_3(z)$$

One way of realizing H_2(z) and H_3(z) is by using the direct form 2 as shown in Figure with a_2 = 0, b_2 = 0, since H_2(z) and H_3(i) are first-order transfer functions. This approach yields the following parallel realization.

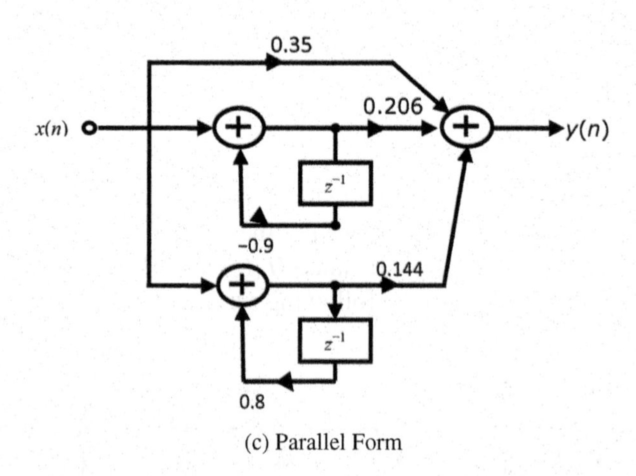

(c) Parallel Form

7.3.5 Lattice Structure for IIR Filter

We begin with all pole system with transfer function, all pole transfer function has no zeros polynomial.

$$H(z) = \frac{Y(z)}{X(z)} = \frac{1}{A_N(z)} = \frac{1}{1 + \sum_{k=1}^{N} a_N(k)\, z^{-k}} \tag{7.37}$$

Therefore the above equation can also be written as

$$Y(z) = H(z)X(z) \tag{7.38}$$

We obtain by cross multiplying

$$Y(z)[1 + a_N(1)z^{-1} + a_N(2)z^{-2} + \ldots + a_N(N)z^{-N}] = X(z) \tag{7.39}$$

$$Y(z) = X(z) - [a_N(1)z^{-1} + a_N(2)z^{-2} + \ldots + a_N(N)z^{-N}]\, Y(z)$$

which yields

$$y(n) = x(n) - a_N(1)\, y(n-1) - a_N(2)\, y(n-2) - \ldots - a_N(N) y(n-k) \quad (7.40)$$

$$y(n) = x(n) - \sum_{k=1}^{N} a_N(k)\, y(n-k) \quad (7.41)$$

it is interesting to note that if we interchange the role of input and output [i.e., interchange $x(n)$ with $y(n)$], we obtain

$$x(n) = y(n) - \sum_{k=1}^{N} a_N(k)\, x(n-k) \quad (7.42)$$

or equivalently

$$y(n) = x(n) + \sum_{k=1}^{N} a_N(k)\, x(n-k). \quad (7.43)$$

We note that Equation (7.43) describes a FIR system having the system function $H(z) = A_N(z)$, while the system is described by the difference equation represents an IIR system with system function

$$H(z) = \frac{1}{A_N(z)}. \quad (7.44)$$

System can be obtained from the other simply by interchanging the role of the input and output.

To demonstrate the set of equations, let us consider the case for $N = 1$.

$$\begin{aligned} f_1(n) &= x(n) \text{ and } y(n) = f_0(n) \\ f_0(n) &= f_1(n) - k_1 g_0(n-1) \end{aligned} \quad (7.45)$$

$$y(n) = x(n) - k_1 y(n-1)$$

$$\begin{aligned} g_1(n) &= k_1 f_0(n) + g_0(n-1) \\ g_1(n) &= k_1 y(n) + y(n-1) \end{aligned}$$

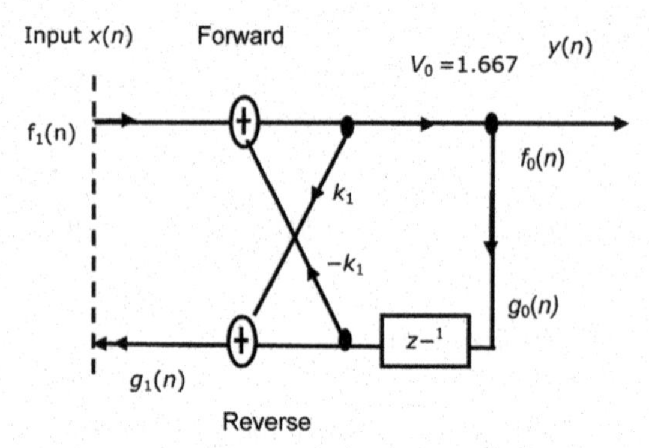

Figure 7.13 Single pole lattice filter structure.

$$\begin{aligned} f_N(n) &= x(n) \\ f_{m-1}(n) &= f_m(n) - k_m g_{m-1}(n-1) \quad m = N, N-1, \ldots\ldots, 1 \\ g_m(n) &= k_m f_{m-1}(n) - g_{m-1}(n-1) \quad m = N, N-1, \ldots\ldots, 1 \\ y(n) &= f_0(n) = g_0(n). \end{aligned} \tag{7.46}$$

To demonstrate the set of equations, let us consider the case for $N = 2$

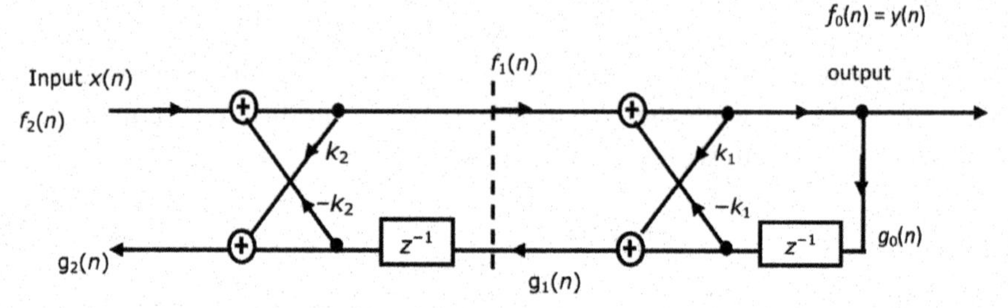

Figure 7.14 Two pole lattice filter structure.

$$y(n) = f_0(n) = g_0(n).$$

The equations corresponding to the structure are

$$\begin{aligned} f_2(n) &= x(n) \\ f_1(n) &= f_2(n) - k_2 g_1(n-1) \\ g_2(n) &= k_2 f_1(n) + g_1(n-1) \end{aligned} \tag{7.47}$$

From (7.45)

$$\begin{aligned} f_0(n) &= f_1(n) - k_1 g_0(n-1) \\ g_1(n) &= k_1 f_0(n) + g_0(n-1). \end{aligned}$$

After some simple substitution and manipulations

$$y(n) = f_0(n) = f_1(n) - k_1 g_0(n-1)$$
$$y(n) = [f_2(n) - k_2 g_1(n-1)] - k_1 g_0(n-1)$$

Because $g_1(n-1) = k_1 y(n-1) + y(n-2)$ and $y(n) = g_0(n)$

$$\begin{aligned} y(n) &= x(n) - k_2\{k_1 y(n-1) + y(n-2)\} - k_1 y(n-1) \\ y(n) &= -k_1(1+k_2)\, y(n-1)] - k_2 y(n-2) + x(n). \end{aligned} \tag{7.48}$$

In a similar way

$$g_2(n) = k_2 y(n) + k_1(1+k_2)\, y(n-1) + y(n-2). \tag{7.49}$$

The difference Equation (7.48) represents a two-pole IIR system, and the relation in Equation (7.49) is the input-output equation for a two-zero FIR system except that they occur in reverse order.

In general, the conclusion holds for any value of N. Indeed, with the definition of $A_m(z)$

$A_m(z) = \frac{F_m(z)}{X(z)} = \frac{F_m(z)}{F_0(z)}$ .... FIR filter

$H_a(z) = \frac{Y(z)}{X(z)} = \frac{F_0(z)}{F_m(z)} = \frac{1}{A_m(z)}$ .... All pole IIR filter

It is interesting to note that all pole lattice structure has an all zero path with input $g_0(n)$ and the output $g_N(n)$, which is identical to the counter part all-zero path in the all-zero lattice structure.

The polynomial $B_m(z)$, which represents the system function and all-zero path common to both lattice structures, it is usually called the backward system function, because it provides the backward path in the all pole lattice structure.

To develop the appropriate ladder structure for denominator polynomial of IIR filter, let us consider a system using Gray and Markel method using recursive equation.

$$v_q(k) = c_q(k) - \sum_{j=k+1}^{q} v_q(j)\alpha_j(j-k).$$

It is to be noted that the method used in calculation of FIR filter lattice coefficients is valid for the all pole filter and for ladder net work we use the recursive formula which is explained in next section.

7.3.5.1 Gray–Markel method of IIR lattice structure for ladder coefficients

The method of realizing an IIR filter transfer function which has been used in Matlab is summarized as below. (The recursion formula for ladder coefficients is included here).

$$H(z) = \frac{C_m(z)}{A_m(z)} = \frac{[c_0 + c_1\, z^{-1} + c_2\, z^{-2} + c_3\, z^{-3} + \ldots]}{[1 + \alpha_1\, z^{-1} + \alpha_2\, z^{-2} + \alpha_3\, z^{-3} + \ldots]}.$$

Using recursion

$$\mathrm{v_q}(k) = \mathrm{c_q}(k) - \sum_{j=k+1}^{q} v_q(j)\alpha_j(j-k)$$

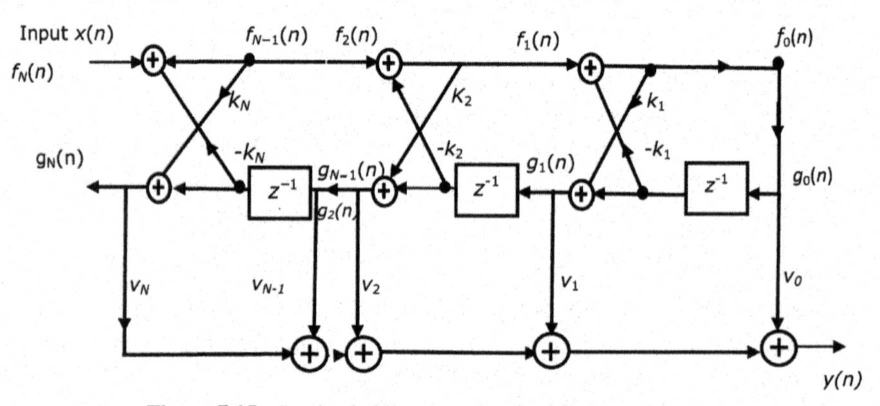

Figure 7.15 Lattice ladder structure of a pole-zero system.

Example 7.6

Determine the lattice ladder structure for the system

$$H(z) = \left[\frac{1 - 0.8\, z^{-1} + 0.15 z^{-2}}{1 + 0.1\, z^{-1} - 0.72\, z^{-2}}\right]$$

Solution 7.6

$H(z) = \frac{C_m(z)}{A_m(z)} = \frac{[c_0 + c_1\, z^{-1} + c_2\, z^{-2} + c_3\, z^{-3} + \ldots]}{[1 + \alpha_1\, z^{-1} + \alpha_2\, z^{-2} + \alpha_3\, z^{-3} + \ldots]}$. Considering the transfer function for lattice–ladder network

$$H(z) = \left[\frac{1 - 0.8\, z^{-1} + 0.15 z^{-2}}{1 + 0.1\, z^{-1} - 0.72\, z^{-2}}\right].$$

For all pole system, we have

$$H(z) = \frac{1}{A(z)}$$

Comparing $A(z) = 1 + 0.1z^{-1} - 0.72z^{-2}$

$$\alpha_2(1) = k_1(1 + k_2) = 0.1 \quad \text{and} \quad \alpha_2(2) = k_2 = -0.72$$

$$k_2 = \alpha_2(2) = -0.72 \quad k_1 = \frac{\alpha_2(1)}{1 + \alpha_2(2)} = 0.357.$$

We get the value of the coefficients for ladder network as shown in Figure 7.16

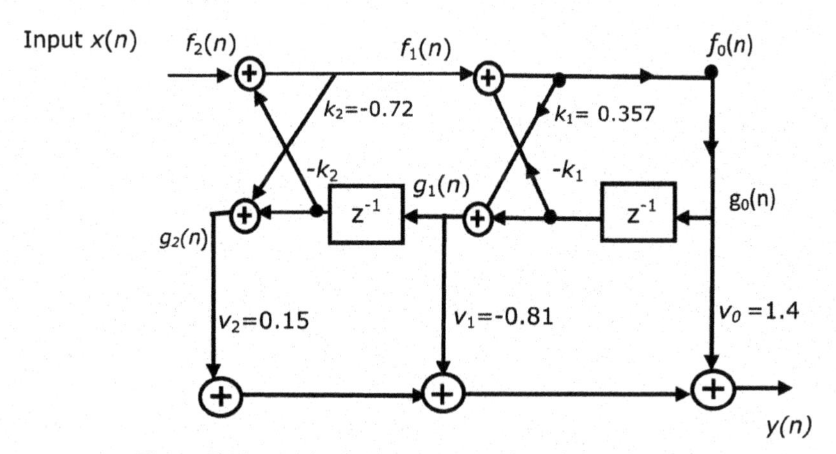

Figure 7.16 Lattice ladder structure of a pole-zero system.

$$H(z) = \frac{C_m(z)}{A_m(z)} = \frac{[c_0 + c_1\ z^{-1} + c_2\ z^{-2} + c_3\ z^{-3} + \ldots]}{[1 + \alpha_1\ z^{-1} + \alpha_2\ z^{-2} + \alpha_3\ z^{-3} + \ldots]}.$$

$$H(z) = \left[\frac{1 - 0.8\ z^{-1} + 0.15\ z^{-2}}{1 + 0.1\ z^{-1} - 0.72\ z^{-2}}\right]$$

7.3.5.2 Calculation of ladder coefficients using Gray–Markel method

Using recursion

$$\mathrm{v_q}(k) = \mathrm{c_q}(k) - \textstyle\sum_{j=k+1}^{q} v_q(j)\alpha_j(j - k).$$

Because the filter transfer function is of second order, therefore, $q = 2$, the value of k will be taken in descending order, i.e., 2, 1, 0

$$\begin{aligned}
v_2(2) &= c_2(2)\\
v_2(1) &= c_2(1) - v_2(2)\,\alpha_2(1)\\
v_2(0) &= c_2(0) - v_2(2)\,\alpha_2(2) - v_2(1)\,\alpha_1(1)
\end{aligned}$$

$$\begin{aligned}
v_2(2) &= c_2(2) = 0.15\\
v_2(1) &= c_2(1) - v_2(2)\,\alpha_2(1) = (-0.8) - (0.15)(0.1) = -0.815\\
v_2(0) &= c_2(0) - v_2(2)\,\alpha_2(2) - v_2(1)\,\alpha_2(1) = 1 - (0.15)(-0.72)\\
&\quad -(-0.815)(0.357) = 1.3989.
\end{aligned}$$

It is to be noted here that to calculate the value of ladder networks the method given in Matlab is used here.

7.4 Implementation of Filters

Having calculated the filter coefficients, chosen a suitable structure, and verified that the filter degradation, after quantizing the coefficients and filter variables to the selected word-lengths, is acceptable, the difference equation must be implemented as a software routine or in hardware.

Whatever the method of implementation, the output of the filter must be computed, for each sample, in accordance with the difference equation (assuming a time-domain implementation).

As the examination of any difference equation will show the computation of $y(n)$ (the filter output) involves only multiplications, additions/subtractions, and delays.

Thus to implement a filter, we need the following basic building blocks:

(a) Memory (for example ROM) for storing filter coefficients
(b) Memory (such as RAM) for storing the present and past inputs and outputs, that is $\{x(n), x(n-1), \ldots\}$ and $\{y(n), y(n-1), \ldots\}$
(c) Hardware or software multiplier(s)
(d) Adder or arithmetic logic unit

Example 7.7

(a) Discuss the five main steps involved in the design of digital filters, using the following design problem to illustrate your answer.
(b) A digital filter is required for real-time noise reduction. The filter should meet the following amplitude response specification:

Passband frequency	0–10 Hz
Stopband frequency	20–64 Hz
Sampling frequency	128 Hz
Maximum passband deviation	<0.036 dB
Stopband attenuation	>30 dB

Other important requirements are that

(i) minimal distortion of the harmonic relationships between the components of the in-band signals is highly desirable,
(ii) the time available for filtering is limited, the filter being part of a larger process, and
(iii) the filter will be implemented using the Texas Instruments TMS32010 DSP processor with the analogue input digitized to 12 bits.

Solution 7.7

This filter was designed and used in a certain biomedical signal processing project. Given here only an is an outline discussion of the design, postponing detailed discussion to Chapter 7 where FIR filter design methods are fully covered.

(i) Requirement specification. As discussed previously, the designer must give the exact role and performance requirements for the filter together with any important constraints. These have already been given for the example.
(ii) Calculation of suitable coefficients. The requirements of minimal distortion and limited processing time are best achieved with a linear phase FIR filter, with coefficients obtained using the optima method.
(iii) Selection of filter structure. The transversal structure will lead to the most efficient implementation using the processor either floating or fixed point.
(iv) Analysis of finite word-length effects. In the processor fixed point arithmetic is used with each coefficient represented by 16 bits (after rounding) for efficiency. FIR filter degradation may result from input signal quantization, coefficient quantization, roundoff and overflow errors.

A check should be made to ensure that the world-lengths are sufficiently long. Analysis of finite word-length effects for this case showed that the input quantization noise and deviation in the frequency response due to coefficient quantization are both insignificant.

The use of the processor having 32-bit accumulator to sum the coefficient data products, rounding only the final sum, would reduce roundoff errors to negligible levels. To avoid overflow each coefficient should be divided by$\sum_{k=0}^{N-1}|h(k)|$before quantizing to 16 bits.

(v) *Implementation.* Design and configure the processor is based hardware (if it does not already exist) with the necessary input/output interfaces. Then write a program for the processor to handle the I/O protocols and calculate filter output,

(vi) $y(n) = \sum_{k=0}^{N-1} h(k)x(n-k)$, for each new input, $x(n)$

7.5 Problems and Solutions

Problem 7.1

The transfer function for an FIR filter is given by

$$H(z) = 1 - 1.3435\, z^{-1} + 0.9025\, z^{-2}$$

Draw the realization block diagram for each of the following cases:

(1) Transversal structure;
(2) Two-stage lattice structure.

Calculate the values of the coefficients for the lattice structure.

Solution 7.1

(1) From the transfer function, the diagram for the transversal structure is given in Figure (a). The input and output of the transversal structure are given by

$$y(n) = x(n) + h(1)\, x(n-1) + h(2)\, x(n-2) \tag{1}$$

(2) A two-stage lattice structure for the filter is given in Figure (b). The outputs of the structure are related to the input as

$$y_2(n) = y_1(n) + K_2 w_1(n-1)$$

$$y_2(n) = x(n) + K_1 w_1(1 + K2)x(n-1) + K_2 x(n-2) \tag{2}$$

$$w_2(n) = K_2 x(n) + K_2(1 + K_2)x(n-1) + x(n-2) \tag{3}$$

Comparing Equations (1) and (2), and equating coefficients, we have

$$K_1 = \frac{h(1)}{1 + h(2)}; \quad K_2 = h(2)$$

from which $K_2 = 0.9025$ and $K_1 = -1.3435/(1 + 0.9025) = -0.7062$.

Notice that the coefficients of $y_2(n)$ and $w_2(n)$ in Equations (2) and (3) are identical except that one is written in reverse order. This is a characteristic feature of the FIR lattice structure.

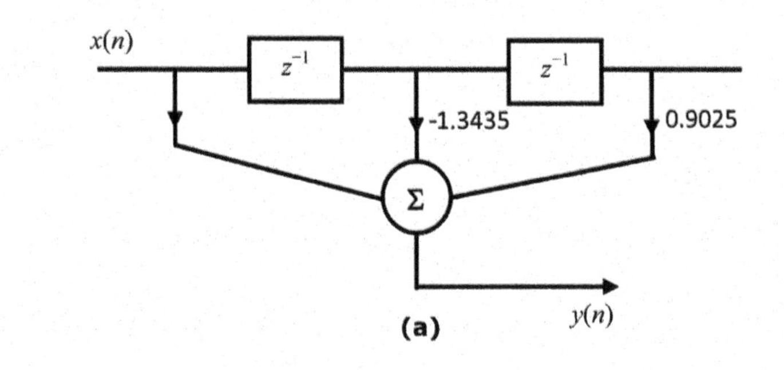

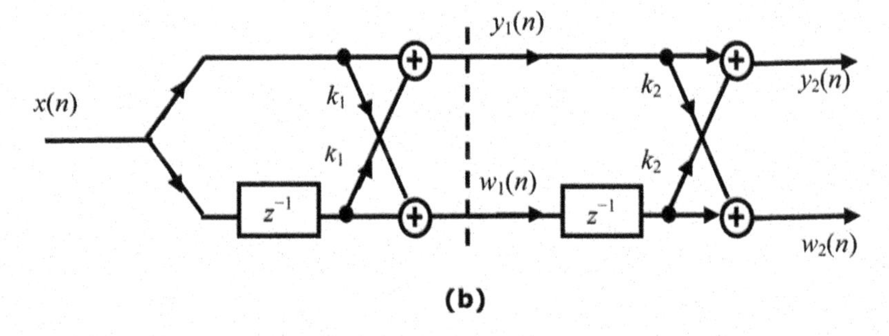

Problem 7.2

Determine the parameter $\{k_m\}$ of the lattice filter corresponding to the FIR filter described by the system function

$$H(z) = A_2(z) = 1 + 2\, z^{-1} + \frac{1}{3}\, z^{-2}$$

Solution 7.2

We note that $k_2 = \alpha_2(2) = 1/3$

$$B_2(z) = \frac{1}{3} + 2\, z^{-1} + \, z^{-2}.$$

The step-down relationship with $m = 3$ yields

$$A_{m-1}(z) = \frac{A_m(z) - k_m\, B_m(z)}{1 - k_m^2}$$

$$A_1(z) = \frac{A_2(z) - k_2\, B_2(z)}{1 - k_2^2}$$

$$A_1(z) = \frac{[1 + 2\, z^{-1} + \frac{1}{3}\, z^{-2}] - \frac{1}{3}[\frac{1}{3} + 2\, z^{-1} + \, z^{-2}]}{1 - \frac{1}{9}}$$

$$A_1(z) = 1 + \frac{3}{2} z^{-1}$$

$$k_1 = \alpha_2(1) = 3/2$$

Problem 7.3

Determine the impulse response of an FIR lattice with parameter $k_1 = 0.6$, $k_2 = 0.3,\ k_3 = 0.5,$ and $k_4 = 0.9$.

Solution 7.3

$$A_0(z) = B_0(z) = 1$$

$$A_m(z) = A_{m-1}(z) \ + k_m\, z^{-1}\, B_{m-1}(z) \quad m = 0, 1, 2. \ldots, M - 1$$

for $m = 1$

$$A_1(z) = \ A_0(z) + k_1\, z^{-1}\, B_0(z) = 1 + 0.6\, z^{-1}$$

$k_1 = 0.6$

since $B_m(z)$ is the reverse polynomial of $A_m(z)$

$$B_1(z) = \ 0.6 + \ z^{-1}$$

we add second stage $m = 2$

$$A_2(z) = \ A_1(z) + k_2\, z^{-1}\, B_1(z)$$
$$A_2(z) = 1 + 0.78\, z^{-1} + 0.3\, z^{-2}$$

also

$$B_2(z) = 0.3 + 0.78\, z^{-1} + \ z^{-2}$$

Finally, we add second stage $m = 3$

$$A_3(z) = \ A_2(z) + k_3\, z^{-1}\, B_2(z)$$
$$A_3(z) = 1 + 0.93\, z^{-1} + 0.69\, z^{-2} + 0.5\, z^{-3}$$

$$B_3(z) = 0.5 + 0.69\, z^{-1} + 0.93\, z^{-2} + z^{-3}$$
$$A_4(z) = \ A_3(z) + k_4\, z^{-1}\, B_3(z)$$
$$A_4(z) = 1 + 1.38\, z^{-1} + 1.311\, z^{-2} + 1.337\, z^{-3} + 0.9\, z^{-4}$$

$\alpha_4(0) = 1$, $\alpha_4(1) = 1.38$, $\alpha_4(2) = 1.311$, $\alpha_4(3) = 1.337$, and $\alpha_4(4) = 0.9$

Problem 7.4

Given the second-order transfer function

$$H(z) = \frac{0.7(z^2 - 0.36)}{z^2 + 0.1z - 0.72}.$$

Obtain the following realization.

(d) The direct form II (or canonical).
(e) Series form in terms of first-order sections.
(f) Parallel form in terms of first-order sections.

Solution 7.4

Direct Form II

(a) We rewrite $H(z)$ as

$$H(z) = \frac{0.7(1 - 0.36z^{-2})}{1 + 0.1z^{-1} - 0.72z^{-2}} = \frac{0.7(a_0 + a_1z^{-1} + a_2z^{-2})}{1 + b_1z^{-1} + b_2z^{-2}}.$$

From figure below, it is clear that the canonical form realization is as shown in the following sketch.

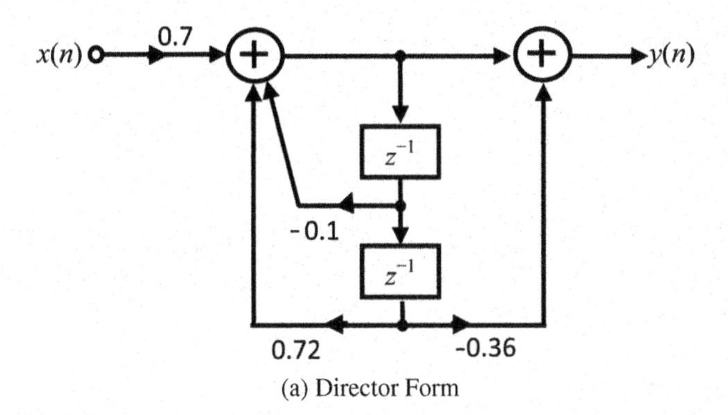

(a) Director Form

Series Form

(b) To obtain the series form realization, we factor the numerator and denominator polynomials of $H(z)$ to obtain

$$H(z) = \frac{0.7(1 + 0.6z^{-1})(1 - 0.6z^{-1})}{(1 + 0.9z^{-1})(1 - 0.8z^{-1})}$$

There is nothing unique about how one combines the first-order polynomials above to obtain corresponding first-order transfer functions. For example, one choice is as follows: $H_1(z) = \frac{1+0.6z^{-1}}{1-0.8z^{-1}}$

$$H_2(z) = \frac{1-0.6z^{-1}}{1+0.9z^{-1}}$$

The above transfer functions in can be realized in terms of either of the two direct forms (i.e., 1 or 2). To illustrate, we use the direct form 2 realization in Figure given below with $a_2 = b_2 = 0$, since $H_1(z)$ and $H_2(z)$ are first-order transfer functions. This results in the following series form realization.

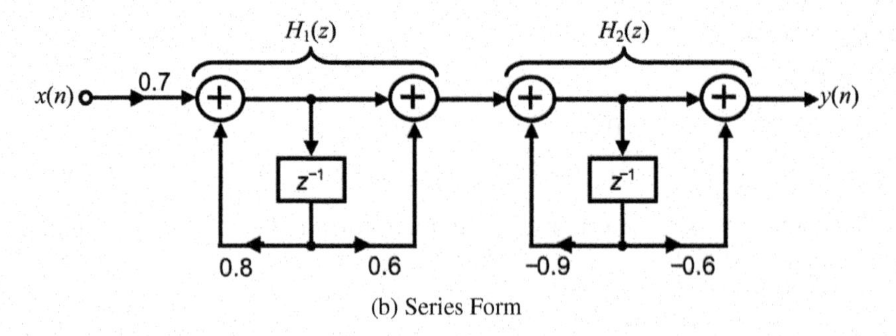

(b) Series Form

Parallel Form

(c) Next we seek a parallel form realization, and hence obtain a PFE of $H(z)/z$ is given as below,

$$H(z) = \frac{H(z)}{z} = \frac{0.7(z+0.6)(z-0.6)}{z(z+0.9)(z-0.8)}$$

$$= \frac{A}{z} + \frac{B}{z+0.9} + \frac{C}{z-0.8}$$

where $A = \widehat{H}(z)|_{z=0} = 0.35$; $B = (z+0.9)H(z)|_{z=-0.9} = 0.206$, and $C = (z-0.8)\widehat{H}(z)|_{z=0.8} = 0.144$.

Thus $H(z) = 0.35 + \frac{0.206z}{z+0.9} + \frac{0.144z}{z-0.8}$ $H(z) = 0.35 + \frac{0.206}{1+0.9z^{-1}} + \frac{0.144}{1-0.8z^{-1}}$

$$= H_1(z) + H_2(z) + H_3(z).$$

One way of realizing $H_2(z)$ and $H_3(z)$ is by using the direct form 2 as shown in Figure with $a_2 = 0$, $b_2 = 0$, since $H_2(z)$ and $H_3(z)$ are first-order transfer functions. This approach yields the following parallel realization.

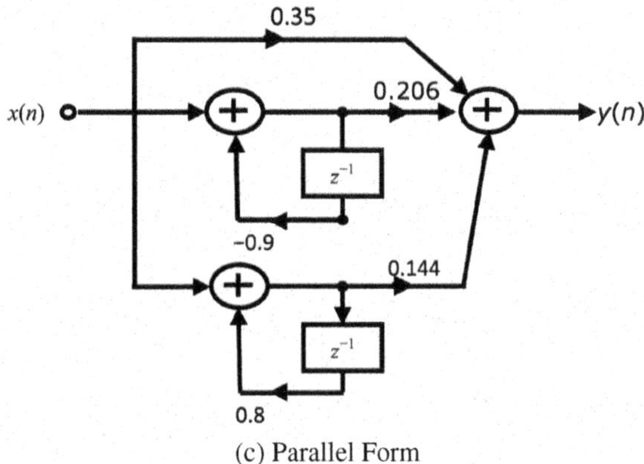

(c) Parallel Form

Problem 7.5

Given the second-order transfer function

$$H(z) = \frac{0.7(z^2 - 0.36)}{z^2 + 0.1z - 0.72}$$

Obtain the following realization.

(a) The direct form II (or canonical).
(b) Series form in terms of first-order sections.
(c) Parallel form in terms of first-order sections.

Solution 7.5

Direct Form II

(a) We rewrite $H(z)$ as

$$H(z) = \frac{0.7(1 - 0.36z^{-2})}{1 + 0.1z^{-1} - 0.72z^{-2}} = \frac{0.7(a_0 + a_1 z^{-1} + a_2 z^{-2})}{1 + b_1 z^{-1} + b_2 z^{-2}}$$

(a) Director Form II

From figure below, it is clear that the canonical form realization is as shown in the following sketch.

Series Form

(b) To obtain the series form realization, we factor the numerator and denominator polynomials of $H(z)$ to obtain

$$H(z) = \frac{0.7(1 + 0.6z^{-1})(1 - 0.6z^{-1})}{(1 + 0.9z^{-1})(1 - 0.8z^{-1})}$$

There is nothing unique about how one combines the first-order polynomials above to obtain corresponding first-order transfer functions. For example, one choice is as follows:

$$H_1(z) = \frac{1 + 0.6z^{-1}}{1 - 0.8z^{-1}} \qquad H_2(z) = \frac{1 - 0.6z^{-1}}{1 + 0.9z^{-1}}$$

The above transfer functions in can be realized in terms of either of the two direct forms (i.e., 1 or 2). To illustrate, we use the direct form 2 realization in Figure given below with $a_2 = b_2 = 0$, since $H_1(z)$ and $H_2(z)$ are first-order transfer functions. This results in the following series form realization.

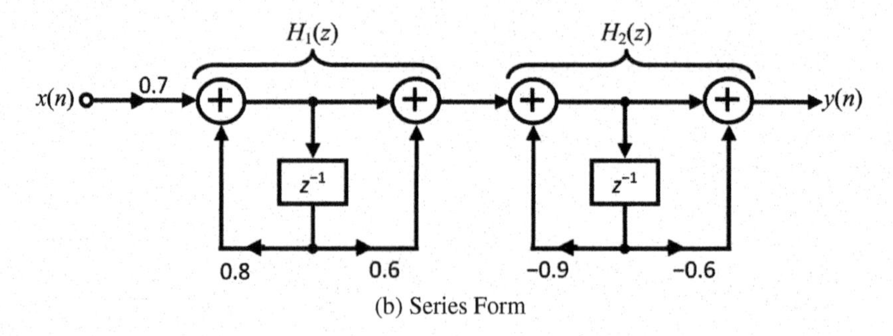

(b) Series Form

Parallel Form

(c) Next we seek a parallel form realization, and hence obtain a PFE of $H(z)/z$ is given as below,

$$H(z) = \frac{H(z)}{z} = \frac{0.7(z + 0.6)(z - 0.6)}{z(z + 0.9)(z - 0.8)}$$

$$= \frac{A}{z} + \frac{B}{z + 0.9} + \frac{C}{z - 0.8}$$

where A = $\widehat{H}(z)|_{z=0} = 0.35$ $B = (z+0.9)\widehat{H}(z)|_{z=-0.9} = 0.206$
$C = (z-0.8)\widehat{H}(z)|_{z=0.8} = 0.144$

Thus $H(z) = 0.35 + \frac{0.206z}{z+0.9} + \frac{0.144z}{z-0.8}$ $H(z) = 0.35 + \frac{0.206}{1+0.9z^{-1}} + \frac{0.144}{1-0.8z^{-1}}$

$$= H_1(z) + H_2(z) + H_3(z)$$

One way of realizing $H_2(z)$ and $H_3(z)$ is by using the direct form 2 as shown in Figure with $a_2 = 0$, $b_2 = 0$, since $H_2(z)$ and $H_3(z)$ are first-order transfer functions. This approach yields the following parallel realization.

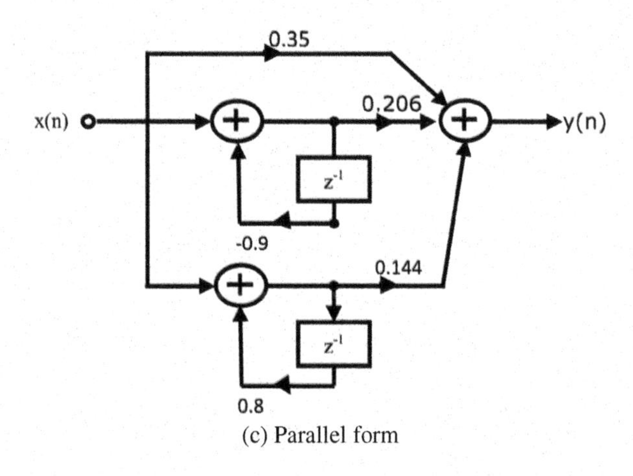

(c) Parallel form

Problem 7.6

Consider a causal IIR filter with system function

$$H(z) = \frac{1+2z^{-1}+3z^{-2}+2z^{-3}}{1+0.9z^{-1}-0.8z^{-2}+0.5z^{-3}}$$

Determine the equivalent lattice ladder structure for the system

Solution 7.6

$$H(z) = \frac{1+2z^{-1}+3z^{-2}+2z^{-3}}{1+0.9z^{-1}-0.8\,z^{-2}+0.5z^{-3}} = \frac{C_3(z)}{A_3(z)}$$

For all pole system, we have

$$H(z) = \frac{1}{A(z)}$$

Comparing $A_3(z) = 1 + 0.9\,z^{-1} - 0.8\,z^{-2} + 0.5\,z^{-3}$

$$B_3(z) = 0.5 - 0.8\,z^{-1} + 0.9\,z^{-2} + z^{-3}$$

From above equation, $k_3 = 0.5$

$$A_{m-1}(z) = \frac{A_m(z) - k_m\,B_m(z)}{1 - k_m^2} \qquad m = 0, 1, 2. \ldots, M - 1$$

The equations mentioned earlier and again given above are valid for all pole system,

$$A_2(z) = \frac{A_3(z) - k_3\,B_3(z)}{1 - k_3^2}$$

$$A_2(z) = \frac{[1 + 0.9\,z^{-1} - 0.8\,z^{-2} + 0.5\,z^{-3}]}{1 - (0.5)^2}$$

$$\frac{-0.5[0.5 - 0.8\,z^{-1} + 0.9\,z^{-2} + z^{-3}]}{1 - (0.5)^2}$$

$$A_2(z) = 1 + 1.73\,z^{-1} - 1.67\,z^{-2}$$

$$B_3(z) = -1.67 + 1.73z^{-1} + z^{-2}$$

From above equation, $k_2 = -1.67$

$$A_1(z) = \frac{A_2(z) - k_2\,B_2(z)}{1 - k_2^2}$$

$$A_1(z) = \frac{[1 + 1.73\,z^{-1} - 1.67\,z^{-2}] - (-1.67)[-1.67 + 1.73\,z^{-1} + z^{-2}]}{1 - (-1.67)^2}$$

$$A_1(z) = 1 - 2.58\,z^{-1}$$

$$B_1(z) = -2.58 + z^{-1}$$

From above equation, $k_1 = -2.58$

Solution for ladder coefficients using Gray-Markel Method

$$H(z) = \frac{1 + 2\,z^{-1} + 3z^{-2} + 2z^{-3}}{1 + 0.9\,z^{-1} - 0.8\,z^{-2} + 0.5z^{-3}}$$

$$H(z) = \frac{C_m(z)}{A_m(z)} = \frac{[c_0 + c_1\,z^{-1} + c_2\,z^{-2} + c_3\,z^{-3} + \ldots]}{[1 + \alpha_1\,z^{-1} + \alpha_2\,z^{-2} + \alpha_3\,z^{-3} + \ldots]}.$$

In the above solution k_1, k_2 and k_3 has already been calculated as follows:

$$\alpha_1(1) = k_1 = -2.58 \; \alpha_2(2) = k_2 = -1.67 \;\; \alpha_3(3) = k_3 = -0.5$$

It is to be noted here that to calculate the value of ladder networks the method given in Matlab is used here.

Now using Gray–Markel method, calculation of ladder coefficients are as follows Using recursion

$$v_q(k) = c_q(k) - \sum_{j=k+1}^{q} v_q(j)\alpha_j(j-k)$$

Because the filter transfer function is of second order therefore $q = 3$, the value of k will be taken in descending order i.e. 3, 2, 1, 0

$$\begin{aligned}
v_3(3) &= c_3(3) = 2 \\
v_3(2) &= c_3(2) - v_3(3)\,\alpha_3(1) = 3 - 2(0.9) = 1.2 \\
v_3(1) &= c_3(1) - v_3(3)\,\alpha_3(2) - v_3(2)\,\alpha_2(1) = 2 - 2(-0.8) \\
&\quad - (1.2)(1.73) = 1.524 \\
v_3(0) &= c_3(0) - v_3(3)\,\alpha_3(3) - v_3(2)\,\alpha_2(2) - v_3(1)\,\alpha_1(1) \\
&= 1 - 2(0.5) - 1.2(-1.67) - 1.524(-2.58) = 5.93592
\end{aligned}$$

It is to be noted that value of ladder networks are calculated using Gray and Markel method. The values of the ladder coefficients by this method are same as calculated using Matlab. We get the value of the coefficients for ladder network as shown in Figure.

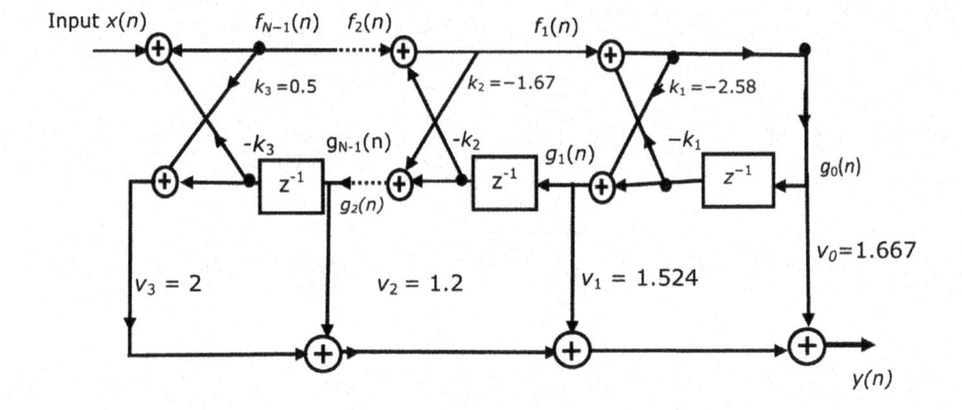

Problem 7.7

Determine the direct form II realization for each of the following LTI system

(a) $2\,y(n) + y(n-1) - 4\,y(n-3) = x(n) + 3\,x(n-5)$
(b) $y(n) = x(n) - x(n-1) + 2\,x(n-2) - 3\,x(n-4)$

Solution 7.7

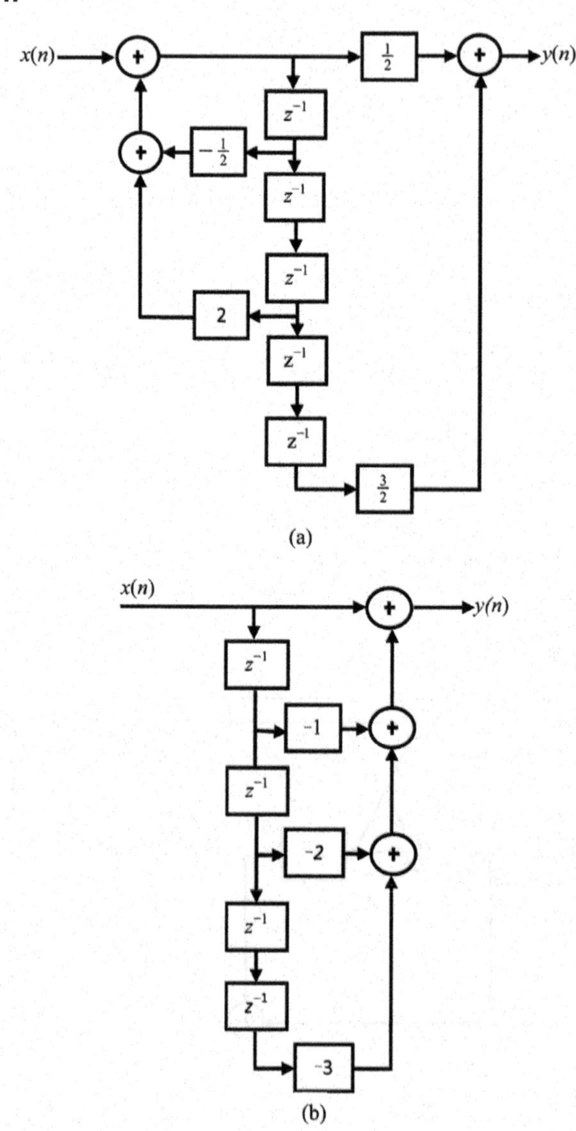

Problem 7.8

A discrete time system is realized by the structure shown below. Determine realization for its inverse system, that is, the system which produces $x(n)$ as an output when $y(n)$ is used as an input.

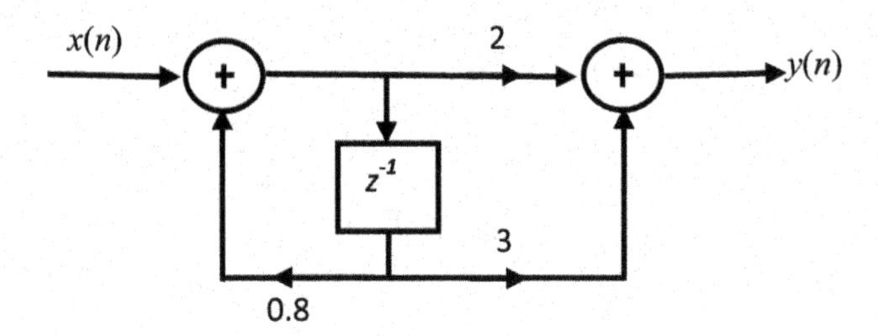

Solution 7.8

The inverse system is characterized by the difference equation

$$x(n) = -1.5\,x(n-1) + y(n) - 0.4\,y(n-1)$$

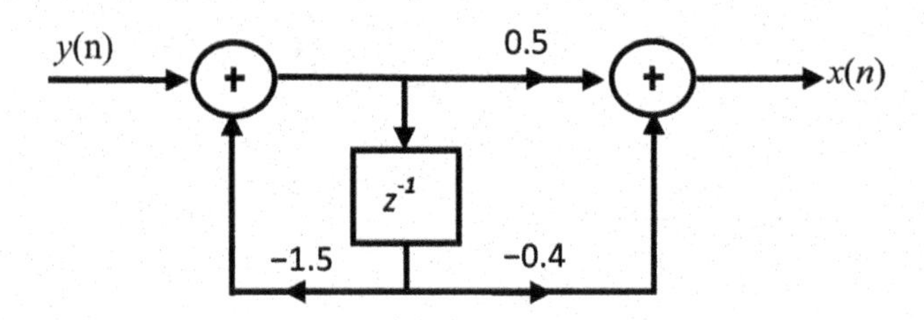

Problem 7.9

Two signals $s(n)$ and $v(n)$ are related through the following difference equations

$$s(n) + a_1 s(n-1) + \ldots + a_N s(n-N) = b_o v(n)$$

Design the block diagram realization of:

(a) The system that generates $s(n)$ when excited by $v(n)$.
(b) The system that generates $v(n)$ when excited by $s(n)$.
(c) What is the impulse response of the cascade interconnection of systems in parts (a) and (b)?

Solution 7.9

(a) $s(n) = -a_1 s(n-1) \; - \; a_2 s(n-2) \; - \; \ldots \; - \; a_N s(n-N) \; + \; b_0 v(n)$

(b) $v(n) = \frac{1}{b_0}[s(n) \; + \; a_1 s(n-1) + \; \ldots \; + \; a_N s(n-N)]$

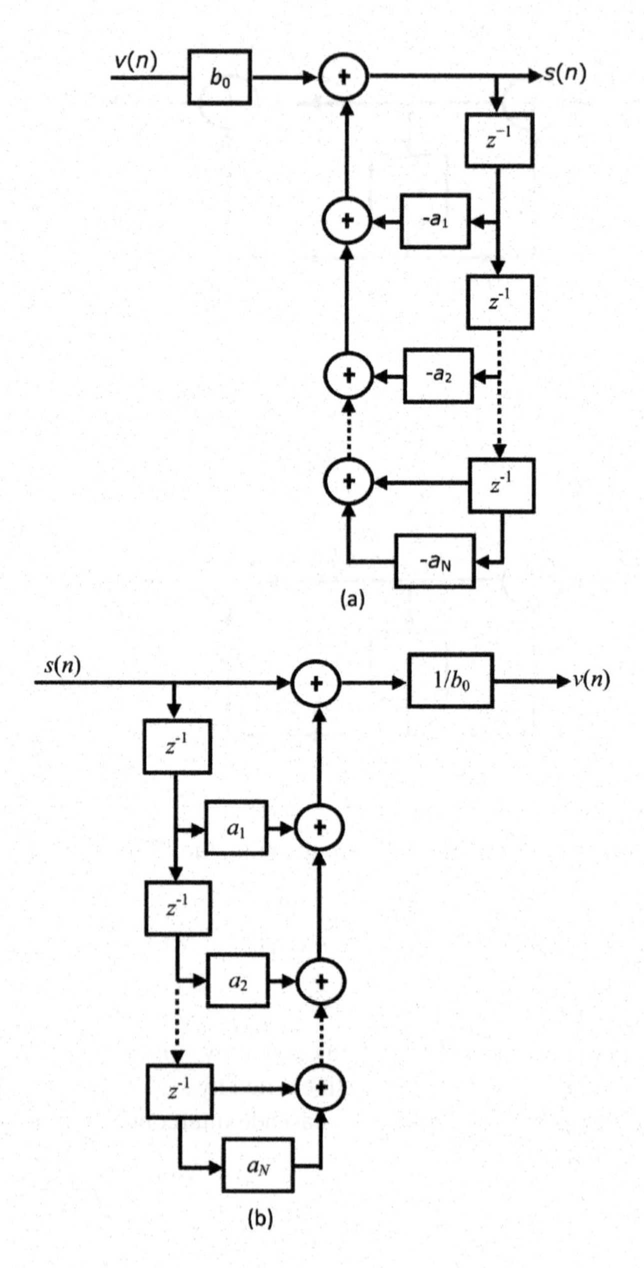

Problem 7.10

Determine and sketch the impulse response of the following systems for $n = 0, 1, \ldots, 9$.

(a) Figures (a), (b), and (c)
(b) Classify the systems below as FIR or IIR.
(c) Find an explicit expression for the impulse response of the system in part (c).

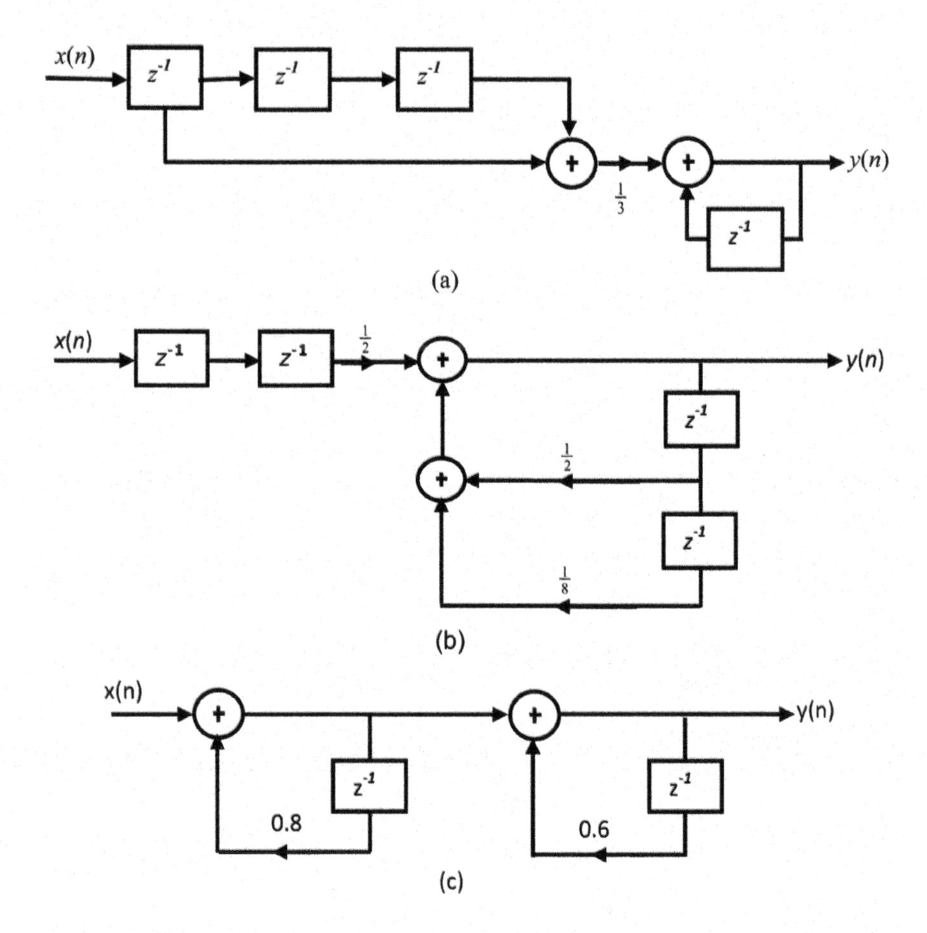

Solution 7.10

(a)

(i) $y(n) = \frac{1}{3}x(n) + \frac{1}{3}x(n-3) + y(n-1)$

for $x(n) = \delta(n)$, we have

$$h(n) = \{\tfrac{1}{3}, \tfrac{1}{3}, \tfrac{1}{3}, \tfrac{2}{3}, \tfrac{2}{3}, \tfrac{2}{3}, \tfrac{2}{3}, \ldots\}$$

(ii) $y(n) = \frac{1}{2}y(n-1) + \frac{1}{8}y(n-2) + \frac{1}{2}x(n-2)$
with $x(n) = \delta(n)$, and $y(-1) = y(-2) = 0$

$$h(n) = \{0, 0, \tfrac{1}{2}, \tfrac{1}{4}, \tfrac{3}{16}, \tfrac{1}{8}, \tfrac{11}{128}, \tfrac{15}{256}, \tfrac{41}{1024}, \ldots\}$$

(iii) $y(n) = 1.4y(n-1) - 0.48y(n-2) + x(n)$
For $x(n) = \delta(n)$, $y(-1) = 0$, $y(-2) = 0$, we obtain
$h(n) = \{1,\ 1.4,\ 1.48,\ 1.4,\ 1.2496,\ 1.0774,\ 0.9086,\ \ldots\}$

(b) All three systems are IIR
(c) $y(n) = 1.4y(n-1) - 0.48y(n-2) + x(n)$
The characteristic equation is $\lambda^2 - 1.4\lambda + 0.48 = 0$

Hence $\lambda = 0.8,\ 0.6$ and $y_h(n) = C_1(0.8)^n + C_2(0.6)^n$
For $x(n) = \delta(n)$, we have $C_1 + C_2 = 1$ and $0.8\ C_1 + 0.6\ C_2 = 1.4$

Therefore $C_1 = 4$, $C_2 = -3$, yields the final solution

$$y_h(n) = [4\ (0.8)^n - 3\ (0.6)^n]\ u(n)$$

Problem 7.11

Obtain the direct form II for the following systems.

(a) $y(n) = \frac{3}{4}y(n-1) - \frac{1}{8}y(n-2) + x(n) + \frac{1}{3}x(n-1)$
(b) $y(n) = -0.1\ y(n-1) + 0.72\ y(n-2) + 0.7\ x(n) - 0.252x(n-2)$
(c) $y(n) = -0.1\ y(n-1) + 0.2\ y(n-2) + 3\ x(n) + 3.6\ x(n-1) + 0.6\ x(n-2)$

Solution 7.11

(a) Taking the Z-transform of both the sides of the given equation (a), we get:

$$Y(z) = \frac{3}{4}z^{-1}Y(z) - \frac{1}{8}z^{-2}Y(z) + X(z) + \frac{1}{3}z^{-1}X(z)$$

$$\frac{Y(z)}{X(z)} = \frac{1 + \frac{1}{3}z^{-1}}{1 - \frac{3}{4}z^{-1} + \frac{1}{8}z^{-2}}$$

$$H(z) = \frac{1 + \frac{1}{3}z^{-1}}{1 - \frac{3}{4}z^{-1} + \frac{1}{8}z^{-2}}$$

Direct Form II

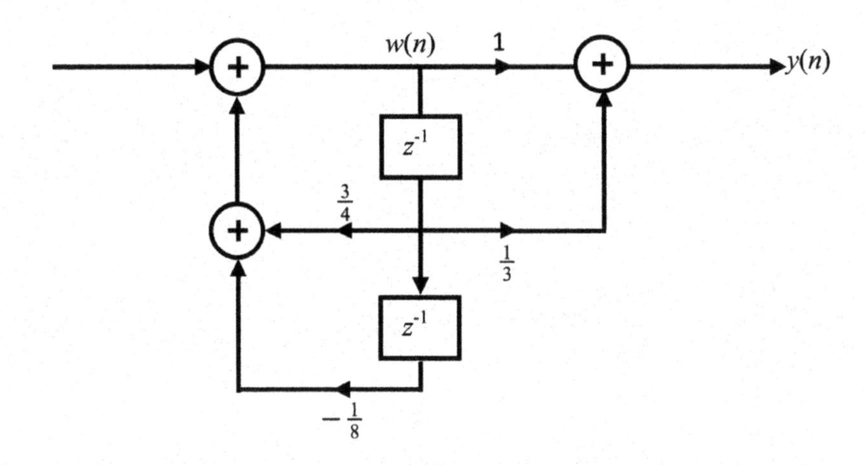

(b)

$$Y(z) = -0.1\,z^{-1}Y(z) + 0.72\,z^{-2}Y(z) + 0.7\,X(z)$$
$$-0.252\,z^{-2}X(z)$$

$$\frac{Y(z)}{X(z)} = \frac{0.7 - 0.252z^{-2}}{1 + 0.1z^{-1} - 0.72z^{-2}}$$

$$H(z) = \frac{0.7 - 0.252z^{-2}}{1 + 0.1z^{-1} - 0.72z^{-2}}$$

Direct Form II

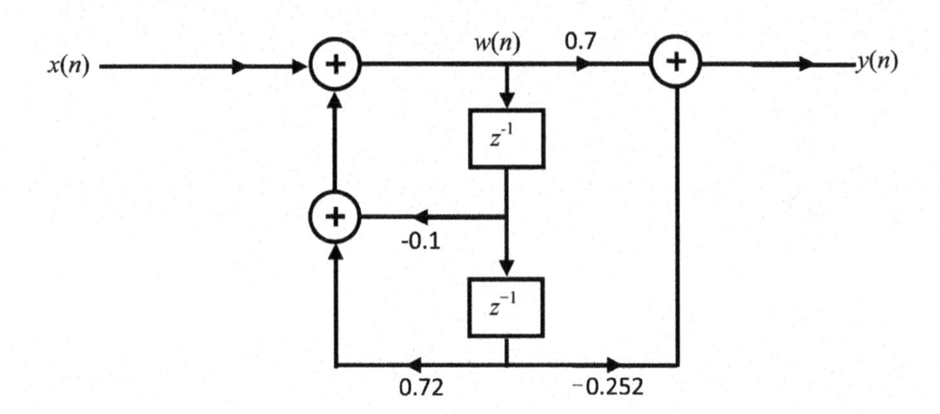

(c)

$$Y(z) = -0.1\,z^{-1}Y(z) + 0.2\,z^{-2}Y(z) + 3X(z) + 3.6z^{-1}X(z) + 0.6\,z^{-2}X(z)$$

$$\frac{Y(z)}{X(z)} = \frac{3 + 3.6z^{-1} + 0.6z^{-2}}{1 + 0.1z^{-1} - 0.2z^{-2}}$$

$$H(z) = \frac{3 + 3.6z^{-1} + 0.6z^{-2}}{1 + 0.1z^{-1} - 0.2z^{-2}}$$

Direct Form II

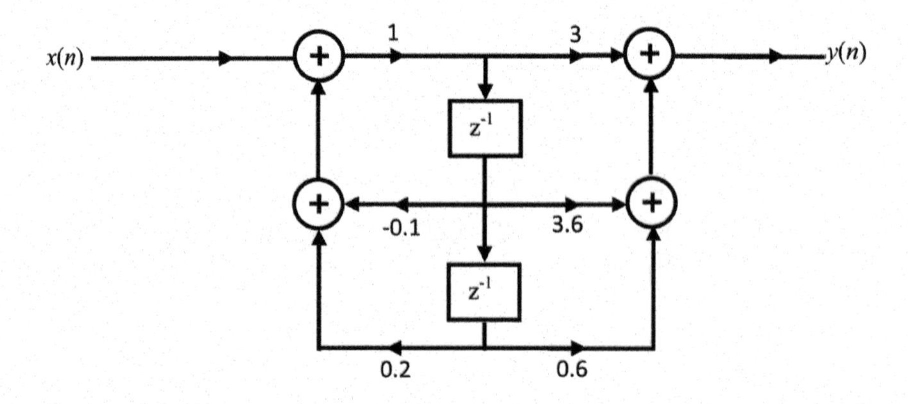

Problem 7.12

Obtain the direct form II for the following systems.

$$H(z) = \frac{2(1 - z^{-1})(1 + \sqrt{2}\,z^{-1} + z^{-2})}{(1 - 0.9z^{-1} + 0.81z^{-2}) + 0.5z^{-1}(1 - 0.9z^{-1} + 0.81z^{-2})}$$

Solution 7.12

$$= \frac{2 + 2\sqrt{2}\,z^{-1} + 2z^{-2} - 2z^{-1} - 2\sqrt{2}\,z^{-2} - 2z^{-3}}{1 - 0.9z^{-1} + 0.81z^{-2} + 0.5z^{-1} - 0.45z^{-2} + 0.405z^{-3}}$$

$$H(z) = \frac{2 + 0.828\,z^{-1} - 0.828z^{-2} - 2z^{-3}}{1 - 0.4z^{-1} + 0.36z^{-2} + 0.405z^{-3}}$$

Direct Form II

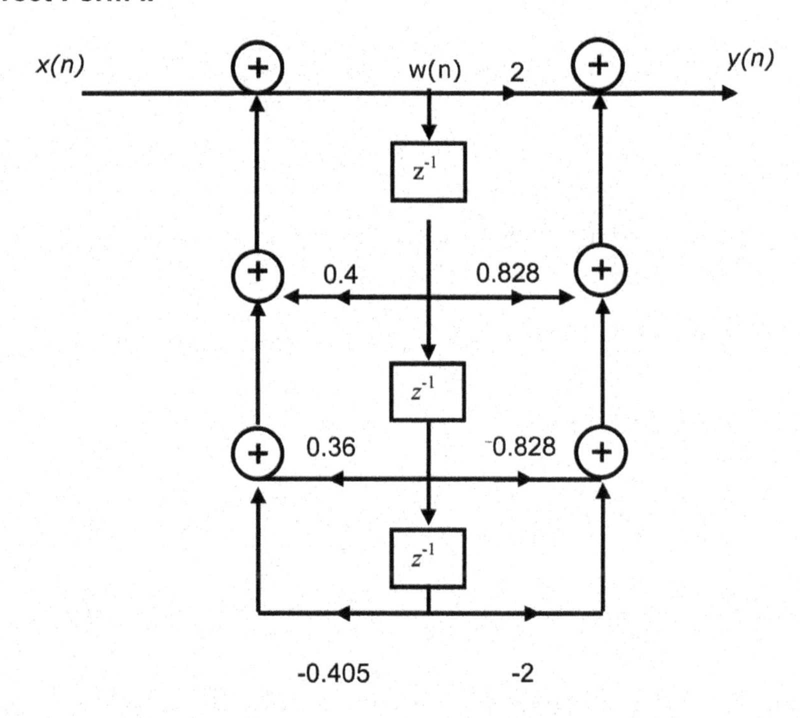

Problem 7.13

Obtain the direct form II for the following systems

(a) $y(n) = \frac{1}{2}y(n-1) + \frac{1}{4}y(n-2) + x(n) + x(n-1)$

(b) $y(n) = y(n-1) + \frac{1}{2}y(n-2) + x(n) + x(n-2) - x(n-1)$

Solution 7.13

(a) Converting the difference equation into Z-transform form

$$Y(z) = \frac{1}{2}z^{-1}Y(z) + \frac{1}{4}z^{-2}Y(z) + X(z) + z^{-1}X(z)$$

$$\frac{Y(z)}{X(z)} = \frac{1 + z^{-1}}{1 - \frac{1}{2}z^{-1} - \frac{1}{4}z^{-2}}$$

$$H(z) = \frac{1 + z^{-1}}{1 - \frac{1}{2}z^{-1} - \frac{1}{4}z^{-2}}$$

Direct Form II

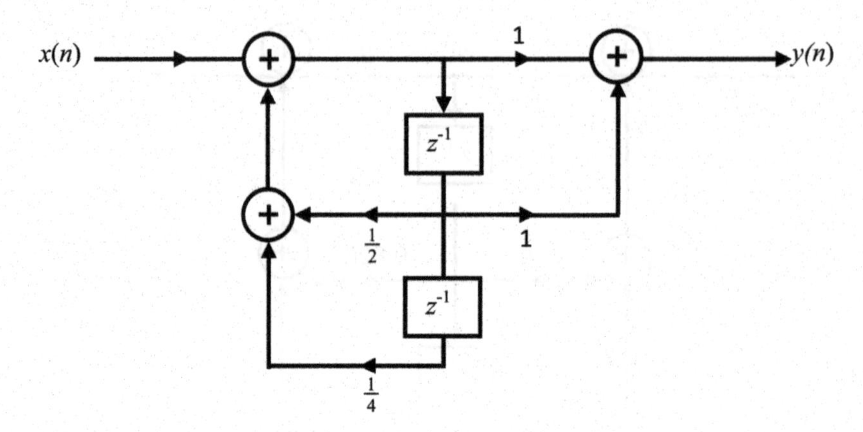

(b) Converting the difference equation into Z-transform form, the drawing the direct form II structure.

$$y(n) = y(n-1) + \frac{1}{2}y(n-2) + x(n) - x(n-1) + x(n-2)$$

$$Y(z) = z^{-1}Y(z) + \frac{1}{2}z^{-2}Y(z) + X(z) - z^{-1}X(z) + z^{-2}X(z)$$

$$H(z) = \frac{Y(z)}{X(z)} = \frac{1 - z^{-1} + z^{-2}}{1 - z^{-1} - \frac{1}{2}z^{-2}}$$

Direct Form II

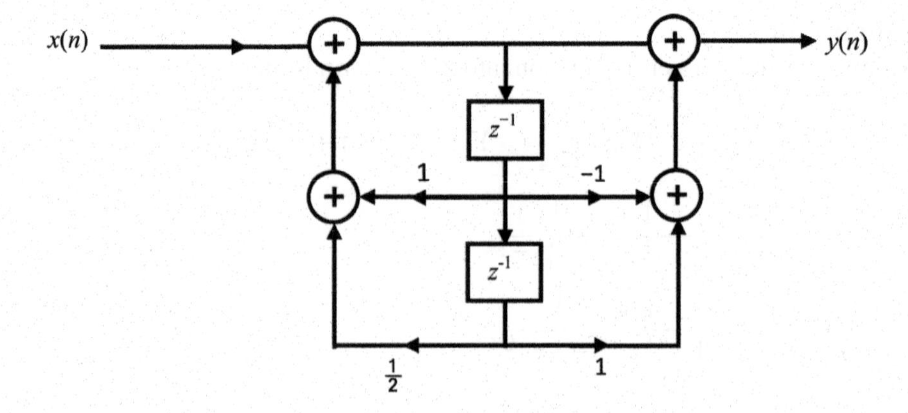

Problem 7.14

Develop different realizations of the third-order IIR transfer function:

$$H(z) = \frac{0.44z^2 + 0.362z + 0.02}{z^3 + 0.4z^2 + 0.18z - 0.2} = \frac{0.44z^{-1} + 0.362z^{-2} + 0.02z^{-3}}{1 + 0.4z^{-1} + 0.18z^{-2} - 0.2z^{-3}}$$

Solution 7.14

By factoring the numerator and the denominator polynomials of $H(z)$ as given in transfer function, we obtain

$$H(z) = \frac{0.44z^2 + 0.362z + 0.02}{(z^2 + 0.8z + 0.5)(z - 0.4)}$$

$$= \left(\frac{0.44 + 0.362z^{-1} + 0.02z^{-2}}{1 + 0.8z^{-1} + 0.5z^{-2}}\right)\left(\frac{z^{-1}}{1 - 0.4z^{-1}}\right)$$

From the above, we arrive at a cascade realization of $H(z)$, as shown in another cascade realization is obtained by using a different pole-zero paring:

$$H(z) = \left(\frac{z^{-1}}{1 + 0.8z^{-1} + 0.5z^{-2}}\right)\left(\frac{0.44 + 0.362z^{-1} + 0.02^{-2}}{1 - 0.4z^{-1}}\right)$$

whose realization is left as an exercise. However, it should be noted that the realization based on the above factored form will be non-canonic since it would require four delays instead of three delays employed. The total number of multipliers in both realizations remains the same.

Next, we make a partial-fraction expansion of $H(z)$ of the form of equation, resulting in

$$H(z) = -0.1 + \frac{0.6}{1 - 0.4z^{-1}} + \frac{-0.5 - 0.2z^{-1}}{1 + 0.8z^{-1} + 0.5z^{-2}}$$

which leads to the parallel form I realization indicated in Figure (a).

Finally, a direct partial-fraction expansion of $H(z)$ expressed as a ratio of polynomials in z is given by

$$H(z) = \frac{0.24}{z - 0.4} + \frac{0.2z + 0.25}{z^2 + 0.8z + 0.5} = \frac{0.24z^{-1}}{1 - 0.4z^{-1}}$$

$$+ \frac{0.2z^{-1} + 0.25z^{-2}}{1 + 0.8z^{-1} + 0.5z^{-2}}$$

resulting in the parallel form II realization sketched in Figure (b).

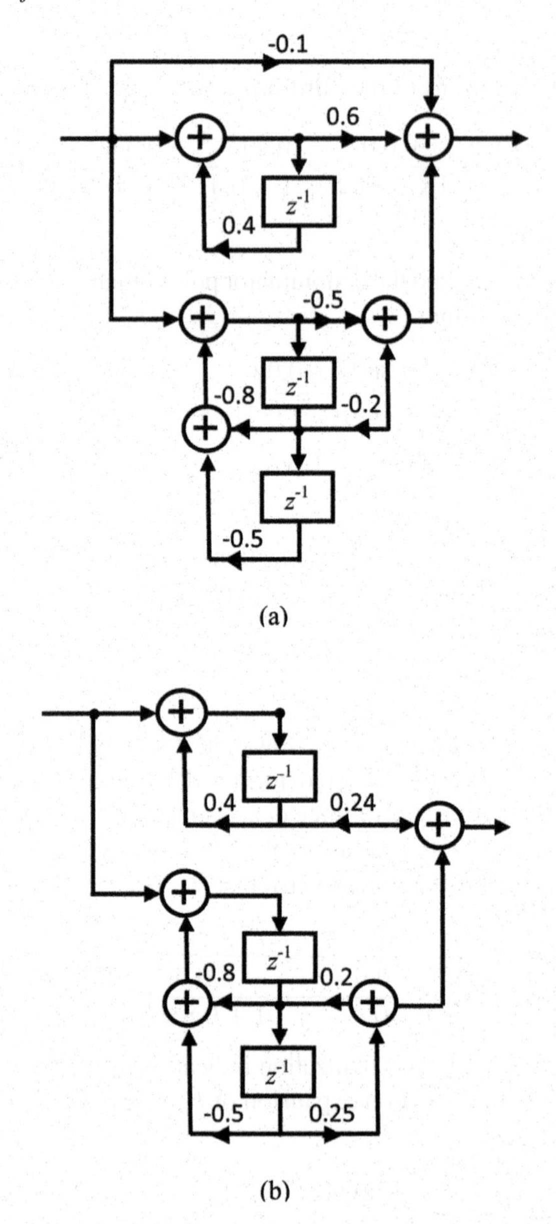

Practice Problem 7.15

Assume that the six methods of calculating filter coefficients given are all available. State and justify which of the methods you should use in each of the following applications:

(a) Phase (delay) equalization for a digital communication system;
(b) Simulation of analogue systems;
(c) A high throughput noise reduction system requiring a sharp magnitude frequency response filter;
(d) Image processing;
(e) High quality digital audio processing;
(f) Real-time biomedical signal processing with minimal distortion.

Practice Problem 7.16

The following transfer function represents two different filters meeting identical amplitude frequency response specifications:

$$(1)\quad H(z) = \frac{b_0 + b_1 z^{-1} + b_2 z^{-2}}{1 + a_1 z^{-1} + a_2 z^{-2}} \times \frac{b_3 + b_4 z^{-1} + b_5 z^{-2}}{1 + a_3 z^{-1} + a_4 z^{-2}}$$

$$\begin{aligned}
a(0) &= 3.13632 \times 10\ \exp(-1)\\
a(1) &= -5.45665 \times 10\ \exp(-2)\\
a(2) &= 4.635728 \times 10\ \exp(-1)\\
a(3) &= -5.45665 \times 10\ \exp(-2)\\
a(4) &= 3.136362 \times 10\ \exp(-1)\\
b(1) &= -8.118702 \times 10\ \exp(-1)\\
b(2) &= -3.339228 \times 10\ \exp(-1)\\
b(3) &= -2.79457 \times 10\ \exp(-1)\\
b(4) &= -3.030631 \times 10\ \exp(-1)
\end{aligned}$$

$$(2)\quad H(z) = \sum_{k=0}^{22} h_k z^{-k}$$

(a) State whether it is an FIR or IIR filter;
(b) Represent the filtering operation in a block diagram form and write down the difference equation;
(c) Determine and comment on the computational and storage requirements.

Problem 7.17

An analogue filter has been converted into an equivalent digital filter that will operate at a sampling frequency of 256 Hz. The converted digital filter has the transfer function:

$$H(z) = \frac{0.1432\ (1 + 3\ z^{-1} + 3\ z^{-2} + z^{-3})}{1 - 0.1801\ z^{-1} + 0.3419\ z^{-2} - 0.0165\ z^{-3}}$$

(a) Assuming that the digital filter is to be realized using the cascade structure, draw a suitable realization block diagram and develop the difference equations.

(b) Repeat (a) for the parallel structure.

Solution 7.17

(a) For the cascade realization, $H(z)$ is factorized using partial fractions

$$H(z) = 0.1432 \frac{1 + 2z^{-1} + z^{-2}}{1 - 0.1307z^{-1} + 0.3355z^{-2}} \times \frac{1 + z^{-1}}{1 - 0.0490z^{-1}}$$

The block diagram representation and the corresponding set of difference equations are given in Figures (a) and (b) as follows:

$$\begin{aligned} w_1(n) &= 0.1432\ x(n) + 0.1307\ w_1(n-1) - 0.3355\ w_1(n-2) \\ y_1(n) &= w_1(n) + 2w_1(n-1) + w_1(n-2) \\ w_2(n) &= y_1(n) + 0.049w_2(n-1) \\ y_2(n) &= w_2(n) + w_2(n-1) \end{aligned}$$

(b) For the parallel realization, $H(z)$ is expressed using partial fractions as

$$H(z) = \frac{1 + 2916 - 0.08407z^{-1}}{1 - 0.1317z^{-1} + 0.3355z^{-2}} + \frac{7.5268}{1 - 0.0490z^{-1}} - 8.6753$$

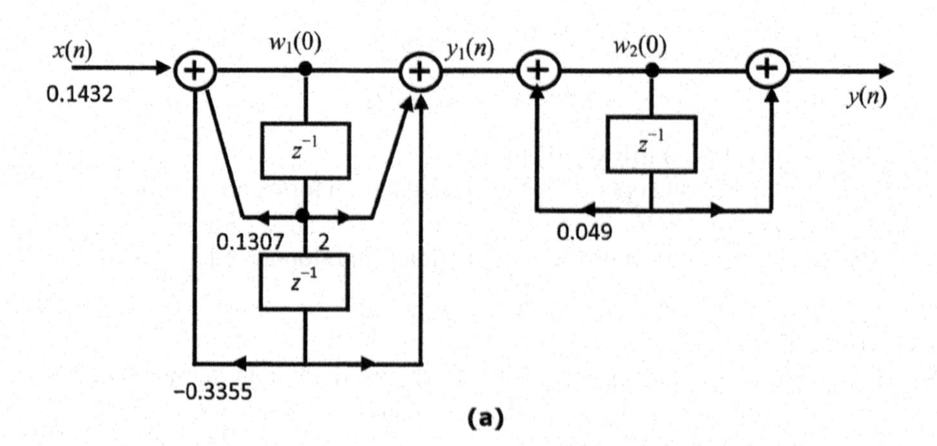

(a)

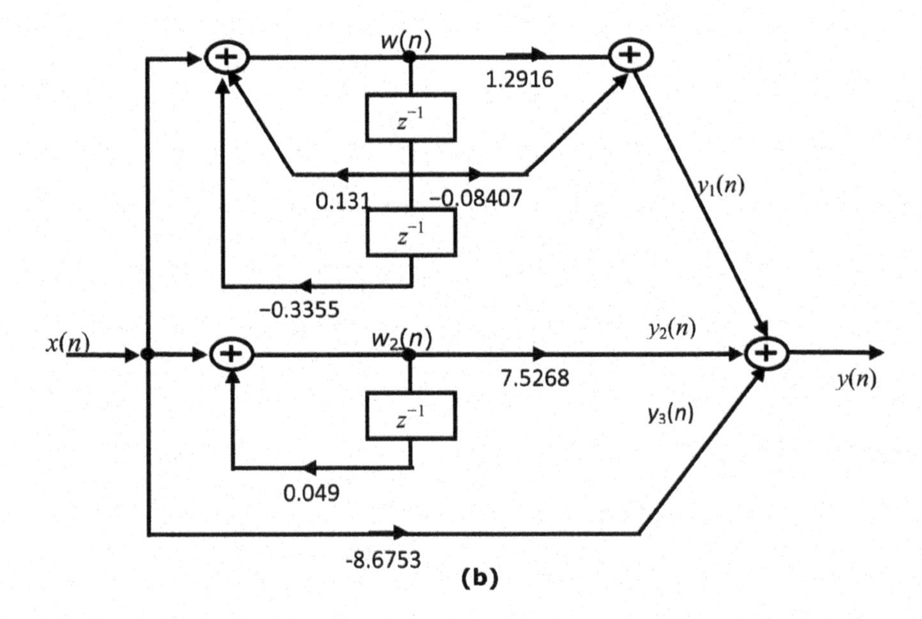

(b)

The parallel realization diagram and its corresponding set of difference equations as given in Figure (a) and Figure (b) are as follows:

$$
\begin{aligned}
w_1(n) &= x(n) + 0.131\, w_1(n-1) - 0.3355\, w_1(n-2) \\
y_1(n) &= 1.2916\, w_1(n) - 0.08407\, w_1(n-1) \\
w_2(n) &= x(n) + 0.049\, w_2(n-1) \\
y_2(n) &= 7.5268\, w_2(n) \\
y_3(n) &= -8.6753\, x(n) \\
y(n) &= y_1(n) + y_2(n) + y_3(n)
\end{aligned}
$$

Problem 7.18

The following transfer functions represent two different filters meeting identical amplitude–frequency response specifications:

(a) $H(z) = \dfrac{a_0 + a_1 z^{-1} + a_2 z^{-2}}{1 + b_1 z^{-1} + b_2 z^{-2}}.$

where

$$
\begin{aligned}
a_0 &= 0.4981819 \\
a_1 &= 0.9274777 \\
a_2 &= 0.4981819 \\
b_1 &= -0.6744878 \\
b_2 &= -0.3633482
\end{aligned}
$$

(b) $H(z) = \sum_{k=0}^{11} h(k)z^{-k}$

where

$$\begin{aligned}
h(0) &= 0.54603280 \times 10^{-2} = h(11) \\
h(1) &= -0.45068750 \times 10^{-1} = h(10) \\
h(2) &= 0.69169420 \times 10^{-1} = h(9) \\
h(3) &= -0.55384370 \times 10^{-1} = h(8) \\
h(4) &= -0.63428410 \times 10^{-1} = h(7) \\
h(5) &= 0.57892400 \times 10^{-0} = h(6)
\end{aligned}$$

For each filter,

(i) State whether it is an FIR or IIR filter,
(ii) Represent the filtering operation in a block diagram form and write down the difference equation,
(iii) Determine and comment on the computational and storage requirements.

Solution 7.18

(i) Filters (a) and (b) are IIR and FIR as per given transfer respectively.
(ii) The block diagram for filter (1) is given in Figure 7.13(a). The corresponding set of difference equations are

$$\begin{aligned}
w(n) &= x(n) - b_1 w(n-1) - b_2 w(n-2) \\
y(n) &= a_0 w(n) + a_1 w(n-1) + b_2 w(n-2)
\end{aligned}$$

The block diagram for filter (2) is given in Figure (b). The corresponding difference equation is

$$y(n) = \sum_{k=0}^{11} h(k)x(n-k)$$

(iii) From examination of the two difference equations the computational and storage requirements for both filters are summarized below:

	FIR	IIR
Number of multiplications	12	5
Number of additions	11	4
Storage locations (coefficients and data)	24	8

It is evident that the IIR filter is more economical in both computational and storage requirements than the FIR filter. However, we could have exploited the symmetry in the FIR coefficients to make the FIR filter more efficient, although at the expense of its obvious implementation simplicity. A point worth making is that, for the same amplitude response specifications, the number of FIR filter coefficients (12 in this example) is typically six times the order (the highest power of z in the denominator) of the IIR transfer function (2 in this case).

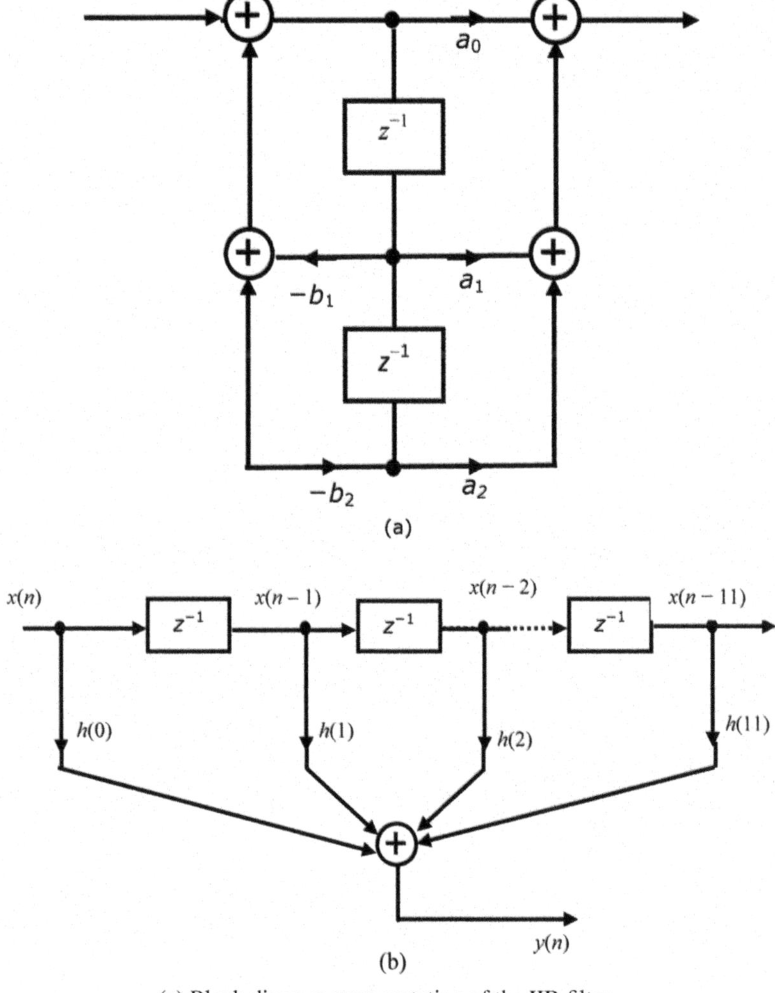

(a) Block diagram representation of the IIR filter.
(b) Block diagram representation of the FIR filter.

8

Introduction to Digital Filters

This chapter covers: Introduction to digital filters, Criteria for selection of digital filters, Filter design steps, Advantage and disadvantages of FIR and IIR filters, and Problems and solutions.

8.1 Introduction

The major emphasis in this chapter is on digital filter. However, some basic information analog passive and active filters will be presented first. An understanding of their design and implementation will make it easier to appreciate the advantages of digital filters. Let's take a look at what filters accomplish. Briefly, they separate frequencies. Common filtering objectives are to improve the quality of a signal (for example, to remove or reduce noise), to extract information from signals or to separate two or more signals previously combined to make, for example, efficient use of an available communication channel. A filter is essentially a system or network that selectively changes the wave shape, amplitude-frequency and/or phase-frequency characteristics of a signal in a desired manner. So, the input signal to a filter might consist of four frequencies and the filter output might contain three, two, or only one of those frequencies.

Filtering is a process by which the frequency spectrum of signal can be modified, reshaped or manipulated to achieve some desired objectives.

8.1.1 Types of Filters

Basically, filters are of two types depending on what type of signal being processed as under:

(i) Analog filters
(ii) Digital Filters

8.1.2 Classification of Filters Development Wise

First before the existence of digital computer, analog filters were being used, after the invention of the digital electronics digital filters came into existence. There are three basic ways to build or realize a filter:

8.1.3 Analog Filters

Passive and Active filters comes under the category of analog filters.

Passive: Use resistors, inductors and capacitors (or at least two of three).

Active: Use resistors, capacitors and transistors, or operational amplifiers.

Analog filters may be defined as a system in which both the input and the output are continuous time signals.

8.1.4 Types of Analog Filter

As a matter of fact, there are five kinds of analog filters which may be listed as under:

(i) Low pass
(ii) High pass
(iii) Band pass
(iv) Band stop
(v) All pass

An ideal filter is often called a brick wall filter. The low-pass passes all the lower frequencies, while high-pass passes all the high frequencies, passband includes all those frequencies that go through the filter with no attenuation (the amplitude is maximum). The stopband includes those frequencies that do not get through the filter (the amplitude is zero and attenuation is infinite). The attenuation from passband to stopband is immediate. Or, to say it another way, the transition width is zero. It is not possible to realize ideal filters, but digital filters can come very close.

8.2 Digital Filters

Digital: Use computers or dedicated processors. Digital filter may be defined as a system in which both the input and the output are discrete time signals.

A digital filter is a mathematical algorithm implemented in hardware and/or software that operate on a digital input signal to produce a digital output signal for the purpose of achieving a filtering objective. The term digital filter refers to the specific hardware or software routine that performs the filtering algorithm. Digital filters often operate on digitized analog signals or just numbers, representing some variables stored in a computer memory.

8.3 Importance and Advantages

Digital filters play very important role in DSP. Compared with analog filters they are preferred in a number of applications (for example data compression, biomedical signal processing, speech processing, image processing, data transmission, digital audio, and telephone echo cancellation) because of one or more of the following advantages.

Digital filters can have characteristics that are not possible with analog filters, such as a truly linear phase response. Unlike analog filters, the performances of digital filters do not vary with environmental changes, for example thermal variations. This eliminates the need to calibrate periodically. The frequency response of a digital filter can be automatically adjusted if it is implemented using a programmable processor, that is why they are widely used in adaptive filters. Several input signals or channels can be filtered by one digital filter without the need to replicate the hardware. Both filtered and unfiltered data can be saved for further use.

Advantage can be readily taken of the tremendous advancements in VLSI technology to fabricate digital filters and to make them small in size, to consume low power, and to keep the cost down. In practice, the precision achievable with analog filters is restricted; for example, typically a maximum of only about 60–70 dB stop-band attenuation is possible with active filters designed with off-the-shelf components. With digital filters the precision is limited only by the word length used. The performance of digital filters is repeatable from unit to unit. Digital filters can be used at very low frequencies, found in many biomedical applications for example, where the use of analog filters is impractical. Also, digital filters can be made to work over a wide range of frequencies by a mere change to the sampling frequency.

8.4 Disadvantages

8.4.1 Speed Limitation

The maximum bandwidth of signals that digital filters can handle, in real time, is much lower than for analog filters. In real-time situations, the analog–digital–analog conversion processes introduce a speed constraint on the digital filter performance. The conversion time of the ADC and the settling time of the DAC limit the highest frequency that can be processed.

Further, the speed of operation of a digital filter depends on the speed of the digital processor used and on the number of arithmetic operations that must be performed for the filtering algorithm, which increases as the filter response is made tighter.

8.4.2 Finite Word-Length Effects

When the digital filters are implemented in hardware, there are some effects of using finite register length to represent all relevant filter parameters.

Some of the effects are:

(a) A/D conversion noise
(b) Round off noise (Uncorrelated)
(c) Coefficient accuracy
(d) Correlated roundoff noise or limit cycles

There are three types of arithmetic used in filter algorithms.

(a) Fixed point
(b) Floating point
(c) Block floating point

All the binary number representations that we have studied use a finite number of bits to represent numerical values. These numerical values could be filter coefficients, data samples, or computational results. Depending on the binary representation used confined to 16, 24, 32, and 64 bits. The finite length of the number of bit results in errors known as finite word-length effects.

Digital filters are subject to ADC noise resulting from quantizing a continuous signal, and to roundoff noise incurred during computation. With higher order recursive filters, the accumulation of roundoff noise could lead to instability.

8.4.3 Limit Cycles

When the input to a filter is constant, either zero or non-zero, it is expected that the output of the filter should also be a constant. In some case this does not happen, and the filter output tends to oscillate, even though the input signal does not change. These undesired oscillations are known as limit cycles. Limit cycles are a problem in audio applications when the oscillation frequency occurs in the range of interest.

Limit cycles occur due to either quantization or overflow. The quantization error is due to the finite number of bits used to represent a binary number. Limit cycles due to quantization errors are relatively small, but still undesirable. Overflow occurs if the internal mathematical operations result in a number too large to fit within the range of number representation. The limit cycle oscillations due to overflow are generally larger in amplitude and more severe.

Several methods are employed to reduce the possibility of limit cycles. These are:

(a) Use of larger word lengths (more bits) for intermediate calculations. This helps to avoid overflow.
(b) The use of special filter structures that prevent the formation of limit cycles.
(c) Prevention of overflow by saturating the result of calculation. Saturation means that if the result of adding two positive numbers is greater than the largest positive number that can be represented, the result is clamped or fixed to the largest positive number. Similarly if the result of adding two negative numbers is more negative than the largest negative number that can be represented, the result is clamped to the largest negative number.

8.4.4 Long Design and Development Times

The design and development times for digital filters, especially hardware development, can be much longer than for analog filters. However, once filter is developed the hardware and/or software can be used for other filtering or DSP tasks with little or no modifications.

Good computer-aided design (CAD) support can make the design of digital filters an enjoyable task, but some expertise is required to make full and effective use of such design aids.

8.5 Types of Digital Filters

8.5.1 FIR (Finite Impulse Response) Filters

Finite impulse response filter response decays to zero after the impulse passes through the system. Its impulse response stays for a finite time. The input and output signals for the filters are related by convolution sum, which is given in the Equation (8.1) for FIR filter.

$$y(n) = \sum_{k=0}^{N-1} h(k)\, x(n-k) \tag{8.1}$$

8.5.2 IIR (Infinite Impulse Response) Filters

Infinite impulse response filter response never quite reaches zero because the feedback makes the decay exponential. It impulse response stays for an infinite time (time taken by its impulse response is much larger than finite impulse response). A practical filter has to eventually settle to zero or it would be useless, unless one needs an oscillator. IIR filters can oscillate if they are improperly designed. All feedback systems have the potential to become unstable and oscillate.

The input and output signals to the filter are related by convolution sum, which is given in Equation (8.2) for IIR filter.

$$y(n) = \sum_{k=0}^{\infty} h(k)\, x(n-k) \tag{8.2}$$

All IIR filters can be implemented in digital form. IIR filter use feedback to sharpen the filter response. There are two sets of filter coefficients. The coefficients '*a*' are called the feed-forward coefficients, the coefficient '*b*' are the feedback coefficients.

The IIR filter equation is expressed in a recursive form by Equations (8.3) and (8.4)

$$y(n) = \sum_{k=0}^{\infty} h(k)x(n-k) = \sum_{k=0}^{N} a_k x(n-k) - \sum_{k=1}^{M} b_k\, y(n-k) \tag{8.3}$$

$$H(z) = \frac{a_0 + a_1 z^{-1} + \ldots + a_N z^{-N}}{1 + b_1 z^{-1} + \ldots + b_M z^{-M}} = \frac{\sum\limits_{k=0}^{N} a_k z^{-k}}{1 + \sum\limits_{k=1}^{M} b_k z^{-k}} \tag{8.4}$$

8.6 Choosing between FIR and IIR Filters

Finite impulse response filters can have an exactly linear phase response. The implication of this is that no phase distortion is introduced into the signal by the filter. This is an important requirement in many applications, for example data transmission, biomedicine, digital audio, and image processing. The phase responses of IIR filters are nonlinear, especially at the band edges. FIR filters realized non-recursively, that is by direct evaluation of difference equation are always stable. The stability of IIR filters cannot always be guaranteed. The effects of using a limited number of bits to implement filters such as roundoff

Table 8.1 A comparison of the characteristics of FIR and IIR filters

Characteristics	FIR	IIR
Efficiency	Low	High
Speed	Slow	Fast
Overflow (Finite word-length Effect)	Not likely	Likely
Stability	Guaranteed	Design issue
Phase response	Generally linear	Generally nonlinear
Analog modeling	Not directly	Yes
Design and noise analysis	Straight forward	Complex
Arbitrary filters	Straight forward	Complex

noise and coefficient quantization errors are much less severe in FIR than in IIR.

FIR requires more coefficients for sharp cut-off filters than IIR. Thus for a given amplitude response specification, more processing time and storage will be required for FIR implementation. Analog filters can be readily transformed into equivalent IIR digital filters meeting similar specifications. This is not possible with FIR filters as they have no analog counterpart. However, with FIR it is easier to synthesize filters of arbitrary frequency responses. Table 8.1 shows a comparison of the characteristics of FIR and IIR Filters.

8.7 Tolerance Scheme of FIR and IIR Filters

8.7.1 FIR Filters

The amplitude–frequency response of an FIR filter is often specified in the form of a tolerance scheme, Figure 8.1 depicts such a scheme for a low-pass filter. The shaded horizontal lines indicate the tolerance limits. In the passband, the magnitude response has a peak deviation of δ_p and, in the stopband; it has a maximum deviation of $\delta_{s.}$

The characteristics of digital filters are often specified in the frequency domain. For frequency selective filters, such as low-pass and band-pass filters, the specifications are often in the form of tolerance schemes for FIR filter is shown in Figure 8.1.

Referring to the figure, the following parameters are of interest:

δ_P = Peak passband deviation (or ripples)

δ_s = Stopband deviation (or ripples)

f_P = Passband edge frequency

f_s = Stopband edge frequency

F_s = Sampling frequency

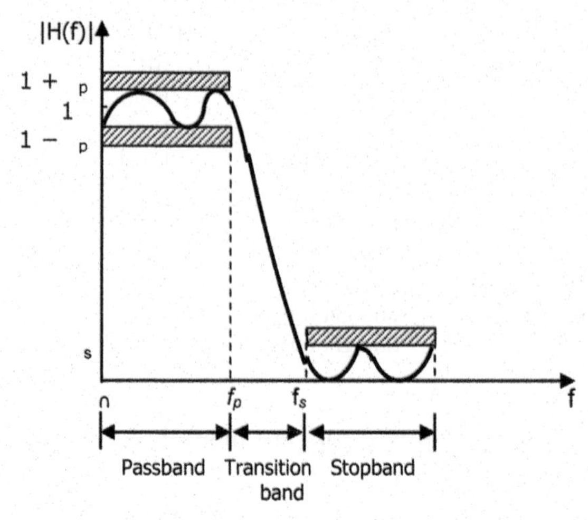

Figure 8.1 Tolerance scheme for a low-pass FIR filter.

Passband and stopband deviations may be expressed as an ordinary number or in decibels. Thus the passband ripple and minimum stop band attenuation, respectively, for FIR filter is given by,

$$A_s \text{ (stopband attenuation) } = -20\log_{10}\delta_s \tag{8.5}$$

$$A_P \text{ (passband ripple) } = 20\log_{10}(1 + \delta_p)$$

For FIR filters, the passband ripple is the difference between the ideal response and the maximum (or minimum) deviation in the passband.

Example 8.1

The peak pass band ripple and the minimum stopband attenuation of a FIR digital filter are, respectively, 0.1 and 35 dB. Determine the corresponding peak passband ripple δ_P and δ_s stopband deviation (or ripples).

Solution 8.1

The pass band attenuation is given by

$$A_P = 20\log(1 + \delta_P)$$

$$\text{or } 0.1 = 20\log_{10}(1 + \delta_P) \quad \log_{10}(1 + \delta_P) = 1/20 = 0.005$$

$$1 + \delta_P = 10^{0.005} \quad \delta_P = 0.01157$$

The stopband attenuation is given by

$$A_s = -20\log\delta_s$$

$$\text{or } 35 = -20\log_{10}\delta_s \quad \log_{10}\delta_s = -35/20 = -1.75 \quad \delta_s = 0.017782$$

8.7.2 IIR Filters

The amplitude-frequency response of an IIR filter is often specified in the form of a tolerance scheme, Figure 8.2 depicts such a scheme for a bandpass filter.

The shaded horizontal lines indicate the tolerance limits for an IIR filter. The following parameters are normally used to specify the frequency response. Referring to the figure, the following parameters are of interest:

ϵ^2 Passband ripple parameters
δ_P Peak passband deviation (or ripples)
δ_s Stopband deviation (or ripples)
f_{P1} and f_{P2} Passband edge frequencies
f_{s1} and f_{s2} Stopband edge frequencies
F_s Sampling frequency

The band edge frequencies are sometimes given in normalized form, which is a fraction of the sampling frequency (f/F_S)

$$A_P = 10\log(1+\varepsilon^2) = -20\log(1-\delta_P) \text{ and } A_s = -20\log(\delta_s). \quad (8.6)$$

Thus, for IIR passband ripple is meant is the peak-to-peak passband ripple.

Example 8.2

The peak pass band ripple and the minimum stopband attenuation of a IIR digital filter are, respectively, 0.1 and 35 dB. Determine the corresponding peak passband ripple δ_P and δ_s stopband deviation (or ripples).

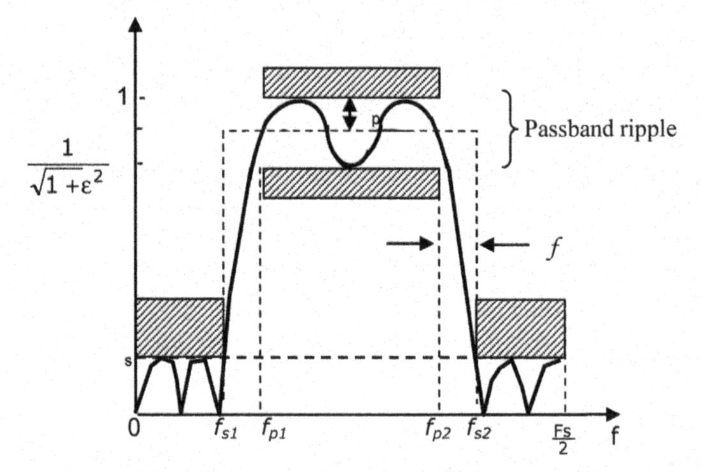

Figure 8.2 Tolerance scheme for an IIR band-pass filter.

Solution 8.2

The pass band attenuation is given by

$$\begin{aligned} A_P &= -20\log(1-\delta_P) \\ \text{or } 0.1 &= -20\log_{10}(1-\delta_P) \quad \log_{10}(1-\delta_P) = 0.1/20 = -0.005 \\ 1-\delta_P &= 10^{-0.005} \quad \delta_P = 0.0114469 \end{aligned}$$

The stopband attenuation is given by

$$\begin{aligned} A_s &= -20\log\delta_s \\ \text{or } 35 &= -20\log_{10}\delta_s \quad \log_{10}\delta_s = -35/20 = -1.75 \quad \delta_s = 10^{-0.005} \\ \delta_s &= 0.988553 \end{aligned}$$

8.8 FIR and IIR Filter Design Stages

The design of a digital filter involves five steps. The five steps are not necessarily independent; nor are they always performed in the order given.

8.8.1 Filter Specification

Requirement specifications include the following:

(a) This may include stating the type of filter, for example low-pass filter, the desired amplitude and/or phase responses and the tolerances (if any) we are prepared to accept, the sampling frequency, and the wordlength of the input data.

(b) Signal characteristics (types of signal source and sink, I/O interface, data rates and width, and highest frequency of interest).

(c) The characteristics of the filter, the desired amplitude and/or phase responses and their tolerances (if any), the speed of operation and modes of filtering (real time or batch).

(d) The manner of implementation (for example, as a high level language routine in a computer or as a DSP processor-based system, choice of signal processor), and other design constraints (for example, the cost of the filter). The designer may not have enough information to specify the filter completely at the outset, but as many of the filter requirements as possible should be specified to simplify the design process.

8.8.2 Coefficient Calculation

At this step, we determine the coefficients of a transfer function, $H(z)$, which will satisfy the specifications given in Figures (8.1) and (8.2). Our choice of coefficient calculation method will be influenced by several factors, the most important of which are the critical requirements as in step in Section 8.8.

8.8.3 Realization

This involves converting the transfer function obtained as in step given in Section 8.8.2 into a suitable filter network or structure as discussed in previous chapter.

8.8.4 Analysis of Finite Word-Length Effects

Here, we analyze the effect of quantizing the filter coefficients and the input data as well as the effect of carrying out the filtering operation using fixed word-lengths on the filter performance.

8.8.5 Implementation

This involves producing the software code and/or hardware and performing the actual filtering. In fact techniques are now available that combine the second and aspects of the third and fourth steps. However, the approach discussed here gives a simple step-by-step guide that will ensure a successful design.

To arrive at an efficient filter, it may be necessary to iterate a few times between the steps, especially if the problem specification is not watertight, as is often the case, or if the designer wants to explore other possible designs.

8.9 Filters Coefficient Calculation Methods

Selection of one of the approximation methods to calculate the values of the coefficients, $h(k)$, for FIR, or a_K and b_K for IIR, such that the filter characteristics are satisfied.

The method used to calculate the filter coefficients depends on whether the filter is IIR or FIR type. Calculations of IIR filter coefficients are traditionally based on the transformation of known analog filter characteristics into equivalent digital filters.

8.9.1 FIR Filters Coefficient Calculation Methods

There are several methods of calculating the coefficients of FIR filters. The three methods are the **window, optimal**, and the **frequency sampling**.

The **window method** offers a very simple and flexible way of computing FIR filter coefficients, but it does not allow the designer adequate control over the filter parameters.

With the availability of an efficient and easy-to-use program, the **optimal method** is now widely used in industry and, for most applications, will yield the desired FIR filters.

The main attraction of the **frequency sampling method** is that it allows a recursive realization of FIR filters which can be computationally very efficient. However, it lacks flexibility in specifying or controlling filter parameters.

Thus, for FIR filters, the optimal method should be the method of first choice unless the particular application dictates otherwise or a CAD facility is unavailable.

8.9.2 IIR Filters Coefficient Calculation Methods

The two basic methods used are the **impulse invariant** and the **bilinear transformation** methods, which is widely used, the third method is the **Pole placement** method which is used for lower order filter specially notch filter.

The **pole–zero placement method** offers an alternative approach to calculate the coefficients of IIR filters. It is an easy way of calculating the coefficients of very simple filters. However, for filters with good amplitude response it is not recommended as it relies on trial and error, shuffling of the pole and zero positions.

With the **impulse invariant method,** after digitizing the analog filter, the impulse response of the original analog filter is preserved, but not its magnitude–frequency response. Because of inherent aliasing, the method is inappropriate for highpass or bandstop filters.

The bilinear transformation method, on the other hand, yields very efficient filters and is well suited to the calculation of coefficients of frequency selective filters. It allows the design of digital filters with known classical characteristics such as Butter-worth, Chebyshev and Elliptic. Digital filters resulting from the bilinear transform method will, in general, preserve the magnitude response characteristics of the analog filter but not the time domain properties. The impulse invariant method is good for simulating analog systems, but the bilinear method is best for frequency selective IIR filters.

8.9.3 Summary of Filters Coefficient Calculation Methods

There are several methods of calculating filter coefficients of which the following are the most widely used:

1. Window (FIR)
2. Optimal (FIR)
3. Frequency sampling (FIR)
4. Pole-zero placement (IIR)
5. Impulse invariant (IIR)
6. Bilinear transformation (IIR)

The choice is influenced by several factors, the most important of which are the critical requirements in the specifications. In general, the crucial choice is really between FIR and IIR. In most cases, if the FIR properties are vital then a good candidate is the optimal method, whereas, if IIR properties are desirable, then the bilinear method will in most cases suffice.

9

Step-by-Step Design of Digital FIR Filters

This chapter covers the characteristic features of FIR filters, the types of linear-phase FIR filters, linear-phase response and its implications, FIR filter design for both non-causal and causal, methods of filter coefficient calculations by the window method for low-pass, high-pass, band-pass, band-stop filters, advantage and disadvantage of these methods, and Problems and Solutions.

9.1 Introduction

The purpose of filter design lies in constructing a transfer function of a filter that meets the prescribed frequency response specifications. The Finite Impulse Response (FIR) filters decay to zero after the impulse passes through the system. In a non-recursive filter, the current output sample, $y(n)$, is a function only of past and present values of the input, $x(n)$; so, the basic FIR filter is characterized by the following equation:

$$y(n) = \sum_{k=0}^{N-1} h(k)\, x(n-k) \tag{9.1}$$

where $h(k)$ is a finite duration sequence defined over the interval $0 \leq k \leq$, $N - 1$, where N **is the length of the filter**. It means that the order of the FIR filter is always $N - 1$.

9.2 Why is it Called the Finite Impulse Response Filter?

A finite impulse response (FIR) filter is a type of a signal-processing filter whose impulse response (or response to any finite length input) is of *finite* duration, because it settles to zero in finite time. This is in contrast to infinite impulse response (IIR) filters, which have internal feedback and may continue to respond indefinitely (usually decaying). The impulse response of an

Nth-order discrete-time FIR filter (i.e. with a Kronecker delta impulse input) lasts for $N + 1$ sample, and then dies to zero.

FIR filters can be discrete-time or continuous-time, and digital or analog.

Let us further investigate the filter defined in Equation (9.1)

$$y(n) = h(0)x(n) + h(1)x(n-1) + h(2)x(n-2) + n \ldots\ldots\ldots(N-1)x(n-(N-1)) \quad (9.2)$$

The Z-transform of Equation (9.2) is

$$Y(z) = h(0)\,X(z) + h(1)z^{-1}X(z) + h(2)z^{-2}X(z) + h(N-1)\,z^{-(N-1)}X(z) \quad (9.3)$$

$$H(z) = \frac{Y(z)}{X(z)} = h(0) + h(1)z^{-1} + h(2)z^{-2} + h(N-1)\,z^{-(N-1)} \quad (9.4)$$

Now to find its impulse response, we put $x(n) = \delta(n)$ and $X(z) = Z\{\delta(n)\} = 1$

$$H(z) = Y(z) = h(0) + h(1)z^{-1} + h(2)z^{-2} + h(N-1)\,z^{-(N-1)} \quad (9.5)$$

$$H(z) = \sum_{n=0}^{N-1} h(n)\,z^{-n} \quad (9.6)$$

Taking the impulse response of Equation (9.6), we get

$$h(n) = \{h(0), h(1), h(2), h(3) \ldots\ldots h(N-1)\} \quad (9.7)$$

To implement the filter practically, the coefficients $h(n) = \{h(0), h(1) \ldots\ldots h(N-1)\}$ must all be real numbers. Since the time factor $(N-1)$ contained in the impulse response is of finite value, which means that the impulse response is of finite duration. $H(z)$ is the transfer function of the filter and N is the filter length, which is the number of filter coefficients. $H(z)$ provides a means of analyzing the filter, for example, when evaluating the frequency response.

9.2.1 Features of FIR Filters

FIR filters can have an exactly linear-phase response. FIR filters are very simple to implement. All DSP processors available have architectures that are suited to FIR filtering. Non-recursive FIR filters suffer less from the effects of finite word-length than IIR filters. Recursive FIR filters also exist and may

offer significant computational advantages, but these areas of study extend beyond the boundaries of this book.
The Fourier transform of $h(n)$ is periodic with a period of 2π

$$H(\omega) = \sum_{n=0}^{N-1} h(n)\, e^{-j\omega n} \quad H(\omega) = \sum_{n=0}^{N-1} h(n)\, e^{-j\,(\omega n+2\pi m)}$$

$$m = 0, 1, 2, \ldots \tag{9.8}$$

In magnitude and phase

$$H(\omega) = \pm|H(\omega)|e^{j\theta(\omega)} \tag{9.9}$$

$$|H(\omega)| = |H(-\omega)|, \quad \pi < \omega < \pi \quad \theta(\omega) = \theta(-\omega) \tag{9.10}$$

Equation (9.10) implies that the magnitude of the Fourier transform is a symmetric function and the phase is anti-symmetric in nature.

Let us assume that the phase is linear, i.e. $(\omega) = -\alpha\omega$, $-\pi < \omega < \pi$ where α is a constant phase delay in samples.

$$H(\omega) = \sum_{n=0}^{N-1} h(n)\, e^{-j\omega n} = \pm|H(\omega)|\, e^{-j\alpha\omega} \tag{9.11}$$

Equating the real and imaginary parts of the components in Equation (9.11) gives the following two equations:

$$\pm|H(\omega)|\ \cos(\alpha\omega) = \sum_{n=0}^{N-1} h(n)\ \cos(\omega n) \tag{9.12}$$

$$\pm|H(\omega)|\ \sin(\alpha\omega) = \sum_{n=0}^{N-1} h(n)\ \sin(\omega n) \tag{9.13}$$

Dividing Equation (9.13) by Equation (9.12), we get

$$\frac{\sin(\alpha\omega)}{\cos(\alpha\omega)} = \tan(\alpha\omega) = \frac{\sum_{n=0}^{N-1} h(n)\ \sin(\omega n)}{\sum_{n=0}^{N-1} h(n)\ \cos(\omega n)} \tag{9.14}$$

$$\text{Hence, we have} \tan(\alpha\omega) = \frac{\sum_{n=0}^{N-1} h(n)\ \sin(\omega n)}{h(0) + \sum_{n=1}^{N-1} h(n)\ \cos(\omega n)} \tag{9.15}$$

There exist two possible solutions of Equations (9.14) and (9.15). The first possibility is that $\alpha = 0$, which is suggested in (9.15)

$$0 = \frac{\sum_{n=0}^{N-1} h(n)\ \sin(\omega n)}{h(0) + \sum_{n=1}^{N-1} h(n)\ \cos(\omega n)} \tag{9.16}$$

For which the only solution is that $h(0)$ is arbitrary and $h(n) = 0$, $n \neq 0$, i.e. the impulse response of the filter is an impulse, which is not a useful result.

Hence, the other possible case is when $\alpha \neq 0$. This is shown in Equation (9.17)

$$\sum_{n=0}^{N-1} h(n)\ \cos(\omega n)\ \sin(\alpha\omega) - \sum_{n=0}^{N-1} h(n)\ \sin(\omega n)\ \cos(\alpha\omega) = 0 \tag{9.17}$$

$$\sum_{n=0}^{N-1} h(n)\ [\cos(\omega n)\ \sin(\alpha\omega) - \sin(\omega n)\ \cos(\alpha\omega)] = 0 \tag{9.18}$$

$$\sum_{n=0}^{N-1} h(n)\ \ \sin[(\alpha - n)\omega] = 0 \tag{9.19}$$

The one possible solution of Equation (9.19) is

$$\alpha = (N-1)/2 \tag{9.20}$$

$$h(n) = h(N-n-\ 1),\ \ 0 \leq n \leq N-1 \tag{9.21}$$

Equations (9.20) and (9.21) represent the necessary and sufficient condition for an FIR filter, i.e. its phase is linear.

Example 9.1

Given the following transfer function, calculate its complex frequency response at 250 Hz if the system sampling frequency is 1000 Hz.

$$H(z) = 0.0935 + 0.3027z^{-1} + 0.40z^{-2} + 0.3027z^{-3} + 0.0935z^{-4}$$

Solution 9.1

The frequency response in the z plane is that value of the transfer function, which lies on the unit circle in the z plane at an angle of digital frequency ω. Thus, $\omega = 2\pi\frac{f}{F_s} = 2\pi\frac{250}{1000} = 0.7854$

The value of z on the unit circle at this angle is $z = e^{j\omega} = e^{j0.7854} = 0.7071 + j\,0.7071$

Therefore, the complex frequency is

$$H(\omega) = 0.0935 + 0.3027e^{-j0.7854} + 0.40e^{-j2(0.7854)} + 0.3027e^{-j3(0.7854)} + 0.0935e^{-j4(0.7854)}$$

$$H(\omega) = 0.0000 - j0.8281 = 0.82815e^{-j1.5708}$$

This result means that the magnitude frequency response at 250 Hz is 0.8282 and the phase response is –1.5708 rad.

9.2.2 Linear-Phase Implications

For the phase response, we need to determine whether positive symmetry or negative symmetry is required (assuming linear phase). The ability to have an exactly linear-phase response is one of the most important properties of FIR filters.

When a signal passes through a filter, it is modified in amplitude and/or phase. The nature and extent of the modification of the signal is dependent on the amplitude and phase characteristics of the filter.

The phase delay or group delay of the filter provides a useful measure of how the filter modifies the phase characteristics of the signal, considering that the signal consists of several frequency components (such as a speech waveform or a modulated signal).

9.2.2.1 Effect of phase distortion on signals

Because the frequency of a filter H(ω) is, in general, a complex number for a specific value of ω, the filter changes both the amplitude and the phase angle of an input sinusoid with a frequency of ω. As we will show, a phase angle change is associated with a delay in the signal as it passes through a filter. A potential problem exists when the delay is different for different frequencies in the pass band of the filter. This delay alters the phase relationship among the frequency components of a signal consisting of many frequencies, resulting in the phenomenon of phase distortion.

9.2.2.2 Phase delay

The phase delay of the filter is the amount of time delay each frequency component of the signal suffers in going through a filter. Mathematically, the phase delay is the negative of the phase angle divided by the frequency.

$$T_P = -\mathrm{d}\theta(\omega)/d\omega \tag{9.22}$$

The requirement for no phase distortion is that the phase delay, T_P, must be constant for all frequencies. A filter with a nonlinear-phase characteristic will cause a phase distortion in the signal that passes through it. This is because the frequency components in the signal will each be delayed by an amount not proportional to the frequency. Such distortion is undesirable in many applications, such as music, data transmission, video, biomedicine, etc. It can be avoided by using filters with linear-phase characteristics over the frequency bands of interest.

Example 9.2

The linear-phase condition results in zero-phase distortion, but adds a phase angle to each sinusoid that is proportional to the frequency of the sinusoids. The signal also picks up a constant phase delay of 5 msec. Find out the phase delay taken by signals.

Solution 9.2

$$\begin{aligned} T_P &= -0.005 = -\tfrac{d\vartheta}{\omega} = -\tfrac{-kf}{2\pi f} = \tfrac{k}{2\pi} \\ k &= -0.001\pi \quad \text{or} \\ T_p &= -0.01\pi f \end{aligned}$$

The phase change in radians for the sinusoids of frequency f would be $T_P = -0.01\pi f$.

9.2.2.3 Group delay

A parameter used to test the phase change with respect to frequency is called the group delay. It is the average time delay the composite signal suffers at each frequency. Mathematically, the group delay is the negative of the derivative of the phase with respect to frequency:

$$T_q = -d\theta(\omega)/d\omega \tag{9.23}$$

If only the constant group delay is desired, a second type of linear-phase filter is defined in which

$$H(\omega) = |H(\omega)|\, e^{j(\beta-\alpha)\omega} \tag{9.24}$$

The only possible solution of $h(n)$, with $\alpha = (N-1)/2$ and $\beta = \pm\,\pi/2$, is

$$h(n) = -h(N-n-1), \;\; 0 \le n \le N-1$$

A filter is said to have a linear-phase response if its phase response satisfies one of the following relationships:

$$\theta(\omega) = -\alpha\omega \tag{9.25}$$

$$\theta(\omega) = \beta - \alpha\omega \tag{9.26}$$

where α and β are constants.

If a filter satisfies the condition given in Equation (9.25), i.e. $\theta(\omega) = -\alpha\omega$, it will have both constant group and phase-delay responses. It can be shown that for the condition $\theta(\omega) = -\alpha\omega$ to be satisfied, the impulse response of the filter must have a positive symmetry.

The phase response in this case is simply a function of the filter length:

$$\begin{aligned} h(n) &= h(N-1-n), \\ n &= 0, 1, \ldots, (N-1)/2 \ (N \, \text{odd}) \\ n &= 0, 1, \ldots, \ (N/2) - 1 \ (N \text{ even}) \end{aligned}$$

Example 9.3

Given the following transfer function, calculate its complex frequency response at 1000 Hz if the system sampling frequency is 1000 Hz.

$$H(z) = 1 + z^{-1}$$

Solution 9.3

The frequency response in the z plane is the value of the transfer function on the unit circle in the z plane at an angle of digital frequency ω.

$$\omega = 2\pi \frac{f}{F_s} = 2\pi \frac{1000}{1000} = 2\pi$$

The value of z on the unit circle at this angle is $z = e^{j\omega} = e^{j2\pi}$

Therefore, the complex frequency is

$$H(\omega) = 1 + e^{-j2\pi}$$

$$H(\omega) = 1 + \cos(2\pi) - j\sin(2\pi) = 1 + 1 - j(0) = 2$$

$$H(\omega) = 2$$

This result means that the magnitude frequency response at 1000 Hz is 2.

Example 9.4

Given the following transfer function, calculate its complex frequency response at 125 Hz if the system sampling frequency is 1000 Hz.

$$H(z) = 0.0935 + 0.3027z^{-1} + 0.40z^{-2} + 0.3027z^{-3} + 0.0935z^{-4}$$

Solution 9.4

The frequency response in the z plane is the value of the transfer function on the unit circle in the z plane at an angle of digital frequency ω. Thus,

$$\omega = 2\pi \frac{f}{F_s} = 2\pi \frac{125}{1000} = \frac{\pi}{4} = 0.7854$$

The value of z on the unit circle at this angle is $z = e^{j\omega} = e^{j0.7854}$

Therefore, the complex frequency is

$$H(\omega) = 0.0935 + 0.0327e^{-j0.7854} + 0.40e^{-j2(0.7854)} + 0.3027e^{-j3(0.7854)} + 0.0935e^{-j4(0.7854)}$$

$$H(\omega) = 0.0935 + 0.0327\{\cos 0.7854 - j \sin 0.7854\} + 0.40\{\cos 1.5708 - j \sin 1.5708\} + 0.3027\{\cos 2.3562 - j \sin 2.3562\} + 0.0935 \{\cos 3.1416 - j \sin 3.1416\}$$

$$H(\omega) = 0.0935 + 0.0327(.7071 - j0.7071) + 0.40(-j) + 0.3027 (-0.7071 - j0.7071) + 0.0935(1)$$

$$H(\omega) = 0.0935 + 0.0231 - j0.0231 - 0.40j - 0.0231 - j0.0231 + 0.0935$$

$$H(\omega) = 0.187 - j0.4462 = 0.4838^{-j1.1739}$$

This result means that the magnitude frequency response at 125 Hz is 0.4838 and the phase response is –1.1739 rad.

Example 9.5

Given the following transfer function, calculate its complex frequency response at 500 Hz if the system sampling frequency is 1000 Hz.

$$H(z) = 0.50z^{-1} + 0.40z^{-2} + 0.60z^{-3}$$

Solution 9.5

The frequency response in the z plane is the value of the transfer function on the unit circle in the z plane at an angle of digital frequency ω.

$$\omega = 2\pi \frac{f}{F_s} = 2\pi \frac{500}{1000} = \pi$$

The value of z on the unit circle at this angle is $z = e^{j\omega} = e^{j\pi}$

Therefore, the complex frequency is

$$H(\omega) = 0.5e^{-j\pi} + 0.40e^{-j2(\pi)} + 0.3027e^{-j3(\pi)}$$

$$H(\omega) = 0.5(-1) + 0.40(1) + 0.6(-1) = -0.7$$
$$H(\omega) = 0.7^{j\pi}$$

This result means that the magnitude frequency response at 500 Hz is 0.7 and the phase response is π rad.

9.3 Type of FIR Filters

9.3.1 Type-1 FIR Filter (Length of the filter *N* is odd)

1. Symmetrical impulse response (order of filter $N - 1$ is even)
2. Its impulse response $h(n)$ possesses the symmetry property of $h(n) = h(N - 1 - n)\ 0 \le n \le N - 1$

For example, for an FIR Filter, which has $N - 1 = 8$,

$$H(z) = \sum_{n=0}^{8} h(n)\, z^{-n}$$

$$H(z) = h(0) + h(1)z^{-1} + h(2)z^{-2} + h(3)z^{-3} + h(4)z^{-4} + h(5)z^{-5} + h(6)z^{-6} + h(7)z^{-7} + h(8)z^{-8}$$

$$h(n) = h(N - 1 - n),\ 0 \le n \le N - 1,\ h(n) = h(8 - n),\ 0 \le n \le 8$$

$h(0) = h(8),\ h(1) = h(7),\ h(2) = h(6),\ h(3) = h(5),\ h(4)$ is left alone without any equivalence.

$$H(z) = h(0)[1 + z^{-8}] + h(1)[z^{-1} + z^{-7}] + h(2)[z^{-2} + z^{-6}] + h(3)[z^{-3} + z^{-5}] + h(4)z^{-4}$$

$$H(\omega) = 2h(0)\left[\frac{1 + e^{-j\omega 8}}{2}\right] + 2h(1)\left[\frac{e^{-j\omega} + e^{-j\omega 7}}{2}\right] + 2h(2)\left[\frac{e^{-j2\omega} + e^{-j6\omega}}{2}\right] + 2h(3)\left[\frac{e^{-j3\omega} + e^{-j5\omega}}{2}\right] + h(4)e^{-4j\omega}$$

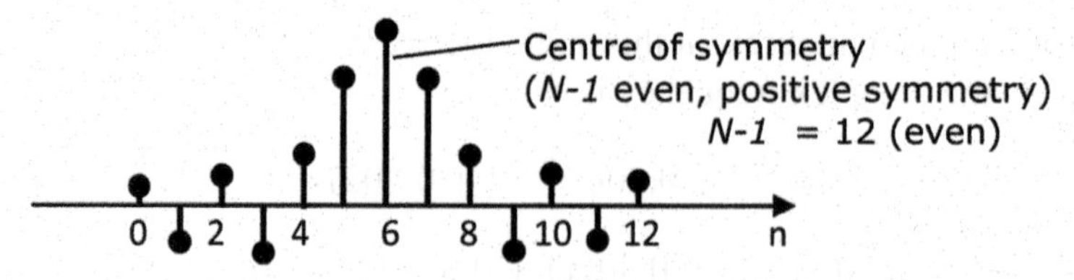

Figure 9.1 Type-I FIR filter: Length of the filter N is odd, positive symmetry.

$$H(\omega) = e^{-4j\omega}\left\{h(4) + 2h(0)\left[\frac{e^{j\omega 4} + e^{-j\omega 4}}{2}\right] + 2h(1)\left[\frac{e^{j3\omega} + e^{-j\omega 3}}{2}\right]\right.$$
$$\left. + 2h(2)\left[\frac{e^{-j2\omega} + e^{-j2\omega}}{2}\right] + 2h(3)\left[\frac{e^{-j\omega} + e^{-j\omega}}{2}\right]\right\}$$

Giving the above expression, a new format is obtained using the following transformation relationship

$$a(0) = h\left[\frac{N-1}{2}\right]; \ a(n) = 2h\left[\left(\frac{N-1}{2}\right) - n\right] \ \ n = 1, 2, \ldots, (N-1)/2$$

$$a(0) = h[4]; \quad a(n) = 2h[4-n] \quad n = 1, \ 2, \ \ldots, \ (N-1)/2$$

Similarly $a(0) = h[4]$; $a(1) = 2h[3]$ $a(2) = 2h[2]$ $a(3) = 2h[1]$ $a(4) = 2h[0]$

The following expression for Type-I filter is expressed as

$$H(\omega) = e^{-4j\omega}a(0) + e^{-4j\omega}a(1)\cos(\omega) + e^{-4j\omega}a(2)\cos(2\omega) + e^{-j4\omega}a(3)$$
$$\cos(3\omega) + e^{-j4\omega}a(4)\cos(4\omega)$$

$$H(\omega) = e^{-4j\omega}\{a(0) + a(1)\cos(\omega) + a(2)\cos(2\omega) + a(3)\cos(3\omega)$$
$$+ a(4)\cos(4\omega)\}$$

$$H(\omega) = e^{-j\omega(4)}\sum_{n=0}^{4} a(n)\cos(\omega n)$$

In generalized format, the above equation for type I FIR filter (Figure 9.1) is expressed as

$$H(\omega) = e^{-j\omega(N-1)/2}\sum_{n=0}^{(N-1)/2} a(n)\cos(\omega n) \tag{9.27}$$

9.3.2 Type-II FIR Filter (Length of the filter *N* is even)

1. Symmetrical impulse response, order of the filter $N - 1$ is odd
2. Its impulse response $h(n)$ possesses the symmetry property of $h(n) = h(N - 1 - n)\ 0 \leq n \leq (N/2) - 1$, i.e. $0 \leq n \leq (8/2) - 1\ 0 \leq n \leq 3$

For example, for an FIR Filter, which has $N - 1 = 7$,

$$H(z) = \sum_{n=0}^{N-1} h(n)\, z^{-n} \quad H(z) = \sum_{n=0}^{7} h(n)\, z^{-n}$$

$$H(z) = h(0) + h(1)z^{-1} + h(2)z^{-2} + h(3)z^{-3} + h(4)z^{-4} + h(5)z^{-5} + h(6)z^{-6} + h(7)z^{-7}$$

The necessary and sufficient condition that an FIR filter should have linear-phase response is $h(n) = h(N - 1 - n)$, $0 \leq n \leq N - 1$, $h(n) = h(7 - n)$, $0 \leq n \leq (N/2) - 1$, $0 \leq n \leq 3$

$$h(0) = h(7),\ h(1) = h(6),\ h(2) = h(5),\ h(3) = h(4).$$

$$H(z) = h(0)[1 + z^{-7}] + h(1)[z^{-1} + z^{-6}] + h(2)[z^{-2} + z^{-5}] + h(3)[z^{-3} + z^{-4}]$$

$$H(\omega) = 2h(0)\left[\frac{1 + e^{-j\omega 7}}{2}\right] + 2h(1)\left[\frac{e^{-j\omega} + e^{-j\omega 6}}{2}\right] + 2h(2)\left[\frac{e^{-j2\omega} + e^{-j5\omega}}{2}\right] + 2h(3)\left[\frac{e^{-j3\omega} + e^{-j4\omega}}{2}\right]$$

$$H(\omega) = e^{-7/2j\omega}\left\{2h(0)\left[\frac{e^{j\omega 7/2} + e^{-j\omega 7/2}}{2}\right] + 2h(1)\left[\frac{e^{j\omega 5/2} + e^{-j\omega 5/2}}{2}\right] + 2h(2)\left[\frac{e^{-j3/2\omega} + e^{-j3/2\omega}}{2}\right] + 2h(3)\left[\frac{e^{j\omega/2} + e^{-j\omega/2}}{2}\right]\right\}$$

$$b(n) = 2h\left[\frac{N}{2} - n\right] \quad n = 1,\ 2,\ \ldots,\ N/2 b(n) = 2h[4 - n] \quad n = 1, 2, 3$$

$$b(1) = 2h[4 - n];\ b(1) = 2h[4]\ \ b(2) = 2h[2]\ \ b(3) = 2h[1]\ \ b(4) = 2h[0]$$

Giving the above expression, a new format obtained using the following transformation relationship is

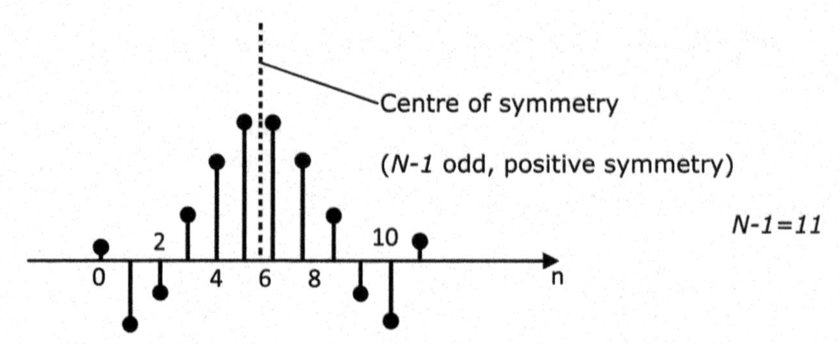

Figure 9.2 Type-II FIR filter: Length of the filter N is even, positive symmetry.

$$H(\omega) = e^{-7/2j\omega}\left\{ b(1)\left[\frac{e^{j\omega 1/2} + e^{-j\omega 1/2}}{2}\right] + b(3)\left[\frac{e^{j\omega 3/2} + e^{-j\omega 3/2}}{2}\right] + b(2)\left[\frac{e^{j\omega 5/2} + e^{-j\omega 5/2}}{2}\right] + b(4)\left[\frac{e^{j\omega 7/2} + e^{-j\omega 7/2}}{2}\right]\right\}$$

The following expression for Type-I filter is expressed as

$$H(\omega) = e^{-j\omega(7/2)}\left\{ b(1)\cos\left(\frac{\omega}{2}\right) + b(2)\cos\left(\frac{3}{2}\omega\right) + b(3)\cos\left(\frac{5}{2}\omega\right) + b(4)\cos\left(\frac{7}{2}\omega\right)\right\}$$

$$H(\omega) = e^{-j\omega(7/2)}\sum_{n=1}^{4} b(n)\cos\left[\omega\left(n - \frac{1}{2}\right)\right]$$

The following expression for Type-II filter (Figure 9.2) is expressed as

$$H(\omega) = e^{-j\omega(N-1)/2}\sum_{n=1}^{N/2} b(n)\cos\left[\omega\left(n - \frac{1}{2}\right)\right] \tag{9.28}$$

9.3.3 Type III-FIR Filter (Length of the filter *N* is odd)

1. Anti-symmetrical impulse response (order of the filter $N - 1$ is even)
2. Its impulse response $h(n)$ possesses the symmetry property of $h(n) = -h(N - 1 - n)$ $0 \leq n \leq N - 1$

For example, for an FIR Filter, which has $N - 1 = 8$, say,

$$h(n) = -h(N-1-n),\ 0 \le n \le N-1,\ h(n) = -h(8-n),\ 0 \le n \le 8$$

The expression for Type-III filter (Figure 9.3) is as following:

$$H(z) = \sum_{n=0}^{N-1} h(n)\, z^{-n} H(z) = \sum_{n=0}^{8} h(n)\, z^{-n}$$

$$H(z) = h(0) + h(1)z^{-1} + h(2)z^{-2} + h(3)z^{-3} + h(4)z^{-4} + h(5)z^{-5} + h(6)z^{-6} + h(7)z^{-7} + h(8)z^{-8}$$

$h(0) = -h(8),\ h(1) = -h(7),\ h(2) = -h(6),\ h(3) = -h(5)$. We also find $h(4) = -h(4)$ and this is possible only if $h(4) = 0$.

$$H(z) = h(0)[1-z^{-8}]+h(1)[z^{-1}-z^{-7}]+h(2)[z^{-2}-z^{-6}]+h(3)[z^{-3}-z^{-5}] + h(4)z^{-4}$$

Converting z format into ω format, we obtain the following expression:

$$H(\omega) = h(0)+h(1)e^{-j\omega}+h(2)e^{-2j\omega}+h(3)e^{-j3\omega}+h(4)e^{-j4\omega}+h(5)e^{-j5\omega} + h(6)e^{-j6\omega} + h(7)e^{-j7\omega} + h(8)e^{-j8\omega}$$

$$H(\omega) = h(0)[1-e^{-j8\omega}] + h(1)[z^{-j\omega}-z^{-j7\omega}] + h(2)[e^{-j2\omega}-e^{-j6\omega}] + h(3)[e^{-j3\omega} - e^{-j5\omega}] + h(4)e^{-j4\omega}$$

$$H(\omega) = e^{-j4\omega}\Bigg\{h(0)\, 2j\left[\frac{e^{j4\omega} - e^{-j4\omega}}{2j}\right] + h(1)2j\left[\frac{z^{3j\omega} - z^{-j3\omega}}{2j}\right] + h(2)2j\left[\frac{e^{-j2\omega} - e^{-j6\omega}}{2j}\right] + h(3)2j\left[\frac{e^{-j3\omega} - e^{-j5\omega}}{2j}\right] + h(4)\Bigg\}$$

Giving the above expression, a new format using the following transformation relationship is obtained:

$$a(0) = h\left[\frac{N-1}{2}\right];\ a(n) = 2h\left[\left(\frac{N-1}{2}\right) - n\right]\quad n = 1,\ 2, \ldots,\ (N-1)/2$$

$$a(0) = h[4];\quad a(n) = 2h[4-n]\quad n = 1,\ 2, \ldots,\ (N-1)/2$$

Similarly, $a(0) = h[4];\quad a(1) = 2h[3]\quad a(2) = 2h[2]\quad a(3) = 2h[1]$
$a(4) = 2h[0]$

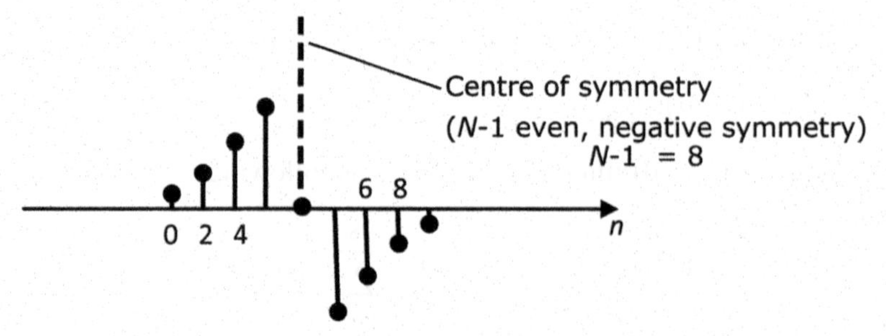

Figure 9.3 Type-III FIR filter: Length of the filter N is odd, negative symmetry.

$$H(\omega) = e^{-j4\omega} e^{j\pi/2} \left\{ h(4) + 2h(3) \left[\frac{e^{j\omega} - e^{-j\omega}}{2j} \right] + 2h(2) \right.$$

$$\left. \left[\frac{e^{-j2\omega} - e^{-j2\omega}}{2j} \right] + 2h(1) \left[\frac{e^{-j3\omega} - e^{-j5\omega}}{2j} \right] + 2h(0) \left[\frac{e^{j4\omega} - e^{-j4\omega}}{2j} \right] \right\}$$

$$H(\omega) = e^{-j[\omega(4)]} e^{j\pi/2} \{a(0)\ \sin(\omega.0) + a(1)\ \sin(\omega.1) + a(2)\ \sin(\omega.2) + a(3)\ \sin(\omega.3) + a(4)\ \sin(\omega.4)\}$$

$$H(\omega) = e^{-j[\omega(4)]} e^{j\pi/2} \sum_{n=0}^{4} a(n) \sin(\omega n)$$

$$H(\omega) = e^{-j[\omega(N-1)/2 - \pi/2]} \sum_{n=0}^{(N-1)/2} a(n) \sin(\omega n) \qquad (9.29)$$

9.3.4 Type-IV FIR Filter (Length of the filter *N* is even)

1. Anti-symmetrical impulse response (order of the filter $N - 1$ is odd)
2. Its impulse response $h(n)$ possesses the symmetry property of $h(n) = -h(N - 1 - n)\ 0 \leq n \leq N/2 - 1$

For example, for an FIR Filter, which has $N - 1 = 7$, say,

The expression for Type-IV filter is given as follows:

$$H(z) = \sum_{n=0}^{N-1} h(n)\, z^{-n} H(z) = \sum_{n=0}^{7} h(n)\, z^{-n}$$

$$H(z) = h(0)+h(1)z^{-1}+h(2)z^{-2}+h(3)z^{-3}+h(4)z^{-4}+h(5)z^{-5} + h(6)z^{-6} + h(7)z^{-7}$$

$$h(n) = -h(N-1-n),\ 0 \leq n \leq N-1,\ h(n) = -h(7-n),\ 0 \leq n \leq 7$$
$$h(0) = -h(7),\ h(1) = -h(6),\ h(2) = -h(5),\ h(3) = -h(4).$$

$$H(z) = h(0)[1-z^{-7}]+h(1)[z^{-1}-z^{-6}]+h(2)[z^{-2}-z^{-6}]+h(3)[z^{-3}-z^{-4}]$$

$$H(\omega) = h(0)[1-e^{-7j\omega}] + h(1)[e^{-j\omega} - e^{-j6\omega}] + h(2)[e^{-j2\omega} - e^{-j5\omega}] + h(3)[e^{-j3\omega} - e^{-j4\omega}]$$

$$H(\omega) = e^{-j\omega(7/2)}\left\{2h(0)\left[\frac{e^{-7/2j\omega} - e^{7/2j\omega}}{2j}\right] + 2h(1)\left[\frac{e^{j(5/2\omega} - e^{-j(5/2)\omega}}{2j}\right] + h(2)\left[\frac{e^{j(3/2)\omega} - e^{-j(3/2)\omega}}{2j}\right] + h(3)\left[\frac{e^{j(1/2)\omega} - e^{-j(1/20\omega}}{2j}\right]\right\}$$

Giving the above expression, a new format using the following transformation relationship is derived,

$$b(n) = 2h\left[\frac{N}{2} - n\right] \quad n = 1, 2, \ldots,\ N/2 b(n) = 2h[4-n] \quad n = 1, 2, 3, 4$$

$$b(1) = 2h[4-n];\ b(1) = 2h[4] \quad b(2) = 2h[2] \quad b(3) = 2h[1] \quad b(4) = 2h[0]$$

$$H(\omega) = e^{-7/2j\omega}\left\{b(4)\left[\frac{e^{j\omega 7/2} - e^{-j\omega 7/2}}{2j}\right] + b(3)\left[\frac{e^{j\omega 5/2} - e^{-j\omega 5/2}}{2j}\right] + b(2)\left[\frac{e^{-j3/2\omega} - e^{-j3/2\omega}}{2j}\right] + b(1)\left[\frac{e^{j\omega/2} - e^{-j\omega/2}}{2j}\right]\right\}$$

$$H(\omega) = e^{-7/2j\omega}\left\{b(1)\left[\frac{e^{j\omega 1/2} + e^{-j\omega 1/2}}{2j}\right] + b(3)\left[\frac{e^{j\omega 3/2} + e^{-j\omega 3/2}}{2j}\right] + b(2)\left[\frac{e^{j\omega 5/2} + e^{-j\omega 5/2}}{2j}\right] + b(4)\left[\frac{e^{j\omega 7/2} + e^{-j\omega 7/2}}{2j}\right]\right\}$$

The following expression for Type-I filter is expressed as

$$H(\omega) = e^{-j\omega(7/2)}\left\{b(1)\sin\left(\frac{\omega}{2}\right) + b(2)\sin\left(\frac{3}{2}\omega\right) + b(3)\sin\left(\frac{5}{2}\omega\right)\right.$$
$$\left. + b(4)\sin\left(\frac{7}{2}\omega\right)\right\}$$

$$H(\omega) = e^{-j\omega(7/2)}.e^{j(\pi/2)}\sum_{n=1}^{4} b(n)\sin\left[\omega\left(n-\frac{1}{2}\right)\right]$$

The expression for Type-IV filter (Figure 9.4) is given as follows:

$$H(\omega) = e^{-j[\omega(N-1)/2-\pi/2]}\sum_{n=1}^{N/2} b(n)\sin\left[\omega\left(n-\frac{1}{2}\right)\right] \tag{9.30}$$

$$b(n) = 2h\left[\frac{N}{2}-n\right] \quad n = 1,\ 2,\ldots,\ N/2$$

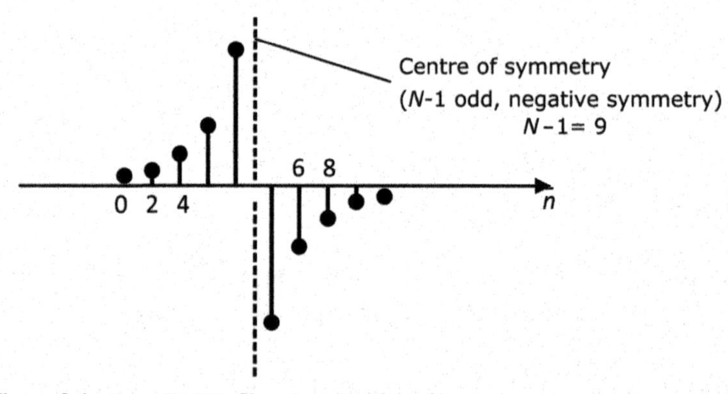

Figure 9.4 Type-IV FIR filter: Length of the filter N is even, negative symmetry.

Example 9.6

Which of the DSP systems given by the following transfer functions show linear phase?

(a) $H(z) = 1 + 4z^{-1} + 3z^{-2} + 4z^{-3} + z^{-4}$
(b) $H(z) = -3 + 2z^{-1} + z^{-2} + 4z^{-3} + z^{-4} - 4z^{-5} + 3z^{-6}$
(c) $H(z) = 1 + 3z^{-1} - z^{-2}$

Solution 9.6

The impulse responses for the three systems mentioned above are:

(a) $h(n) = \{1,\ 4,\ 3,\ 4,\ 1\}$
(b) $h(n) = \{-3,\ 2,\ 1,\ 4,\ 1,\ -4,\ 3\}$
(c) $h(n) = \{1,\ 3,\ -1\}$

All three systems have odd number coefficients. Both (a) and (c) are symmetrical about the centre value 3, and hence, both have linear-phase responses. (b) Lacks symmetry and, therefore, also lacks a centre value. (a) is the case of positive symmetry, (b) is not symmetrical, while (c) falls under odd symmetry.

Example 9.7

Prove that the filter with the following response has a linear-phase response

$$h(n) = \{4,\ 1,\ 1,\ 4\}$$

Solution 9.7

The phase shift of the system function varies linearly. This is called the linear-phase response. The condition for the linear-phase response is given by

$h(n) = h(N - 1 - n)$, where N is the length of the impulse response

In this example,

$$h(n) = \{4,\ 1,\ 1,\ 4\}$$

We have $h(n) = h(4-1-n), h(n) = h(3-n), h(0) = h(3), h(1) = h(2)$, hence this filter has a linear-phase response.

Example 9.8

A filter has the following impulse response

$$h(n) = \{-4,\ 2,\ 1,\ -2,\ 4\}$$

Is it a linear-phase filter?

Solution 9.8

The condition for linear-phase response is given by

$h(n) = h(N - 1 - n)$, where N is the length of the impulse response

In this example, $N = 5$

$$h(n) = \{-4\ ,\ 2,\ 1,\ -2,\ 4\}$$

We have $h(n) = h(5 - 1 - n), h(n) = h(4 - n), h(0) = h(4);\ h(1) = h(3);\ h(2)$.

Hence, this filter has a linear-phase response and its impulse response is odd symmetric.

Example 9.9

(i) Discuss briefly the conditions necessary for a realizable digital filter to have a linear-phase characteristic, and the advantages of filters with such characteristics.

(ii) An FIR digital filter has an impulse response, $h(n)$, defined over the interval $0 \leq n \leq N - 1$. Show that if $N = 5$ and $h(n)$ satisfy the symmetry condition $h(n) = h(N - 1 - n)$, then the filter has a linear-phase characteristic.

(iii) Repeat (ii) for $N = 4$.

Solution 9.9

(i) The necessary and sufficient condition for a filter to have linear-phase response is that its impulse response must be symmetrical.

$$h(n) = h(N - 1 - n) \quad \text{or} \quad h(n) = -h(N - 1 - n)$$

For non-recursive FIR filters, the storage space for coefficients and the number of arithmetic operations are reduced by a factor of approximately 2. For recursive FIR filters, the coefficients can be made to be simple integers, leading to an increased speed of processing. In linear-phase filters, all frequency components experience the same amount of delay through the filter, which has no phase distortion.

(ii) Using the symmetry condition, we find that for $N - 1 = 4$:

$$h(n) = h(N - 1 - n);\ h(0) = h(4);\ h(1) = h(3);$$

The frequency response, $H(\omega)$, for the filter is given by

$$H(\omega) = \sum_{n=0}^{N-1} h(n)\, e^{-j\omega n}$$

$$H(\omega) = \sum_{n=0}^{4} h(n) e^{-j\omega n}$$

$$H(\omega) = h(0) + h(1)e^{-j\omega} + h(2)e^{-j2\omega} + h(3)e^{-j3\omega} + h(4)e^{-j4\omega}$$

$$H(\omega) = e^{-j2\omega}[h(0)e^{j2\omega} + h(1)e^{j\omega} + h(2) + h(3)e^{-j\omega} + h(4)e^{-j2\omega}]$$

Using the symmetry and necessary and sufficient condition, $h(0) = h(4)$; $h(1) = h(3)$; we can group terms whose coefficients are numerically equal.

$$H(\omega) = e^{-j2\omega}[h(0)(e^{j2\omega} + e^{-j2\omega}) + h(1)(e^{j\omega} + e^{-j\omega}) + h(2)]$$
$$= e^{-j2\omega}[2h(0)\cos(2\omega) + 2h(1)\cos(\omega) + h(2)]$$

$$a(0) = h\left[\frac{N-1}{2}\right]; a(n) = 2h\left[\left(\frac{N-1}{2}\right) - n\right] \quad \text{n} = 1,\ 2,\ \ldots,\ (N-1)/2$$

$a(0) = h(2);\quad a(n) = 2h[2-n] \quad n = 1, 2, \ldots, (N-1)/2$

Similarly, $a(0) = h[2]; a(1) = 2h[1] \quad a(2) = 2h[0]$

then $H(\omega)$ can be written in the following form:

$$H(\omega) = e^{-j2\omega}[a(2)(\tfrac{e^{j2\omega}+e^{-j2\omega}}{2}) + a(1)(\tfrac{e^{j\omega}+e^{-j\omega}}{2}) + a(0)]$$
$$H(\omega) = e^{-j2\omega}[a(0) + a(1)\cos(\omega) + a(2)\cos(2\omega)]$$

$$H(\omega) = e^{-j2\omega}\sum_{n=0}^{2} a(n)\cos(\omega n) = e^{j\theta(\omega)}|H(\omega)|$$

where $|H(\omega)| = \sum_{n=0}^{2} a(n)\cos(\omega n);$ and $\theta(\omega) = -2\omega$

Clearly, $\theta(\omega) = -2\omega$ and the phase response is linear.

(iii) In this case, using the symmetry condition, we find that for $N = 4$:

$h(0) = h(3); h(1) = h(2);$

The frequency response, $H(\omega)$, for the filter is given by

$$H(\omega) = \sum_{n=0}^{N-1} h(n)\, e^{-j\omega n}$$

$$H(\omega) = \sum_{n=0}^{3} h(n)e^{-j\omega n}$$
$$H(\omega) = h(0) + h(1)e^{-j\omega} + h(2)e^{-j2\omega} + h(3)e^{-j3\omega}$$

The necssary and sufficient condition that an FIR filter should have linear-phase response: $h(n) = h(N-1-n)$, $0 \leq n \leq N-1$, $h(n) = h(3-n)$, $0 \leq n \leq (N/2) - 1$, $0 \leq n \leq 1$

$$H(\omega) = h(0)[1 + e^{-j\omega 3}] + h(1)[e^{-j\omega} + e^{-j\omega 2}]$$

$$H(\omega) = 2h(0)\left[\frac{1 + e^{-j\omega 3}}{2}\right] + 2h(1)\left[\frac{e^{-j\omega} + e^{-j\omega 2}}{2}\right]$$

$$H(\omega) = e^{-3/2j\omega}\left\{2h(0)\left[\frac{e^{j\omega 3/2}+e^{-j\omega 3/2}}{2}\right]+2h(1)\left[\frac{e^{j\omega/2}+e^{-j\omega/2}}{2}\right]\right\}$$

$$b(n) = 2h\left[\frac{N}{2}-n\right] \quad n = 1, 2, \ldots, N/2$$

$$b(n) = 2h[2-n] \quad n = 1,\ 2$$

$$b(n) = 2h[2-n]; b(1) = 2h[2] \quad b(2) = 2h[0]$$

Giving the above expression, a new format obtained using the following transformation relationship is

$$H(\omega) = e^{-3/2j\omega}\left\{b(1)\left[\frac{e^{j\omega/2}+e^{-j\omega/2}}{2}\right]+b(2)\left[\frac{e^{j\omega 3/2}+e^{-j\omega 3/2}}{2}\right]\right\}$$

The following expression for Type-I filter is expressed as

$$H(\omega) = e^{-j\omega(3/2)}\left\{b(1)\cos\left(\frac{\omega}{2}\right)+b(2)\cos\left(\frac{3}{2}\omega\right)\right\}$$

$$H(\omega) = e^{-j\omega(3/2)}\sum_{n=1}^{2} b(n)\cos\left[\omega\left(n-\frac{1}{2}\right)\right]$$

The expression for Type-II filter is given as following:

$$H(\omega) = e^{-j\omega(N-1)/2}\sum_{n=1}^{N/2} b(n)\cos\left[\omega\left(n-\frac{1}{2}\right)\right] \qquad (9.31)$$

$$H(\omega) = \pm|H(\omega)|e^{j\theta(\omega)}$$

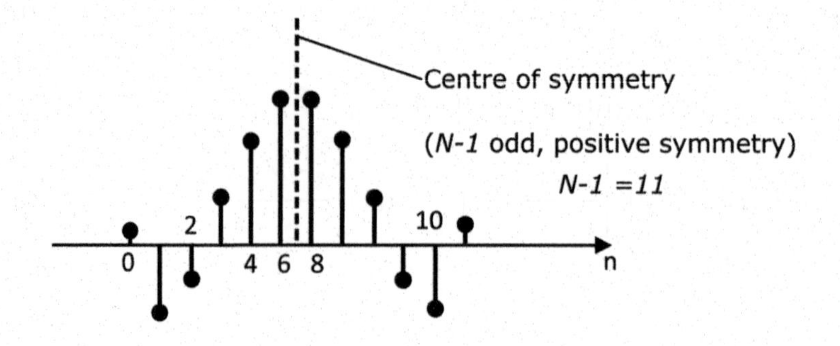

$$|H(\omega)| = \sum_{n=1}^{2} b(n)\cos\left\{\omega\left(n-\frac{1}{2}\right)\right\}; \theta(\omega) = -(3/2)\omega$$

Clearly, $\theta(\omega) = -(3/2)\omega$ and the phase response is linear.

Table 9.1 A summary of the key points of the four types of FIR filters

Impulse response Symmetry	Number of Coefficients, N	Frequency Response $H(\omega)$	Type of Linear Phase
Positive symmetry,	Odd	$e^{-j\omega(N-1)/2} \sum_{n=0}^{(N-1)/2} a(n) \cos(\omega n)$	1
$h(n) = h(N-1-n)$	Even	$e^{-j\omega(N-1)/2} \sum_{n=1}^{N/2} b(n) \cos[\omega(n-\frac{1}{2})]$	2
Negative symmetry,	Odd	$e^{-j[\omega(N-1)/2-\pi/2]} \sum_{n=0}^{(N-1)/2} a(n)\sin(\omega n)$	3
$h(n) = -h(N-1-n)$	Even	$e^{-j[\omega(N-1)/2-\pi/2]} \sum_{n=1}^{N/2} b(n) \sin[\omega(n-\frac{1}{2})]$	4

$a(0) = h[\frac{N-1}{2}]$; $a(n) = 2h[\frac{N-1}{2} - n]$ $b(n) = 2h[\frac{N}{2} - n]$

Example 9.10

For a linear-phase filter, positive symmetry has a phase response of $e^{-j3\omega}$. What is the order of the filter?

Solution 9.10

We know that the phase of a linear-phase filter is given by the phase function for the positive symmetry

$$H(\omega) = e^{-j\omega(N-1)/2} \sum_{n=1}^{N-1/2} a(n)\cos[\omega(n)] \quad \text{or}$$

$$H(\omega) = e^{-j\omega(N-1)/2} \sum_{n=1}^{N/2} b(n)\cos[\omega(n-\tfrac{1}{2})]$$

$$H(\omega) = e^{-j\omega(N-1)/2} \sum_{n=0}^{(N-1)/2} a(n)\cos(\omega n)$$

Comparing the given phase response $e^{-j3\omega}$ with $e^{-j\omega(N-1)/2}$, we obtain

$$\frac{N-1}{2} = 3, \quad N = 7$$

This is a seventh-order filter.

Example 9.11

The frequency response of a Type-I positive symmetry FIR filter is given by the following expression:

$$H(\omega) = e^{-j3\omega}\{2 + 1.8\cos 3\omega + 1.2\cos 2\omega + 0.5\cos\omega\}$$

Determine the impulse response *h*(*n*) of the filter in the form of a sequence.

Solution 9.11

Comparing the phase factor $e^{-j\omega(N-1)/2}$ with that of $H(\omega)$, we obtain $\frac{N-1}{2} = 3, \quad N = 7$ (the length of the filter coefficients)

The frequency response of an FIR for a causal filter is given by the following expression:

$$H(\omega) = \sum_{n=0}^{6} h(n)\, e^{-j\omega n}$$

$$H(\omega) = \sum_{n=0}^{6} h(n) e^{-j\omega n}$$

$$\begin{aligned} H(\omega) &= h(0) + h(1)e^{-j\omega} + h(2)e^{-j2\omega} + h(3)e^{-j3\omega} + h(4)e^{-j4\omega} \\ &\quad + h(5)e^{-j5\omega} + h(6)e^{-j6\omega} \end{aligned}$$

The necessary and sufficient condition for a filter to be causal

$$\begin{aligned} &h(n) = h(6-n) \\ &h(0) = h(6);\; h(1) = h(5);\; h(2) = h(4); h(3) = h(3) \end{aligned}$$

Replacing the above expression using the necessary and sufficient condition, we obtain

$$\begin{aligned} H(\omega) &= h(0){+}h(1)e^{-j\omega}{+}h(2)e^{-j2\omega}{+}h(3)e^{-j3\omega}{+}h(2)e^{-j4\omega}{+}h(1)e^{-j5\omega} \\ &\quad + h(0)e^{-j6\omega} \end{aligned}$$

$$\begin{aligned} H(\omega) = e^{-j\omega 3} \sum_{n=0}^{3} &\{h(0)e^{3j\omega} + h(0)e^{-j3\omega}\} + \{h(1)e^{j2\omega} + h(1)e^{-j2\omega}\} \\ &+ \{h(2)e^{j\omega} + h(2)e^{-j\omega} + h(3)\} \end{aligned}$$

Expanding the formula

$$H(\omega) = e^{-j\omega(3)/}\left\{\sum_{n=0}^{\frac{N-1}{2}} a(n)\cos[\omega(n)]\right\}$$

$$H(\omega) = e^{-j\omega 3}\{a(0) + a(1)\ \cos[\omega] + a(2)\ \cos[2\omega] + a(3)\ \cos[3\omega]\}$$
$$H(\omega) = e^{-j3\omega}\{0.5\cos\omega + 1.2\cos 2\omega + 1.8\cos 3\omega + 2\}$$
$$H(\omega) = e^{-j3\omega}\{h(3) + 2h(0)\cos 3\omega + 2h(1)\cos 2\omega + 2h(2)\cos\omega\}$$

Converting the equation into a coefficient form, we get

$$a(0) = h\left[\frac{N-1}{2}\right];\ a(n) = 2h\left[\frac{N-1}{2} - n\right] \quad n = 1, 2,\ \ldots,\ (N-1)/2$$

$$a(0) = h[3];\ a(n) = 2h[3-n] \quad n = 1, 2, \ldots, (N-1)/2$$
$$a(1) = 2\ h(2);\ a(2) = 2\ h(1);\ h(3)\ = 2\ h(0)$$

$$a(0) =\ 2 = h(3);\ a(1) = 0.5 = 2\ h(2);\ a(2)\ = 1.2\ = 2\ h(1);$$
$$a(3) = 1.8 =\ 2\ h(0)$$

Hence, the required impulse response will be

$$h(n) = \{0.9,\ 0.6,\ 0.25,\ 2,\ 0.25,\ 0.6,\ 0.9\}$$

Example 9.12

The frequency response of a Type-II positive symmetry FIR filter is given by the following expression

$$H(\omega) = e^{-j(3/2)\omega}\{1.5\cos 2\omega + 1.2\cos\omega\}$$

Determine the impulse response *h(n)* of the filter in the form of a sequence.

Solution 9.12

Comparing the phase factor $e^{-j\omega(N-1)/2}$ with that of $H(\omega)$, we obtain $\frac{N-1}{2} = \frac{3}{2}$, $N = 4$ (the length of the filter).

The frequency response of an FIR for a causal filter is given by the following expression:

$$H(\omega) = \sum_{n=0}^{3} h(n)e^{-j\omega n}$$
$$H(\omega)\ = h(0) + h(1)e^{-j\omega} + h(2)e^{-j2\omega} + h(3)e^{-j3\omega}$$

Using the symmetry condition, we find that for $N - 1 = 3$:

$$h(n) = h(N - 1 - n)\ h(0) = h(3);\ h(1) = h(2);$$

$$H(\omega) = e^{-j\omega 3/2}[h(0)(e^{j3\omega/2} + e^{-j3\omega/2}) + h(1)(e^{j\omega/2} + e^{-j\omega/2})]$$

Converting the equation into b(n) coefficient form, let

$$b(n) = 2h\left(\frac{N}{2} - n\right) \quad n = 1, 2, \ldots, N/2$$

$$e^{-j\omega(N-1)/2}\sum_{n=1}^{N/2} b(n)\cos[\omega(n - \tfrac{1}{2})]$$

Expanding the term

$$H(\omega) = e^{-j\omega 3/2}[b(1)\cos(1\omega/2) + b(2)\cos(3\omega/2)]$$

$$b(n) = 2h[2 - n] \quad b(1) = 2h(1) = 1.5) \quad h(1) = 0.75;$$

$$b(2) = 2h(0) = 1.2 \quad h(0) = 0.6$$

$H(\omega) = e^{-j(3/2)\omega}\{1.5\cos 2\omega + 1.2\cos\omega\}$ comparing
Hence, the required impulse response will be

$$h(n) = \{0.6,\ 0.75,\ 0.75,\ 0.6\}$$

9.4 Basic Principle of FIR Filter Design

FIR filters are designed by assuming that the magnitude of the transfer function $H(\omega)$ is unity. In other words, we assume $H(\omega) = 1$.

$$\frac{|Y(\omega)|}{|X(\omega)|} = |H(\omega)|$$

X(ω) is the input and Y(ω) is the output in the frequency domain

$$|Y(\omega)| = |X(\omega)| \tag{9.32}$$

Equation (9.32) tells us that the output in the frequency domain = the input in the frequency domain.

From Equation (9.32), we arrive at a very important concept. An FIR filter does not introduce any losses to signals that get transmitted through it. It must be noted here that all types of IIR filters are designed by choosing a suitable and finite value of $|H(\omega)|$ which is less than 1. FIR filters, on the other hand, are characterized by the following equations:

$$y(n) = \sum_{k=0}^{N-1} h(k)\, x(n-k);$$

$$H(z) = \sum_{n=0}^{N-1} h(n)\, z^{-n}; \; H(\omega) = \sum_{n=0}^{N-1} h(n)\, e^{-j\omega n}$$

The sole objective of most FIR coefficient calculation (or approximation) methods is to obtain values of $h(n)$, so that the resulting filter meets the design specifications, such as amplitude-frequency response as per requirements.

The Fourier series and the Fourier transform methods belong to the category which are used to design FIR filters.

Several methods are available for obtaining $h(n)$. The window, optimal, and frequency sampling methods, however, are the most commonly used. All three can lead to linear-phase FIR filters. The window method is discussed in this chapter.

9.4.1 Windows Used in FIR Filters

The basic assumption behind the DFT is that the input domain signal is periodic. The assumption results in the signal, which is created by repeating the windowed signal, over and over again. This signal is not quite the same as the original time domain signal. The sudden changes require a large number of frequencies. To create a signal that closely approaches these discontinuities, we require a very large number of sinusoids.

It is the discontinuities, or sudden changes, that result in a large number of frequencies. These transitions result in the spreading of energy in the frequency domain referred to earlier as spectral leakage. If we could somehow reduce the abruptness of these transitions, we could reduce the spectral leakage. In practice, it is very difficult to avoid these transitions.

Most practical signals consist of many frequencies, and there is no way to know in advance what these frequencies will be. So we cannot determine the window duration such that we will always obtain an integer number of cycles of all the constituent sinusoids. It is not even possible to do with complex signals such as music or speech. The best method of tackling this concern is to reduce the abruptness of the transitions. This is what windows accomplish.

9.4.1.1 Windowing a signal

FIR filters exhibit poor gain characteristics for ideal filters, which have a jump between one and zero. If the number of coefficients is increased, the transition region of the FIR filter decreases and the ripples in the pass band and stop band are reduced. However, there is a limit to the reduction in the amplitude of the ripple as the number of coefficients used is increased. This is called the Gibbs effect. This effect is due to the jump between zero and one of the ideal gain curves. In order to reduce this effect, the actual values of FIR coefficients are reduced near where they start and end. This is called windowing.

It is important to note that each of the transform assumes that the signal exists for infinite time in the past to infinite time in the future. Of course, we cannot measure a signal for such a long time. In practice, the measurement is made for finite duration that could typically be as small as a few milliseconds to as large as several hours. The duration over which the signal is measured is known as a window because it is as if you are looking at a very small signal through a small window that restricts your field of vision.

9.4.1.2 Rectangular window

When a signal is measured for a finite duration, it is being windowed. Windowing is equivalent to multiplying the signal of interest by another signal, referred to as a window and denoted by w. If any of the windows are not specified, the rectangular window is used which is also known as the uniform window.

The rectangular window has a value of 1 for a certain time interval and a value of zero elsewhere. If the windowed signal has N samples, the equation for the corresponding rectangular window is

$$\omega(n) = 1.0 \tag{9.33}$$

where n ranges from 0 to $N - 1$.

What happens when you multiply a signal by a rectangular window? You only get a part of the signal that exists when the values of the rectangular window are nonzero.

9.4.1.3 Hanning window

Sudden transitions at the ends of the windowed signals result in spectral leakage. Without any particular window, the rectangular window is used by default. However, the rectangular window does not do anything to reduce the amplitudes of the transitions at the beginning and end of the measurement interval. There are other windows that smooth the transition and reduce spectral leakage. This is achieved by making the values of the signal almost the same at the two ends of the measurement intervals. This value is usually zero or close to zero. One such popular window is the Hanning window, which is used for general purpose applications.

If you are not sure which window to use, you can start with the Hanning window. The shape of the Hanning window is that of a cosine wave, but it is given a dc offset so that it never goes negative. If the window signal has N samples, the equation for the corresponding Hanning window is where n ranges from 0 to $N - 1$ (Figures 9.5 and 9.6).

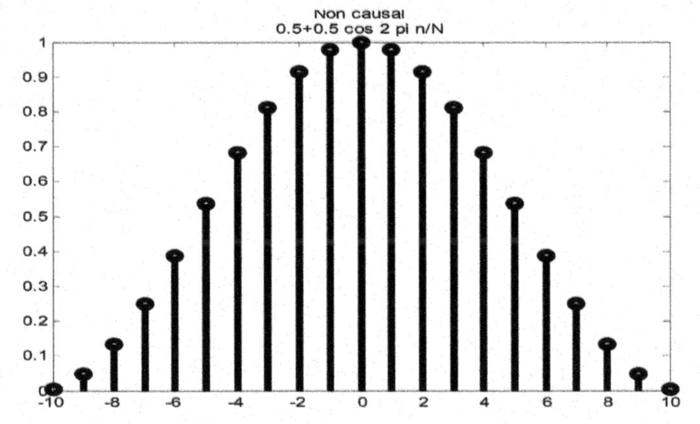

Figure 9.5 Hanning expression non-causal $\omega(n) = 0.5(1 + \cos 2n\pi/N - 1)$.

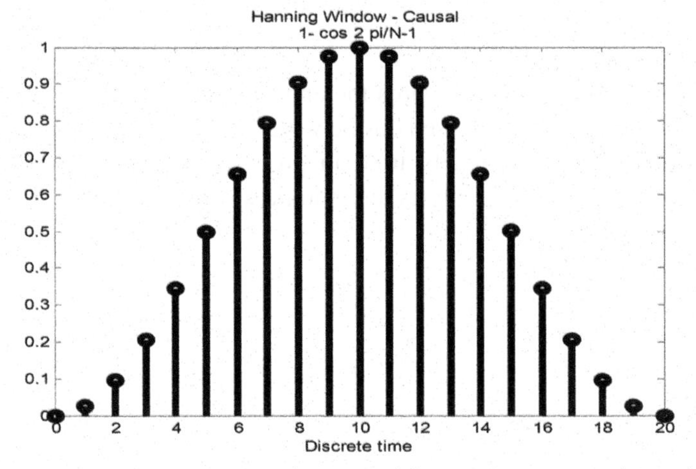

Figure 9.6 Hanning expression causal $\omega(n) = 0.5(1 - \cos 2n\pi/N - 1)$.

$$\omega[n] = 0.5 - 0.5 \cos \frac{2\pi n}{N-1} \quad 0 \leq n \leq N-1 \tag{9.34}$$

Equation (9.34) has been taken as a standard window for designing a causal filter.

$$\omega[n] = 0.5 + 0.5 \cos \frac{2\pi n}{N-1} \quad -\frac{N-1}{2} \leq n \leq \frac{N-1}{2} \tag{9.35}$$

Equation (9.35) has been taken as a standard window for designing a non-causal filter.

Hanning Window			
Coefficients Value (n)	$\omega[n] = 0.5 + 0.5 \cos \frac{2\pi n}{N-1}$ $\left(-\frac{N-1}{2}\right) \leq n \leq \left(\frac{N-1}{2}\right)$ $N-1 = 10$	Coefficients Value (n)	$\omega[n] = 0.5 - 0.5 \cos \frac{2\pi n}{N-1}$ $0 \leq n \leq (N-1)$ $N-1 = 10$
–5	0.0000	0	0.0000
–4	0.0955	1	0.0955
–3	0.3455	2	0.3455
–2	0.6545	3	0.6545
–1	0.9045	4	0.9045
0	1.0000	5	1.0000
1	0.9045	6	0.9045
2	0.6545	7	0.6545
3	0.3455	8	0.3455
4	0.0955	9	0.0955
5	0.0000	10	0.0000

9.4.1.4 Hamming window

The amplitude of the windows towards the ends can be different from zero. This window is also called the raised cosine window. The Hamming window is one such example. It is widely used in processing speech signals for applications such as spectral analysis and computer voice response systems (Figures 9.7 and 9.8). Its equation is

$$\omega[n] = 0.54 - 0.46 \cos \frac{2\pi n}{N-1} \quad 0 \leq n \leq N-1 \tag{9.36}$$

Equation (9.36) has been taken as a standard window for designing a causal filter.

$$\omega[n] = 0.54 + 0.46 \cos \frac{2\pi n}{N-1} \quad -\frac{N-1}{2} \leq n \leq \frac{N-1}{2} \tag{9.37}$$

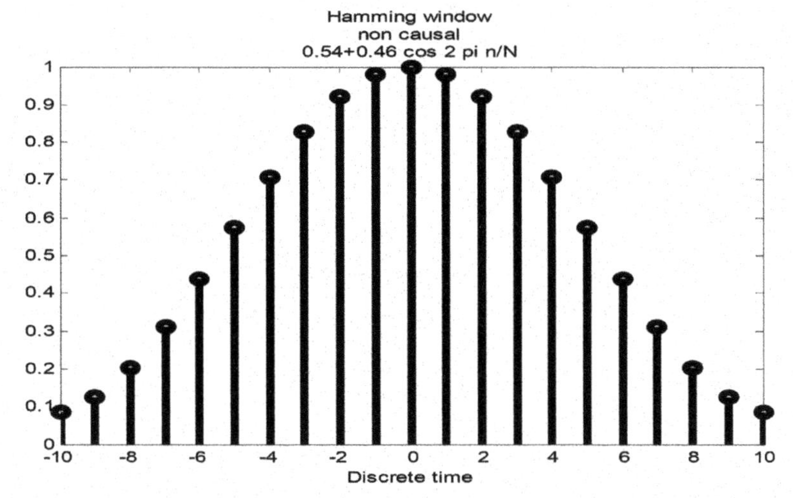

Figure 9.7 Hamming expression non-causal $\omega(n) = 0.54 + 0.46 \cos 2\pi n/(N - 1)$.

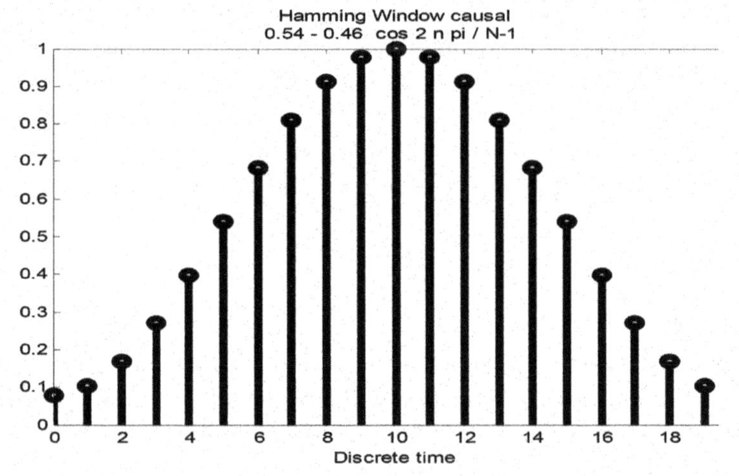

Figure 9.8 Hamming expression causal $\omega(n) = 0.54 - 0.46 \cos 2\pi n/(N - 1)$.

Equation (9.37) has been taken as a standard window for designing a non-causal filter.

Equation (9.36) has taken a standard window of $n = 0$ to $n = N - 1$ for the window to be causal for a low-pass filter as shown in Figure 9.9. It is worthwhile to spend a few minutes here to decide which formula is going to be fit for calculation of coefficients $h(n)$ of FIR filters of different passes such as high pass, band pass, and band stop.

Hamming Window			
Coefficient Value (n)	$\omega[n] = 0.54 + 0.46\ \cos\frac{2\pi n}{N-1}$ $(-\frac{N-1}{2}) \le n \le (\frac{N-1}{2})$ $N-1 = 10$	Coefficient Value (n)	$\omega[n] = 0.54 - 0.46\ \cos\frac{2\pi n}{N-1}$ $0 \le n \le (N-1)$ $N-1 = 10$
–5	0.0800	0	0.0800
–4	0.1679	1	0.1679
–3	0.3978	2	0.3978
–2	0.6821	3	0.6821
–1	0.9121	4	0.9121
0	1.0000	5	1.0000
1	0.9121	6	0.9121
2	0.6821	7	0.6821
3	0.3978	8	0.3978
4	0.1679	9	0.1679
5	0.0800	10	0.0800

9.5 Design of FIR Filter using the Window Method

The window method is one of the simplest methods of designing FIR digital filters. It is well suited for designing filters with simple frequency response shapes, such as ideal low-pass filters. Some typical filter shapes that can be designed are shown in Figure 9.9.

In this method, the frequency response of a filter, $H_{\mathrm{D}}(\omega)$, and the corresponding impulse response, $h_{\mathrm{D}}(n)$, are related by the inverse Fourier transform:

$$h_D(n) = \frac{1}{2\pi}\int_0^{2\pi} H_{\mathrm{D}}(\omega)\ e^{j\omega n} d\omega \tag{9.38}$$

The subscript D is used to distinguish between the ideal and practical impulse responses. The need for this distinction will soon become clear. If we know $H_{\mathrm{D}}(\omega)$, $h_{\mathrm{D}}(n)$ can be obtained by evaluating the inverse Fourier transform of Equation (9.38).

As an illustration, suppose we wish to design a low-pass filter. We could start with the ideal low-pass response shown in Figure 9.9 (a), where ω_{C} is the cutoff frequency and the frequency scale is normalized: $G = 1$. Letting the response from $-\omega_{\mathrm{C}}$ to ω_{C}, the impulse response is given by

$$h_D(n) = \frac{G}{2\pi}\int_{-\pi}^{\pi} 1 \times e^{j\omega n} d\omega = \frac{G}{2\pi}\int_{-\omega_c}^{\omega_c} e^{j\omega n} d\omega$$

$$h_D(n) = \frac{G}{2\pi}\int_{-\omega_c}^{\omega_c} e^{j\omega n} d\omega = G\frac{e^{j\omega_c n} - e^{-j\omega_c n}}{2j\pi\ n} = \frac{G}{\pi\ n}\ \sin\omega_c n, G = 1$$

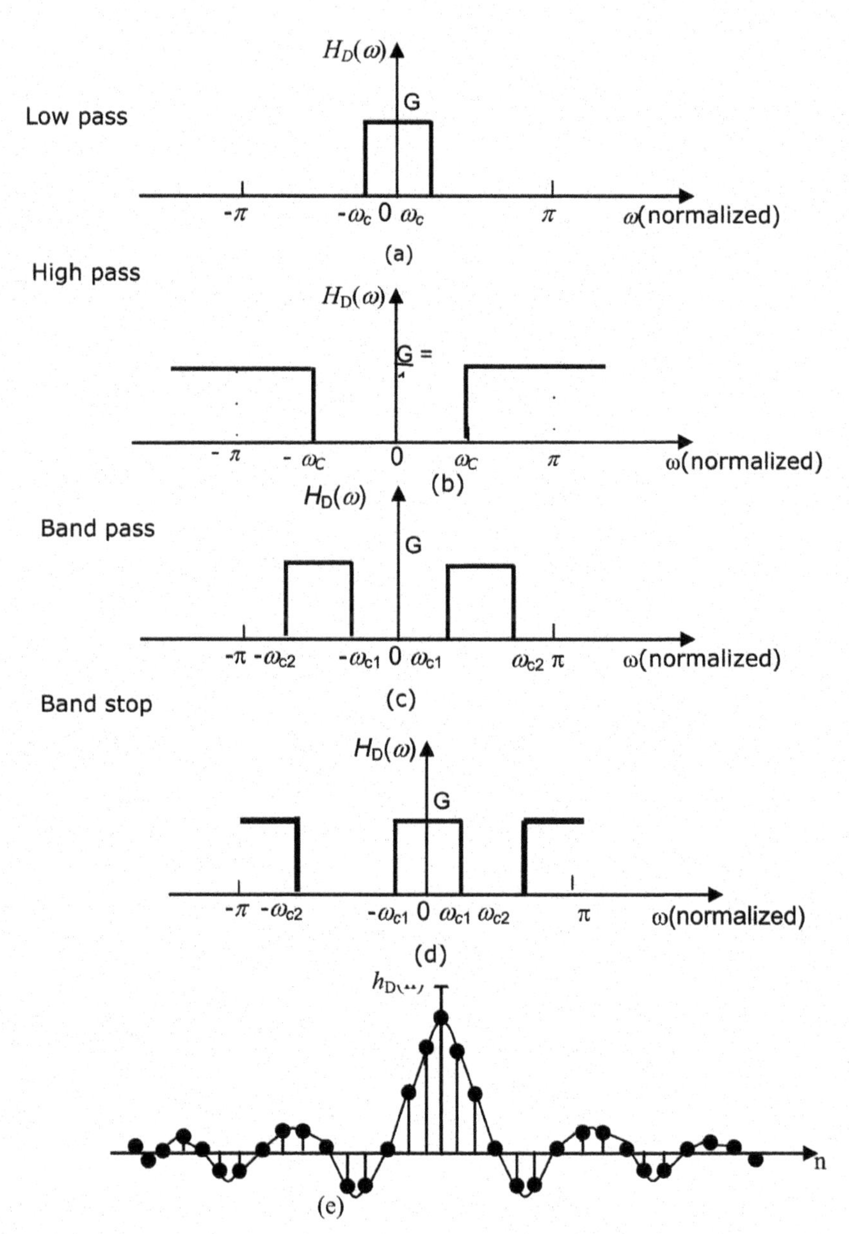

Figure 9.9 (a, b, c, d) Ideal frequency response of low-pass, high-pass, band-pass and bandstop filters. (e) Impulse response of an ideal low-pass filter.

$$\text{low-pass filter } h_D(n) = \frac{1}{n\pi} \sin n\omega_c, \quad n = 0, \pm 1, \pm 2, \ldots.. \tag{9.39}$$

$\frac{\omega_c}{\pi} = 2f_C$ when $n = 0$ (using L' Hopital's rule)
Similarly, we find for the high-pass, band-pass, and band-stop filter of Figure 9.9 defined over $-\infty\, n < \infty$

$$\text{high-pass filter } h_D(n) = \delta(n) - \frac{1}{n\pi} \sin n\omega_c \tag{9.40}$$

$$\text{band-pass filter } h_D(n) = \frac{\sin n\omega_{c2} - \sin n\omega_{c1}}{n\pi}. \tag{9.41}$$

$$\text{band-stop filter } h_D(n) = \delta(n) - \frac{\{\sin n\omega_{C2} - \sin n\omega_{C1}\}}{n\pi}. \tag{9.42}$$

Note that the same values of the cutoff frequencies ω_C, ω_{C1}, ω_{C2} low-pass/high-pass and band-pass/band-stop filters are complementary, that is their impulse responses add up to a unit impulse $\delta(n)$ and their frequency responses add up to unity.

$$h_{\text{LP}}(n) + h_{\text{HP}}(n) = \delta(n) \;\Leftrightarrow\; H_{\text{LP}}(\omega) + h_{\text{HP}}(\omega) = 1 \tag{9.43}$$

$$h_{BP}(n) + h_{BP}(n) = \delta(n) \;\Leftrightarrow\; H_{BP}(\omega) + h_{BS}(\omega) = 1$$

The impulse responses for the ideal high-pass, band-pass, and band-stop filters obtained from the low-pass and high-pass cases of Equations (9.39) and (9.40) are summarized in Table 9.2. The impulse response for the low-pass filter is plotted in Figure 9.9 (e) from which we note that $h_D(n)$ is symmetrical about $n = 0$. That is $h_D(n) = h_D(-n)$, so that the filter will have a linear (in this case, zero)-phase response. Several practical problems with this simple approach are apparent.

Table 9.2 Summary of ideal responses for standard frequency selective filters

Filter Type	Impulse Response $h_D(n)$	$h_D(0)$
Low-pass	$\frac{\sin(n\omega_c)}{n\pi}$	$\frac{\omega_c}{\pi}$
High-pass	$\delta(n) - \frac{\sin(n\omega_c)}{n\pi}$	$1 - \frac{\omega_c}{\pi}$
Band-pass	$\frac{\sin(n\omega_{c2}) - \sin(n\omega_{c1})}{n\pi}$	$\frac{\omega_{c2} - \omega_{c1}}{\pi}$
Band-stop	$\delta(n) - \frac{\{\sin(n\omega_{c2}) - \sin(n\omega_{c1})\}}{n\pi}$	$1 - \frac{\omega_{c2} - \omega_{c1}}{\pi}$

The most important of these is that, although $h_D(n)$ decreases as we move away from $n = 0$, it nevertheless carries on, theoretically, to $n = \pm\infty$.

ω_c, ω_{c1}, and ω_{c2} are the normalized pass-band or stop-band edge frequencies; N is the length of the filter.

9.5.1 To Find the Filter Coefficients using Window

Once either in the case of non-causal or causal filter using different window has to be designed, the respective coefficients have to be multiplied by respective window coefficients. For example, if $h_D(n)$ has to be calculated by taking inverse Fourier Transform, then it should be multiplied by the respective window magnitude $\omega(n)$ obtained to find the final coefficient value of the filter $h(n)$; therefore, the final value becomes $h(n) = h_D(n).\ \omega(n)$.

9.5.2 Filter Design Steps for Non-causal Filters

We start the filter design first by taking non-causal filters. A design procedure for a digital filter is now described in the following way, which is nearly the same as that of causal filters.

(a) Determine the normalized cut-off frequency.
(b) Determine the unit sample response $h_D(n)$ that will produce the desired frequency response. In other words, find the value of $h_D(n)$ for the given requirements.
(c) Determine the transfer function from the calculated impulse response sequence.

Example 9.12

Design a **low-pass, non-causal,** positive symmetry FIR filter. Find the coefficients according to their filter lengths for the following specifications.

Cut-off frequency = 500 Hz; Sampling Frequency = 2000 Hz
Order of the Filter = 4; Filter Length required = 5, $G = 1$

Solution 9.12

Normalized Cut-off frequency is

$$\omega_c = 2\pi\frac{f_c}{F_S} = 2\pi\frac{500}{2000} = \frac{\pi}{2}$$

For determination of the filter coefficients, we substitute various values of n and find the corresponding value of $h_D(n)$ using $h_D(n) = \frac{1}{n\pi}\sin n\omega_c$.

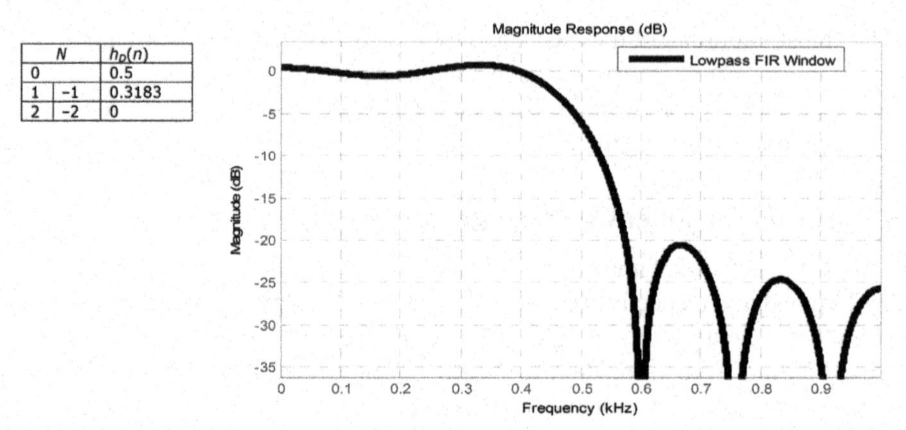

N		$h_D(n)$
0		0.5
1	-1	0.3183
2	-2	0

Figure 9.10 (a) Coefficients of a low-pass FIR filter, non-causal. (b) Magnitude response of a low-pass FIR filter.

$$h_D(1) = h_D(-1) = \frac{1}{\pi} \sin \frac{1.\pi}{2} = 3.183$$

$$h_D(2) = h_D(-2) = \frac{1}{2.\pi} \sin \frac{2.\pi}{2} = 0$$

$h_D(0)$ can be calculated in a different way using the relationship obtained by

$$h_D(0) = \frac{1}{\pi}\omega_c = \frac{1}{\pi}\frac{\pi}{2} = \frac{1}{2} = 0.5$$

It means for the length of the filter, *N* is odd. The centre one value will be the highest value. The filter coefficients can be verified using the FDA Tool (Figure 9.10).

9.5.3 Filter Design Steps for Causal Filters

A design procedure for digital filters will now be discussed. The method is similar to that of causal filters, except for Equations (9.39) to (9.42).

(a) Determine the normalized cut-off frequency, if it is unknown.
(b) Determine the unit sample response $h_D(n)$ that will produce the desired frequency response.
(c) Determine the transfer function from the impulse response sequence.

The expression for finding the coefficients of a non-causal filter is defined as

$$h_D(n) = \frac{G}{n\pi} \sin n\omega_c$$

We modify the expression for the non-causal filter to causal filter by shifting the coefficients to the right side. Normally, in all the cases discussed here, G has been taken as 1.

$$h_D(n) = \frac{1}{\left(\frac{N-1}{2} - n\right)\pi} \sin\left(\frac{N-1}{2} - n\right)\omega_c$$

Example 9.13

Design a **low-pass, causal,** positive symmetry FIR filter. Find the coefficients according to their filter lengths for the following specifications.

Cut-off frequency = 500 Hz; Sampling Frequency = 2000 Hz
Order of the Filter = 4; Filter Length required (N) = 5

Solution 9.13

Normalized Cut-off frequency

$$\omega_C = 2\pi \frac{f_C}{F_S} = 2\pi \frac{500}{2000} = \frac{\pi}{2}$$

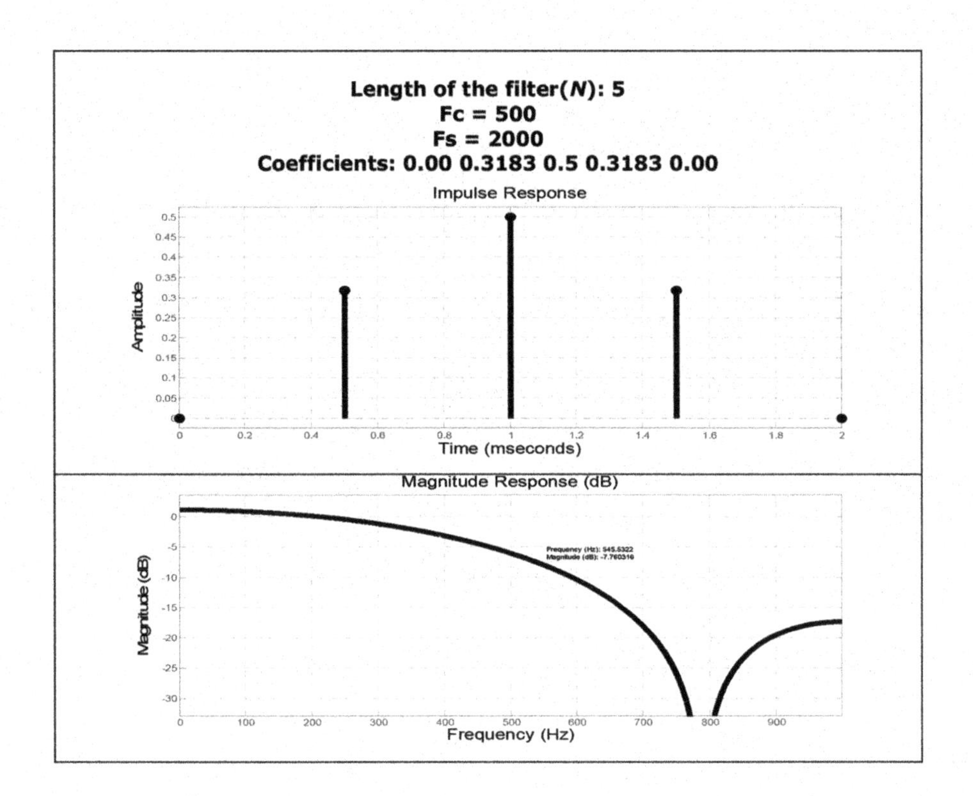

For determination of the filter coefficients, we substitute various values of n and determine the corresponding values of $h_D(n)$ using $h_D(n) = \frac{1}{n\pi} \sin n\omega_c$.

$$h_D(n) = \frac{1}{(\frac{N-1}{2} - n)\pi} \sin\left(\frac{N-1}{2} - n\right)\omega_c$$

$$h_D(n) = \frac{1}{(2-n)\pi} \sin(2-n)\omega_c$$

Because the filter is causal, it follows the relationship $h_D(n) = h_D(N-1-n)$

$$h_D(n) = \frac{1}{(2-n)\pi} \sin(2-n)\omega_c$$

$n = 0 \quad h_D(0) = h_D(4) = \frac{1}{2\pi} \sin \frac{2.\pi}{2} = 0.00$
$n = 1 \quad h_D(1) = h_D(3) = \frac{1}{.\pi} \sin \frac{.\pi}{2} = 0.3183$
$n = 2 \quad h_D(2) = \frac{1}{\pi} . \frac{\pi}{2} = 0.50$

We stop the computation at $h(2)$, since the required length of the filter is 5. The filter coefficients can be verified using the FDA Tool.

Example 9.14

Design a **low-pass, causal,** positive symmetry FIR filter. Find the coefficients according to their filter lengths for the following specifications.

Cut-off frequency = 500 Hz; Sampling Frequency = 2000 Hz
Order of the Filter = 5; Filter Length required **(N) = 6**

Solution 9.14

Normalized Cut-off frequency

$$\omega_C = 2\pi \frac{f_c}{F_S} = 2\pi \frac{500}{2000} = \frac{\pi}{2}$$

Determination of the filter coefficients, we substitute various values of n and determine the corresponding values of $h_D(n)$ using $h_D(n) = \frac{1}{n\pi} \sin n\omega_c$.

$$h_D(n) = \frac{1}{(\frac{N-1}{2} - n)\pi} \sin\left(\frac{N-1}{2} - n\right)\omega_c$$

$$h_D(n) = \frac{1}{(2.5-n)\pi} \sin(2.5-n)\omega_c$$

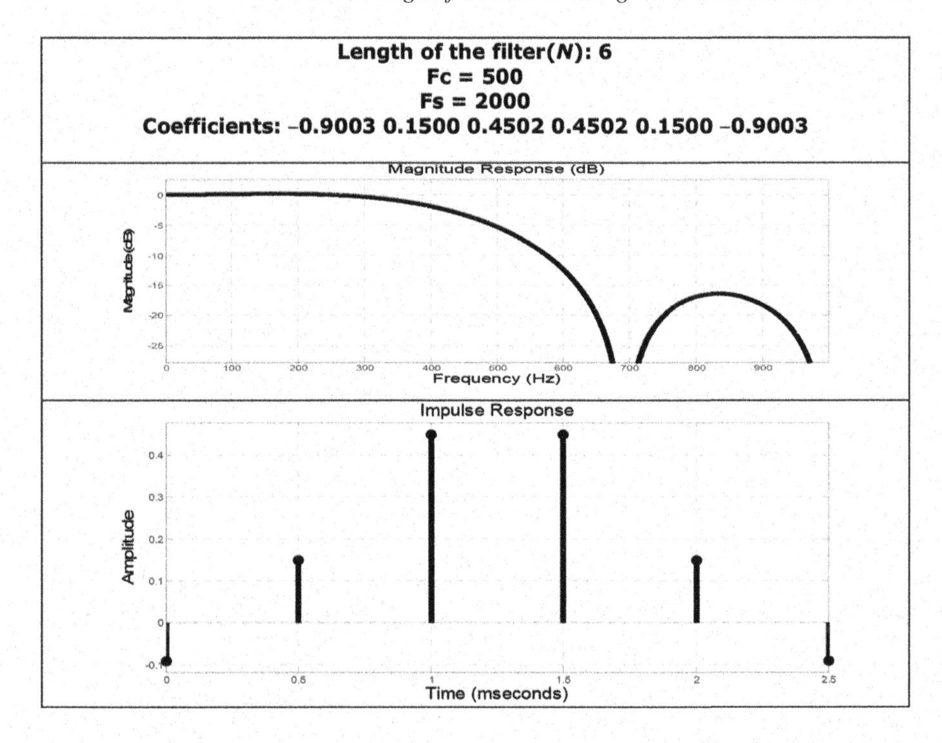

Because the filter is causal, it follows the relationship $h_D(n) = h_D(N-1-n)$

$$h_D(n) = \frac{1}{(2.5-n)\pi}\sin(2.5-n)\omega_c$$

$$\begin{aligned}
n = 0;\quad & h_D(0) = h_D(5) = \tfrac{1}{2.5\pi}\sin\tfrac{2.5\pi}{2} = -0.9003\\
n = 1;\quad & h_D(1) = h_D(4) = \tfrac{1}{1.5.\pi}\sin\tfrac{1.5\pi}{2} = 0.1500\\
n = 2;\quad & h_D(2) = h_D(3) = \tfrac{1}{0.5\pi}\sin\tfrac{0.5\pi}{2} = 0.4502
\end{aligned}$$

We stop the computation at $h(2)$, $h(3)$ since the required length of the filter is 6. The filter coefficients can be verified using the FDA Tool.

9.5.4 Designing Other Types of FIR Filters

For FIR filters, there are individual procedures as discussed for a low-pass filter. Now the other pass filters such as high-pass, band-pass, and band-stop filters procedures appear to be different at first sight, on closer observation, and we can see that they all follow more or less the same design steps; the major difference in these procedures is the use of appropriate limits of integration in each case for finding the impulse response.

The impulse response for a low-pass filter is given by

$$\text{low-pass filter } h_D(n) = \frac{1}{n\pi}\sin n\omega_c, \quad n = 0, \pm 1, \pm 2, \ldots. \tag{9.44}$$

Similarly, for the high-pass, band-pass, and band-stop filter of Figure 9.9 defined over $-\pi < n < \pi$, we have

$$\text{high-pass filter } h_D(n) = \delta(n) - \frac{1}{n\pi}\sin n\omega_c \tag{9.45}$$

$$\text{band-pass filter } h_D(n) = \frac{\sin n\omega_{C2} - \sin n\omega_{C1}}{n\pi} \tag{9.46}$$

$$\text{band-stop filter } h_D(n) = \delta(n) - \frac{\sin n\omega_{C2} - \sin n\omega_{C1}}{n\pi} \tag{9.47}$$

Example 9.15

Develop an expression for a **causal high-pass**, positive symmetry Finite Impulse Response (FIR) filter length with $N = 5$, $f_c = 500$ Hz and $f_s = 2000$ Hz

Solution 9.15

It should be kept in mind that the required filter is an FIR causal filter, so the impulse response formula must first be modified as shown below.

$$h_D(n) = \frac{1}{n\pi}\sin n\omega_c$$

$$h_D(n) = \delta(n) - \frac{1}{\left(\frac{N-1}{2} - n\right)\pi}\sin\left(\frac{N-1}{2} - n\right)\omega_c$$

The FIR coefficients, therefore given by

$$h_D(n) = -\frac{1}{(2-n)\pi}\sin(2-n).\frac{\pi}{2}$$

$n = 0 \quad h_D(0) = h_D(4) = -\frac{1}{2\pi}\sin 2.\frac{\pi}{2} = -0.00$

$n = 1 \quad h_D(1) = h_D(3) = \frac{1}{1\pi}\sin 1.\frac{\pi}{2} = -0.3183$

$n = 2 \quad h_D(2) = 1 - \frac{1}{\pi}.\frac{\pi}{2} = 0.5$

The expression for an FIR filter can be expressed using the Hamming window

$$H(z) = -0.3183\,z^{-1} + 0.50\,z^{-2} - 0.3183\,z^{-3}$$

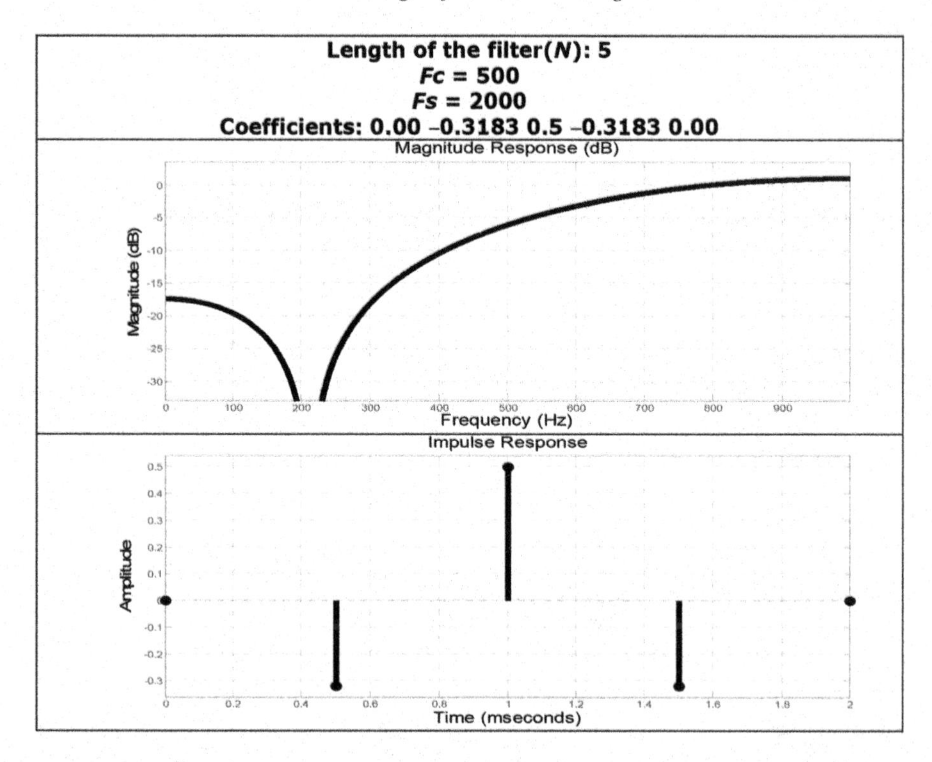

Example 9.16

Obtain the impulse response for a **causal band-pass** FIR filter for the following specification, and write the expression in transfer function form of $h(z)$.

Cut-off frequency = 300 Hz and 600 Hz

Sampling frequency = 2000 Hz

Order of the filter = 5

Filter length required (N) = **6**

Solution 9.16

Using the design equation of a band-pass filter

band-pass filter $h_D(n) = \frac{\sin n\omega_{C2} - \sin n\omega_{C1}}{n\pi}$ and modifying this expression for a causal filter.

$$\text{band-pass filter } h_D(n) = \frac{\sin(\frac{N-1}{2} - n)\omega_{C2} - \sin(\frac{N-1}{2} - n)\omega_{C1}}{(\frac{N-1}{2} - n)\pi}.$$

Normalize the cut-off frequencies

$$\omega_{C1} = 2\pi \frac{f_{C1}}{F_s} = 2\pi \frac{300}{2000} = 0.3\pi \quad \omega_{C2} = 2\pi \frac{f_{C2}}{F_s} = 2\pi \frac{600}{2000} = 0.6\pi$$

$$\text{band-pass filter } h_D(n) = \frac{\sin\left(\frac{N-1}{2} - n\right)(0.6\pi) - \sin\left(\frac{N-1}{2} - n\right)(0.3\pi)}{\left(\frac{N-1}{2} - n\right)\pi}.$$

$$n = 0,5;\ h_D(0) = \frac{\sin(2.5)(0.6\pi) - \sin(2.5)(0.3\pi)}{(2.5)\pi} = \frac{\sin(1.5\pi) - \sin(0.75\pi)}{(2.5)\pi} = -0.217$$

$$n = 1,4;\ h_D(1) = \frac{\sin(0.9\pi) - \sin(0.45\pi)}{(1.5)\pi} = \frac{0.3090 - 0.9876}{1.5\pi} = \frac{-0.6786}{1.5\pi} = -0.1440$$

$$n = 2,3;\ h_D(2) = \frac{\sin(0.5)(0.6\pi) - \sin(0.5)(0.3\pi)}{(0.5)\pi} = 0.226$$

The expression for an FIR filter can be written as

$$H(z) = -0.217 - 0.144\, z^{-1} + 0.226 z^{-2} + 0.226\, z^{-3} - 0.144 z^{-4} - 0.217\, z^{-5}$$

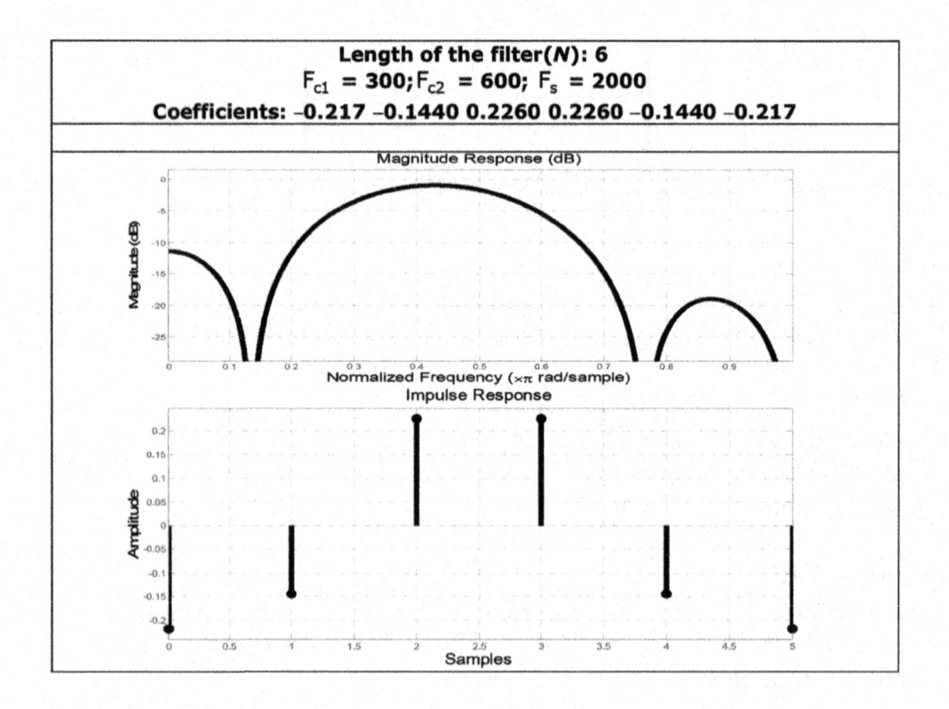

Example 9.17

Find the impulse response of a **causal band-pass** FIR filter for the following specification:

Cut-off frequency = 300 Hz and 600 Hz
Sampling frequency = 2000 Hz
Order of the filter = 6
Filter length required (N) = 7

Solution 9.17

Using the design equation of a band-pass filter
band-pass filter $h_D(n) = \frac{\sin n\omega_{C2} - \sin n\omega_{C1}}{n\pi}$ and modifying this expression for a causal filter.

$$\text{band-pass filter } h_D(n) = \frac{\sin\left(\frac{N-1}{2} - n\right)\omega_{C2} - \sin\left(\frac{N-1}{2} - n\right)\omega_{C1}}{\left(\frac{N-1}{2} - n\right)\pi}$$

Normalize the cut-off frequencies

$$\omega_{C1} = 2\pi\frac{f_{C1}}{F_s} = 2\pi\frac{300}{2000} = 0.3\pi$$

$$\omega_{C2} = 2\pi\frac{f_{C2}}{F_s} = 2\pi\frac{600}{2000} = 0.6\pi$$

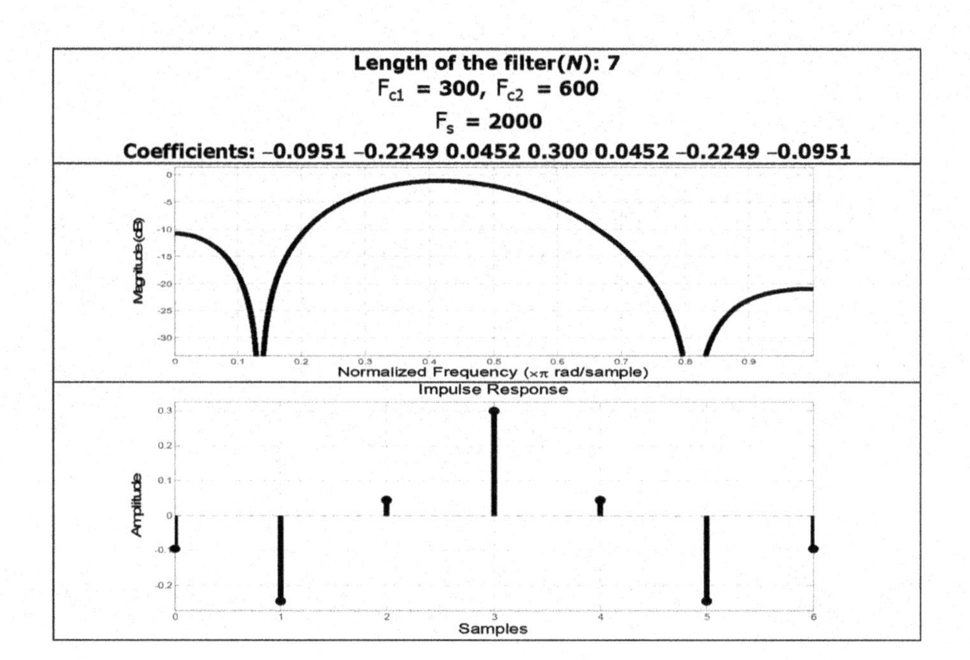

$$\text{band-pass filter } h_D(n) = \frac{\sin\left(\frac{N-1}{2} - n\right)(0.6\pi) - \sin\left(\frac{N-1}{2} - n\right)(0.3\pi)}{\left(\frac{N-1}{2} - n\right)\pi}$$

$n = 0{,}6;\ h_D(0) = \frac{\sin(3)(0.6\pi) - \sin(3)(0.3\pi)}{(3)\pi} = \frac{\sin(1.8\pi) - \sin(0.9\pi)}{(3)\pi} = -0.0951$

$n = 1{,}5;\ h_D(1) = \frac{\sin(2)(0.6\pi) - \sin(2)(0.3\pi)}{(2)\pi} = \frac{\sin(1.2\pi) - \sin(0.6\pi)}{(2)\pi} = -0.2249$

$n = 2{,}4;\ h_D(2) = \frac{\sin(1)(0.6\pi) - \sin(1)(0.3\pi)}{(1)\pi} = \frac{\sin(0.6\pi) - \sin(0.3\pi)}{(1)\pi} = 0.0452$

$n = 3;\ \ h_D(3) = \frac{(0.6\pi) - (0.3\pi)}{\pi} = 0.3$

The expression for an FIR filter can be expressed as

$$H(z) = -0.0951 \ - 0.2249z^{-1} + 0.0451z^{-2} + 0.3z^{-3} + 0.04520z^{-4}$$
$$- 0.2249z^{-5} - 0.0951z^{-5}$$

Example 9.18

Find the impulse response of a **causal band-stop** FIR filter for the following specification:

Cut-off frequency = 300 and 600 Hz

Sampling frequency = 2000 Hz

Order of the filter = 6

Filter length required (N) = 7

Solution 9.18

Using the design equation of a band-pass filter

band-stop filter $h_D(n) = \delta(n) - \frac{\sin n\omega_{C2} - \sin n\omega_{C1}}{n\pi}$ and modifying this expression for a causal filter.

$$\text{band-stop filter } h_D(n) = \delta(n) - \frac{\sin\left(\frac{N-1}{2} - n\right)\omega_{C2} - \sin\left(\frac{N-1}{2} - n\right)\omega_{C1}}{\left(\frac{N-1}{2} - n\right)\pi}$$

Normalize the cut-off frequencies

$$\omega_{C1} = 2\pi\frac{f_{C1}}{F_s} = 2\pi\frac{300}{2000} = 0.3\pi$$

$$\omega_{C2} = 2\pi\frac{f_{C2}}{F_s} = 2\pi\frac{600}{2000} = 0.6\pi$$

$$\text{band-stop filter } h_D(n) = \frac{\sin\left(\frac{N-1}{2} - n\right)(0.6\pi) - \sin\left(\frac{N-1}{2} - n\right)(0.3\pi)}{\left(\frac{N-1}{2} - n\right)\pi}.$$

$$n=0,6;\ h_D(0) = -\frac{\sin(3)(0.6\pi)-\sin\,(3)(0.3\pi)}{(3)\pi} = -\frac{\sin(1.8\pi)-\sin\,(0.9\pi)}{(3)\pi} = 0.0951$$
$$n=1,5;\ h_D(1) = -\frac{\sin(2)(0.6\pi)-\sin\,(2)(0.3\pi)}{(2)\pi} = -\frac{\sin(1.2\pi)-\sin\,(0.6\pi)}{(2)\pi} = 0.2249$$
$$n=2,4;\ h_D(2) = -\frac{\sin(1)(0.6\pi)-\sin(1)(0.3\pi)}{(1)\pi} = -\frac{\sin(0.6\pi)-\sin(0.3\pi)}{(1)\pi} = -0.0452$$
$$n=3;\quad h_D(3) = 1-\frac{(0.6\pi)-(0.3\pi)}{\pi} = 0.7$$

The expression for an FIR filter can be expressed as

$$H(z) = 0.0951 \ + 0.2249z^{-1} - 0.0451z^{-2} + 0.7z^{-3} - 0.04520z^{-4} + 0.2249z^{-5} + 0.0951z^{-5}$$

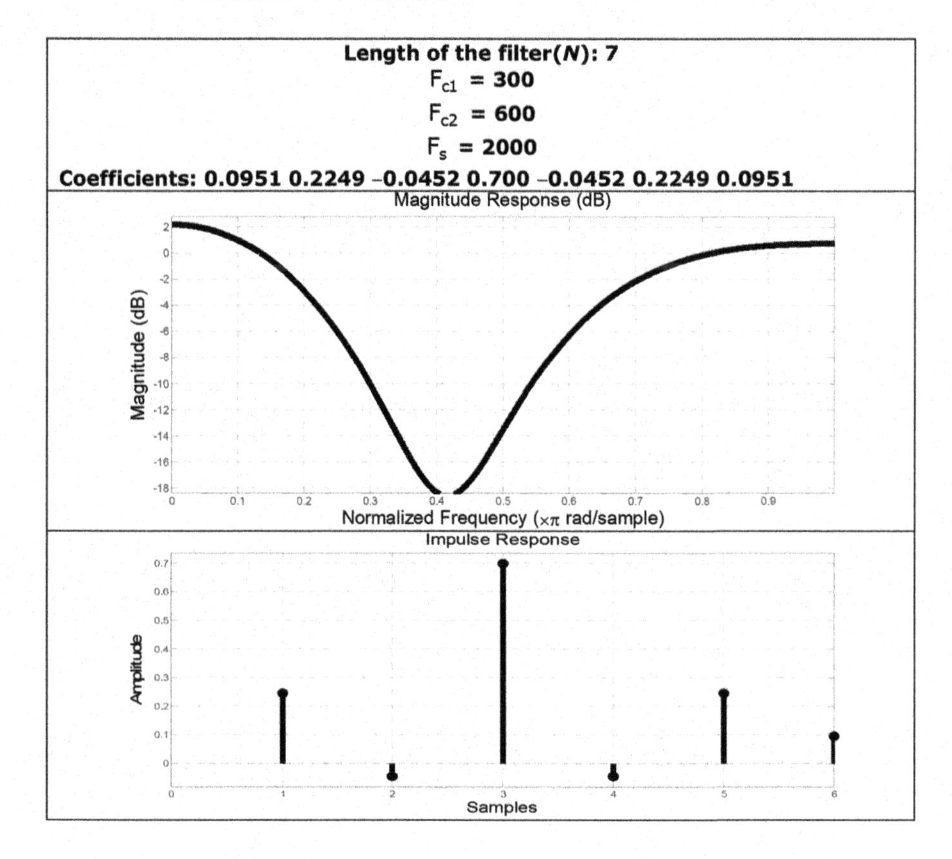

9.5.5 Smearing Effect

Because of the smearing effect of the window on the filter response, the cutoff frequency of the resulting filter will be different from that given in

Table 9.3 Summary of important features of common window functions

Name of Window Function	Transition Width (Hz) Δf (normalized)	Pass-Band Ripple (dB)	Main lobe Relative to Side Lobe (dB)	Stop-band Attenuation (dB) (maximum)	Window Function $\omega(n)$, $\lvert n\rvert \leq (N-1/2)$
Rectangular	0.9/*N*	0.7416	13	21	1
Hanning	3.1/*N*	0.0546	31	44	$0.5 - 0.5\cos\left(\frac{2\pi n}{N-1}\right)\ 0 \leq n \leq N-1$
					$0.5 + 0.5\cos\left(\frac{2\pi\, n}{N-1}\right)\ -(\frac{N-1}{2}) \leq n \leq (\frac{N-1}{2})$
Hamming	3.3/*N*	0.0194	41	53	$0.54 - 0.46\cos\left(\frac{2\pi n}{N-1}\right)\ 0 \leq n \leq N-1$
Kaiser	2.93/*N* ($\beta = 4.54$)	0.0274		50	$0.54 + 0.46\cos\left(\frac{2\pi\, n}{N-1}\right)\ -(\frac{N-1}{2}) \leq n \leq (\frac{N-1}{2})$
					$\frac{I_0\ (\alpha)}{I_o(\beta)}$
	4.32/N ($\beta = 6.76$)	0.00275		70	
					$\alpha = \beta\,[\{1 - [2n/(N-1)]^2\}^{1/2}]$
	5.71/*N* ($\beta = 9.96$)	0.000275		90	

the specifications for designing. To account for this, we use an f'_c that is centered on the transition band:

$f'_c = f_c$ + Transition width/2 and
Δf = Transition width/Sampling frequency.

In Table 9.3 different window function transfer function, transition width, the pass-band and stop-band ripple are shown, which helps out in designing of FIR filters.

In above table when $\alpha = 0$, the Kaiser window corresponds to the rectangular window, and the resulting window is very similar, though not identical, to the Hamming window. The value of α is determined by the stop-band attenuation requirements and may be estimated from one of the empirical relationships stated above.

Example 9.19

Obtain the coefficients for **non-causal low-pass** positive symmetry FIR filter to meet the specifications given below using the window method.

Pass-band edge frequency 1.5 kHz
Transition width 0.5 kHz
Stop-band attenuation >50 dB
Sampling frequency 8 kHz

Solution 9.19

From Table 9.2, we select $h_D(n)$ for a low-pass filter which is given by

$$\begin{aligned} h_D(n) &= \tfrac{\sin(n\omega_c)}{n\pi} \quad & n \neq 0 \\ h_D(n) &= \tfrac{1}{\pi}\omega_c \quad & n = 0 \end{aligned}$$

Table 9.2 indicates that the Hamming, Blackman or Kaiser window will satisfy the stop-band attenuation requirements. We will use the Hamming window for simplicity

$$\omega[n] = 0.54 + 0.46 \cos \frac{2\pi n}{N-1} \quad -\frac{N-1}{2} \leq n \leq \frac{N-1}{2}$$

Now $\Delta f = 0.5/8 = 0.0625$. From Table 9.3, $N = 3.3/\Delta f = 3.30/0.0625 = 52.8$, let $N = 53$ (Note: here N is the order of the filter)

The filter coefficients are obtained from

$$h_D(n)w(n) \qquad -26 \leq n \leq 26$$

$$h_D(n) = \frac{\sin(n\omega_c)}{n\pi} \qquad n \neq 0$$
$$h_D(n) = \frac{\omega_c}{\pi} \qquad n = 0$$
$$\omega(n) = 0.54 + 0.46\cos(2\pi n/52) \qquad -26 \leq n \leq 26$$

Because of the smearing effect of the window on the filter response, the cutoff frequency of the resulting filter will be different from that given in the specifications. To account for this, we will use an f_c that is centered on the transition band:

$$f_C' = f_c + \text{Transition width}/2 = (1.5 + 0.25)\text{kHz}$$
$$f_C' = 1.75\text{ kHz} \equiv 1.75/8 = 0.21875$$

Noting that $h(n)$ is symmetrical; we need only compute values for $h(0)$, $h(1)$, …, $h(26)$ and then use the symmetry property to obtain the other coefficients with $n = 0$;

$$h_D(n) = 2\,f_c' = 2\,x\,0.21875 = 0.4375$$
$$\omega(0) = 0.54 + 0.46\cos(0) = 1$$
$$h(0) = h_D(0)\omega(0) = 0.4375$$

with $n = 1$;

$$h_D(1) = \frac{2 \times 0.21875}{2\pi \times 0.21875}\sin(2\pi \times 0.21875)$$
$$= \frac{\sin(360 \times 0.21875)}{\pi} = 0.31219$$

$$= \omega(1) = 0.54 + 0.46\cos(2\pi/53) = 0.54 + 0.46\cos(360/53) = 0.98713$$
$$= h(1) = h(-1) = h_D(1)\omega(1) = 0.31119$$

with $n = 2$;

$$h_D(2) = \frac{2 \times 0.21875}{2 \times 2\pi \times 0.21875}\sin(2 \times 2\pi \times 0.21875)$$
$$= \frac{\sin(157.5°)}{2\pi} = 0.06091$$
$$\omega(2) = 0.54 + 0.46\cos(2\pi \times 2/52)$$
$$= 0.54 + 0.46\cos(720°/52) = 0.98713$$
$$h(2) = h(-2) = h_D(1)\omega(2) = 0.06012$$

$n = 26$:

$$h_D(26) = \frac{2 \times 0.21875}{26 \times 2\pi \times 0.15}\frac{\sin(26 \times 2\pi \times 0.21875)}{2\pi}$$

$$= 0.01131$$

$$\begin{aligned}\omega(26) &= 0.54 + 0.46\cos(2\pi \times 26/53)\\ &= 0.54 + 0.46\cos(720^{\circ}/53) = 0.08081\\ h(26) &= h(-26) = h_D(26)w(26) = 0.000913\end{aligned}$$

We note that the indices of the filter coefficients run from –26 to 26. To make the filter causal (necessary for implementation) we add 26 to each index so that the indices start from zero.

9.5.6 Kaiser Window

To design a good window, it has become necessary to choose a time limited function whose Fourier transform best approximates a band limited function i.e. a time limited function that has the minimum energy outside some selected interval. The solution for the case of continuous time functions is the set of functions called "prolate spheroidal wave function".

The relative simple approximation of the Kaiser window is given by the weighting function

$$w[n] = \frac{I_0\left[\beta\sqrt{1-[2n/(N-1)]^2}\right]}{I_o(\beta)}; -\left(\frac{N-1}{2}\right) \leq n \leq \left(\frac{N-1}{2}\right) \tag{9.48}$$

$$w[n] = \frac{I_0(\alpha)}{I_o(\beta)} \quad \alpha = \left[\beta\sqrt{1-[2n/(N-1)]^2}\right] \tag{9.49}$$

where α is a constant that specifies a frequency response trade-off between the peak height of the side lobe ripple and the width of the energy of the main lobe and $I_0(\beta)$ is the modified zeroth-order Bessel function.

$$I_o[\beta] = 1 + \sum_{k=1}^{L}\left[\frac{(\beta/2)^{-k}}{k!}\right]^2 \tag{9.50}$$

Typically the value of $L = 25$

The number of filter coefficients, N, is given by

$$N \geq \frac{A - 7.95}{14.36\ \Delta f} \tag{9.51}$$

where Δf is the normalized transition width.

No optimum window designed filter exists in spite of optimum windows, since the actual filter response is being obtained by the "convolution" of the windows with desired ideal response.

9.5.6.1 Procedure to FIR filter design using Kaiser windows

(i) Determine $h(n)$ using the Fourier series approach assuming a idealized frequency response.

$$\begin{aligned} H(\omega) &= 1 \quad |\omega| \leq \omega_C \\ H(\omega) &= 0 \end{aligned} \tag{9.52}$$

$$\text{where} \quad \omega_C = \frac{1}{2}(\omega_P + \omega_s)$$

and ω_P is the pass-band frequency and ω_S is the stop-band frequency in radians/sec.

(ii) Choose δ after calculating δ_P and δ_S

$$\text{Pass-band Attenuation } A_P = 20 \log(1 + \delta_P) \tag{9.53}$$

$$\text{Pass-band Attenuation } A_S = -20 \ \log \ \delta_s$$

A suitable value δ is calculated from the minimum value of δ_P or δ_S. δ_p is the desired pass-band ripple and δ_s is the desired stop-band ripple

$$\delta = \min(\delta_S, \ \delta_P) \tag{9.54}$$

$$\text{where } \delta_S = 10^{-0.05A_S} \quad \text{and} \quad \delta_P = 10^{-0.05Ap} - 1 \tag{9.55}$$

(iii) Calculate A_s again using (9.52)

(iv) Choose the parameter β (ripple control factor) as follows:

$$\begin{aligned} b &= 0 && \text{if } A_S \leq 21 \text{ dB} \\ b &= 0.5842(A_S - 21)^{0.4} && \\ &\quad + 0.07886(A_S - 21) && \text{if } 21 \text{ dB} < A_S < 50 \text{ dB} \\ b &= 0.1102(A_S - 9.7) && \text{if } A_S \geq 50 \text{ dB} \end{aligned} \tag{9.56}$$

(v) Form $\omega(n)$ using equation

$$\omega(n) = \frac{I_0(\alpha)}{I_0(\beta)} \quad \text{for} \quad |\, n \,| \leq \frac{N-1}{2} \omega(n) = 0 \text{ otherwise} \tag{9.57}$$

Where α is the independent parameter and

$$\alpha = \beta \sqrt{1 - \left[\frac{2n}{N-1}\right]^2} \tag{9.58}$$

where $I_0(\beta)$ is the zero-order Bessel function of the first kind. This can be evaluated to any desired degree of accuracy by using the rapidly converging series.

$$I_o[\beta] = 1 + \sum_{k=1}^{L} \left[\frac{(\beta/2)^k}{k!} \right]^2 \tag{9.59}$$

(vi)

$$H(z) = z^{\frac{-(N-1)}{2}} H_1(z)$$

where

$$H(Z) = Z[\omega(n)h(n)] \tag{9.60}$$

9.5.7 Comparison of Window Methods

Rectangular	If the window is not specified, the rectangular window is used which is also known as uniform window. The spectral side lobes are down only about 14 dB from the main lobe peak.
Hamming	The amplitudes of the windows toward ends can be different from zero. The Hamming window is one such example. It is widely used in processing speech signals for applications such as spectral analysis and computer voice response systems. The main lobe of the frequency response of the Hamming windows is twice that of a rectangular window. For $\alpha = 0.54$, 99.96% of the spectral energy is the main lobe and the peak side lobe ripple is down about 40 dB from the main lobe peak.
Hanning	One such popular window is the Hanning window, which is used for general purpose applications. If you are not sure which window to use, you can start with the Hanning window.
Kaiser	Kaiser window is an optimum window in the sense that it is a finite duration sequence that has the minimum spectral energy beyond some specified frequency.

9.5.8 Analysis of Different FIR Filter Types

There are interesting results for the response of FIR filters, and students should note that the filter in all four types cannot be bulid up. An analysis has been carried out with different values of N; positive and negative symmetry are given below, which clearly shows that positive symmetry filters are widely used for designing FIR filter.

9.5.9 Conclusion for the Types of FIR Filter

It is evident from the analysis diagram done on Matlab that type I, positive symmetry is the most versatile for designing all types of filters; type II are designed only for odd order filters, and types III and IV are often used to design differentiators.

Length of the Filter N is Odd	Length of the Filter N is Even	
(type I)	(type II)	Positive Symmetry
Low-Pass, High-Pass, Band-Pass, and Band-Stop Filter	Low-Pass and Band-Pass Filter	
(type III) No filter	(type IV) High-Pass filter	Negative Symmetry

9.5.10 Advantages/Disadvantages of the Window Method

(i) An important advantage of the window method is its simplicity: it is simple to apply and understand. It involves a minimum amount of computational effort, even for the more complicated Kaiser window.

(ii) The major disadvantage is its lack of flexibility. Both the peak pass-band and stop-band ripples are approximately equal, so the designer may end up with either too small pass-band ripple or too large a stop-band attenuation.

(iii) Because of the effect of convolution on the spectrum of the window function and the desired response, the pass-band and stop-band edge frequencies cannot be precisely specified.

9.6 Problems and Solutions

Problem 9.1

Given the following transfer function, calculate its complex frequency response at 100 Hz if the system sampling frequency is 500 Hz.

$$H(z) = 0.6 + 0.3z^{-1} + 0.9z^{-2}$$

Solution 9.1

The frequency response in the z plane is the value of the transfer function, which lies on the unit circle in the z plane at an angle of digital frequency ω. Thus, $\omega = 2\pi \frac{f}{F_s} = 2\pi \frac{100}{500} = 0.4\pi$

The value of z on the unit circle at this angle is: $z = e^{j\omega} = e^{j0.4\pi}$

Therefore, the complex frequency is:

$$H(\omega) = 0.6 + 0.3e^{-j0.4\pi} + 0.9e^{2(-j0.4\pi)}$$

$$H(\omega) = 0.6 + 0.3\{\cos 0.4\pi - j\sin 0.4\pi\} + +0.9\{\cos 0.8\pi - j\sin 0.8\pi\}$$

$$H(\omega) = 0.6 + 0.3(0.3090 - j0.9510) + 0.9(-0.8090 - j0.5877)$$

$$H(\omega) = 0.6 + 0.0927 - j0.2853 - 0.7281 - j0.5289$$

$$H(\omega) = -0.0353 - j0.8142 = 0.8149^{-j1.6142}$$

This result means that the magnitude frequency response at 100 Hz is 0.8149 and the phase response is –1.6142 rad.

Problem 9.2

The linear-phase condition results in zero phase distortion, but adds a phase angle to each sinusoid that is proportional to the frequency of the sinusoids. The signal also picks up a constant phase delay of 10 msec. Find out the phase delay taken by signals.

Solution 9.2

$$T_P = -0.010 = -\tfrac{d\vartheta}{\omega} = \tfrac{-kf}{2\pi f} = \tfrac{k}{2\pi}$$
$$k = -.02\pi \quad \text{or}$$
$$T_p = -0.02\pi f$$

The phase change in radians for the sinusoids of frequency f would be $T_P = -0.02\pi f$.

Problem 9.3

Given the following transfer function, calculate its complex frequency response at 250 Hz if the system sampling frequency is 1000 Hz.

$$H(z) = 0.0935 + 0.3027z^{-1} + 0.40z^{-2} + 0.3027z^{-3} + 0.0935z^{-4}$$

Solution 9.3

The frequency response in the z plane is the value of the transfer function on the unit circle in the z plane at an angle of digital frequency ω. Thus,

$$\omega = 2\pi\frac{f}{F_s} = 2\pi\frac{250}{1000} = 1.5707$$

The value of z on the unit circle at this angle is: $z = e^{j\omega}$

Therefore, the complex frequency is

$$H(\omega) = 0.0935 + 0.0327e^{-j1.5707} + 0.40e^{-j2(1.5707)} + 0.3027e^{-j3(1.5707)} + 0.0935e^{-j4(1.5707)}$$

$$H(\omega) = 0.0.2135e^{-j3.14}$$

This result means that the magnitude frequency response at 250 Hz is 0.2135 and the phase response is –3.14 rad.

Problem 9.4

A linear-phase filter has a phase response of $e^{-j6\omega}$. What is the order of the filter?

Solution 9.4

We know that the phase of a linear-phase filter is given by the phase function for positive symmetry

$$H(\omega) = e^{-j\omega(N-1)/2} \sum_{n=1}^{N/2} a(n) \cos[\omega(n)]$$

$$H(\omega) = e^{-j\omega(N-1)/2} \sum_{n=0}^{(N-1)/2} a(n) \cos(\omega n)$$

Comparing the given phase response $e^{-j6\omega}$ with $e^{-j\omega(N-1)/2}$

$$\frac{N-1}{2} = 6, \quad N = 13$$

This is a 13th order filter.

Problem 9.5

The frequency response of a Type-I positive symmetry FIR filter is given by the following expression:

$$H(\omega) = e^{-j3\omega}\{2 + 1.8\cos 3\omega + 1.2\cos 2\omega + 0.5\cos\omega\}$$

Determine the impulse response *h*(*n*) of the filter in the form of a sequence.

Solution 9.5

Comparing the phase factor $e^{-j\omega(N-1)/2}$ with that of $H(\omega)$, we obtain

$$\frac{N-1}{2} = 3, \quad N = 7$$

The frequency response of an FIR for a causal filter is given by the following expression:

$$H(\omega) = \sum_{n=0}^{6} h(n)\, e^{-j\omega n}$$

$$H(\omega) = \sum_{n=0}^{6} h(n) e^{-j\omega n}$$

$$\begin{aligned} H(\omega) &= h(0) + h(1)e^{-j\omega} + h(2)e^{-j2\omega} + h(3)e^{-j3\omega} + h(4)e^{-j4\omega} \\ &+ h(5)e^{-j5\omega} + h(6)e^{-j6\omega} \end{aligned}$$

The necessary and sufficient condition for a filter to be causal

$$\begin{aligned} &h(n) = h(6-n) \\ &h(0) = h(6)\ ;\ h(1)\ = h(5);\ h(2) = h(4); h(3) = h(3) \end{aligned}$$

Replacing the above expression using the necessary and sufficient condition

$$\begin{aligned} H(\omega)\ &= h(0) + h(1)e^{-j\omega} + h(2)e^{-j2\omega} + h(3)e^{-j3\omega} + h(2)e^{-j4\omega} \\ &+ h(1)e^{-j5\omega}\ +\ h(0)e^{-j6\omega} \end{aligned}$$

$$\begin{aligned} H(\omega) &= e^{-j\omega 3} \sum_{n=0}^{3} \{h(0)e^{3j\omega} + h(0)e^{-j3\omega}\} + \{h(1)e^{j2\omega} + h(1)e^{-j2\omega}\} \\ &+ \{h(2)e^{j\omega} + h(2)e^{-j\omega}\} \end{aligned}$$

Expanding the formula

$$H(\omega) = e^{-j\omega(3)/} \left\{ \sum_{n=0}^{\frac{N-1}{2}} a(n) \cos[\omega(n)] \right\}$$

$$H(\omega) = e^{-j\omega 3}\{a(0) + a(1)\ \cos[\omega] + a(2)\ \cos[2\omega] + a(3)\ \cos[3\omega]\}$$

$$H(\omega) = e^{-j3\omega}\{2 + 1.8\cos 3\omega + 1.2\cos 2\omega + 0.5\cos\omega\}$$

$$H(\omega) = e^{-j3\omega}\{h(3) + 2\,h(0)\cos 3\omega + 2h(1)\cos 2\omega + 2h(2)\cos\omega\}$$

Converting the equation into a coefficient form

$a(0) = h[\frac{N-1}{2}]; a(n) = 2h[\frac{N-1}{2} - n]$ n = 1, 2, ..., $(N-1)/2$

$a(0) = h[3];\ a(n) = 2h[3-n]$ n = 1, 2, ..., $(N-1)/2$

$a(1) = 2\,h(2);\ a(2) = 2\,h(1);\ a(3) = 2\,h(0)$

$a(0) = 2 = h(3);\ a(1) = 0.5 = 2\,h(2)\ ;\ a(2) = 1.2 = 2\,h(1);$
$a(3) = 1.8 = 2\,h(0)$

$h(n) = h(6-n)$
$h(0) = h(6);\ h(1) = h(5);\ h(2) = h(4);\ h(3) = h(3)$

Hence, the required impulse response will be

$$h(n) = \{0.9,\ 0.6,\ 0.25,\ 2,\ 0.25,\ 0.6,\ 0.9\}$$

Problem 9.6

The impulse response of a Type-I positive symmetry FIR filter is given by the following expression:

$$h(n) = \{0.6,\ 0.25,\ 2,\ 0.25,\ 0.6\}$$

Determine the frequency response $H(\omega)$ of the filter in the form of a sequence.

Solution 9.6

The impulse response as given in the question is five, which means the length of the FIR filter is $N = 5$. The frequency response of an FIR for a causal filter is given by the following expression:

$$H(\omega) = \sum_{n=0}^{N-1} h(n)e^{-j\omega n} = \sum_{n=0}^{4} h(n)e^{-j\omega n}$$
$$H(\omega) = h(0) + h(1)e^{-j\omega} + h(2)e^{-j2\omega} + h(3)e^{-j3\omega} + h(4)e^{-j4\omega}$$

The necessary and sufficient condition for a filter to be causal

$$h(n) = h(4-n)$$
$$h(0) = h(4);\ h(1) = h(3);\ h(2) = h(2)$$

Replacing the above expression using the necessary and sufficient condition

$$H(\omega) = h(0) + h(1)e^{-j\omega} + h(2)e^{-j2\omega} + h(1)e^{-j3\omega} + h(0)e^{-j4\omega}$$

$$H(\omega) = e^{-j\omega 2}[h(2) + \sum_{n=0}^{1} \{h(0)e^{2j\omega} + h(0)e^{-j2\omega}\} + \{h(1)e^{j\omega} + h(1)e^{-j\omega}\}]$$

$$H(\omega) = e^{-j\omega 2}[h(2) + \sum_{n=0}^{1} h(0)\{e^{2j\omega} + e^{-j2\omega}\} + h(1)\{e^{j\omega} + e^{-j\omega}\}]$$

$$a(0) = h[\tfrac{N-1}{2}]; a(n) = 2h[\tfrac{N-1}{2} - n] \quad \text{n} = 1,\ 2,\ \ldots,\ (N-1)/2$$

$$a(0) = h[2]; a(n) = 2h[2-n] \quad \text{n} = 1,\ 2,\ \ldots,\ (N-1)/2$$

$$a(0) = h(2) = 2;\ a(1) = 2\ h(1) = 2(0.25) = 0.5;$$

$$a(2) = 2\ h(0) = 2(0.6) = 1.2$$

$$H(\omega) = e^{-j\omega 2}\{a(0) + a(1)\ \cos[\omega] + a(2)\ \cos[2\omega]\}$$

$$H(\omega) = e^{-j\omega 2}\{2 + 0.5\ \cos[\omega] + 1.2\ \cos[2\omega]\}$$

Problem 9.7

Derive the impulse responses of an ideal, zero-phase **non-causal, low pass**, and positive symmetry digital filter.

Cut-off frequency = 250 Hz; Sampling Frequency = 2000 Hz
Order of the Filter = 20; Filter Length required = 21 [N = 21].

Solution 9.7

The value of ω_c is calculated from the cut-off frequency and the sampling frequency

$$\omega_c = 2\pi\frac{f_c}{F_s} = 2\pi\frac{250}{2000} = \frac{\pi}{4}$$

The finite numbers of terms are required to implement a low-pass digital filter design. So, we calculate the expression for $h_D(n)$ at some reasonable value of N = 21.

Using inverse Fourier transform $h_D(n) = \frac{G}{n\pi}\sin n\omega_c,\ n = 0, \pm 1, \pm 2, \ldots$

$h_D(0)$ can be calculated in a different way using the relationship obtained by $h_D(0) = \frac{1}{\pi}\omega_c$

The coefficient values have been calculated and tabulated using an even function symmetric condition, $h(n) = h(-n)$.

Causal filter positive symmetric condition $h(0)$; $h(1) = h(-1)$; $h(2) = h(-2)$; $h(3) = h(-3)$; $h(4) = h(-4)$; $h(5) = h(-5)$; $h(6) = h(-6)$; $h(7) = h(-7)$; $h(8) = h(-8)$; $h(9) = h(-9)$; $h(10) = h(-10)$.

$$h_D(0) = \frac{1}{\pi}\omega_c = \frac{1}{\pi}\frac{\pi}{4} = 0.25$$

$$h_D(1) = h_D(-1) = \frac{1}{\pi}\sin\frac{1.\pi}{4} = 0.225$$

$$h_D(2) = h_D(-2) = \frac{1}{2.\pi}\sin\frac{2.\pi}{4} = 0.159$$

$$h_D(3) = h_D(-3) = \frac{1}{\pi}\sin\frac{3.\pi}{4} = 0.075$$

$$h_D(4) = h_D(-4) = \frac{1}{\pi}\sin\frac{4.\pi}{4} = 0.0$$

$$h_D(5) = h_D(-5) = \frac{1}{\pi}\sin\frac{5.\pi}{2} = -0.045$$

$$h_D(6) = h_D(-6) = \frac{1}{\pi}\sin\frac{6.\pi}{2} = -0.053$$

$$h_D(7) = h_D(-7) = \frac{1}{\pi}\sin\frac{7.\pi}{2} = -0.032$$

$$h_D(8) = h_D(-8) = \frac{1}{\pi}\sin\frac{8.\pi}{2} = 0.0$$

$$h_D(9) = h_D(-9) = \frac{1}{\pi}\sin\frac{9.\pi}{2} = 0.025$$

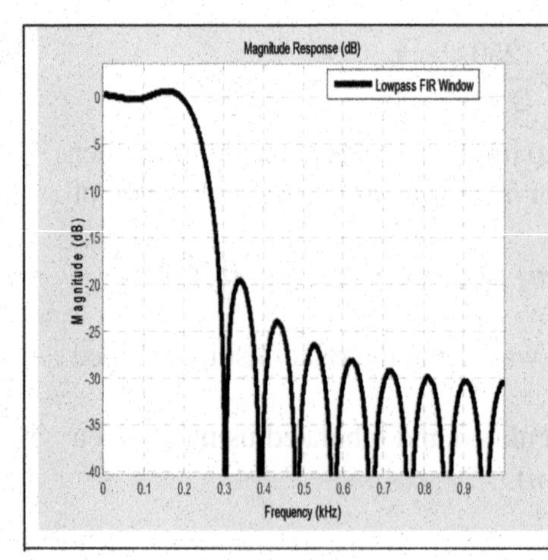

$h_D(n)$	Coefficient's value
$h[0]$	0.25
$h[1] = h[-1]$	0.225
$h[2] = h[-2]$	0.159
$h[3] = h[-3]$	0.075
$h[4] = h[-4]$	0
$h[5] = h[-5]$	-0.045
$h[6] = h[-6]$	-0.053
$h[7] = h[-7]$	-0.032
$h[8] = h[-8]$	0
$h[9] = h[-9]$	0.025
$h[10] = h[-10]$	0.032

(a) Magnitude response of the low-pass filter (b) Coefficients values of the impulse response

$$h_D(10) = h_D(-10) = \frac{1}{\pi} \sin \frac{10.\pi}{2} = 0.032$$

The impulse and magnitude response have been generated using the FDA tool in Matlab.

Problem 9.8

Find the impulse responses of an ideal, linear-phase, **causal low-pass** digital filter with a cutoff frequency $\omega_C = \pi/4$ and G = 1. The length of the filter = 21.

Solution 9.8

We simply truncate the expression for $h_D(n)$ at some reasonable value of $(N - 1)/2$.

A linear-phase low-pass filter means that the filter is causal. The necessary and sufficient condition, i.e. $h(n) = h(N - 1 - n)$, is used to calculate the filter coefficients. The impulse response formula $h_D(n) = 1/([n]\pi) \sin[n]\omega_C, \quad n = 0,\ 1,\ 2, 3, \ldots..N - 1$ is modified by shifting as either subtracting n by $(N - 1)/2$ or subtracting n from $(N - 1)/2$, normally we use the first case to avoid any minus sign in the impulse response formula.

$$h_D(n) = \frac{1}{\left[\frac{N-1}{2} - n\right]\pi} \sin\left[\frac{N-1}{2} - n\right] \omega_C, \quad n = 0, 1, 2, 3, \ldots..N - 1$$

$$h_D(n) = \frac{1}{\left[n - \frac{N-1}{2}\right]\pi} \sin\left[n - \frac{N-1}{2}\right] \omega_C, \quad n = 0, 1, 2, 3, \ldots..N - 1$$

The values of shifted coefficients are calculated and tabulated for $\omega_C = \pi/4$ and $G = 1$. The impulse response value table is generated below for 21 terms. Using the formula

$$h_D(n) = \frac{1}{\left[\frac{N-1}{2} - n\right]\pi} \sin\left[\frac{N-1}{2} - n\right] \omega_C, \quad n = 0, 1, 2, 3, \ldots..N - 1$$

For $N = 21$, the above impulse response formula for no causal system reduces to

$$h_D(n) = \frac{1}{[10 - n]\pi} \sin[10 - n]\, \omega_C, \quad n = 0, 1\ ,\ 2, 3, \ldots..N - 1$$

For $n = 0$ or $n = 20$. $h_D(0) = \frac{1}{[10]\pi} \sin[10]\, \omega_C$ $\qquad h_D(0) = \frac{1}{[10]\pi} \sin[10]\, \frac{\pi}{4} = 0.032$

For $n = 1$ or $n = 19$. $h_D(1) = \frac{1}{[9]\pi} \sin[9]\, \omega_C$ $\qquad h_D(1) = \frac{1}{[9]\pi} \sin[9]\, \frac{\pi}{4} = 0.025$

For $n = 2$ or $n = 19$. $h_D(2) = \frac{1}{[8]\pi}\sin[8]\,\omega_C \quad h_D(2) = \frac{1}{[8]\pi}\sin[8]\,\frac{\pi}{4} = 0.0$

For $n = 3$ or $n = 17$. $h_D(3) = \frac{1}{[7]\pi}\sin[7]\,\omega_C \quad h_D(3) = \frac{1}{[7]\pi}\sin[7]\,\frac{\pi}{4} = -0.032$

For $n = 4$ and $n = 16$. $h_D(4) = \frac{1}{[6]\pi}\sin[6]\,\omega_C \quad h_D(4) = \frac{1}{[6]\pi}\sin[6]\,\frac{\pi}{4} = -0.053$

For $n = 5$ and $n = 15$. $h_D(5) = \frac{1}{[5]\pi}\sin[5]\,\omega_C \quad h_D(5) = \frac{1}{[5]\pi}\sin[5]\,\frac{\pi}{4} = -0.045$

For $n = 6$ and $n = 14$. $h_D(6) = \frac{1}{[4]\pi}\sin[4]\,\omega_C \quad h_D(6) = \frac{1}{[4]\pi}\sin[4]\,\frac{\pi}{4} = 0.0$

For $n = 7$ and $n = 13$. $h_D(7) = \frac{1}{[3]\pi}\sin[3]\,\omega_C \quad h_D(7) = \frac{1}{[3]\pi}\sin[3]\,\frac{\pi}{4} = 0.075$

For $n = 8$ and $n = 12$. $h_D(8) = \frac{1}{[2]\pi}\sin[2]\,\omega_C \quad h_D(8) = \frac{1}{[8]\pi}\sin[8]\,\frac{\pi}{4} = 0.159$

For $n = 9$ and $n = 11$. $h_D(9) = \frac{1}{[1]\pi}\sin[1]\,\omega_C \quad h_D(9) = \frac{1}{[1]\pi}\sin[1]\,\frac{\pi}{4} = 0.225$

or $n = 10$, $\quad h_D(10) = \frac{1}{\pi}\,\omega_C \quad h_D(10) = \frac{1}{\pi}\,\frac{\pi}{4} = 0.25$

N	$h_D(n)$
0,20	0.032
1,19	0.025
2,18	0
3,17	–0.032
4,16	–0.053
5,15	–0.045
6,14	0
7,13	0.075
8,12	0.159
9,11	0.225
10	0.25

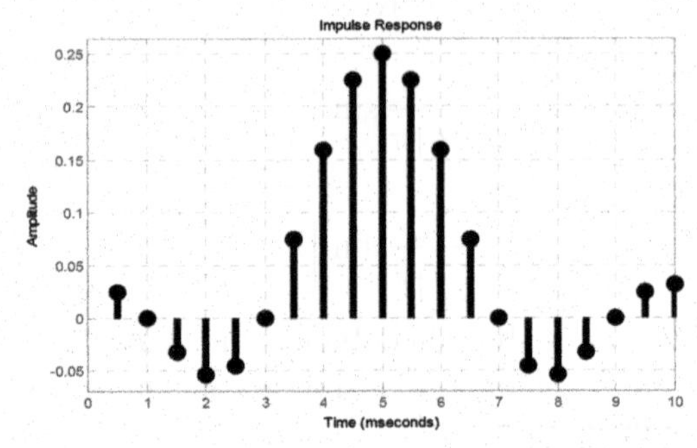

(a) Coefficients values of the FIR causal filter (b) Impulse response for $N = 21$ for a Causal Low-pass

Problem 9.9

Find and sketch the impulse responses of an ideal, zero-phase, **non-causal low-pass**, positive symmetry, digital filter which has filter length $N = 17$ with cut-off frequencies (a) $\omega = \pi/5$; (b) $\omega = \pi/2$, where $G = 1$.

Solution 9.9

The coefficient values have been calculated and tabulated using even function symmetric condition $h(n) = h(-n)$. Causal filter positive symmetric condition

$h(0)$; $h(1) = h(-1)$; $h(2) = h(-2)$; $h(3) = h(-3)$; $h(4) = h(-4)$; $h(5) = h(-5)$; $h(6) = h(-6)$; $h(7) = h(-7)$; $h(8) = h(-8)$.

(a) Using the equation, $h_D(n) = \frac{G}{n\pi} \sin n\omega_c, \quad n = 0, \pm 1, \pm 2, \ldots..$

The coefficient $h[0]$ cannot be calculated in the straight forward manner carried out for the other coefficients, because the numerator and denominator are both zero. Hence, here we will resort to L'Hopital's rule

$$h[0] = \frac{\omega_c}{\pi} = \frac{\pi/5}{\pi} = 0.2$$

$$h_D(1) = h_D(-1) = \frac{1}{\pi} \sin \frac{1.\pi}{5} = 0.18709$$

$$h_D(2) = h_D(-2) = \frac{1}{2.\pi} \sin \frac{2.\pi}{5} = 0.15136$$

$$h_D(3) = h_D(-3) = \frac{1}{3\pi} \sin \frac{3.\pi}{5} = 0.1009$$

$$h_D(4) = h_D(-4) = \frac{1}{3\pi} \sin \frac{4.\pi}{5} = 0.0467$$

$$h_D(5) = h_D(-5) = \frac{1}{5\pi} \sin \frac{5.\pi}{5} = 0.0$$

$$h_D(6) = h_D(-6) = \frac{1}{6\pi} \sin \frac{6.\pi}{2} = -0.0312$$

$$h_D(7) = h_D(-7) = \frac{1}{7\pi} \sin \frac{7.\pi}{2} = -0.0432$$

$$h_D(8) = h_D(-8) = \frac{1}{\pi} 8 \sin \frac{8.\pi}{2} = -0.0378$$

Coefficient value (n)	(a) $h_D(n)$ $\omega_c = \pi/5$	(b) $h_D(n)$ $\omega_c = \pi/2$
[0]	0.2	0.5
$h[1] = h[-1]$	0.1870	0.3183
$h[2] = h[-2]$	0.1 513	0
$h[3] = h[-3]$	0.1 009	-0.1 061
$h[4] = h[-4]$	0.0 467	0
$h[5] = h[-5]$	0	0.0 636
$h[6] = h[-6]$	-0.0 3118	0
$h[7] = h[-7]$	-0.0 4324	-0.0 454
$h[8] = h[-8]$	-0.0 3784	0

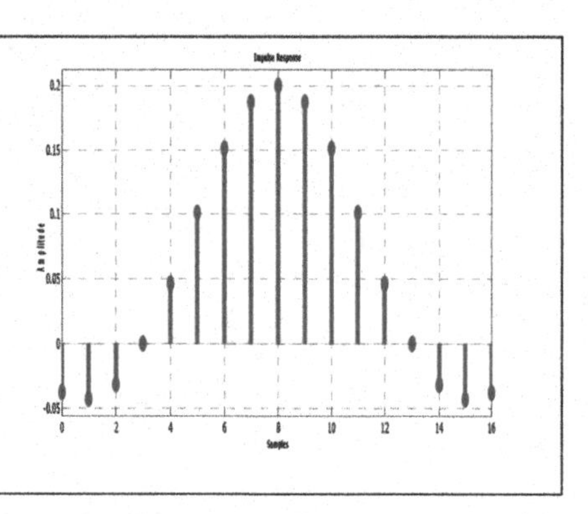

(b) Using equation $h_D(n) = \frac{G}{n\pi} \sin n\omega_c, \quad n = 0, \pm 1, \pm 2, \ldots\ldots$

$$\omega_c = \frac{\pi}{2}\; h[n] = \frac{1}{n\pi} \sin \frac{n\pi}{2}$$

$$h_D(0) = \frac{1}{\pi}\omega_c = \frac{1}{\pi}\frac{\pi}{2} = 0.5$$

$$h_D(1) = h_D(-1) = \frac{1}{\pi} \sin \frac{1.\pi}{2} = 0.3183$$

$$h_D(2) = h_D(-2) = \frac{1}{2.\pi} \sin \frac{2.\pi}{2} = 0.159$$

$$h_D(3) = h_D(-3) = \frac{1}{3\pi} \sin \frac{3.\pi}{2} = 0.0$$

$$h_D(4) = h_D(-4) = \frac{1}{4\pi} \sin \frac{4.\pi}{2} = 0.0$$

$$h_D(5) = h_D(-5) = \frac{1}{5\pi} \sin \frac{5.\pi}{2} = -0.1061$$

$$h_D(6) = h_D(-6) = \frac{1}{6\pi} \sin \frac{6.\pi}{2} = 0.0$$

$$h_D(7) = h_D(-7) = \frac{1}{7\pi} \sin \frac{7.\pi}{2} = -0.0454$$

$$h_D(8) = h_D(-8) = \frac{1}{8\pi} \sin \frac{8.\pi}{2} = 0.0$$

The impulse response is drawn over the range $-8 \leq n \leq 8$ in Figure 9.12

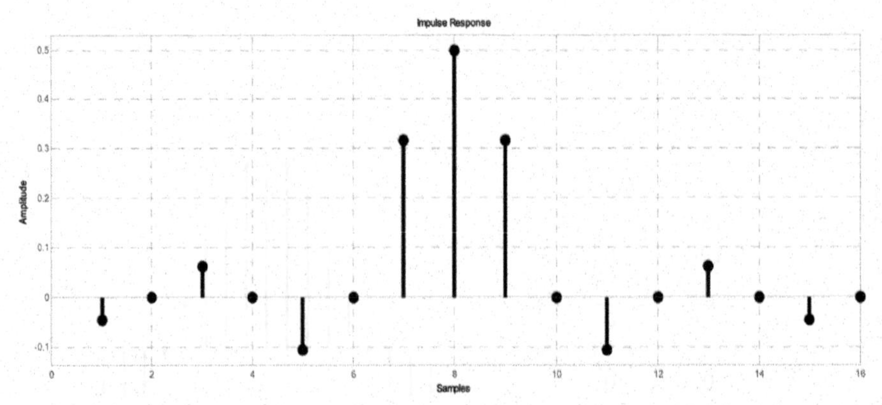

(b) Ideal frequency response of a zero-phase low-pass filter

Problem 9.10

Find and plot the impulse response sequence, positive symmetry Finite Impulse Response (FIR) for a **low-pass causal filter** with 21 terms and a gain of 1. A sampling frequency of 1000 Hz, and a cut-off frequency of 250 Hz is to be used.

Solution 9.10

The value of ω_c is calculated from the cut-off frequency and the sampling frequency

$$\omega_c = 2\pi \frac{f_c}{F_s} = 2\pi \frac{250}{1000} = \frac{\pi}{2}$$

The finite number of terms are required to implement a low-pass digital filter design for some reasonable value of $N = 21$.

For finding coefficient values of causal FIR filters, the formula of the impulse response obtained from the inverse Fourier transform should be modified. Below is the impulse response of a low-pass filter, which has been modified to cater to the case of causal filters.

The necessary and sufficient condition of $h(n) = h(N - 1 - n)$ is used to calculate the filter coefficients. The impulse response formula $h_D(n) = (G/([n]\pi)) \sin[n]\ \omega_C, n = 0, 1, 2, 3, \ldots\ldots N - 1$ is modified by shifting as follows:

Either subtracting n by $(N - 1)/2$ or subtracting n from $(N - 1)/2$, normally we use the first case to avoid any minus sign in the impulse response formula.

$$h_D(n) = \frac{G}{\left[\frac{N-1}{2} - n\right]\pi} \sin\left[\frac{N-1}{2} - n\right] \omega_C, \quad n = 0, 1, 2, 3, \ldots\ldots N - 1$$

$$h_D(n) = \frac{G}{\left[n - \frac{N-1}{2}\right]\pi} \sin\left[n - \frac{N-1}{2}\right] \omega_C, \quad n = 0, 1, 2, 3, \ldots\ldots N - 1$$

The value of shifted coefficients are calculated and tabulated for $\omega_C = \pi/4$ and $G = 1$. The impulse response value table is generated below for 21 terms for the case of a causal filter. Using the formula

$$h_D\left(\frac{N-1}{2} - n\right) = \frac{G}{\left[\frac{N-1}{2} - n\right]\pi} \sin\left[\frac{N-1}{2} - n\right] \omega_C,$$

$$n = 0, 1, 2, 3, \ldots\ldots N - 1$$

For $N = 21$, the above impulse response formula for no causal system reduces to

$$h_D(10-n) = \frac{1}{[10-n]\pi}\sin[10-n]\,\omega_C, \quad n = 0, 1, 2, 3, \ldots . N-1$$

For $n = 0$ or $n = 20$. $h_D(10) = \frac{1}{[10]\pi}\sin[10]\,\omega_C$

$h_D(10) = \frac{1}{[10]\pi}\sin[10]\,\frac{\pi}{2} = 0.0$

For $n = 1$ or $n = 19$. $h_D(9) = \frac{1}{[9]\pi}\sin[9]\,\omega_C$

$h_D(9) = \frac{1}{[9]\pi}\sin[9]\,\frac{\pi}{2} = 0.025$

For $n = 2$ or $n = 19$. $h_D(8) = \frac{1}{[8]\pi}\sin[8]\,\omega_C$

$h_D(8) = \frac{1}{[8]\pi}\sin[8]\,\frac{\pi}{2} = 0.0$

For $n = 3$ or $n = 17$ $h_D(7) = \frac{1}{[7]\pi}\sin[7]\,\omega_C$

$h_D(7) = \frac{1}{[7]\pi}\sin[7]\,\frac{\pi}{2} = -0.032$

N	$h_D(n)$
0,20	0
1,19	0.0354
2,18	0
3,17	–0.0455
4,16	0
5,15	0.0637
6,14	0
7,13	–0.1061
8,12	0
9,11	0.3183
10	0.5

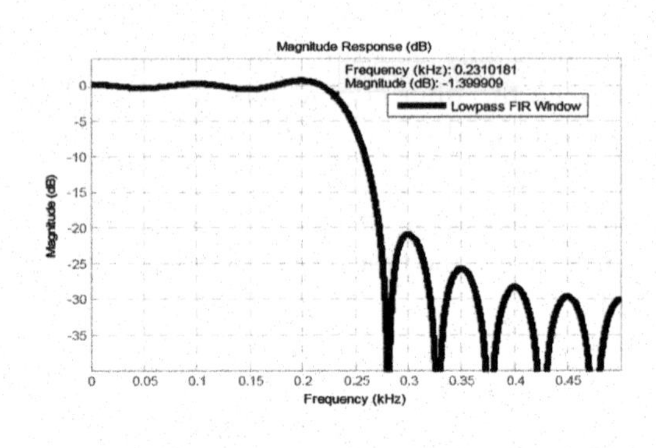

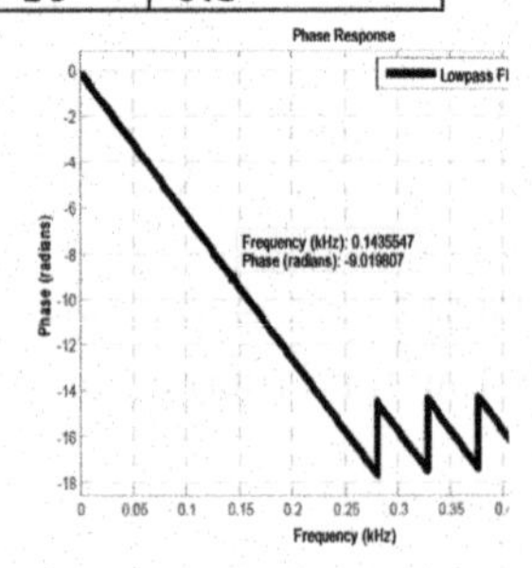

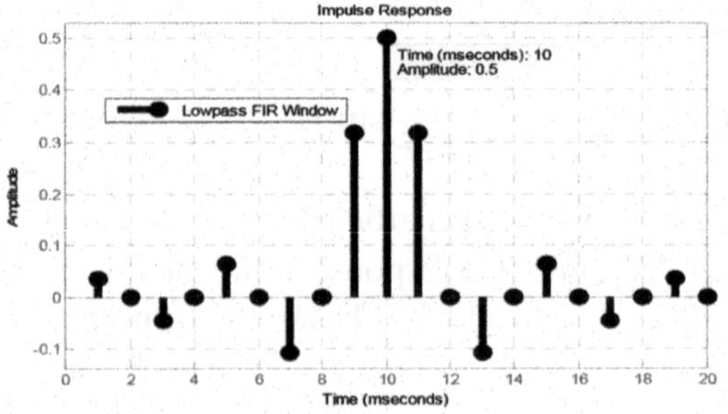

(a) Coefficient value and Phase Response of a low-pass filter (b) Magnitude and Impulse response of a Low-pass FIR filter

For $n = 4$ and $n = 16$. $h_D(6) = \frac{1}{[6]\pi} \sin[6]\ \omega_C$

$h_D(6) = \frac{1}{[6]\pi} \sin[6]\ \frac{\pi}{2} = -0.053$

For $n = 5$ and $n = 15$. $h_D(5) = \frac{1}{[5]\pi} \sin[5]\ \omega_C$

$h_D(5) = \frac{1}{[5]\pi} \sin[5]\ \frac{\pi}{2} = -0.045$

For $n = 6$ and $n = 14$. $h_D(4) = \frac{1}{[4]\pi} \sin[4]\ \omega_C$

$h_D(4) = \frac{1}{[4]\pi} \sin[4]\ \frac{\pi}{2} = 0.0$

For $n = 7$ and $n = 13$. $h_D(3) = \frac{1}{[3]\pi} \sin[3]\ \omega_C$

$h_D(3) = \frac{1}{[3]\pi} \sin[3]\ \frac{\pi}{2} = 0.075$

For $n = 8$ and $n = 12$. $h_D(2) = \frac{1}{[2]\pi} \sin[2]\ \omega_C$

$h_D(10) = \frac{1}{[8]\pi} \sin[8]\ \frac{\pi}{2} = 0.159$

For $n = 9$ and $n = 11$. $h_D(1) = \frac{1}{[1]\pi} \sin[1]\ \omega_C$

$h_D(1) = \frac{1}{[1]\pi} \sin[1]\ \frac{\pi}{2} = 0.383$

For $n = 10$, $h_D(0) = \frac{1}{\pi}\ \omega_C \quad h_D(10) = \frac{1}{\pi}\ \frac{\pi}{2} = 0.5$

Problem 9.11

Develop an expression for a **causal low-pass** Finite Impulse Response (FIR) filter having a length of terms (***N* = 7**) and a gain of 1. A sampling frequency of 1000 Hz, and a cutoff frequency of 250 Hz is to be used in designing.

(a) Rectangular
(b) Hanning
(c) Hamming

Solution 9.11

The generalized expression of the FIR filter having filter length $N = 7$ is

$$H(z) = \sum_{n=0}^{N-1} h(n)\, z^{-n}$$

$$H(z) = h(0)+h(1)z^{-1}+h(2)z^{-2}+h(3)z^{-3}+h(2)z^{-4}+h(1)z^{-5}+h(0)z^{-6}$$

Using the necessary and sufficient condition, $h(n) = h(N - 1 - n)$, to convert the above expression into another readymade formula for causal FIR, we get

For $N = 7$, $h(n) = h(N-1-n)$ means $h(n) = h(6-n)$: the calculated values are required as follows: $h(0) = h(6); h(1) = h(5); h(2) = h(4); h(3)$.

It should be kept in mind that the required filter is an FIR causal filter, so the impulse response formula must first be modified as shown below.

$$h_D(n) = \frac{1}{n\pi} \sin n\omega_c$$

The necessary and sufficient condition, i.e. $h(n) = h(N - 1 - n)$, is used to calculate the filter coefficients. The impulse response formula $h_D(n) = \frac{G}{[n]\pi} \sin[n]\ \omega_C, n = 0,\ 1,\ 2,\ 3, \ldots\ldots N-1$ is modified by shifting as follows:

either subtracting n by $(N-1)/2$ or subtracting n from $(N-1)/2$, normally we use the first case to avoid the minus sign in the impulse response formula.

$$h_D(n) = \frac{G}{\left[\frac{N-1}{2} - n\right]\pi} \sin\left[\frac{N-1}{2} - n\right] \omega_C, \quad n = 0, 1, 2, 3, \ldots\ldots N-1$$

or

$$h_D(n) = \frac{G}{\left[n - \frac{N-1}{2}\right]\pi} \sin\left[n - \frac{N-1}{2}\right] \omega_C, \quad n = 0, 1, 2, 3, \ldots\ldots N-1$$

The values of shifted coefficients are calculated and tabulated for $\omega_C = \pi/2$ and $G = 1$. The impulse response value table is generated below for 7 terms for the case of causal filters. Using this formula,

$$h_D(n) = \frac{G}{\left[\frac{N-1}{2} - n\right]\pi} \sin\left[\frac{N-1}{2} - n\right] \omega_C, \quad n = 0, 1, 2, 3, \ldots\ldots N-1$$

For $N = 7$, the above impulse response formula for no causal system reduces to

$$h_D(n) = \frac{1}{[3-n]\pi} \sin[3-n]\ \omega_C, \quad n = 0, 1, 2, 3, \ldots\ldots N-1$$

$$\omega_C = \frac{2\pi f_c}{F_s} = \frac{2\pi 250}{1000} = \frac{\pi}{2}$$

For $n = 0$ or $n = 6$,

$$h_D(0) = \frac{1}{[3]\ \pi} \sin\,[3]\ \omega_C \quad h_D(0) = \frac{1}{[3]\ \pi} \sin\,[3]\ \frac{\pi}{2} = 0 - .10610$$

For $n = 1$ or $n = 5$,

$$h_D(1) = \frac{1}{[2]\pi} \sin[2]\ \omega_C \quad h_D(1) = \frac{1}{[2]\pi} \sin[2]\ \frac{\pi}{2} = 0.0$$

For $n = 2$ or $n = 4$,

$$h_D(2) = \frac{1}{[1]\pi}\sin[1]\,\omega_C \quad h_D(2) = \frac{1}{[1]\pi}\sin[1]\,\frac{\pi}{2} = 0.3183$$

For $n = 3$,

$$h_D(3) = \frac{1}{\pi}\,\omega_C \quad h_D(3) = \frac{1}{\pi}\frac{\pi}{2} = 0.5$$

N		$h_D(n)$
0	6	−0.10610
1	5	0.00000
2	4	0.31830
3		0.5

(a) For a rectangular window,

$\omega(n) = 1.0$, it means $h_D(n) = h(n)$

Similarly, the following chart shows that $h(n)$ is the same as the impulse response $h_D(n)$, because the magnitude of the rectangular window is 1.0 at every instant.

n		$h_D(n)$	$\omega(n)$	$h(n) = h_D(n).\omega(n)$
0	6	−0.10610	1.0	−0.10610
1	5	0.00000	1.0	0.00000
2	4	0.31830	1.0	0.31830
3		0.5		0.5

The expression for FIR filters can be given using rectangular windows as

$$H\ (z) = -0.10610 + 0.3183\ z^{-2} + 0.5\ \ z^{-3} + 0.3183\ \ z^{-4} + -0.10610\ z^{-6}$$

(b) For a Hanning window,

$$\omega[n] = 0.5 - 0.5\ \cos\frac{2\pi n}{N-1};\ 0 = n = 6$$

$$\omega[0] = \omega[6] = 0.00\omega[1] = \omega[5] = 0.5 - 0.5\ \cos\frac{\pi}{3} = 0.25$$

$$\omega[2] = \omega[4] = 0.5 - 0.5\ \cos\frac{2\pi}{3} = 0.75$$

$$\omega[3] = 0.5 - 0.5\ \cos\pi = 1.0$$

The expression for FIR filter can be obtained using the Hanning window as

$$H(z) = 0.238725\ z^{-2} + 0.5\ z^{-3} + 0.238725\ z^{-4}$$

n		$h_D(n)$	$\omega(n)$	$h(n) = h_D(n)*\omega(n)$
0	6	−0.10610	0	0
1	5	0.00000	0.25	0.000
2	4	0.31830	0.75	0.238725
3		0.5	1.0	0.5

(c) For a Hamming window,

$\omega[n] = 0.54 - 0.46 \cos \frac{2\pi n}{6}; 0 = n = 6$

$$\omega[0] = \omega[6] = 0.08$$

$$\omega[1] = \omega[5] = 0.54 - 0.46 \cos \frac{\pi}{3} = 0.31$$

$$\omega[2] = \omega[4] = 0.54 - 0.46 \cos \frac{2\pi}{3} = 0.77$$

$$\omega[3] = 0.54 - 0.46 \cos \pi = 1$$

The FIR coefficients are given by

n		$h_D(n)$	w(n)	$h(n) = h_D(n)*\text{w}(n)$
0	6	−0.10610	0.08	−0.008488
1	5	0.00000	0.31	0.00000
2	4	0.31830	0.77	0.24509
3		0.5	1.0	0.50000

The expression for FIR filters can be written using the Hamming window as

$$H(z) = -0.008488 \ + 0.24509z^{-2} + 0.31831z^{-3} + 0.24509z^{-4}$$
$$-0.008488z^{-6}$$

The frequency responses of the FIR low-pass filters for three windows are shown below

Problem 9.12

Write the expression for a causal low-pass filter with a frequency response

$$\begin{aligned} H(\omega) &= 1 \quad \text{for } |\omega| \leq \omega_C \\ H(w) &= 0 \quad \text{for } \omega_C < \omega \leq \omega_s/2 \end{aligned}$$

where ω_s is the sampling frequency

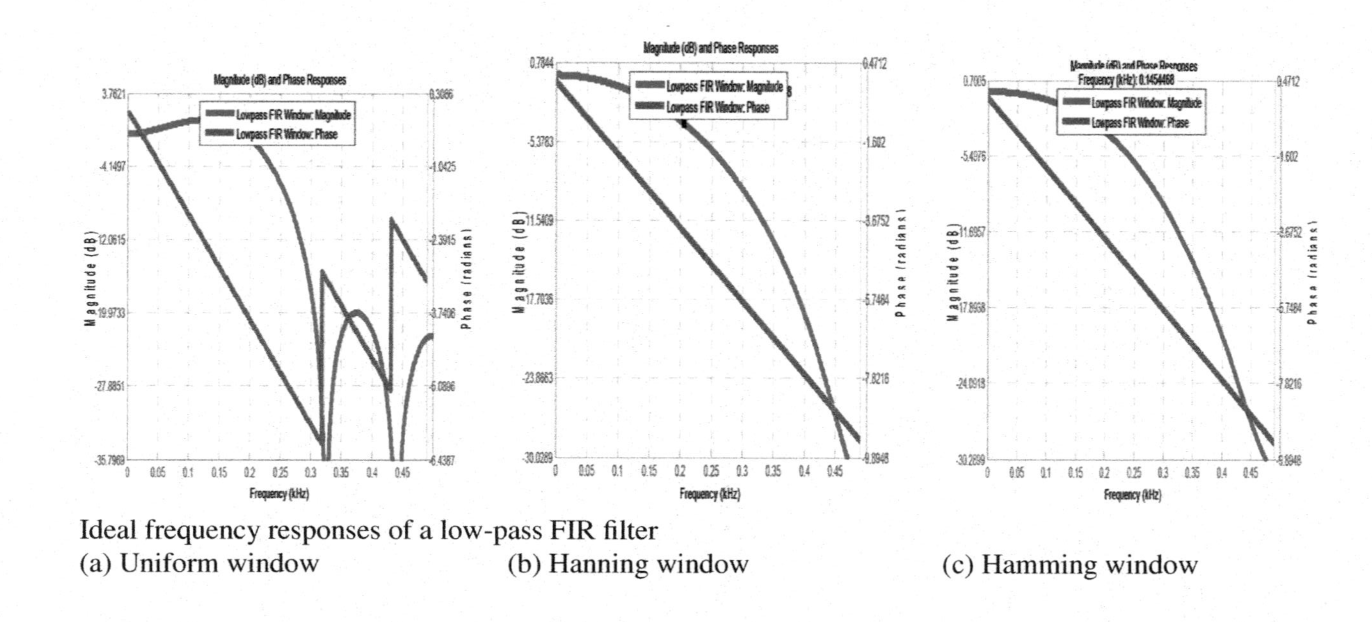

Ideal frequency responses of a low-pass FIR filter
(a) Uniform window (b) Hanning window (c) Hamming window

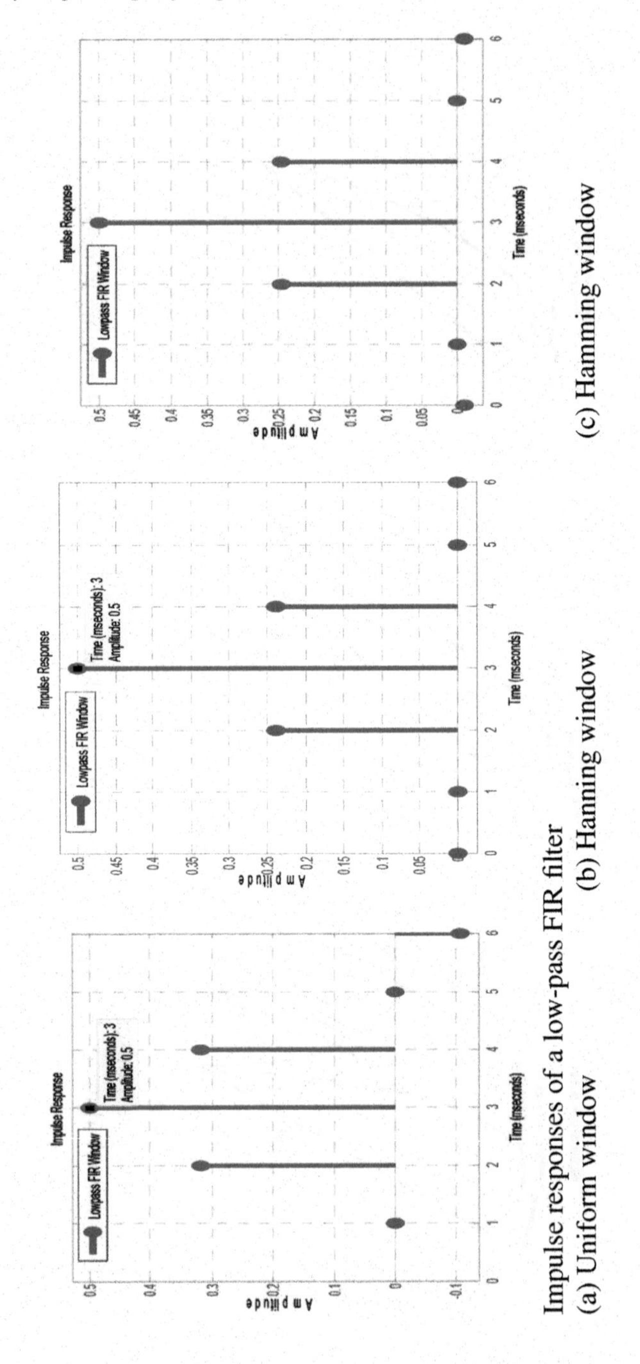

Impulse responses of a low-pass FIR filter
(a) Uniform window (b) Hanning window (c) Hamming window

Solution 9.12

The frequency response for linear-phase, positive symmetry FIR filters is given by

$$h_D(n) = \frac{1}{\omega_S}\int_{-\omega\pi}^{\omega c} 1 \times e^{j\omega n} d\omega = \frac{1}{n\pi}\sin \omega_c n$$

Hence,

$$H(z) = z^{\frac{-(N-1)}{2}} \sum_{n=0}^{(\frac{N-1}{2})} \tfrac{a(n)}{2}(z^n + z^{-n}) \text{ where}$$

$$a(0) = h\left[\frac{N-1}{2}\right] \quad a(n) = 2h\left[\frac{N-1}{2} - n\right]$$

Problem 9.13

A low-pass digital filter is required for physiological noise reduction. The filter should meet the following specifications:

Pass-band edge frequency 10 Hz
Stop-band edge frequency < 20 Hz
Stop-band attenuation > 30 dB
Pass-band attenuation < 0.026 dB
Sampling frequency 256 Hz

Solution 9.13

Important requirements in this application are: (i) the filter should introduce as little distortion as possible to the in-band signals and (ii) the length of the filter should be as low as possible and should not exceed 37.

Problem 9.14

Find the coefficients of an ideal **high-pass non-causal**, positive symmetry, filter that has a pass band for frequencies greater than 10 kHz, where the system sampling frequency = 50 kHz. Let the pass band gain be unity for filter length, $N = 21$.

Solution 9.14

The frequency response for positive symmetry FIR filters is given by

$$\text{digtal cutoff frequency} = \tfrac{fc}{F_s} = \tfrac{10\times(10^3)}{50\times(10^3)} = \tfrac{1}{5}$$
$$\omega_C = 2(\pi)\ f_c = 2(\pi)\ (1/5) = 0.4\pi \text{ rad}$$

$$h_D(n) = \delta(n) - \frac{G}{n\pi}\sin n\omega_c \quad \begin{array}{l} h_D(n) = \delta(n)\frac{1}{n\pi}\sin\ n(0.4\pi) \\ h_D(0) = 1 - 2(0.2) = 0.6 \end{array}$$

The values of high-pass FIR coefficients are tabulated below for $\omega_C = 0.4\pi$ and $G = 1$.

Coefficients value	
h[0]	0.5999
h[1] = h[−1]	−0.3027
h[2] = h[−2]	−0.0934
h[3] = h[−3]	0.0623
h[4] = h[−4]	0.0756
h[5] = h[−5]	0.0000
h[6] = h[−6]	−0.0504
h[7] = h[−7]	−0.0267
h[8] = h[−8]	−0.0234
h[9] = h[−9]	0.0336
h[10] = h[−10]	0.0000

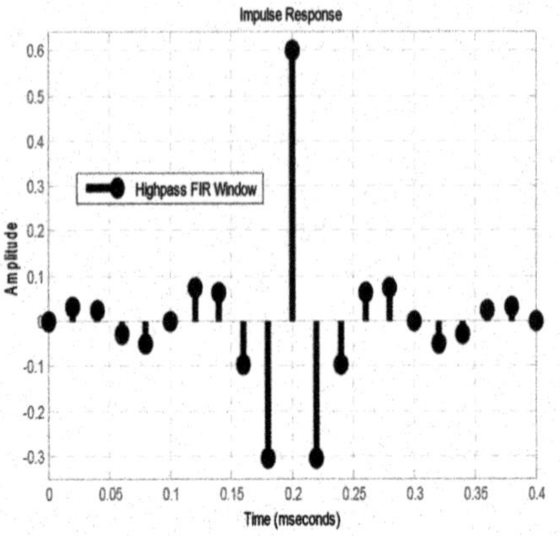

(a) Coefficient value of the high-pass filter (b) Impulse response of the filter

Problem 9.15

Design a **causal high-pass**, positive symmetry, Finite Impulse Response (FIR) filter with $N = 7$ and $\omega_C = 2$ rad/sec using following windows. Develop the transfer function $H(z)$ of the filters.

(a) Rectangular
(b) Hanning

Solution 9.15

For finding coefficient values of the causal FIR filter, the formula for the impulse response obtained from inverse Fourier transform should be modified. Below is the impulse response of a high-pass filter, which has been modified to cater to the case of causal filters.

$$h_D(n) = \delta(n) - \frac{G}{n\pi} \sin n\omega_c$$

$$h_D(n) = \delta\left(\frac{N-1}{2} - n\right) - \frac{G}{\left[\frac{N-1}{2} - n\right]\pi} \sin\left[\frac{N-1}{2} - n\right]\omega_c$$

$$h_D(n) = \delta(3-n) - \frac{G}{[3-n]\pi} \sin[3-n]\omega_c$$

$G = 1$.

For $n = 0$ or $n = 6$.

$$h_D(0) = \delta(3) - \frac{1}{[3]\pi}\sin[3]\,\omega_C \quad h_D(0) = 0 - \frac{1}{[3]\pi}\sin[3]\,2 = 0.0296$$

For $n = 1$ or $n = 5$.

$$h_D(1) = \delta(2) - \frac{1}{[2]\pi}\sin[2]\,\omega_C \quad h_D(1) = 0 - \frac{1}{[2]\pi}\sin[2]\,2 = 0.1205$$

For $n = 2$ or $n = 4$.

$$h_D(2) = \delta(1) - \frac{1}{[1]\pi}\sin[1]\,\omega_C \quad h_D(1) = 0 - \frac{1}{[1]\pi}\sin[1]\,2 = -0.2894$$

For $n = 3$.

$$h_D(3) = 1 - \frac{1}{\pi}\,\omega_C \quad h_D(3) = 1 - \frac{1}{\pi}\,2 = 0.3634$$

n		$h_D(n)$
0	6	0.0296
1	5	0.1205
2	4	−0.2894
3		0.3634

For calculating coefficient values of causal FIR filters, we look at the impulse response of a high-pass filter, which has been modified using a rectangular window.

(a) By using a rectangular window, the magnitude is the same as that of the impulse response

n		$h_D(n)$	$\omega(n)$	$h(n) = h_D(n)*\omega(n)$
0	6	0.0296	1.0	0.0296
1	5	0.1205	1.0	0.1205
2	4	−0.2894	1.0	−0.2894
3		0.3634	1.0	0.3634

The expression for an FIR filter can be given as

$$H(z) = 0.0296 + 0.1205\,z^{-1} - 0.2894\,z^{-2} + 0.3634\,z^{-3} - 0.2894\,z^{-4} + 0.1205\,z^{-5} + 0.0296\,z^{-6}$$

(b) Using a Hanning window, the filter coefficients are obtained. For finding coefficient values of a causal FIR filter, we look at the impulse response of a high-pass filter, which has been modified using a Hanning window.

N		$h_D(n)$	w(n)	$h(n) = h_D(n)$*w(n)
0	6	0.0296	0	0
1	5	0.1204	0.25	0.031
2	4	−0.2894	0.75	−0.2171
	3	0.3634	1.0	0.3634

The expression for FIR filters can be expressed by a Hanning window as

$$H(z) = 0.031z^{-1} - 0.2171z^{-2} + 0.3634z^{-3} - 0.2171z^{-4} + 0.031z^{-5}$$

Problem 9.16

A 13-coefficient causal non-recursive **high-pass, causal** positive symmetry, filter is to be designed using the Fourier series approach to approximate the frequency response characteristics with $\omega_C = \pi/2$ rad/sec. Calculate the coefficient values for the FIR.

Solution 9.16

A general expression for the unit sample response (filter coefficients) for the causal high-pass filter can be written from the above expression as

Coefficients value	
h[0] = h[12]	0.0000
h[1] = h[11]	-0.0636
h[2] = h[10]	0.0000
h[3] = h[9]	0.1061
h[4] = h[8]	0.0000
h[5] = h[7]	-0.318
h[6]	0.5

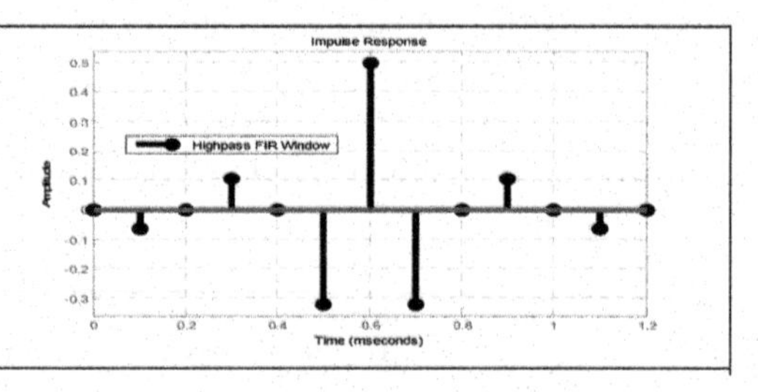

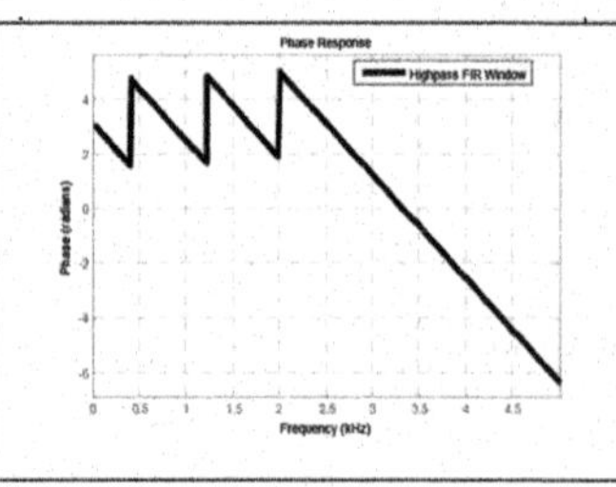

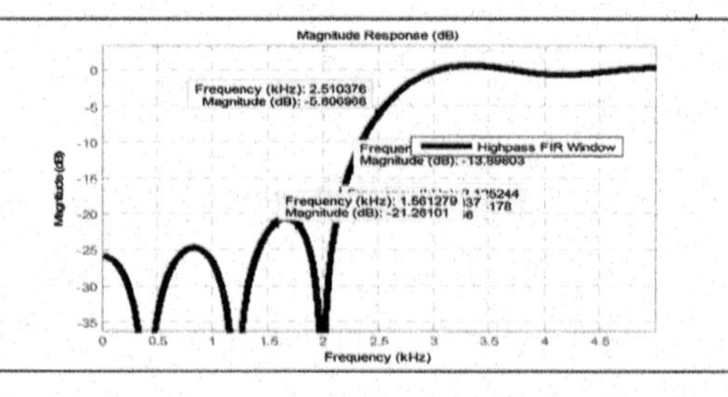

(a) Coefficients of the FIR high-pass filter (b) Impulse response of FIR high-pass filter (c) Phase response (d) Magnitude response

$$h_D(n) = \delta\left(\frac{N-1}{2} - n\right) - \frac{(1)}{\left[\frac{N-1}{2} - n\right]\pi} \sin\left[\frac{N-1}{2} - n\right]\omega_C$$

$$h_{D\ HP}(0) = 1 - \frac{\omega_C}{\pi}$$

Problem 9.17

A 7-coefficient **causal high-pass** filter is to be designed using the Fourier series approach to approximate the frequency response characteristics shown in the figure below.

(a) Find the transfer function of the filter.
(b) Find the transfer function of the filter if a Hamming window is used.

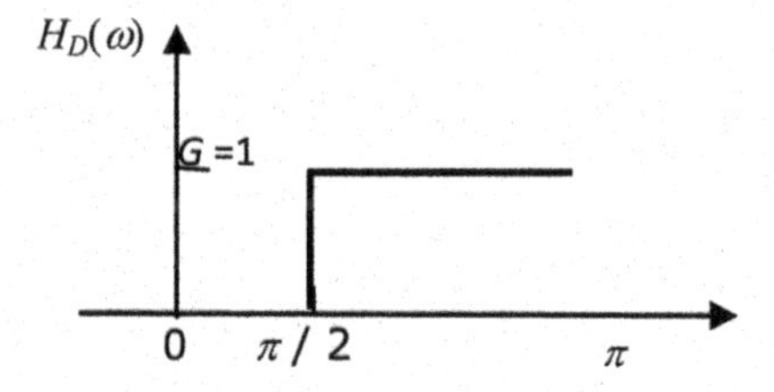

Frequency response of a high-pass filter

Solution 9.17

A general expression for the unit sample response (filter coefficients) for the causal high-pass filter can be written from the above expression as

$$h_D(n) = \delta\left(\frac{N-1}{2} - n\right) - \frac{1}{\left[\frac{N-1}{2} - n\right]\pi} \sin\left[\frac{N-1}{2} - n\right]\omega_C$$

$$h_{D\ \mathrm{HP}}(0) = 1 - 2\, f_c = 1 - \frac{\omega_C}{\pi} \quad h_D(n) = \delta(3-n) - \frac{\sin(3-n)\omega_C}{\pi(3-n)}$$

(a) The filter coefficients are obtained.

n		$h_D(n)$
0	6	0.1061
1	5	0
2	4	−0.3183
3		0.5

$$H(z) = \sum_{n=0}^{6} h(n)\, z^{-n}$$

$$H(z) = h(0) \; + h(1)z^{-1} + h(2)z^{-2} + h(3)z^{-3} + h(4)z^{-4}$$
$$+ \, h(5)z^{-5} \; + h(6)z^{-6}$$

$$H(z) = 0.1061 \;\; -0.3183z^{-2} + 0.5z^{-3} - 0.3183z^{-4} + 0.1061z^{-6}$$

(b) Using hamming windows, the filter coefficients are obtained.

$$\omega[n] = 0.54 - 0.46 \, \cos \frac{2\pi n}{N-1} \qquad 0 \leq n \leq N-1$$

n		$h_D(n)$	$\omega(n)$	$h(n) = h_D(n)*\omega(n)$
0	6	0.1061	0.08	0.0085
1	5	0	0.31	0
2	4	–3.183	0.77	–0.2451
3		0.5	1.0	0.5

The expression for FIR filters can be written by the hanning window as

$$H(z) = 0.0085 \;\; - 0.2451 \; z^{-2} + 0.5 \; z^{-3} - 0.2451 \; z^{-4}$$
$$+ \, 00.0085 \, z^{-6}$$

The filter frequency response magnitude (dB) and phase (degree) plots are shown in figures.

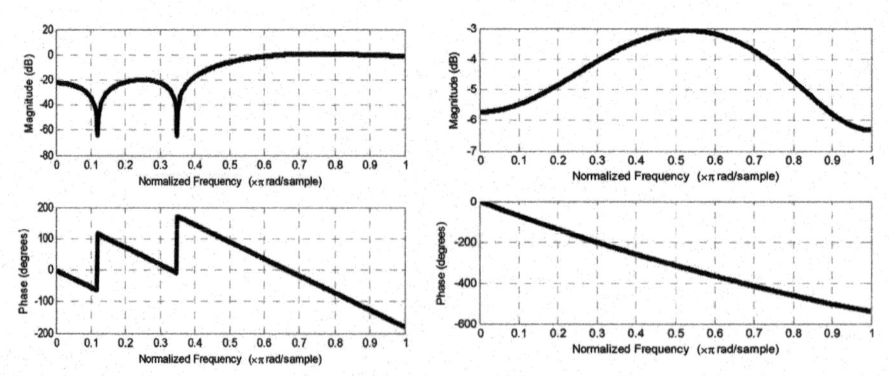

Frequency response of a FIR filter with $N = 7$ terms. (a) Uniform window (b) Hamming window

Problem 9.18

Design a **band-pass non-causal** FIR filter for the following specification:

Cut-off frequency = 250 Hz and 500 Hz
Sampling frequency = 2000 Hz
Order of the filter = 6
Filter length required = 7

Solution 9.18

Using the design equation of a band-pass filter,

$$\text{band-pass filter } h_D(n) = \frac{\sin n\omega_{c2} - \sin n\omega_{c1}}{n\pi}$$

Normalize the cut-off frequencies

$$\omega_{C1} = 2\pi\frac{f_{C1}}{F_s} = 2\pi\frac{250}{2000} = 0.25\pi \quad \omega_{C2} = 2\pi\frac{f_{C2}}{F_s} = 2\pi\frac{500}{2000} = 0.5\pi$$

$$h_D(0) = \frac{(0.5\pi) - (0.25\pi)}{\pi} = 0.25 \; h_D(1) = h_D(-1)$$

$$= \frac{\sin(0.5\pi) - \sin\,(0.25\pi)}{\pi} = 0.0932$$

$$h_D(2) = h_D(-2) = \frac{\sin 2(0.5\pi) - \sin\ 2(0.25\pi)}{2\pi} = -0.15915$$

$$h_D(3) = h_D(-3) = \frac{\sin 3(0.5\pi) - \sin\ 3(0.25\pi)}{3\pi} = -0.1811$$

The above equation is used to find out the filter coefficients. We then use the appropriate window to obtain the desired impulse response and the modified filter coefficients.

Problem 9.19

Design a **band-pass non-causal** FIR filter for the following specification:

Cut-off frequency = 600 Hz and 1200 Hz
Sampling frequency = 3000 Hz
Order of the filter = 6
Filter length required = 7

Solution 9.19

Using the design equation of a band-pass filter

$$\text{band-pass filter } h_D(n) = \frac{\sin n\omega_{C2} - \sin n\omega_{C1}}{n\pi}.$$

Normalize the cut-off frequencies

$$\omega_{C1} = 2\pi\frac{f_{C1}}{F_s} = 2\pi\frac{600}{3000} = \frac{2\pi}{5} = 0.4\pi$$

$$\omega_{C2} = 2\pi\frac{f_{C2}}{F_s} = 2\pi\frac{1200}{3000} = \frac{4\pi}{5} = 0.8\pi$$

$$h_D(0) = \frac{(8\pi) - (0.4\pi)}{\pi} = 0.4 \quad h_D(1) = h_D(-1)$$

$$= \frac{\sin(0.8\pi) - \sin(0.4\pi)}{\pi} = -0.1156$$

$$h_D(2) = h_D(-2) = \frac{\sin 2(0.8\pi) - \sin 2(0.4\pi)}{2\pi} = -0.2449$$

$$h_D(3) = h_D(-3) = \frac{\sin 3(0.8\pi) - \sin 3(0.4\pi)}{3\pi} = -0.1632$$

The above equation is used to find out the filter coefficients. We then use the appropriate window to obtain the desired impulse response and the modified filter coefficients.

Problem 9.20

Design a **band-pass causal** FIR filter for the following specification:

Cut-off frequency = 400 Hz and 800 Hz
Sampling frequency = 2000 Hz
Order of the filter = 10
Filter length required = 11

Solution 9.20

Using the design equation of a band-pass filter

band-pass filter $h_D(n) = \frac{\sin n\omega_{C2} - \sin n\omega_{C1}}{n\pi}$ and modifying this expression for a causal filter.

$$\text{band-pass filter } h_D(n) = \frac{\sin(\frac{N-1}{2} - n)\omega_{C2} - \sin(\frac{N-1}{2} - n)\omega_{C1}}{(\frac{N-1}{2} - n)\pi}$$

Normalize the cut-off frequencies

$$\omega_{C1} = 2\pi\frac{f_{C1}}{F_s} = 2\pi\frac{400}{2000} = 0.4\pi \quad \omega_{C2} = 2\pi\frac{f_{C2}}{F_s} = 2\pi\frac{800}{2000} = 0.8\pi$$

$$\text{band-pass filter } h_D(n) = \frac{\sin\left(\frac{N-1}{2} - n\right)(0.8\pi) - \sin\left(\frac{N-1}{2} - n\right)(0.4\pi)}{\left(\frac{N-1}{2} - n\right)\pi}.$$

$n = 0{,}10$

$$h_D(0) = \frac{\sin(5)(0.8\pi) - \sin(5)(0.4\pi)}{(5)\pi} = \frac{\sin(4\pi) - \sin(2\pi)}{(5)\pi} = 0.00$$

$n = 1{,}9$

$$h_D(1) = \frac{\sin(4)(0.8\pi) - \sin(3)(0.4\pi)}{(4)\pi} = \frac{-0.58778 + 0.95}{4\pi}$$

$$= \frac{-0.58778 + 0.95}{4\pi} = 0.0289$$

$n = 2{,}8$

$$h_D(2) = \frac{\sin(3)(0.8\pi) - \sin(3)(0.4\pi)}{(3)\pi} = 0.1632$$

$n = 3{,}7$

$$h_D(3) = \frac{\sin(2)(0.8\pi) - \sin(2)(0.4\pi)}{(2)\pi} = -0.2449$$

$n = 4{,}6$

$$h_D(4) = \frac{\sin(1)(0.8\pi) - \sin(1)(0.4\pi)}{(1)\pi} = -0.1156$$

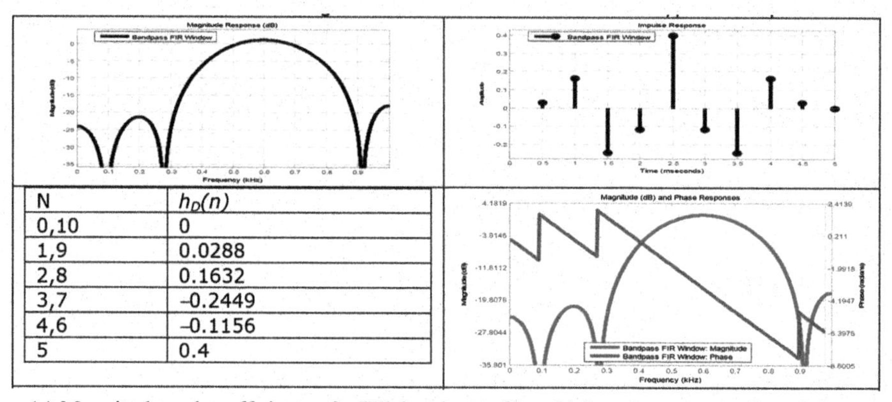

N	$h_D(n)$
0,10	0
1,9	0.0288
2,8	0.1632
3,7	–0.2449
4,6	–0.1156
5	0.4

(a) Magnitude and coefficients of a FIR band-pass filter (b) Impulse, magnitude and phase response of a band-pass FIR filter

$n = 5$

$$h_D(5) = \frac{0.8\pi - 0.4\pi}{\pi} = 0.400$$

The above equation is used to find out the filter coefficients. We then use the appropriate window to obtain the desired impulse response and the modified filter coefficients. These operations follow the same procedures we have adopted in the design of low-pass and high-pass filters. Hence, they are not repeated here. However, the FDA tool has been used to generate coefficients and different types of responses.

Problem 9.21

Design a **band-stop non-causal** FIR filter for the following specification:

Cut-off frequency = 400 Hz and 800 Hz
Sampling frequency = 2000 Hz
Order of the filter = 10
Filter length required = 11

Solution 9.21

Using the design equation of a band-stop filter

$$\text{band-stop filter } h_D(n) = \delta(n) - \frac{\sin n\omega_{C2} - \sin n\omega_{C1}}{n\pi}.$$

Normalize the cut-off frequencies

$$\omega_{C1} = 2\pi\frac{f_{C1}}{F_s} = 2\pi\frac{400}{2000} = 0.4\pi \quad \omega_{C2} = 2\pi\frac{f_{C2}}{F_s} = 2\pi\frac{800}{2000} = 0.8\pi$$

$$\text{band-stop filter } h_D(n) = \delta(n) - \frac{\sin\ n(0.8\pi) - \sin\ n(0.4\pi)}{n\pi}$$

$$h_D(0) = \delta(0) - \frac{\omega_{C2} - \omega_{C1}}{\pi} = 1 - \frac{0.8\pi - 0.4\pi}{\pi} = 1 - \frac{0.4\pi}{\pi} = 0.60$$

$$h_D(1) = h_D(-1) = \delta(1) - \frac{\sin 0.8\pi - \sin 0.4\pi}{\pi} = 0.1156$$

$$h_D(2) = h_D(-2) = \delta(2) - \frac{\sin 2(0.8\pi) - \sin 2(0.4\pi)}{2\pi} = 0.2449$$

$$h_D(3) = h_D(-3) = \delta(3) - \frac{\sin 3(0.8\pi) - \sin 3(0.4\pi)}{3\pi} = -0.1632$$

$$h_D(4) = h_D(-4) = \delta(4) - \frac{\sin 4(0.8\pi) - \sin 4(0.4\pi)}{4\pi} = -0.0289$$

$$h_D(5) = h_D(-5) \ = \delta(5) - \frac{\sin 5(0.8\pi) - \sin 5(0.4\pi)}{5\pi} = 0.0$$

The above equation is used to find out the filter coefficients. We then use the appropriate window to obtain the desired impulse response and the modified filter coefficients. These operations follow the same procedures we have adopted in the design of low-pass, high-pass, and pass-band filters. Hence, they are not repeated here.

Problem 9.22

Design a **band-stop causal** FIR filter for the following specification:

Cut-off frequency = 400 Hz and 800 Hz
Sampling frequency = 2000 Hz
Order of the filter = 10
Filter length required = 11

Solution 9.22

Using the design equation of a band-stop filter

$$\text{band-stop filter } h_D(n) = \delta\left(\frac{N-1}{2} - n\right) - \frac{\sin\left(\frac{N-1}{2} - n\right)\omega_{c1} - \sin\left(\frac{N-1}{2} - n\right)\omega_{c2}}{n\pi}.$$

Normalize the cut-off frequencies

$$\omega_{C1} = 2\pi\frac{f_{c1}}{F_s} = 2\pi\frac{400}{2000} = 0.4\pi$$

$$\omega_{C2} = 2\pi\frac{f_{c2}}{F_s} = 2\pi\frac{800}{2000} = 0.8\pi$$

$$\text{band-stop filter } h_D(n) = \delta(5-n) - \frac{\sin(5-n)\omega_{C1} - \sin(5-n)\omega_{C2}}{n\pi}$$

$n = 0,10$

$$h_D(5) = \delta(5) - \frac{\sin(5)0.8\pi - \sin(5)0.4\pi}{5\pi} = 0.0$$

$n = 1,9$

$$\text{band-stop filter } h_D(4) = \delta(4) - \frac{\sin(4)0.8\pi - \sin(4)0.4\pi}{4\pi} = -0.0289$$

$n = 2,8$

$$\text{band-stop filter } h_D(3) = \delta(3) - \frac{\sin(3)0.8\pi - \sin(3)0.4\pi}{3\pi} = -0.1632$$

$n = 3,7$

$$\text{band-stop filter } h_D(2) = \delta(2) - \frac{\sin(2)0.8\pi - \sin(2)0.4\pi}{2\pi} = 0.2449$$

$n = 4,6$

$$\text{band-stop filter } h_D(1) = \delta(1) - \frac{\sin(1)0.8\pi - \sin(1)0.4\pi}{1\pi} = 0.1156$$

$n = 5$

$$h_D(5) = \delta(0) - \frac{\omega_{C2} - \omega_{C1}}{\pi} = 1 - \frac{0.8\pi - 0.4\pi}{\pi} = 1 - \frac{0.4\pi}{\pi} = 0.6$$

The above equation is used to find out the filter coefficients. We then use the appropriate window to obtain the desired impulse response and the modified filter coefficients. These operations follow the same procedures we have adopted in the design of low-pass, high-pass, and pass- and stop-band filters. Hence, they are not repeated here.

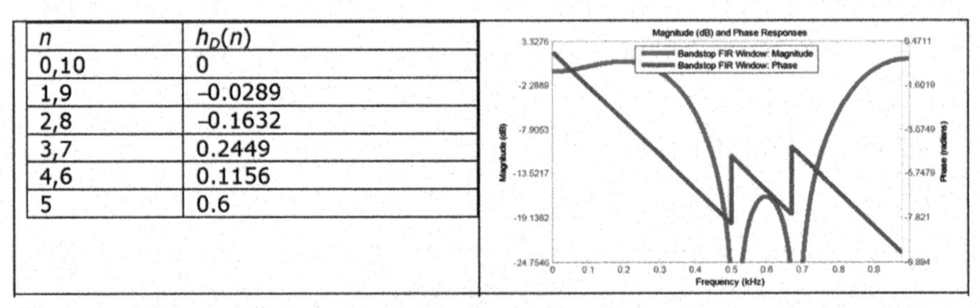

n	$h_D(n)$
0,10	0
1,9	-0.0289
2,8	-0.1632
3,7	0.2449
4,6	0.1156
5	0.6

(a) Magnitude and coefficients of a FIR band-stop filter (b) Impulse, magnitude, and phase response of a band-stop FIR filter

Problem 9.23

An FIR digital filter is to meet the following specifications:

Pass-band 150–250 Hz
Transition width 50 Hz
Pass-band attenuation 0.1 dB
Stop-band attenuation 60 dB
Sampling frequency 1000 Hz

Obtain the filter coefficients and spectrum using the window method.

Solution 9.23

From the specifications, the pass-band and stop-band ripples are

$$20 \log (1 + \delta_p) = 0.1 \text{ dB, giving } \delta_p = 0.0115$$
$$-20 \log (\delta_S) = 60 \text{ dB, giving } \delta_s = 0.001$$

Thus,

$$\delta = \min (\delta_p, \delta_s) = 0.001$$

The attenuation requirements can only be met by the Kaiser or the Blackman window. For the Kaiser window, the numbers of filter coefficients are:

$$N \geq= \frac{A - 7.95}{14.36\Delta F} = \frac{60 - 7.95}{14.36(50/1000)} = 72.49$$

Let $N = 73$. The ripple parameter is given by

$$\beta = 0.1102(60 - 9.7) = 5.65$$

With $N = 73$, $\beta = 5.65$, Matlab is used to compute the values of $\omega(n)$, the ideal impulse response $h_D(n)$, and the filter coefficients.

To account for the smearing effects of the window functions in computing the ideal impulse response, cutoff frequencies of $f_{C1} - \Delta f/2$ and $f_{C2} + \Delta f/2$ were used, that is f_{C1} = 125 Hz and 275 Hz, respectively.

For the Blackman window, an estimate of the number of filter coefficients is obtained as

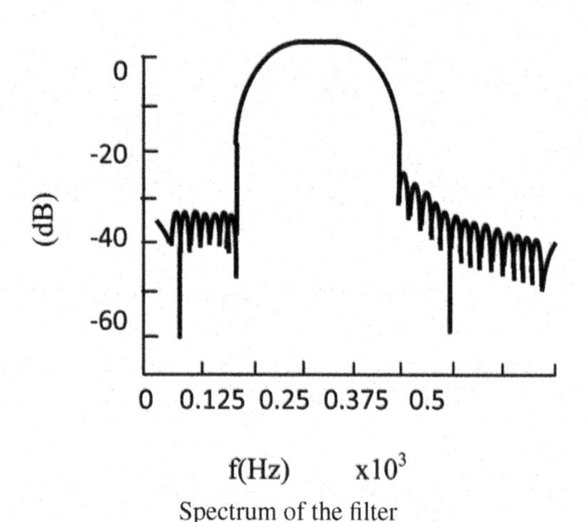

Spectrum of the filter

$N = 5.5/\Delta f = 5.5/(50/1000) \approx 110$

The filter coefficients for the Blackman window are not given here owing to lack of space. It is evident that the Kaiser window is more efficient than the Blackman window in terms of the number of coefficients required to meet the same specifications. In general, the Kaiser window is more efficient compared with the other windows in this respect.

Problem 9.24

Design a low-pass filter with Finite Impulse Response (FIR) using the Kaiser window to satisfy the following specification.

Pass-band ripple: in the frequency range 0 to 1.5 rad/sec ≤ 0.1 dB
Minimum stop-band attenuation: range 2.5 to 5 rad/sec ≥ 40 dB
Sampling frequency: 10 rad/sec.

Obtain the filter coefficients and spectrum using the window method.

Solution 9.24

$$h_D(n) = \frac{1}{n\pi} \sin n\omega_c$$

$$\omega_C = \frac{1}{2}(1.5 + 2.5) = 2 \text{ rad/sec}$$

$$\delta_s = 10^{-0.05(40)} = 0.01$$

$$\delta_p = 10^{0.05(0.1)} - 1 = 0.0115$$

$$\delta = \min(\delta_P,\ \delta_S) = 0.0115$$

$A_S = -20 \log \delta_S = 39.76$

Therefore $N \geq \frac{A-7.95}{14.36\Delta F} = \frac{38.786-7.95}{14.36(50/1000)} = 21.47 = 21$

$$H(z) = z^{-\frac{N-1}{2}} \sum_{n=0}^{\left(\frac{N-1}{2}\right)} \frac{a_n}{2}(z^n + z^{-n})$$

Problem 9.25

Obtain the coefficients of a linear-phase FIR filter using the Kaiser window to satisfy the following amplitude response specification:

Stop-band attenuation 40 dB
Pass-band ripple 0.01 dB
Transition width 500 Hz
Sampling frequency 10 kHz
Ideal cutoff frequency 1 kHz

Solution 9.25

From the specification,

$$20 \log\,(1\ + \delta_p) = 0.01 \text{ dB}, \text{ giving}\, \delta_p =\ 0.00115$$
$$-20 \log\,(\delta_s) = 40 \text{ dB}, \text{ giving}\, \delta_s =\ 0.01$$

Since both the pass-band and stop-band ripples are equal (as they cannot be specified independently) in the window method, we use the smaller of the ripples:

$$\delta = \min(\delta_P,\ \delta_S) = 0.0115$$

This means that the stop-band attenuation is required. In this case,

$$-20\,\log(0.00115) = 59.8 \text{ dB}.$$

The number of filter coefficients required is

$$N = \frac{A - 7.95}{14.36\Delta F} = \frac{58.8 - 7.95}{14.36(500/10000)} \approx 71$$

If the required attenuation specification of 40 dB was used, *N* would have been 45. Thus, the need for δ_p to be equal to δ_s in the window method has led to a higher than necessary number of filter coefficients.

The ripple parameter is obtained as shown below

$$\alpha = 0.5842(59.8\ - 21)^{0.4} + 0.07886(59.8 - 21) = 5.48$$

The FIR coefficients are obtained from $h(n) = h_D(n)w(n)$, where (from Table 9.2)

$$h_D(n) = 2f_c \frac{\sin(n\omega_c)}{n\omega_c} \quad n \neq 0$$
$$h_D(n) = 2f_c \quad n = 0$$

As explained before, the cutoff frequency, f_c, used in calculating $h(n)$ is different to that given in the specifications to account for the smearing effect of the window function.

We select f_c such that it is in the middle of the transition band:

$$f_c' = 1200 + \Delta f/2 = 1450\,\text{Hz}.$$

The following filter parameters are used with the computer program.

Cutoff frequency 1450 Hz

Ripple parameter, β 5.48

Number of filter coefficients 71 and

Sampling frequency 10 kHz

The resulting filter coefficients can be calculated.

Practice Problem 9.26

The frequency response, $H(\omega)$, of a type II, linear-phase FIR filter may be expressed as

$$H(\omega) = e^{-j\omega(N-1)/2} \sum_{n=1}^{N/2} b(n) \cos\left[\omega\left(n - \frac{1}{2}\right)\right]$$

where $b(n)$ is related to the filter coefficients. Explain why filters with the response above are unsuitable as high-pass filters. Use a simple case (such as $N = 4$) to illustrate your answer.

Practice Problem 9.27

An FIR filter has an impulse response, $h(n)$, which is defined over the interval $0 \leq n \leq N-1$. Show that if N is even and $h(n)$ satisfies the positive symmetry condition, that is $h(n) = h(N - n - 1)$, the filter has a linear-phase response. Obtain expressions for the amplitude and phase responses of the filter.

Practice Problem 9.28

Show that the impulse response for an ideal band-pass filter is given by

$$\begin{aligned} h_D(n) &= 2f_2 \frac{\sin n\omega_2}{n\omega_2} - 2f_1 \frac{\sin n\omega_1}{n\omega_1} \quad && n \neq 0 \\ &= 2(f_2 - f_1) && n = 0 \end{aligned}$$

where f_1 is the lower pass-band frequency and f_2 is the upper pass-band frequency.

Practice Problem 9.29

Obtain the coefficients of an FIR low-pass digital filter to meet the following specifications using the window method:

Stop-band attenuation 50 dB
Pass-band edge frequency 3.4 kHz
Transition width 0.6 kHz
Sampling frequency 8 kHz

10

Step-by-Step Design of IIR Filters

This chapter begins with general consideration of analog filters; it further discusses the design stages of IIR Digital Filters, the calculation of IIR Filter coefficient by Bilinear Z-transformation, Impulse Invariant and Pole-Zero Placement methods, and Problems and Solutions.

10.1 Introduction

The filters magnitude response is represented with respect to frequency; therefore, these filters are also named as frequency selective filters. The system transfer functions for an analog filter are denoted by $H(s)$, where $s = \sigma + j\Omega$, where Ω is continuous time angular frequency. The frequency transfer function of an analog filter, denoted by $H(j\Omega)$, is obtained by evaluating the system function H in the s-plane along the frequency axis, or $H(j\Omega) = H(s)_{s=j\Omega}$. The squared magnitude response $|H(j\Omega)|^2$ [power transfer characteristic] of the filter is shown below.

10.2 Analog Prototype Filters

Prior to the widespread use of DSP, high-performance frequency selective filters needed to de designed with analog electronics DSP allowed the development of a new class of filter, the linear-phase FIR filter, for which there is no analog electronics counterpart.

The digital version of analog filters can realize certain filter characteristics with a lower order (fewer coefficients) than a comparable FIR filter. The digital counterparts to classic analog filters are all IIR design.

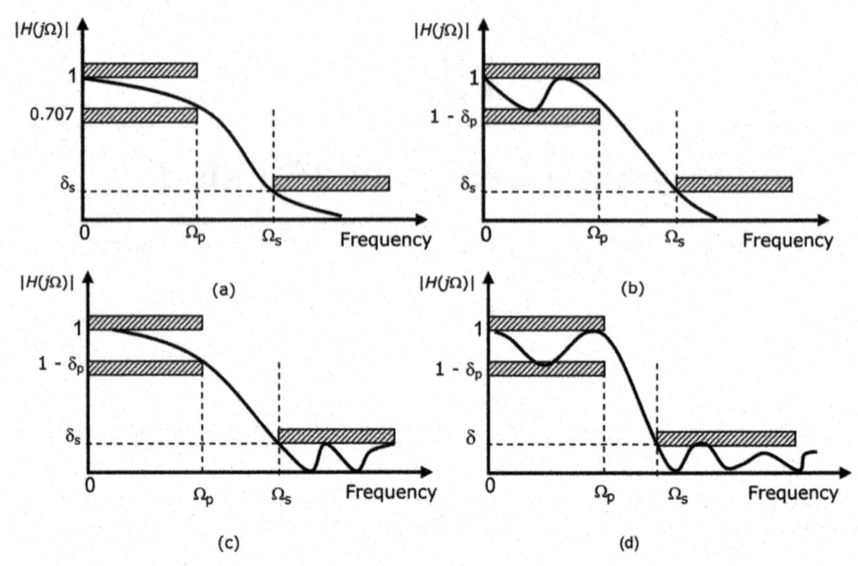

Figure 10.1 Sketches of frequency responses of some classical analog filters (a) Butterworth; (b) Chebyshev type I; (c) Chebyshev type II; and (d) Elliptic.

Analog Filter Type	Pass-Band Ripple	Stop-Band Ripple	Transition Witn
Butterworth	Monotonic	Monotonic	Wide
Chebyshev-I	Equiripple	Monotonic	Narrow
Chebyshev-II	Monotonic	Equiripple	Narrow
Elliptic	Equiripple	Equiripple	Very Narrow

10.2.1 Preview of Butterworth Filter

The low-pass Butterworth filter is characterized by the following magnitude-squared-frequency response, and it should be noted clearly that poles of Butterworth filter lies on the circle.

$$|H(\Omega)|^2 = \frac{1}{1+\left(\frac{\Omega}{\Omega_p^p}\right)^{2N}} \tag{10.1}$$

where N is the filter order and Ω_p^p is the 3-dB cutoff frequency of the low-pass filter (for the normalized prototype filter, Ω_p^p is always equal to 1). The magnitude-frequency response of a typical Butterworth low-pass filter is depicted in Figure 10.1(a), and is seen to be monotonic in both the pass-band and stop-band. The response is said to be maximally flat because of its initial flatness (with a slope of zero at dc).

The Butterworth polynomials can be determined from the following formula, which has been derived for understanding.

$$\left(\frac{\Omega}{\Omega_p^p}\right)^{2N} + 1 = 0; \quad \left(\frac{j\Omega}{j\,\Omega_p^p}\right)^{2N} = -1; \quad \left(\frac{\Omega}{j\,\Omega_p^p}\right)^{2N} = -1$$

$$(s)^{2N} = (-1)(j\,\Omega_p^p)^{2N} \text{ or } (e^{-j\pi}) = -1\left(e^{j\pi/2}\right)^{2N} = \left(e^{j\pi/2}\right) = j$$

As the value of a complex number is unchanged when multiplied by $e^{j(2\pi m)}$ for an integer value of m, i.e. $m = 1, 2, 4, \ldots\ldots..,$

$$(s)^{2N} = (e^{-j\pi})\left(e^{j\pi/2}\right)^{2N} \Omega_p^{p\,2N}\, e^{j(2\pi m)}$$

$$(s)^{2N} = (e^{-j\pi})\left(e^{j\pi/2}\right)^{2N} \Omega_p^{p\,2N}\, e^{j(2\pi m)}$$

$$(s)^{2N} = \Omega_p^{p\,2N}\, e^{j(2m+N-1)\pi}$$

distinct location of the poles are found by

$$s_m = \Omega_p^p\, e^{j\left(\frac{2m+N-1}{2N}\right)\pi)}$$

The transfer function of the normalized analog Butterworth filter, $H(s)$, contains zeros at infinity and poles which are uniformly spaced on a circle of radius $\Omega_p^p = 1$ in the s-plane at the following locations.

$$s_m = e^{j\pi(2m+N-1)/2N} = \cos\left[\frac{(2m+N-1)\pi}{2N}\right] + j\,\sin\left[\frac{(2m+N-1)\pi}{2N}\right] \quad m = 1, 2, \ldots, N \tag{10.2}$$

The poles occur in complex conjugate pairs and lie on the left-hand side of the s-plane.

Using Equation (10.2), the normalized Butterworth polynomial is derived and is given in tabular form for different values of N. Students are advised to verify themselves that how it has been derived.

Order of N	Factors for the Butterworth denominator polynomial
1	$s + 1$
2	$s^2 + 1.414s + 1$
3	$(s + 1)(s^2 + s + 1)$
4	$(s^2 + 0.765s + 1)(s^2 + 1.848s + 1)$

10.2.2 Preview of Chebyshev Filter

The Chebyshev characteristic provides an alternative way of obtaining a suitable analog transfer function, $H(s)$. There are two types of Chebyshev filters, types I and II, (Figure 10.1(b) and (c)): with the following features and again to be noted that poles of the Chebyshev filter lie on the ellipse

- Type I with equal ripple in the pass-band, monotonic in the stop-band;
- Type II, with equal ripple in the stop-band, monotonic in the pass-band.

Type I Chebyshev filters; for example, are characterized by the magnitude-squared response

$$|H(\Omega)|^2 = \frac{K}{1+\varepsilon^2 C_N^2(\Omega/\Omega_p)} \tag{10.3}$$

where ϵ is a constant less than 1, $C_N(\Omega/\Omega_P)$ is a Chebyshev polynomial which exhibits equal ripple in the pass band, *N* is the order of the polynomial as well as that of the filter, and ϵ determines the pass-band attenuation, which in decibels is given by pass-band attenuation $\leq 10\log_{10}(1+\epsilon^2) = -20\log_{10}(1-\delta p)$.

A typical amplitude response of a type I Chebyshev characteristic is shown in Figure 10.1. The transfer function, $H(s)$, for the Chebyshev response depends on the desired pass-band attenuation and the filter order, *N*.

The first method of finding polynomial of the normalized Chebyshev LPF, which lies on an ellipse in the s-plane, can be found with the help of the following coordinates:

$$s_m = \sigma_m + j\Omega_m$$

$$s_m = -\sinh(\alpha)\ \sin[\tfrac{2m-1}{2N}]\pi + j\cosh(\alpha)\cos[\tfrac{2m-1}{2N}]\pi \quad m = 1, 2, \ldots.., N \tag{10.4}$$

where

$$\alpha = \frac{1}{N}\sinh^{-1}\left(\frac{1}{\varepsilon}\right) \tag{10.5}$$

The second method of finding polynomial of the normalized Chebyshev LPF is described by a relation that looks trigonometric in nature, where $C_N(x)$ is actually a polynomial of degree *N*.

$$C_N(x) = \cos\left[N\cos^{-1}(x)\right] \quad \text{where} \quad \varphi = \left[\cos^{-1}(x)\right] \tag{10.6}$$

where $x = (\Omega/\Omega_p)$

$$C_0(x) = \cos\left[0\right] = 1 \quad \text{and} \quad C_1(x) = \cos\left[\cos^{-1}(x)\right] = x \tag{10.7}$$

$$C_N(x) = \cos\left[N\varphi\right]$$

Using trigonometric identities:

$$\cos(\alpha \pm \beta) = \cos\alpha \ \cos\beta \pm \sin\alpha \ \sin\beta$$

$$\cos[(N+1)\varphi] + \cos[(N-1)\varphi] = 2 \ \cos[N\varphi]\cos[\varphi]$$

$$C_{N+1}(x) = 2\cos[N\varphi]\cos[\varphi] - C_{N-1}(x) \text{ for } n \geq 1 \tag{10.8}$$

The normalized Chebyshev polynomial up to order 4 has been calculated using Equations (10.8) and (10.9) and placed below in the tabular form without the ripple factor (ϵ) for different values of *N*. Students are advised to verify themselves that how it has been derived.

Order of N	$C_{\mathrm{N}(x)} = \omega/\omega_p$
0	1
1	x
2	$2x^2 - 1$
3	$4x^3 - 3x$
4	$8x^4 - 8x^2 + 1$

$$|H(\Omega)|^2 = \frac{K}{1 + \varepsilon^2 C_N^2(x)} \tag{10.9}$$

Using Equation (10.9), the normalized Chebyshev polynomial, with the ripple factor (ϵ), is derived and is given in the tabular form for different values of ϵ. The formula for the pass-band ripple (A_P) and the ripple factor (ϵ) is given in Equation (10.10).

$$A_P(\mathrm{dB}) = 10 \ \log(1 + \varepsilon^2) \tag{10.10}$$

$$\varepsilon = \sqrt{10^{A_P} - 1} \tag{10.11}$$

Normalized Chebyshev Polynomial Pass-band 0.5 dB Ripple ($\epsilon = 0.3493$)

Order of N	Factors of Polynomial ($s = j\Omega$)
1	$s + 2.863$
2	$s^2 + 1.426s + 1.5164$
3	$(s + 0.626)(s^2 + 0.626s + 1.42453)$
4	$(s^2 + 0.350s + 1.062881)(s^2 + 0.846s + 0.35617)$

Normalized Chebyshev Polynomial Pass-band 3.0 dB Ripple ($\epsilon = 0.9953$)

Order of N	Factors of Polynomial ($s = j\Omega$)
1	$s + 1.002$
2	$s^2 + 0.2986s + 0.8395$
3	$(s + 0.299)(s^2 + 0.2986s + 0.8395)$
4	$(s^2 + 0.17s + 0.90214)(s^2 + 0.412s + 0.1961)$

10.2.3 Preview of Elliptic filter

The elliptic filter exhibits equal ripple behavior in both the pass band and the stop band; as shown in Figure 10.1(d), it is characterized by the following magnitude-squared response:

$$|H(\Omega')|^2 = \frac{K}{1 + \varepsilon^2 G_N^2(\Omega')}$$

where $G_N(\Omega')$ is a Chebyshev rational function. Unlike the Butterworth and Chebyshev filters, there is no simple expression for the poles of the Elliptic filter. A procedure is available for computing locations of the poles. The zeros of the elliptic low-pass filter are entirely imaginary.

The elliptic characteristic provides the most efficient filters in terms of amplitude response. It yields the smallest filter order for a given set of specifications and should be the method of first choice in IIR filter design, except where the phase response is of concern when the Butterworth response may be preferred.

Tables of the polynomials of $H(s)$ for the Butterworth, Chebyshev, and Elliptic characteristics are available in most analog design books in normalized form and can be used in the bilinear transformation. In practice, however, the computation of $H(z)$ from $H(s)$ is done by the software.

10.3 Basic Structure of IIR Filters

IIR (Infinite Impulse Response) filter response never quite reaches zero because the feedback makes the decay exponential. A practical filter has to eventually settle to zero or it would be useless, unless one needs an oscillator. IIR filters can oscillate if they are improperly designed. All feedback systems have the potential to become unstable and oscillate.

The input and output signals to the filter are related by convolution sum, which is given in Equation (10.12) for IIR filter.

$$y(n) = \sum_{k=0}^{\infty} h(k)\, x(n-k) \tag{10.12}$$

All IIR filters can be implemented in digital form. IIR filter uses feedback to sharpen the filter response. There are two sets of filter coefficients. The coefficient a is called the feed-forward coefficient, the coefficient b is the feedback coefficient and $x(n)$ and $y(n)$ are the input and output to the filter.

The IIR filter equation is expressed in a recursive form using Equations (10.13) and (10.14)

$$y(n) = \sum_{k=0}^{\infty} h(k)\,x(n-k) = \sum_{k=0}^{N} a_k\,x(n-k) - \sum_{k=1}^{M} b_k y(n-k) \quad (10.13)$$

$$H(z) = \frac{a_0 + a_1 z^{-1} + \cdots + a_N z^{-N}}{1 + b_1 z^{-1} + \cdots + b_M z^{-M}} = \frac{\sum_{k=0}^{N} a_k z^{-k}}{1 + \sum_{k=1}^{M} b_k z^{-k}} \quad (10.14)$$

The transfer function of the IIR filter, $H(z)$, which is given in Equation (10.15) can be factored as

$$H(z) = \frac{K(z - z_1)(z - z_2)\cdots(z - z_N)}{(z - p_1)(z - p_2)\cdots(z - p_M)} \quad (10.15)$$

The following methods of calculating the coefficients of IIR filters are discussed here.

(a) Bilinear Z transformation method
(b) Pole-zero placement method
(c) Impulse invariant method

10.4 Bilinear z-Transform (BZT) Method

One of the best digital IIR filter approximations to an analog filter is obtained using the bilinear transform method (BZT). The formula obtained for this method uses the fact that for a Laplace transform $X(s)$ of a signal $x(t)$, its integral without initial conditions is given by $X(s)/s$, thus multiplying the Laplace transform of a signal by $1/s$ is the same as integrating the signal in the time domain.

However, the following difference equation also approximately integrates the signal $x(t)$ by using its sampled values $x(nT) = x(n)$. Remember that the integral of a signal without initial condition is just the area under the signal.

$$y(n) = y(n-1) + \frac{x(n) + x(n-1)}{2}T \quad (10.16)$$

Equation (10.16) just says that the area under a signal after a new input sample $x(n)$ is just the previously computed area $y(n-1)$ plus the average of the new input and the previous input multiplied by the time between samples. Now this is just the previous area plus the average of new area. After Z-transformation of (10.16) difference equation, we get the following equation, which is solved for $T(z)$ once we get transforms of the input $x(z)$ and the output $Y(z)$.

The most important method of obtaining IIR filter coefficients is the BZT method, and the basic operation required to convert an analog filter $H(s)$ into an equivalent digital filter is to replace as follows:

$$Y(z) = z^{-1}Y(z) + 0.5[X(z) + z^{-1}X(z)]T$$

$$Y(z)(1 - z^{-1}) = 0.5\,T\,X(z)[1 + z^{-1}]$$

$$T(z) = \frac{Y(z)}{X(z)} = \frac{T}{2}\frac{1 + z^{-1}}{1 - z^{-1}} \qquad (10.17)$$

Equation (10.17) is a transfer function of a DSP system that approximates the analog transfer function of a system that does integration. The two transforms are equated in the following form and then solved for s after multiplying the numerator and denominator by s.

$$\frac{\mathrm{T}}{2}\;\frac{z+1}{z-1} \equiv \frac{1}{s}$$

The above transformation maps the analog transfer function, $H(s)$, from the s-plane into the discrete transfer function, $H(z)$, in the z-plane as shown in Figure 10.2(a).

Once converting the digital filter given digital frequency into analog frequency, we prewarp one or more critical frequencies before applying to the given analog transfer function and finally to BZT to avoid the warping effect shown in Figure 10.2(b).

$$s = k\frac{(z-1)}{(z+1)} \qquad k = 1 \text{ or } \frac{2}{T}$$

We here describe the prewarping effect: the prewarping effect occurs when the digital frequency is being converted into an analog frequency, and the relationship is not linear; to minimize this effect, the formula has been derived from the BZT transformation.

For example, we often prewarp the cutoff or band edge frequency as follows:

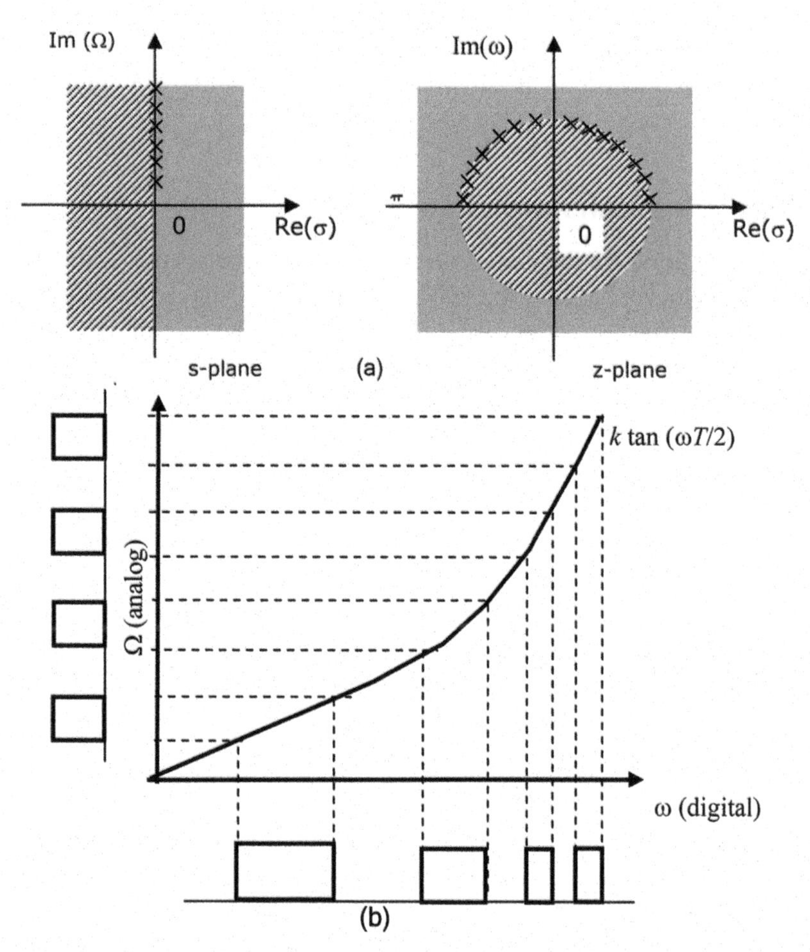

Figure 10.2 (a) An illustration of the s-plane ($s = \sigma + j\Omega$) to z-plane mapping using the bilinear Z-transformation ($z = \sigma + j\omega$). Note that $+j\Omega$ axis maps to the upper half and the negative $j\Omega$ axis maps to the lower half unit circle in the z-plane and (b) Relationship between analog and digital frequencies shows the warping effect. Notice that the equally spaced analog pass bands are pushed together at the high frequency and, after transformation, in the digital domain.

$$s = k\frac{(z-1)}{(z+1)} \quad k = 1 \text{ or } \frac{2}{T}$$

Substituting in this expression $z = e^{j\omega}$

$$j\Omega_p = \frac{2}{T}\frac{(e^{j\omega}-1)}{(e^{j\omega}+1)} = k\frac{(e^{j\omega/2}-e^{-j\omega/2})}{(e^{j\omega/2}+e^{-j\omega/2})} \text{ i.e. } \Omega_p = k \tan\left(\frac{\omega_p T}{2}\right)$$

where ω_p = specified cutoff (pass-band edge) digital frequency

Ω_p = prewarped cutoff analog frequency; $k = 1$ or $\frac{2}{T}$ and T = sampling period.

10.5 Frequency Transformation

It is good to mention here that, once we start the designing of filter using the BZT method, whatever be the type of IIR ever filter, which has to be designed by the frequency transformation, we convert it into respective desired four passes, i.e. from low pass to low pass, low pass to high pass, low pass to band pass, and low pass to band stop. It is also assumed that if the transfer function is given, it is of low-pass filter, or if the transfer function is not known, the frequency transformation is done for a desired filter, then the order of the prototype filter is calculated, and from the calculated transfer function using transformation equation, it is converted again to the respective desired transfer function.

(i) Digital pass-band or cutoff frequency calls this ω_P which is given in specification as the digital frequency of the filter should be converted into analog frequency called Ω_p using the prewarping relationship.

$$\Omega_P = k \tan\left(\frac{\omega_p T}{2}\right) \quad k = 1 \ or \ \frac{2}{T}$$

(ii) Using the relationship of frequency transformation, the analog filter by frequency scaling, using one of the following transformations, depends on the type of filter.

$$\mathrm{s} = \frac{\mathrm{s}}{\Omega_\mathrm{p}} \quad \text{Low-pass to low-pass} \tag{10.18}$$

$$\mathrm{s} = \frac{\Omega_\mathrm{p}}{\mathrm{s}} \quad \text{Low-pass to high-pass} \tag{10.19}$$

$$\mathrm{s} = \frac{\mathrm{s}^2 + \Omega_0^2}{\mathrm{Ws}} \quad \text{Low-pass to band-pass} \tag{10.20}$$

$$\mathrm{s} = \frac{\mathrm{Ws}}{\mathrm{s}^2 + \Omega_0^2} \quad \text{Low-pass to band-stop} \tag{10.21}$$

$$\text{where } \mathrm{W} = \Omega_{p2} - \Omega_{p1} \quad \text{and} \quad \Omega_0^2 = \Omega_{p2}\,\Omega_{p1} \tag{10.22}$$

(iii) Apply the bilinear transformation to obtain the desired digital filter transfer function $H(z)$ by replacing s $= k\frac{(z-1)}{(z+1)}$ $\quad k = 1\ or\ \frac{2}{T}$

It can be seen that factor k is cancelled out, and it would not have been a matter whether k is substituted as 1 or 2/T.

10.6 Design of Filters for Known Transfer Function

In many practical cases, the analog transfer function $H(s)$ is known, from which $H(z)$ is obtained. For standard frequency, selective digital filtering tasks (i.e. involving low-pass, high-pass, band-pass, and band-stop filters), $H(s)$, can be derived from the classical filters with Butterworth, Chebyshev, or Elliptic characteristics. Low-pass filters are considered, since other filter types are normally derived from normalized low-pass filters.

Example 10.1 (Low-Pass filter)

Determine, using the BZT method, the transfer function and difference equation for the digital equivalent of the resistance-capacitance (RC) low-pass filter. Assume a sampling frequency of 150 Hz and a cutoff digital frequency (pass-band edge frequency) of 30 Hz.

Solution 10.1

The normalized transfer function for the RC filter is

$$H(s) = \frac{1}{s+1}$$

The critical frequency for the digital filter is $\omega_p = 2\pi$ x $30 = 60\pi$ rad/sec.

The analog frequency, after prewarping, is $\Omega_p = \tan(\omega_p\ T/2)$, with $T = 1/150$ Hz, $\Omega_p = \tan(\pi/5) = 0.7265$.

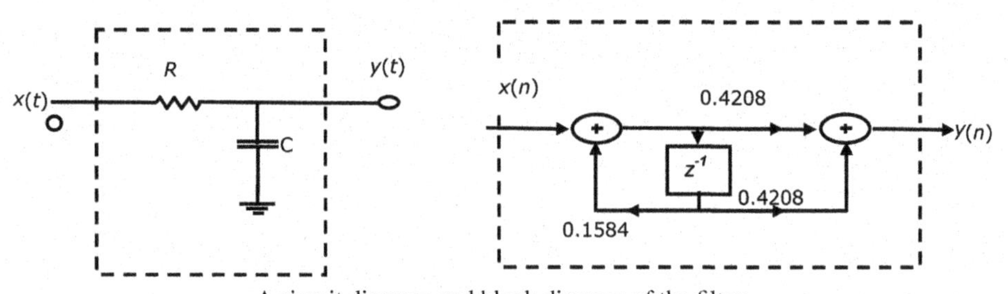

A circuit diagram and block diagram of the filter.

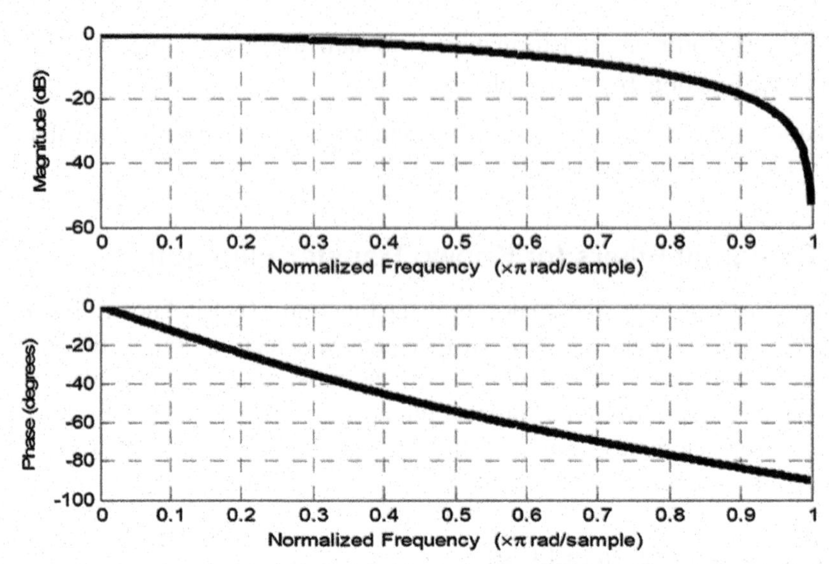

Magnitude and phase response of filter.

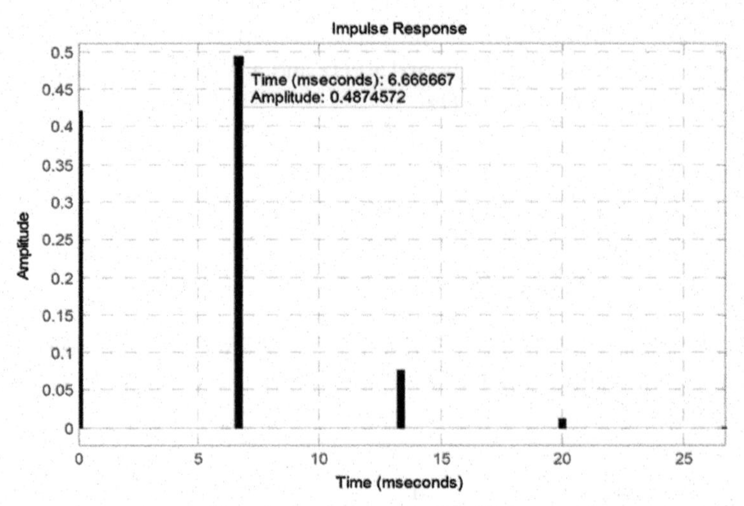

Impulse response of an IIR filter.

The denormalized analog filter transfer function is obtained from $H(s)$ as

$$H'(s) = H(s)|_{s=0.7265} = \frac{1}{s/0.7265+1} = \frac{0.7265}{s+0.7265}$$

$$H(z) = H'(s)|_{s=(z-1)/(z+1)} = \frac{0.7265(1+z)}{(1+0.7265)z+0.7265-1}$$

$$H(z) = \frac{0.7265(1+z)}{1.7265z + 0.7265 - 1}\frac{0.7265(1+z)}{1.7265z + 0.2735} = \frac{0.7265}{1.7265}\frac{(1+z)}{(z+0.0.584)}$$

$$H(z) = 0.4208\frac{(1+z^{-1})}{1-0.1584z^{-1}}$$

The difference equation is

$$y(n) = 0.1584\, y(n-1) + 0.4208[x(n) + x(n-1)]$$

The block diagram representation, with its frequency response and impulse responses shown in figures.

Example 10.2 (High Pass Filter)

Determine, using the BZT method, the transfer function for the digital equivalent of the resistance-capacitance (RC) high-pass filter. Assume a sampling frequency of 2000 Hz and a cutoff digital frequency (pass-band edge frequency) of 8000 Hz.

Solution 10.2

The normalized transfer function for the RC filter is

$$H(s) = \frac{1}{s+1}$$

The critical frequency for the digital filter is $\omega_p = 2\pi \times 2000$ rad/sec.

The analog frequency, after prewarping, is $\Omega_p = \tan(\omega_p\, T/2), \Omega_p = \tan(\pi/4) = 1$

The denormalized analog filter transfer function is obtained from $H(s)$, $\Omega_p = 1$

$$H'(s) = H(s)|_{s=\frac{\Omega_p}{s}} = \frac{1}{1/\mathrm{s}+1} = \frac{\mathrm{s}}{\mathrm{s}+1}$$

$$H(z) = H'(s)|_{\mathrm{s}=(\mathrm{z}-1)/(\mathrm{z}+1)} = \frac{(\mathrm{z}-1)/(\mathrm{z}+1)}{(\mathrm{z}-1)/(\mathrm{z}+1)+1}$$

$$H(z) = \frac{0.5(1-\mathrm{z}^{-1})}{1}$$

$$H(z) = 0.5(1-\mathrm{z}^{-1})$$

Example 10.3

Derive the transfer function for a Digital Butterworth approximation high-pass filter from the following analog filter

(i) A –3-dB cutoff frequency of 2000 Hz, and
(ii) A sampling frequency (F_s) of 8000 Hz.

Solution 10.3

The transfer function of the low-pass filter is given as:

$$H(s) = \frac{1}{s^2 + \sqrt{2}s + 1}$$

This transfer function is now denormalized using the transform given in $s \to \Omega_p/s$, as follows, which converts a low-pass to a high-pass filter.

The prewarped value is

$$\Omega_p = \tan\left(\frac{\omega_p T}{2}\right)$$

$$G(s)\Big|'_{s=\frac{\Omega_p}{s}} = \frac{1}{\left(\frac{\Omega_p}{s}\right)^2 + \sqrt{2}\left(\frac{\Omega_p}{s}\right) + 1}$$

$$\text{i.e.}\quad G(s) = \frac{s^2}{s^2 + \sqrt{2}\Omega_p s + \Omega_p^2}$$

The (Ω_p) radian cutoff frequency of the digital filter is

$\omega_p = 2\pi \times 2000 = 4000\pi$ rad/sec. $F_s = 8000$ Hz

The prewarped value is

$$\Omega_p = \tan\left(\frac{\omega_p T}{2}\right) \quad \Omega_p = \left[\tan\left(\frac{4000\pi \times 1}{2 \times 8000}\right)\right]$$

$\Omega p = 1$ radians per second

Substituting this value in $H(s)$ yields i.e. $G(s) = \frac{s^2}{s^2+\sqrt{2}s+1}$

For the bilinear z-transform, s is replaced with

$$s = \left[\frac{z-1}{z+1}\right]$$

that is

$$H(z) = \frac{\frac{(z-1)^2}{(z+1)^2}}{\left\{\frac{(z-1)^2}{(z+1)^2} + \sqrt{2}\frac{(z-1)}{(z+1)} + 1\right\}}$$

$$H(z) = \frac{(z-1)^2}{\{z^2+2z+1+1.414z^2-1.414+z^2-2z+1)\}}$$

$$H(z) = \frac{z^2-2z+1}{3.414z^2+0.586}$$

$$H(z) = \frac{1-2z^{-1}+z^{-2}}{3.414(1+0.171z^{-2})}$$

$$H(z) = 0.29289\frac{1-2z^{-1}+z^{-2}}{1+0.1715z^{-2}}$$

Example 10.4 (low-pass filter)

The transfer function of a filter is given by

$$H(s) = \frac{5s+1}{s^2+0.4s+1}$$

Design a corresponding discrete-time low-pass resonant filter that resonates at 10 Hz when the sampling frequency is 60 Hz.

Solution 10.4

In this case, the critical frequency is 10 Hz (i.e., the resonant frequency).

Thus, $\omega_P = 2\pi(10)$ radians/second.

With $\omega_p = 2\pi(10)$

$$\Omega_p = \tan\left\{\frac{\omega_p T}{2}\right\} = \tan\left\{\frac{2\pi(10)}{2(60)}\right\} = \tan\frac{\pi}{6} = \frac{1}{\sqrt{3}}$$

$$H'(s) = H(s)|_{s=\sqrt{3}\,s}$$

$$= \frac{5(\sqrt{3})s+1}{3s^2+(\sqrt{3})(0.4)s+1}$$

That is,

$$H'(s) = \frac{8.6603s+1}{3s^2+0.6928s+1}$$

$$= \frac{\left(8.6603\left(\frac{z-1}{z+1}\right)+1\right)(z+1)^2}{3(z-1)^2+0.6928(z-1)(z+1)+(z+1)^2}$$

The transfer function of the resonant filter we seek is given by

$$H(z) = H^{'}(s)|_{s=(z-1)/(z+1)}$$

Evaluating $H^{'}(s)$, we get

$$H(z) = \frac{9.6603 + 2.0z^{-1} - 7.6603z^{-2}}{4.6928 - 4.0z^{-1} + 3.3072z^{-2}}$$

$$= \frac{4.69(2.059 + 0.426z^{-1} - 1.633)z^{-2}}{4.69(1 - 0.85z^{-1} + 0.705z^{-2})}$$

$$H(z) = \frac{2.09 + 0.4262\ z^{-1} - 1.633z^{-2}}{1 - 0.8524z^{-1} + 0.705z^{-2}}$$

$$= 2.09\frac{1 + 0.2039z^{-1} - 0.7813z^{-2}}{1 - 0.8524z^{-1} + 0.705z^{-2}}$$

From the preceding examples, it is apparent that the most tedious part of the design process is that of evaluating $H(z)$ from $H(s)$ via the BZT using $s = (z - 1)/(z + 1)$.

By exploiting some interesting properties of the BZT, it has been shown that the process of evaluating $H(z)$ from $H(s)$ can be carried out by means of an efficient algorithm. For our purposes, it suffices to use a computer program that implements this algorithm. This program accepts the coefficients of a given $H(s)$ and yields the corresponding $H(z)$.

10.7 Design of Filters for Unknown Transfer Function

In cases where the prototype low-pass filter does not exist or the transfer function of the analog filter is not known, the stages of the BZT method are;

(i) Prewarp the band pass or critical frequencies, i.e. $\Omega_{\mathrm{p}}^{\mathrm{P}}$ and $\Omega_{\mathrm{s}}^{\mathrm{P}}$ of the digital filter.

(ii) Find out the value of N from the given specification. Once N is known, the corresponding filters transfer function is written. Find a suitable low-pass prototype analog filter from the classical filter characteristics.

(iii) Using frequency transformation formulae, change the low-pass filter transfer function to the desired filter transfer function. Then apply the BZT to obtain the desired digital filter transfer function, $H(z)$.

The basic concepts for each of the filter types (LP, HP, BP and BS) will be considered here.

10.7.1 Low-Pass Filter – Basic Concepts

The low pass-to-low pass transformation is given by Equation (10.18)

$$s = \frac{s}{\Omega_p} \tag{10.23}$$

If we replace s by jΩ in the equation and denote frequencies for the prototype filter by ω^{P} and those for the low-pass filter to be designed by ω_{LP}, to distinguish between them, then the above equation becomes

$$\mathrm{j}\Omega^{\mathrm{P}} = \mathrm{j}\frac{\Omega_{\mathrm{Lp}}}{\Omega_p}, \quad \text{i.e.}\Omega^{\mathrm{P}} = \frac{\Omega_{\mathrm{Lp}}}{\Omega_p} \tag{10.24}$$

Equation (10.24) defines the relationship between the frequencies in the prototype filter response and those to the denormalized low-pass filter that we wish to design.

Given the critical frequencies for a denormalized low-pass filter, we use Equation (10.3) to determine the critical frequencies for the prototype filter and hence its specifications.

The three key critical frequencies for the prototype filter are: 0, pass-band edge frequency ($\Omega_{\mathrm{p}}^{\mathrm{P}}$). This is, in fact, always 1, and the stop-band edge frequency ($\Omega_{\mathrm{s}}^{\mathrm{P}}$):

(1) $\Omega_{\mathrm{LP}} = 0, \quad \Omega^{P} = 0$
(2) $\Omega_{\mathrm{LP}} = \Omega_p$

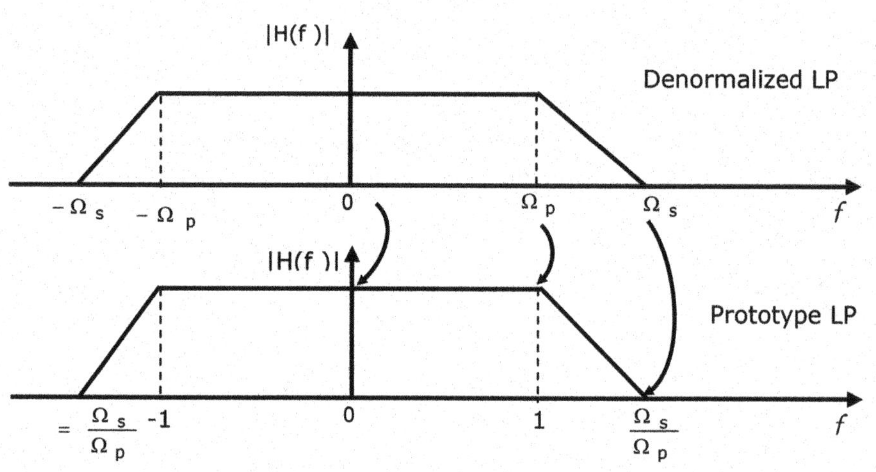

Figure 10.3 Relationships between frequencies in the denormalized LP and prototype LP filters.

(i.e. the pass-band edge frequency), $\Omega^{\mathrm{P}} = \frac{\Omega_{\mathrm{LP}}}{\Omega_p}$ therefore $\Omega_{\mathrm{P}}^{\mathrm{P}} = \frac{\Omega_p}{\Omega_p} = 1$

(3) $\Omega_{\mathrm{LP}} = \Omega_s = \frac{\Omega_s}{\Omega_p} = \Omega_{\mathrm{S}}^{\mathrm{P}}$

Thus, the critical frequencies for the prototype filter are:

$$0,\ 1,\ \frac{\Omega_{\mathrm{s}}}{\Omega_{\mathrm{p}}}$$

The relationships between the frequencies of the denormalized low-pass filter and those of the prototype filter are shown in Figure 10.3.

The order (N) of the Butterworth filter.

10.7.2 The Order (*N*) of the Butterworth Approximation

$$N \geq \frac{\log\left(\dfrac{10^{\frac{A_s}{10}}-1}{10^{\frac{A_p}{10}}-1}\right)}{2\log\left(\frac{\Omega_s^P}{\Omega_p^P}\right)} \tag{10.25}$$

where A_p and A_s are pass-band and stop-band attenuations in dB, $\Omega_{\mathrm{S}}^{\mathrm{P}}$ is the prototype stop-band edge frequency, and $\Omega_{\mathrm{P}}^{\mathrm{P}}$ is the prototype pass-band edge frequency, i.e. cutoff frequency. The subscript is written for pass-band end or stop-band edge frequency or the cutoff frequency in radians/sec; the superscript represents the prototype filters.

$\Omega_{\mathrm{P}}^{\mathrm{P}}$ and $\Omega_{\mathrm{S}}^{\mathrm{P}}$ are calculated for the corresponding desired filters, such as low pass to low pass, low pass to high pass, low pass to band stop, and low pass to band stop.

Example 10.5

A low-pass digital filter with the following specifications is required; assume a Butterworth characteristic for the filter. Determine the order (N) of the prototype high-pass filter.

Pass-band edge frequency: 500 Hz, Stop-band edge frequency: 800 Hz

Pass-band ripple: 3 dB Stop-band attenuation: 20 dB Sampling frequency: 2000 Hz.

Solution 10.5

From the specifications, the prewarped frequencies are

$$\Omega_{\mathrm{p}} = \tan\left\{\frac{2\pi \times 500}{2 \times 1000}\right\} = 1.0 \quad \Omega_s = \tan\left\{\frac{2\pi \times 800}{2 \times 2000}\right\} = 3.077$$

$$\Omega_S^P = \frac{\Omega_s}{\Omega_p} = \frac{3.077}{1} = 3.077$$

Thus, the prewarped pass- and stop-band edge frequencies for HP filter are: 0, 1, and 3.077.

$$10^{A_s/10} - 1 = 10^{20/10} - 1 = 99; \quad 10^{A_P/10} - 1 = 10^{3/10} - 1 = 0.9952;$$

$$\log\left(\frac{99}{0.9952}\right) = 1.9976$$

For the prototype LPF

$$N \geq \frac{\log\left(\frac{10^{\frac{A_s}{10}} - 1}{10^{\frac{A_p}{10}} - 1}\right)}{2\log\left(\frac{\Omega_s^P}{\Omega_p^P}\right)}$$

$$\Omega_P^P = 1.0 \quad \Omega_S^P = 3.077 \quad 2\log\left(\frac{\Omega_S^P}{\Omega_P^P}\right) = 0.967$$

$$N \geq \frac{1.9976}{0.967} = 2.04 \cong N = 2$$

The poles of the prototype filter are from Equation (9.2)

$$s_{P1} = \cos\left[\frac{(2+2-1)\pi}{4}\right] + j\,\sin\left[\frac{(2+2-1)\pi}{4}\right] = -\frac{\sqrt{2}}{2} + j\frac{\sqrt{2}}{2}$$

$$s_{P2} = -\frac{\sqrt{2}}{2} - j\frac{\sqrt{2}}{2}$$

The s-plane transfer function, $H(s)$, is $H(s) = \frac{1}{s^2+\sqrt{2}\,s+1}$

Example 10.6

A low-pass digital filter meeting the following specifications is required; assume a Butterworth characteristic for the filter.

Pass-band frequency	500 Hz
Stop-band frequency	2000 Hz
Pass-band ripple	3 dB
Stop-band attenuation	20 dB
Sampling frequency	8 kHz

Determine the following:

(1) Pass- and stop-band edge frequencies for a suitable analog prototype low-pass filter;
(2) The order (N) of the prototype low-pass filter;
(3) Filter coefficients, and hence the transfer function, of the discrete-time filters using the bilinear z-transformation.

Solution 10.6

From the specifications, the prewarped frequencies are

$$\Omega_p = \tan\left\{\frac{2\pi \times 500}{2 \times 8000}\right\} = 0.198912 \quad \Omega_s = \tan\left\{\frac{2\pi \times 2000}{2 \times 8000}\right\} = 1.0$$

$$\Omega_S^P = \frac{\Omega_s}{\Omega_p} = \frac{1.0}{0.1989} = 5.0273$$

Thus, the prewarped pass- and stop-band edge frequencies for the prototype LP filter are: 0, 1, 5.0273.

Next, Equation (9.3) and the values of the parameters given above are used to determine the order of the filter.

Now,

$$10^{A_s/10} - 1 = 10^{20/10} - 1 = 99; \quad 10^{A_P/10} - 1 = 10^{3/10} - 1 = 0.9952$$
$$\log\left(\tfrac{99}{0.9952}\right) = 1.9976$$

For the prototype LPF

$$N \geq \frac{\log\left(\frac{10^{\frac{A_s}{10}} - 1}{10^{\frac{A_p}{10}} - 1}\right)}{2\log\left(\frac{\Omega_s^P}{\Omega_p^P}\right)}$$

$$\Omega_P^P = 1.0 \quad \Omega_S^P = 5.0273 \quad 2\log\left(\frac{\Omega_S^P}{\Omega_P^P}\right) = 1.4206$$

$$N \geq \frac{1.9976}{1.4026} = 1.424 \cong N = 2$$

The poles of the prototype filter are (from Equation (9.2))

$$s_{P1} = \cos\left[\frac{(2+2-1)\pi}{4}\right] + j\,\sin\left[\frac{(2+2-1)\pi}{4}\right] = -\frac{\sqrt{2}}{2} + j\frac{\sqrt{2}}{2}$$

$$s_{P2} = -\frac{\sqrt{2}}{2} - j\frac{\sqrt{2}}{2}$$

The s-plane transfer function, $H(s)$, is

$$H(s) = \frac{1}{s^2 + \sqrt{2}\,s + 1}$$

The frequency scaled s-plane transfer function is

$$H'(s) = H(s)\Big|_{s=\frac{s}{\Omega_p}} = \frac{1}{\left(\frac{s}{\Omega_p}\right)^2 + \sqrt{2}\left(\frac{s}{\Omega_p}\right) + 1}$$

$$H^{'}(s) = \frac{\Omega_p^2}{s^2 + \sqrt{2}\, s\, \Omega_p + \Omega_p^2}$$

Applying the BZT:

$$H(z) = H^{'}(s)\Big|_{s=\frac{z-1}{z+1}} = \frac{\Omega_p^2}{\left(\frac{z-1}{z+1}\right)^2 + \Omega_p\sqrt{2}\left(\frac{z-1}{z+1}\right) + \Omega_p^2}$$

$$H(z) = \frac{\Omega_p^2(z+1)^2}{(z-1)^2 + \Omega_p\sqrt{2}\,(z-1)(z+1) + \Omega_p^2(z+1)^2}$$

After simplification and dividing top and bottom by z^2, we have

$$H(z) = \frac{\Omega_p^2}{1+\sqrt{2}\ \Omega_p + \Omega_p^2} \times \frac{1 + 2\, z^{-1} + z^{-2}}{1 + \frac{2(\Omega_p^2 - 1)\, z^{-1}}{1+\Omega_p\sqrt{2}+\Omega_p^2} + \frac{(1-\Omega_p\sqrt{2}+\Omega_p^2)\, z^{-2}}{1-\Omega_p\sqrt{2}+\Omega_p^2}}$$

Using the values of the parameters

$$1+\sqrt{2}\ \Omega_p + \Omega_p^2 = 1.3208; \quad \Omega_p^2 - 1 = -0.9604$$
$$1-\sqrt{2}\ \Omega_p + \Omega_p^2 = 0.7582; \quad \Omega_p^2 = 0.0395$$

and substituting in the equation above and simplifying, we have

$$H(z) = 0.02995 \frac{(1 + 2z^{-1} + z^{-2})}{1 - 1.4542z^{-1} + 0.57408z^{-2}}$$

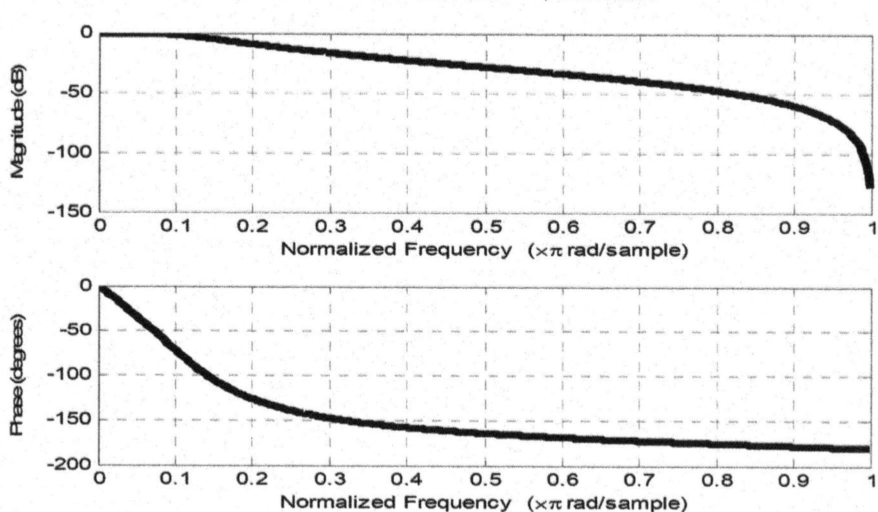

Magnitude and phase responses of low-pass filter.

10.7.3 High-Pass Filter – Basic Concepts

From the low pass-to-high pass transformation, $s = \Omega_p/s$, and denoting the frequencies of the denormalized high-pass filter by ω_{hp} and those of the prototype LP filter by Ω^{P} (as before), the following relationship between the frequencies of the prototype LP filter and the desired high-pass filter is obtained:

$$s = \frac{\Omega_P}{s}; \tag{10.26}$$

$$j\Omega^p = \frac{\Omega_P}{j\Omega_{hp}}, i.e.\ \Omega^p = -\frac{\Omega_P}{\Omega_{hp}}$$

Using Equation (10.19), we can now specify the critical frequencies of the prototype LP filter in terms of those of the desired high-pass filter:

(1) $\Omega_{\mathrm{hP}} = 0,\ \Omega^{\mathrm{P}} = \infty$
(2) $\Omega_{\mathrm{hP}} = \Omega_P$ (i.e. the pass-band edge frequency), $\Omega_{\mathrm{hP}} = -\Omega_P,\ \Omega^{\mathrm{P}}_{\mathrm{p}} = 1$
(3) $\Omega_{\mathrm{hP}} = \Omega_s = -\frac{\Omega_{\mathrm{P}}}{\Omega_{\mathrm{s}}} = \Omega^{\mathrm{P}}_{\mathrm{s}}$
(4) $\Omega_{\mathrm{hP}} = -\Omega_p,\quad \Omega^{\mathrm{P}}_{\mathrm{p}} = 1$
(5) $\Omega_{\mathrm{hP}} = -\Omega_s = \frac{\Omega_{\mathrm{P}}}{\Omega_{\mathrm{s}}} = \Omega^{\mathrm{P}}_{\mathrm{s}}$

Thus, the three key critical frequencies for the prototype low-pass filter for designing the high-pass filter are: 0, 1, $\frac{\Omega_{\mathrm{P}}}{\Omega_{\mathrm{s}}}$, The critical frequencies for the prototype LP filter and their relationships with the frequencies of the denormalized high-pass filter are

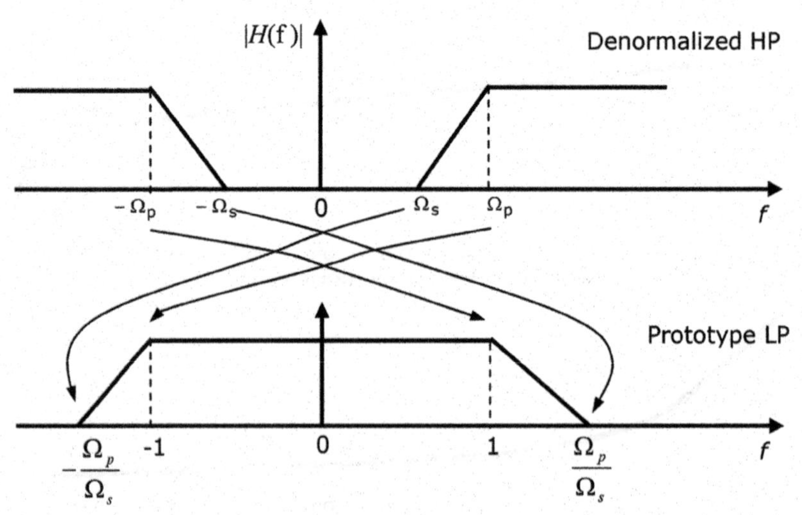

Figure 10.4 Relationships between frequencies in the denormalized HP and prototype LP Filters.

depicted in Figure 10.4. We note that the low pass-to-high pass transformation maps frequencies in the denormalized high-pass filter as follows: it maps zero frequency to infinity, Ω_p to unity, and infinity to zero.

Example 10.7

A high-pass digital filter with the following specifications is required; assume a Butterworth characteristic for the filter. Determine the order (N) of the prototype high-pass filter.

Stop-band frequency 800 Hz Pass-band frequency 3000 Hz Pass-band ripple 3 dB Stop-band attenuation 20 dB Sampling frequency 10000 Hz

Solution 10.7

From the specifications, the prewarped frequencies are

$$\Omega_p = \tan\left\{\frac{2\pi \times 3000}{2 \times 10000}\right\} = 1.3768 \quad \Omega_s = \tan\left\{\frac{2\pi \times 800}{2 \times 10000}\right\} = 0.25675$$

$$\Omega_S^P = \frac{\Omega_p}{\Omega_s} = 5.36$$

Thus, the prewarped pass- and stop-band edge frequencies for HP filter are: 0, 1, 5.36

$$10^{A_s/10} - 1 = 10^{20/10} - 1 = 99; \quad 10^{A_P/10} - 1 = 10^{3/10} - 1 = 0.9952$$

$$\log\left(\frac{99}{0.9952}\right) = 1.9976$$

For the prototype LPF,

$$N \geq \frac{\log\left(\frac{10^{\frac{A_s}{10}} - 1}{10^{\frac{A_p}{10}} - 1}\right)}{2\log\left(\frac{\Omega_s^P}{\Omega_p^P}\right)}$$

$$\Omega_P^P = 1.0 \quad \Omega_S^P = 5.36 \quad 2\log\left(\frac{\Omega_S^P}{\Omega_P^P}\right) = 2\log(5.36) = 1.4584$$

$$N \geq \frac{1.9976}{1.4584} = 1.37 \cong N = 2$$

The poles of the prototype filter are (from (10.2))

$$s_{P1} = \cos\left[\frac{(2+2-1)\pi}{4}\right] + j\,\sin\left[\frac{(2+2-1)\pi}{4}\right] = -\frac{\sqrt{2}}{2} + j\frac{\sqrt{2}}{2}$$

$$s_{P2} = -\frac{\sqrt{2}}{2} - j\frac{\sqrt{2}}{2}$$

The s-plane transfer function $H(s)$ is $H(s) = \dfrac{1}{s^2 + \sqrt{2}\,s + 1}$

Example 10.8

A high-pass digital filter meeting the following specifications is required:

Pass-band frequency	2000 Hz
Stop-band frequency	500 Hz
Pass-band ripple	3 dB
Stop-band attenuation	20 dB
Sampling frequency	8 kHz

Determine the following:

(1) Pass- and stop-band edge frequencies for a suitable analog prototype low-pass filter;
(2) Order (N) of the prototype low-pass filter;
(3) Filters coefficients, and hence the transfer function, of the discrete-time filter using the bilinear z-transformation.

Assume a Butterworth characteristic for the filter.

Solution 10.8

(1) From the specifications, the prewarped frequencies are

$$\Omega_p = \tan\left\{\frac{2\pi\times 2000}{2\times 8000}\right\} = 1.0 \qquad \Omega_s = \tan\left\{\frac{2\pi\times 500}{2\times 8000}\right\} = 0.1989$$
$$\Omega_S^P = \frac{\Omega_p}{\Omega_s} = \frac{1.0}{0.1989} = 5.0273$$

Thus, the pass- and stop-band edge frequencies for the prototype LP filter are: 0, 1, and 5.0273.

(2) Next, we use Equation (10.3) and the values of the parameters above to determine the order of the filter.

Now

$$10^{A_s/10} - 1 = 10^{20/10} - 1 = 99; \quad 10^{A_P/10} - 1 = 10^{3/10} - 1 = 0.9952$$
$$\log\left(\frac{99}{0.9952}\right) = 1.9976$$

For the prototype LPF,

$$N \geq \frac{\log\left(\dfrac{10^{\frac{A_s}{10}} - 1}{10^{\frac{A_p}{10}} - 1}\right)}{2\log\left(\dfrac{\omega_s^p}{\omega_p^p}\right)}$$

$$\Omega_P^P = 1.0 \quad \Omega_S^P = 5.0273 \quad 2\log\left(\frac{\Omega_S^P}{\Omega_P^P}\right) = 1.4206$$

$$N \geq \frac{1.9976}{1.4026} = 1.424 \cong N = 2$$

The poles of the prototype filter are (from Equation (10.2))

$$s_{P1} = \cos\left[\frac{(2+2-1)\pi}{4}\right] + j\ \sin\left[\frac{(2+2-1)\pi}{4}\right] = -\frac{\sqrt{2}}{2} + j\frac{\sqrt{2}}{2}$$

$$s_{P2} = -\frac{\sqrt{2}}{2} - j\frac{\sqrt{2}}{2}$$

The s-plane transfer function, $H(s)$, is

$$H(s) = \frac{1}{(s - s_{P1})(s - s_{P2})} = \frac{1}{s^2 + \sqrt{2}\ s + 1}$$

The frequency scaled s-plane transfer function is

$$H^{'}(s) = H(s)\Big|_{s=\frac{\Omega_p}{s}} = \frac{1}{\left(\frac{\Omega_p}{s}\right)^2 + \sqrt{2}\left(\frac{\Omega_p}{s}\right) + 1}$$

$$H^{'}(s) = \frac{s^2}{s^2 + \sqrt{2}\ s\ \Omega_p + \Omega_p^2}$$

Applying the BZT:

$$H(z) = H^{'}(s)\Big|_{s=\frac{z-1}{z+1}} = \frac{\left(\frac{z-1}{z+1}\right)^2}{\left(\frac{z-1}{z+1}\right)^2 + \Omega_p\sqrt{2}\left(\frac{z-1}{z+1}\right) + \Omega_p^2}$$

$$H(z) = \frac{(z-1)^2}{(z-1)^2 + \Omega_p\sqrt{2}\,(z-1)(z+1) + \Omega_p^2(z+1)^2}$$

$$H(z) = \frac{z^2 - 2z + 1}{z^2 - 2z + 1 + \Omega_p\sqrt{2}\,(z^2 - 1) + \Omega_p^2(z^2 - 2z + 1)}$$

$$H(z) = \frac{z^2 - 2z + 1}{z^2 - 2z + 1 + \Omega_p\sqrt{2}\,z^2 - \Omega_p\sqrt{2} + \Omega_p^2 z^2 - 2\Omega_p^2 z + \Omega_p^2}$$

$$H(z) = \frac{z^2 - 2z + 1}{z^2(1 + \Omega_p\sqrt{2} + \Omega_p^2) - 2(\Omega_p^2 - 1)\ z + 1 - \Omega_p\sqrt{2} + \Omega_p^2}$$

$$H(z) = \frac{z^2 - 2z + 1}{z^2(1 + \Omega_p\sqrt{2} + \Omega_p^2) - 2(\Omega_p^2 - 1)\ z + 1 - \Omega_p\sqrt{2} + \Omega_p^2}$$

After simplification and dividing top and bottom by z^2, we have

$$H(z) = \frac{\dfrac{1 - 2\ z^{-1} + z^{-2}}{1 + \sqrt{2}\,\Omega_p + \Omega_p^2}}{1 + \dfrac{2(\Omega_p^2 - 1)\ z^{-1}}{1 + \Omega_p\sqrt{2} + \Omega_p^2} + \dfrac{(1 - \Omega_p\sqrt{2} + \Omega_p^2)\ z^{-2}}{(1 + \Omega_p\sqrt{2} + \Omega_p^2}}$$

Using the values of the parameters

$$1+\sqrt{2}\ \Omega_p+\Omega_p^2=3.4142; \quad \Omega_p^2-1=0$$
$$1-\sqrt{2}\ \Omega_p+\Omega_p^2=0.5857; \quad \Omega_p^2=1$$

and substituting in the Equation above and simplifying, we have

$$H(z) = \frac{\frac{1-2\,z^{-1}+z^{-2}}{3.1412}}{1+\frac{2(\Omega_p^2-1)\,z^{-1}}{0.5867}+\frac{0.5867\,z^{-2}}{3.1412}} = \frac{0.3183(1-2\,z^{-1}+z^{-2})}{1+\frac{2(1-1)\,z^{-1}}{0.5867}+0.1867\,z^{-2}}$$

$$H(z)=0.3183\frac{(1-2z^{-1}+z^{-2})}{1+0.1867z^{-2}}$$

10.7.4 Band-Pass Filters – Basic Concepts

The low pass-to-band pass transformation is given by

$$s=\frac{s^2+\Omega_o^2}{Ws} \tag{10.27}$$

From the low pass-to-band pass transformation, the frequencies of the band-pass filter ω_{hp}, and those of the prototype LPF, ω^{p}, are related as

$$j\Omega^{\mathrm{p}}=\frac{(j\Omega_{bp})^2+\Omega_o^2}{jW\Omega_{bp}} \tag{10.28}$$

$$\Omega^{\mathrm{p}}=\frac{\Omega_{bp}^2-\Omega_o^2}{W\Omega_{bp}} \tag{10.29}$$

Now, a band-pass filter has four critical or band-edge frequencies and a centre frequency:

Ω_{p1},Ω_{p2} = lower and upper pass-band edge frequencies
Ω_{s1},Ω_{s2} = lower and upper stop-band edge frequencies
Centre frequency $\Omega_0^2=\Omega_{p1}\,\Omega_{p2}$

The band edge frequencies for the prototype LP filter can be found in terms of the band edge frequencies for the band-pass filter:

(1) $\Omega_{bp}=\Omega_{s1}, \quad \Omega^{\mathrm{p}}=\Omega_{s1}^{p}=\frac{\Omega_{s1}^2-\Omega_o^2}{W\Omega_{s1}}$

(2) $\Omega_{bp}=\Omega_{p1}, \quad \Omega_{p1}=\Omega_{p1}^{p}=\frac{\Omega_{p1}^2-\Omega_o^2}{W\Omega_{p1}}=\frac{\Omega_{p1}^2-\Omega_{p1}\Omega_{p2}}{(\Omega_{p2}-\Omega_{p1})\Omega_{p1}}=-1$

(3) $\Omega_{bp}=\Omega_{p2}, \quad \Omega_{p2}=\Omega_{p2}^{p}=\frac{\Omega_{p2}^2-\Omega_o^2}{W\Omega_{p2}}=\frac{\Omega_{p2}^2-\Omega_{p1}\Omega_{p2}}{(\Omega_{p2}-\Omega_{p1})\Omega_{p2}}=1$

(4) $\Omega_{bp}=\Omega_{s2}, \quad \Omega^{\mathrm{p}}=\Omega_{s2}^{p}=\frac{\Omega_{s2}^2-\Omega_o^2}{W\Omega_{s2}}$

(5) $\Omega_{bp}=\Omega_0, \quad \Omega^{\mathrm{p}}=\frac{\Omega_0^2-\Omega_o^2}{W\Omega_0^2}=0$

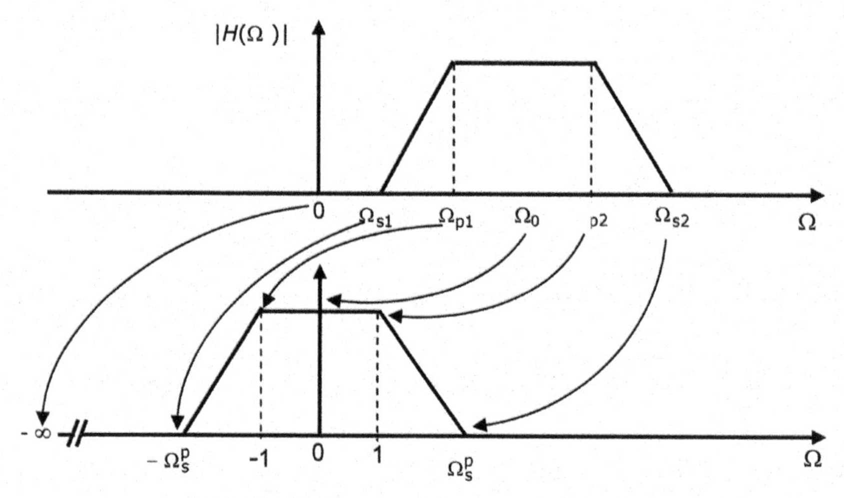

Figure 10.5 Mapping of the prototype LP to BPF.

(6) $\Omega_{\mathrm{s}}^{\mathrm{P}} = \min(|\Omega_{\mathrm{s1}}^{\mathrm{P}}|, |\Omega_{\mathrm{s2}}^{\mathrm{P}}|)$

$$\Omega_p^P = 1$$

$$\Omega_{s1}^P = \frac{\Omega_{s1}^2 - \Omega_o^2}{W\Omega_{s1}} = \frac{\Omega_{s1}^2 - \Omega_o^2}{(\Omega_{p2} - \Omega_{p1})\Omega_{s1}}$$

$$\Omega_{s2}^P = \frac{\Omega_{s2}^2 - \Omega_o^2}{W\Omega_{s2}} = \frac{\Omega_{s2}^2 - \Omega_o^2}{(\Omega_{p2} - \Omega_{p1})\Omega_{s2}}$$

Thus, the critical frequencies of interest for the prototype LP filter are

$$0, 1, \Omega_{\mathrm{s}}^{\mathrm{P}} = \min(|\Omega_{\mathrm{s1}}^{\mathrm{P}}|, |\Omega_{\mathrm{s2}}^{\mathrm{P}}|)$$

The mappings of the frequencies between the band-pass filter and the prototype LP filter are depicted in Figure 10.5. We note, for example, that the centre frequency of the band-pass filter is mapped to zero in the prototype filter, and the upper pass-and and stop-band edge frequencies, Ω_{p2} and Ω_{s2}, respectively, are mapped to the positive pass-band and stop-band edge frequencies in the prototype filter.

On the other hand, the lower pass-band and stop-band edge frequencies, $\Omega_{\mathrm{p1}}$and Ω_{s1}, respectively, are mapped to the negative pass-band and stop-band edge frequencies in the prototype filter.

Example 10.9

A requirement exists for a band-pass digital filter, with a Butterworth magnitude-frequency response, that satisfies the following specification:

Lower pass-band edge frequency 200 Hz
Upper pass-band edge frequency 300 Hz

Lower stop-band edge frequency	50 Hz
Upper stop-band edge frequency	450 Hz
Pass-band ripple	3 dB
Stop-band attenuation	20 dB
Sampling frequency	1 kHz

Determine the following:

(1) Pass- and stop-band edge frequencies of a suitable prototype low-pass filter;
(2) Order, N, of the prototype low-pass filter;
(3) Filter coefficients, and hence the transfer function, of the discrete-time filter using the BZT method.

Solution 10.9

The prewarped critical frequencies for the band-pass filter are

$$\Omega_{p1} = \tan\left\{\frac{2\pi \times 200}{2 \times 1000}\right\} = 0.7265 \qquad \Omega_{p2} = \tan\left\{\frac{2\pi \times 300}{2 \times 1000}\right\} = 1.3763$$

$$\Omega_{s1} = \tan\left\{\frac{2\pi \times 50}{2 \times 1000}\right\} = 0.1584 \qquad \Omega_{s2} = \tan\left\{\frac{2\pi \times 450}{2 \times 1000}\right\} = 6.3138$$

$$\Omega_0 = \sqrt{\Omega_{p2}.\Omega_{p1}} = \sqrt{(1.37638)(0.7265)} = 1$$

$$W = \Omega_{p2}. - \Omega_{p1} = 1.37638 - 0.7265 = 0.6498$$

Thus, the band edge frequencies for the prototype LP filter are (using the relationships above)

$$\Omega^p_{s1} = \frac{\Omega^2_{s1} - \Omega^2_o}{W\Omega_{s1}} = \frac{\Omega^2_{s1} - \Omega^2_o}{(\Omega_{p2} - \Omega_{p1})\Omega_{s1}}$$

$$= \frac{(0.1584)^2 - 1}{(1.3763 - 0.7265)(0.1584)} = \frac{-0.9749}{(0.6498)(0.1584)} = \frac{-0.9749}{0.1029} = -9.4705$$

$$\Omega^p_{s2} = \frac{\Omega^2_{s2} - \Omega^2_o}{W\Omega_{s2}} = \frac{\Omega^2_{s2} - \Omega^2_o}{(\Omega_{p2} - \Omega_{p1})\Omega_{s2}}$$

$$= \frac{(6.3138)^2 - 1}{(1.3763 - 0.7265)(6.3138)} = \frac{39.864 - 1}{(0.6498)(6.3138)} = \frac{38.864}{4.1027} = 9.4727$$

$$\Omega^p_p = 1, \Omega^P_s = \min(|\Omega^P_{s1}|, |\Omega^P_{s2}|)$$

$$\Omega^P_s = 9.4705$$

Thus, we require a prototype LPF with $\Omega^p_p = 1, \Omega^p_s = 9.4705$, $A_p = 3$ dB; $A_s = 20$ dB. From Equation (10.3), the order of the prototype LPF is obtained as

Now

$$10^{A_s/10} - 1 = 10^{20/10} - 1 = 99;$$

$$10^{A_P/10} - 1 = 10^{3/10} - 1 = 0.9952 \quad \log\left(\frac{99}{0.9952}\right) = 1.9976$$

For the prototype LPF

$$N \geq \frac{\log\left(\frac{10^{\frac{A_s}{10}}-1}{10^{\frac{A_p}{10}}-1}\right)}{2\log\left(\frac{\Omega_s^P}{\Omega_p^P}\right)}$$

$$\Omega_P^P = 1.0 \quad \Omega_S^P = 9.4705 \quad 2\log\left(\frac{\Omega_S^P}{\Omega_P^P}\right) = 2\log(9.4705) = 1.9527$$

$$N \geq \frac{1.997697}{1.95279} = 1.0229 \cong N = 1$$

N must be an integer, and this time for simplicity we will use N = 1. The s-plane transfer function for a first-order prototype LP filter is given by

$$H(s) = \frac{1}{s+1}$$

The frequency scaled s-plane transfer function is

$$H^{'}(s) = H(s)\bigg|_{s=\frac{s^2+\Omega_o^2}{W\,\Omega}} = \frac{1}{\left(\frac{s^2+\Omega_o^2}{Ws}\right)+1} = \frac{Ws}{s^2+Ws\ +\Omega_0^2}$$

Applying the BZT:

$$H(z) = H^{'}(s)\bigg|_{s=\frac{z-1}{z+1}} = \frac{W\left(\frac{z-1}{z+1}\right)}{\left(\frac{z-1}{z+1}\right)^2 + W\left(\frac{z-1}{z+1}\right) + \Omega_0^2}$$

$$= \frac{(z-1)(z+1)}{(z-1)^2 + W(z-1)(z+1) + \Omega_0^2(z+1)^2}$$

$$H(z) = H^{'}(s)\bigg|_{s=\frac{z-1}{z+1}} = \frac{W(\frac{z-1}{z+1})}{(\frac{z-1}{z+1})^2+W(\frac{z-1}{z+1})+\Omega_0^2}$$

$$H(z) = \frac{W(z-1)(z+1)}{(z-1)^2+W(z-1)(z+1)+\Omega_0^2(z+1)^2}$$

After simplification, and substituting the values for Ω_0^2 and W,

$$H(z) = \frac{0.6498(z-1)(z+1)}{(z-1)^2 + 0.6498(z-1)(z+1) + 1(z+1)^2}$$

$$H(z) = \frac{0.6498(z^2-1)}{(z^2-2z+1) + 0.6498(z^2-1) + (z^2+2z+1)}$$

$$H(z) = \frac{0.6498(z^2 - 1)}{2.6498z^2 + 1.3502} \qquad H(z) = \frac{0.6498(z^2 - 1)}{2.6498(z^2 + 0.5095)}$$

$$H(z) = 0.2452\frac{(z^2 - 1)}{(z^2 + 0.5095)} \qquad H(z) = \frac{0.2452(1 - z^{-2})}{1 + 0.5095\, z^{-2}}$$

10.7.5 Band-Stop Filters – Basic Concepts

The low pass-to-band stop transformation is given by

$$s = \frac{Ws}{s^2 + \Omega_o^2} \tag{10.30}$$

The band-stop frequency, Ω_{bs}, and those of the prototype LPF, Ω^p, are related as

$$j\Omega^p = \frac{jW\Omega_{bs}}{(j\Omega_{bs})^2 + \Omega_o^2} \tag{10.31}$$

$$\Omega^p = \frac{W\Omega_{bs}}{\Omega_o^2 - \Omega_{bs}^2} \tag{10.32}$$

From the relationship, we can determine the band edge frequencies for the prototype LP filter from those of the desired digital band-stop filter. We recall that a band-stop filter has four band edge frequencies $-\Omega_{P1}$, Ω_{P2} (lower and upper pass-band edge frequencies), Ω_{s1}, Ω_{s2} (lower and upper stop-band edge frequencies), and a centre frequency $(\Omega_0)\Omega_0^2 = \Omega_{P1}.\Omega_{P2}$

(1) $\Omega_{bs} = \Omega_{p1}, \quad \Omega^p = \Omega_{p1}^p = \frac{W\Omega_{p1}}{(\Omega_0^2 - \Omega_{p1}^2)} = \frac{(\Omega_{p2} - \Omega_{p1})\Omega_{p1}}{(\Omega_{p2}.\Omega_{p1} - \Omega_{p1}^2)} = 1$

(2) $\Omega_{bs} = \Omega_{s1}, \quad \Omega^p = \Omega_s^{p(1)} = \frac{W\Omega_{s1}}{\Omega_o^2 - \Omega_{s1}^2}$

(3) $\Omega_{bs} = \Omega_{s2}, \quad \Omega^p = \Omega_s^{p(2)} = \frac{W\Omega_{s2}}{\Omega_o^2 - \Omega_{s2}^2}$

(4) $\Omega_{bs} = \Omega_{p2}, \quad \Omega^p = \Omega_{p2}^p = \frac{W\Omega_{p2}}{\Omega_o^2 - \Omega_{p2}^2} = \frac{(\Omega_{p2} - \Omega_{p1})\Omega_{p2}}{(\Omega_{p1}.\Omega_{p2} - \Omega_{p2}^2)} = -1$

(5) $\Omega_{bp} = \Omega_0, \quad \Omega^p = \frac{W\Omega_o}{\Omega_0^2 - \Omega_0^2} = \infty$

(6) $\Omega_S^p = \min\left(\left|\Omega_s^{p(1)}\right|, \left|\Omega_s^{p(2)}\right|\right)\Omega_p^p = 1$

$$\Omega_s^{p(1)} = \frac{W\Omega_{s1}}{\Omega_o^2 - \Omega_{s1}^2} = \frac{(\Omega_{p2} - \Omega_{p1})\Omega_{s1}}{\Omega_o^2 - \Omega_{s1}^2} \qquad \Omega_s^{p(2)} = \frac{W\Omega_{s2}'}{\Omega_o^2 - \Omega_{s2}^2} = \frac{(\Omega_{p2}' - \Omega_{p1}')\Omega_{s2}'}{\Omega_o^2 - \Omega_{s2}'^2}$$

Thus, the stop-band edge frequency for the prototype LP filter, $\Omega_S^p = \min\left(\left|\Omega_s^{p(1)}\right|, \left|\Omega_s^{p(2)}\right|\right)$, and its pass-band edger frequency is 1. The pass-band ripple and stop-band attenuations are, respectively, A_p and A_s. The mapping between the frequencies of the band-stop filter and those of the prototype low-pass filter is shown in Figure 10.6. We see that the upper band-stop and band-pass edge frequencies in the band-stop filter are mapped to the negative frequencies in the prototype filter,

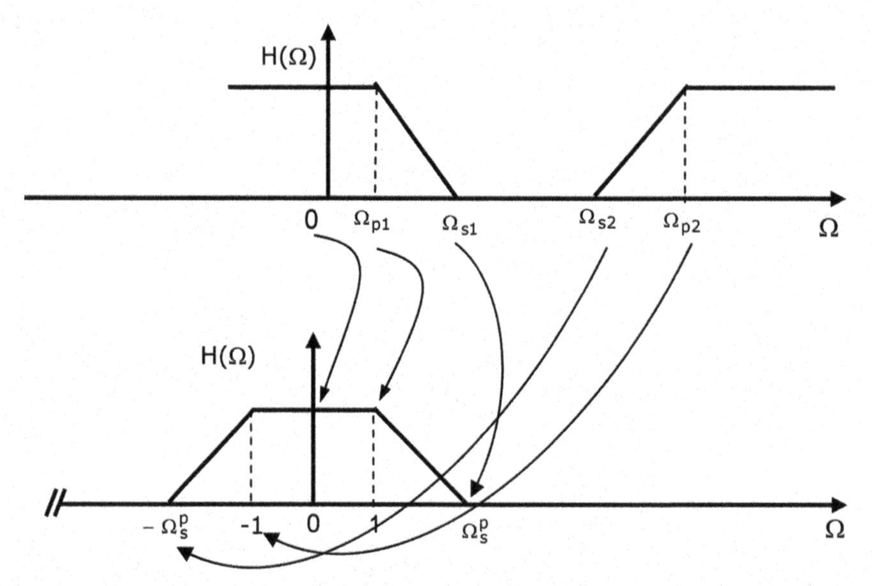

Figure 10.6 Relationship between the frequencies in the denormalized BS and prototype LP filters.

whereas the lower pass-band and stop-band edge frequencies are mapped to the positive frequencies in the prototype filter.

The critical frequencies of interest for the prototype LP filter are

$$0, 1, \ \min\left(\left|\Omega_s^{p(1)}\right|, \left|\Omega_s^{p(2)}\right|\right)$$

From the specifications for the prototype LP filter, we can determine the order, N.

Example 10.10

A requirement exists for a band-stop digital IIR filter, with a Butterworth magnitude-frequency response, that meets the following specifications:

Pass-band frequency	50–450 Hz
Stop-band frequency	200–300 Hz
Pass-band ripple	3 dB
Stop-band attenuation	20 dB
Sampling frequency	1 kHz

Determine the following:

(1) Pass- and stop-band edge frequencies of a suitable prototype low-pass filter;
(2) Order, N, of the prototype low-pass filter;
(3) Filter coefficients, and hence the transfer function, of the discrete-time filter using the BZT method.

Solution 10.10

The prewarped critical frequencies for the band-pass filter are

$$\Omega_{p1} = \tan\left(\frac{\omega_{p1}T}{2}\right) = \tan\left(\frac{2\,\pi\,50}{(2)(1000)}\right) = 0.15838$$

$$\Omega_{p2} = \tan\left(\frac{\omega_{p2}T}{2}\right) = \tan\left(\frac{2\,\pi\,450}{(2)(1000)}\right) = 6.3137$$

$$\Omega_{s1} = \tan\left(\frac{\omega_{s1}T}{2}\right) = \tan\left(\frac{2\,\pi\,200}{(2)(1000)}\right) = 0.7265$$

$$\Omega_{s2} = \tan\left(\frac{\omega_{s2}T}{2}\right) = \tan\left(\frac{2\,\pi\,300}{(2)(1000)}\right) = 1.37638$$

$$\omega_0 = \sqrt{\omega'_{P2}\Omega} = \sqrt{(6.3138)(0.1584)} = 1$$

$$W = \Omega_{p2} - \Omega_{p1} = 6.3138 - 0.1584 = 6.1554$$

$$\begin{aligned}\omega_s^{'p(1)} &= \frac{W\omega'_{s1}}{\omega_o^2 - \omega_{s1}^{'2}} = \frac{(\omega'_{p2}-\omega'_{p1})\omega'_{s1}}{\omega_o^2 - \omega_{s1}^{'2}}\\ &= \frac{(6.3137-0.15838)0.7265}{(0.7265)^2-1} = \frac{(6.15532)(0.7265)}{-0.47219} = -9.4709\end{aligned}$$

$$\begin{aligned}\omega_s^{'p(2)} &= \frac{W\omega'_{s2}}{\omega_o^2 - \omega_{s2}^{'2}} = \frac{(\omega'_{p2}-\omega'_{p1})\omega'_{s2}}{\omega_o^2 - \omega_{s2}^{'2}}\\ &= \frac{(6.3137-0.15838)1.37638}{(1.37638)^2-1} = \frac{(8.47205)}{0.8944} = 9.47209\end{aligned}$$

$$\Omega_s^p = \min\left(\left|\Omega_s^{p(1)}\right|, \left|\Omega_s^{p(2)}\right|\right)$$

Thus, the band edge frequencies for the prototype LP filter are (using the relationships above)

$$\Omega_p^p = 1;\ \Omega_s^p = 9.47$$

We require a prototype LPF with $\Omega_p^p = 1$; $\Omega_s^p = 9.471$, $A_p = 3$ dB; $A_s = 20$ dB. From Equation (9.3), the order of the prototype LPF is obtained as $10^{A_s/10} - 1 = 10^{20/10} - 1 = 99$; $10^{A_P/10} - 1 = 10^{3/10} - 1 = 0.9952$ $\log\left(\frac{99}{0.9952}\right) = 1.9976$. For the prototype LPF

$$N \geq \frac{\log\left(\dfrac{10^{\frac{A_s}{10}}-1}{10^{\frac{A_p}{10}}-1}\right)}{2\log\left(\frac{\Omega_s^p}{\Omega_p^p}\right)}$$

$$\Omega_P^P = 1.0 \quad \Omega_S^P = 9.4705 \quad 2\log\left(\frac{\Omega_S^P}{\Omega_P^P}\right) = 2\log(9.4705) = 1.9527$$

$$N \geq \frac{1.997697}{1.95279} = 1.0229 \cong N = 1$$

N must be an integer, and for simplicity we will use $N = 1$. The s-plane transfer function for a first-order prototype LP filter is given by

$$H(s) = \frac{1}{s+1}$$

Using the low pass-to-band stop transformation for the table, we obtain

The frequency scaled s-plane transfer function is

$$H^{'}(s) = H(s)\Big|_{s=\frac{Ws}{s^2+\omega_o^2}} = \frac{1}{\left(\frac{Ws}{s^2+\omega_o^2}\right)+1}$$

$$H^{'}(s) = \frac{s^2+\Omega_0^2}{s^2+Ws+\Omega_0^2}$$

Applying the BZT:

$$H(z) = H^{'}(s)\Big|_{s=\frac{z-1}{z+1}} = \frac{\left(\frac{z-1}{z+1}\right)^2+\Omega_0^2}{\left(\frac{z-1}{z+1}\right)^2+W\left(\frac{z-1}{z+1}\right)+\Omega_0^2}$$

$$H(z) = \frac{(z-1)^2+\Omega_0^2(z+1)^2}{(z-1)^2+W(z-1)(z+1)+\Omega_0^2(z+1)^2}$$

$$H(z) = \frac{(z-1)^2+(z+1)^2}{(z-1)^2+W(z-1)(z+1)+(z+1)^2}$$

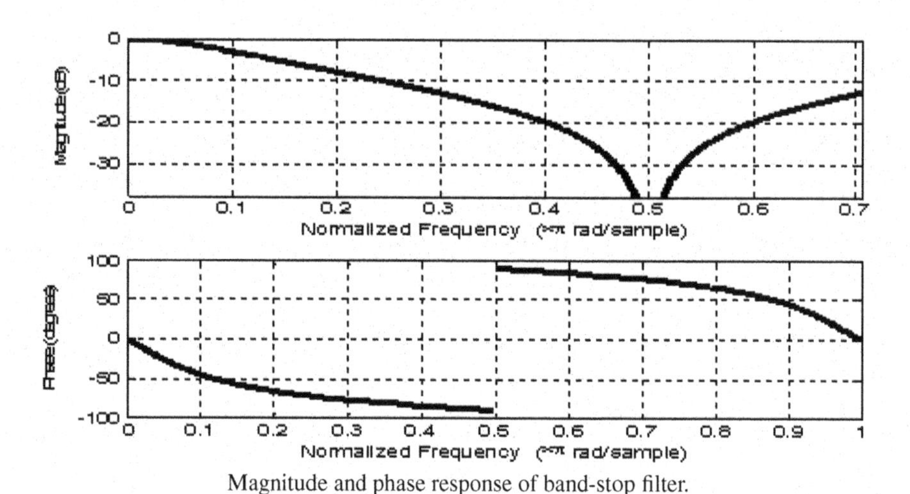

Magnitude and phase response of band-stop filter.

$$H(z) = \frac{(z^2 - 2z + 1) + (z^2 + 2z + 1)}{(z-1)^2 + W(z-1)(z+1) + (z+1)^2}$$

After simplification, and substituting the values for Ω_0^2 and W, we have

$$H(z) = \frac{2z^2 + 2}{(z^2 - 2z + 1) + 6.1554(z^2 - 1) + (z^2 + 2z + 1)}$$

$$H(z) = \frac{2(z^2 + 1)}{8.1554(z^2 - 0.5095)} = \frac{0.2452\,(z^2 + 1)}{(z^2 - 0.5095)} = 0.2452\frac{(1 + z^{-2})}{1 - 0.5095z^{-2}}$$

Example 10.11

Obtain the transfer function of a low-pass digital filter meeting the following specifications and Butterworth characteristic:

Pass-band frequency	0–60 Hz
Stop-band frequency	>85 Hz
Stop-band attenuation	>20 dB
Sampling frequency	256 Hz

Solution 10.11

This example illustrates how two BZT process transformations can be combined into one for computational efficiency, as suggested by Equation

$$s = \cot\left(\frac{\omega_p T}{2}\right)\left[\frac{z-1}{z+1}\right]$$

(1) The critical frequencies for the digital filter are

$$\omega_1 T = \left(\frac{2\,\pi\, f_1}{F_s}\right) = \left(\frac{2\,\pi\, 60}{256}\right) = 2\,\pi(0.2344)$$

$$\omega_2 T = \left(\frac{2\,\pi\, f_2}{F_s}\right) = \left(\frac{2\,\pi\, 85}{256}\right) = 2\,\pi(0.3320)$$

(2) The prewarped equivalent analog frequencies are

$$\Omega_1 = \tan\left(\frac{\omega_1 T}{2}\right) = 0.9063$$
$$\Omega_2 = \tan\left(\frac{\omega_2 T}{2}\right) = 1.7158$$

(3) Next, we need to obtain $H(s)$ with Butterworth characteristics, a 3-dB cutoff frequency of 0.9063, and a response at 85 Hz that is down by 15 dB. For an attenuation of 15 dB and a pass-band ripple of 3 dB, from Equation (8.3), $N = 2.68$. We use $N = 3$, since it must be an integer. A normalized third-order filter is given by

$$H(s) = \frac{1}{(s+1)(s^2+s+1)} = H_1(s)H_2(s)$$

$$\cot\left(\frac{\Omega_1 T}{2}\right) = \cot\left(\frac{2\ \pi\ 0.2344}{2}\right) = 1.1031$$

Performing the transformation in two stages, one for each of the factors of $H(s)$ above, we obtain

$$H_2(z) = H_2(s)\Big|_{s=\cot\left(\frac{\omega_1 T}{2}\right)\left[\frac{z-1}{z+1}\right]}$$
$$= 0.3012\frac{1+2z^{-1}+z^{-2}}{1-0.1307z^{-1}+0.3355z^{-2}}$$

which we have arrived at after considerable manipulation. Similarly, we obtain $H_1(z)$ as

$$H_1(z) = H_1(s)\Big|_{s=\cot\left(\frac{\omega_1 T}{2}\right)\left[\frac{z-1}{z+1}\right]}$$
$$H_1(z) = 0.4754\frac{(1+z^{-1})}{1-0.049\ z^{-1}}$$

$H_1(z)$ and $H_2(z)$ may then be combined to give that desired transfer function $H(z)$,

$$H(z) = H_1(z)H_2(z) = 0.1432\frac{1+3z^{-1}+3z^{-2}+z^{-3}}{1-0.1801z^{-1}+0.3419z^{-2}-0.0165z^{-3}}$$

Magnitude and phase response of filter.

10.8 Pole-Zero Placement Method

When a zero is placed at a given point on the z-plane, the frequency response will be zero at the corresponding point. A pole on the other hand produces a peak at the corresponding frequency point. Poles that are close to the unit circle give rise to large peaks, whereas zeros close to or on the circle produce a trough or minimum peak.

An approximate relationship between r (distance of the pole location), for $r > 0.9$, and bandwidth (bw) is given by:

$$r \approx 1 - (\text{bw}/F_s)\pi \tag{10.33}$$

Example 10.12

A band-pass digital filter is required to meet the following specifications:

Complete signal rejection at dc (zero Hz) and 250 Hz;
Narrow pass-band centered at 125 Hz;
A 3-dB bandwidth of 10 Hz;
Sampling frequency of 500 Hz

Obtain the transfer function of the filter, by suitably placing z-plane poles and zeros, and its difference equations.

Solution 10.12

Complete rejection is required at 0 and 250 Hz, which gives the information that where to place the poles and zeros on the z-plane. These are at angles of 0° and $360° \times 250/500 = 180°$ on the unit circle.

The pass band centered at 125 Hz gives the information to place poles at

$$\pm 360° \times 125/500 = \pm 90°.$$

To ensure that the coefficients are real, it is necessary to have a complex conjugate pole pair. The radius, r, of the poles is determined by the desired bandwidth. An approximate relationship between r, for $r > 0.9$, and bandwidth (bw) is given by

$$r \approx 1 - (\text{bw}/F_s)\pi$$

[bw = 10 Hz and $F_s = 500$ Hz, giving $r = 1 - (10/500)\pi = 0.937$]

The pole-zero diagrams are shown below. From the pole-zero diagrams, the transfer function can be written down by inspection:

$$H(z) = \frac{(z-1)(z+1)}{(z - re^{j\pi/2})(z - re^{-j\pi/2})}$$

$$= \frac{z^2 - 1}{z^2 + 0.877969} = \frac{1 - z^{-2}}{1 + 0.877969z^{-2}}$$

The difference equation is

$$y(n) = -0.877969\, y(n-2) + x(n) - x(n-2)$$

Comparing the transfer function, $H(z)$, with the general IIR, the following coefficients have been found out.

$$a_0 = 1 \qquad b_1 = 0$$
$$a_1 = 0, a_2 = -1 \quad b_2 = 0.877969$$

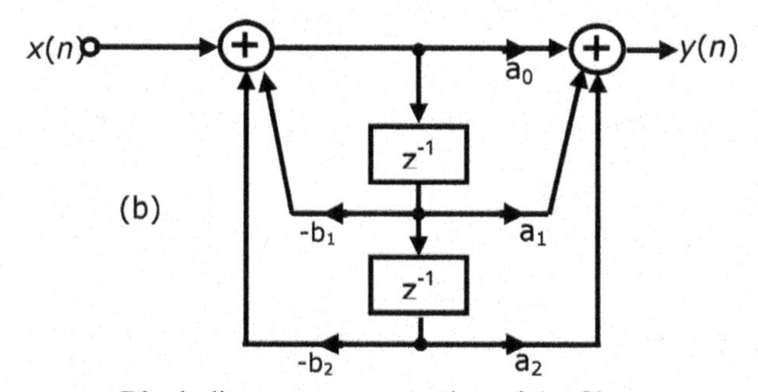

Block diagram representation of the filter.

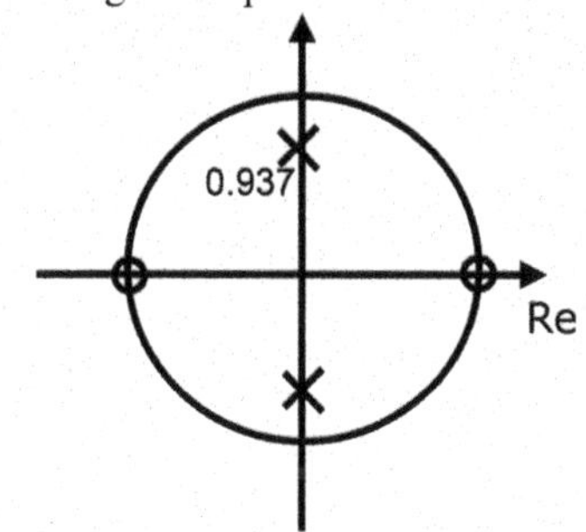

(a) Pole zero diagram.

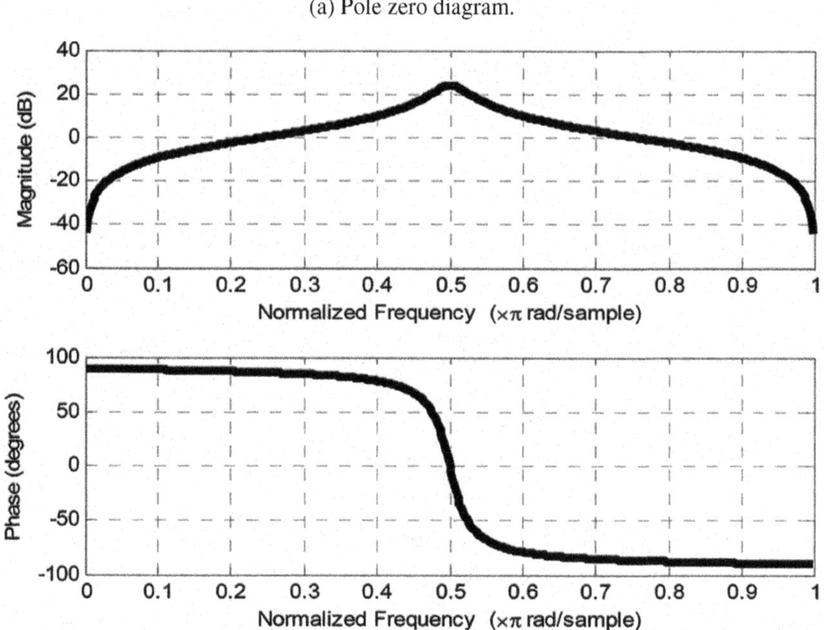

(b) Corresponding magnitude and phase frequency response of a band-pass filter.

Example 10.13

Using the pole-zero placement method, calculate the coefficients of a notch filter. Obtain transfer function and the difference equation of a simple digital notch filter that meets the following specification:

Notch frequency	50 Hz
3-dB width of notch	± 5 Hz
Sampling frequency	500 Hz

Solution 10.13

To reject the component at 50 Hz, we place a pair of complex zeros at points on the unit circle corresponding to 50 Hz, that is at angles of $360° \times 50/500 = \pm 36°$.

To achieve a sharp notch filter and improved amplitude response on either side of the notch frequency, a pair of complex conjugate poles are placed at a radius $r < 1$. The width of the notch is determined by the locations of the poles. The relationship between the bandwidth and the radius of the poles is given by $r \approx 1 - (\mathrm{bw}/F_s)\pi$

[bw = 10 Hz and $F_s = 500$ Hz, giving $r = (1 - 10/500)\,\pi = 0.937$]

The pole-zero diagrams shown below. From the figure the transfer function of the filter is given by

$$H(z) = \frac{[z - e^{-j36°}][z - e^{j36°}]}{[z - 0.937\, e^{-j36°})][z - 0.937\, e^{j36°}]}$$

$$= \frac{z^2 - 1.6180z + 1}{z^2 - 1.874z + 0.8779} = \frac{1 - 1.6180z^{-1} + z^{-2}}{1 - 1.874z^{-1} + 0.8779z^{-2}}$$

The difference equation is

$$y(n) = x(n) - 1.6180\, x(n-1) + x(n-2) + 1.874\, y(n-1) - 0.8779\, y(n-2)$$

Comparing $H(z)$ with Equation (8.16) shows that the coefficients for the notch filter are

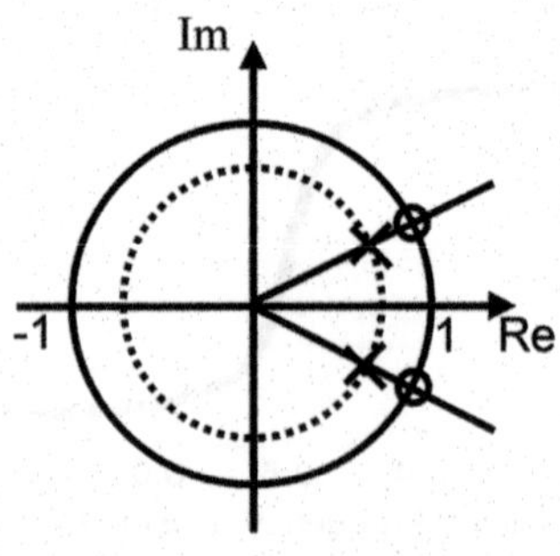

(a) Pole zero diagram.

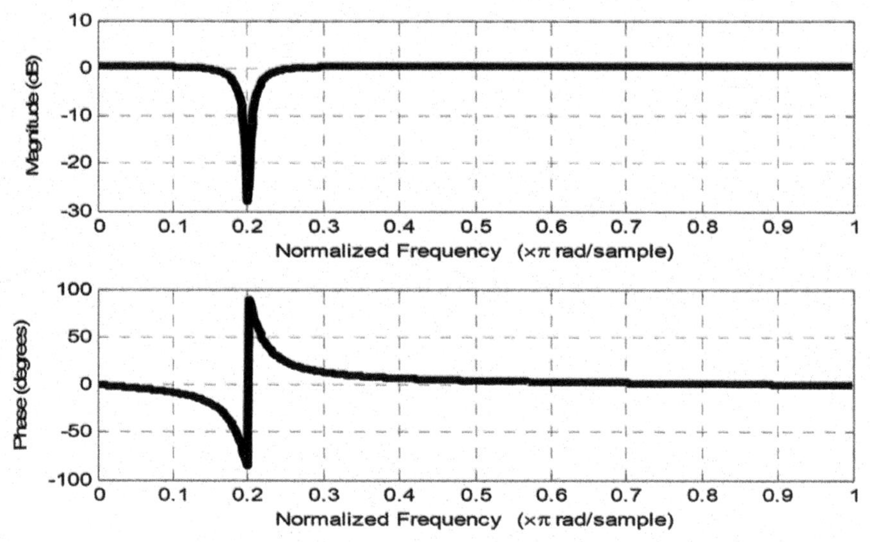

(b) Corresponding magnitude and phase frequency response.

$$
\begin{array}{ll}
a_0 = 1 & b_1 = -1.874 \\
a_1 = -1.6180 & b_2 = 0.8779 \\
a_2 = 1 &
\end{array}
$$

10.9 Impulse Invariant Method

In this method, starting with a suitable analog transfer function, $H(s)$, the impulse response, $h(t)$, is obtained using the Laplace transform. The $h(t)$ so obtained is suitably sampled to produce $h(nT)$, and the desired transfer function $H(z)$, is obtained by Z-transforming $h(nT)$, where T is the sampling interval.

To apply the impulse invariant method to a high-order, IIR filter with simple poles, the transfer function, $H(s)$, is first expanded using partial fractions as the sum of single-pole filters:

$$
\begin{aligned}
H(s) &= \frac{C_1}{s-p_1} + \frac{C_2}{s-p_2} + \ldots. + \frac{C_M}{s-p_M} \qquad (10.34) \\
&= \sum_{K=1}^{M} \frac{C_K}{s-p_K}
\end{aligned}
$$

High-order IIR filters are normally realized as cascades or parallel combinations of standard second-order filter sections. Thus, the case when $M = 2$ is of particular interest. In this case, the transform of Equation (10.34) becomes

$$\frac{C_1}{s-p_1}+\frac{C_2}{s-p_2}\rightarrow\frac{C_1}{1-e^{p_1T}z^{-1}}+\frac{C_2}{1-e^{p_2T}z^{-1}} \tag{10.35}$$

$$H(s)=\frac{C_1(1-e^{p_2T}z^{-1})+C_2(1-C_2e^{p_1T}z^{-1})}{1-(e^{p_1T}+e^{p_2T})z^{-1}+e^{(p_1+p_2)T}z^{-2}}$$

$$H(s)=\frac{(C_1+C_2)-(C_1e^{p_2T}+C_2e^{p_1T})z^{-1}}{1-(e^{p_1T}+e^{p_2T})z^{-1}+e^{(p_1+p_2)T}z^{-2}}$$

If the poles, p_1 and p_2, are complex conjugates, then C*1 and c_1^* will also be complex conjugates and Equation (10.35) reduces to

$$H(s)=\frac{C_1}{1-e^{p_1T}z^{-1}}+\frac{C_1^*}{1-e^{p_1^*T}z^{-1}} \tag{10.36}$$

$$H(s)=\frac{C_1(1-e^{p_1*T}z^{-1})+C_1^*(1-e^{p_1T}z^{-1})}{1-(e^{p_1T}+e^{p_1*T})z^{-1}+e^{(p_1+p_1*)T}z^{-2}}$$

where C_r and C_i are the real and imaginary parts of C_1,
p_r and p_i are the real and imaginary parts of p_1,
Note: here * symbolizes a complex conjugate.

$$C_1+C_1^*=2\,C_r\quad C_1=C_r+jC_i\quad C_1^*=C_r-jC_i$$
$$p_1+p_1^*=2\,p_r\quad p_1=p_r+jp_i\quad p_1^*=p_r-jp_i$$

$$H(z)=\frac{(C_1+C_1^*)+(C_1e^{p_1*T}+C_1^*e^{p_1T})z^{-1}}{1-2\frac{(e^{p_1T}+e^{p_1*T})}{2}z^{-1}+e^{2p_rT}z^{-2}}$$

$$H(z)=\frac{2C_r-[(C_r+jC_i)\{e^{p_rT}e^{-jp_iT}\}+(C_r-jC_i)\{e^{p_rT}e^{jp_iT}\}]z^{-1}}{1-2\,e^{prT}\frac{(e^{p_iT}+e^{p_i*T})}{2}z^{-1}+e^{2p_rT}z^{-2}}$$

$$H(z)=\frac{2C_r-C_re^{p_rT}[\{e^{jp_iT}+e^{-jp_iT}\}-jC_ie^{p_rT}\{e^{jp_iT}-e^{-jp_iT}\}]z^{-1}}{1-2\,e^{prT}\frac{(e^{p_iT}+e^{p_i*T})}{2}z^{-1}+e^{2p_rT}z^{-2}}$$

$$H(z)=\frac{2C_r-2C_re^{p_rT}\cos(p_iT)+2C_ie^{p_rT}\sin(p_iT)]z^{-1}}{1-2e^{p_rT}\cos(p_iT)z^{-1}+e^{2p_rT}z^{-2}}$$

$$H(z)=\frac{2\,C_r-2\,e^{p_rT}[C_r\cos(p_iT)+C_i\sin(p_iT)]z^{-1}}{1-2e^{p_rT}\cos(p_iT)z^{-1}+e^{2p_rT}z^{-2}} \tag{10.37}$$

Example 10.14

Applying the impulse invariant method to filter design, a digital filter to approximate the following normalized analog transfer function is obtained:

$$H(s) = \frac{1}{s^2 + s\sqrt{2} + 1}$$

Using the impulse invariant method, obtain the transfer function, $H(z)$, of the digital filter assuming a 3-dB cutoff frequency of 150 Hz and a sampling frequency of 1280 Hz.

Solution 10.14

Before applying the impulse invariant method, frequency scaling of the normalized transfer function is needed. This is achieved by replacing s by s/ω_p, where $\omega_p = 2\pi \times 150 = 942.4778$, to ensure that the resulting filter has the desired response.

Thus

$$H'(s) = H(s)_{s=s/\Omega_P} = \frac{\Omega_P^2}{s^2 + \sqrt{2}\,\Omega_P\, s + \Omega_P^2} = \frac{C_1}{s+p_1} + \frac{C_2}{s+p_2}$$

$$H'(s) = H(s)_{s=s/\Omega_P} = \frac{1}{\frac{s^2}{\Omega_P^2} + \frac{s}{\Omega_P}\sqrt{2}\ + 1}$$

$$H'(s) = H(s)_{s=s/\omega_P} = \frac{\Omega_P^2}{s^2 + \sqrt{2}\,\Omega_P\, s + \Omega_P^2} = \frac{C_1}{s+p_1} + \frac{C_2}{s+p_2}$$

$$p_{1,2} = \frac{-\Omega_P\,\sqrt{2} \pm \sqrt{2\Omega_P^2 - 4\Omega_P^2)}}{2} = \frac{-\Omega_P\,\sqrt{2} \pm j\Omega_P\sqrt{2}}{2} = \frac{-\Omega_P\,\sqrt{2}(1 \mp j)}{2}$$

where $\Omega_p = 942.4778$

$$p_1 = \frac{-\sqrt{2}\Omega_P(\,1-j)}{2} = -666.4324(1-j);\ p_2 = p_1^*$$

$$H^{'}(s) = \frac{C_1}{s + \frac{\{-\sqrt{2}\Omega_P(1-j)\}}{2}} + \frac{C_1^*}{s + \frac{\{-\sqrt{2}\Omega_P(1+j)\}}{2}}$$

$$C_1 = \lim s = \frac{\{-\sqrt{2}\Omega_P(1-j)\}}{2}[s + \frac{\{-\sqrt{2}\Omega_P(1+j)\}}{2}]$$

$$[\frac{\Omega_P^2}{[s + \frac{\{-\sqrt{2}\Omega_P(1+j)\}}{2}][s + \frac{\{-\sqrt{2}\Omega_P(1-j)\}}{2}]}]$$

$$C_1 = \frac{\Omega_P^2}{\frac{\{-\sqrt{2}\Omega_P(1-j) + \sqrt{2}\Omega_P(1+j)\}}{2}}$$

$$C_1 = -\frac{\Omega_P}{\sqrt{2}j} = -\frac{\Omega_P}{\sqrt{2}}j = -666.43j;\quad C_2 = C_1^*$$

Since the poles are complex conjugates, the transformation in Equation (10.35) to Equation (10.36) are used to obtain the discrete-time transfer function, $H(z)$.

In this problem,
$C_r = 0$, $C_i = -666.4324$, $p_i T = 0.5207$, $e^{p_r T} = 0.5941$, $p_r T = -0.5207$,
$\sin(p_i T) = .4974$, $\cos(p_i T) = 0.8675$,
Substituting these values into Equation (10.23), $H(z)$ is obtained:

$$H(z) = \frac{2C_r - (0.5941)[0 - (-666.4324)(0.4974)]z^{-1}}{1 - 2(0.5941)(0.8675)z^{-1}}$$

$$H(z) = \frac{393.9264z^{-1}}{1 - 1.0308z^{-1} + 0.3530z^{-2}}$$

Such a large gain is a characteristic of the impulse invariant filter. In general, the gain of the transfer function obtained by this method is equal to the sampling frequency that is 1/*T*, and results from sampling the impulse response. To keep the gain down and to avoid overflows, when the filter is implemented, it is a common practice to multiply $H(z)$ by *T* (or equivalently to divide it by the sampling frequency). Thus, for the problem, the transfer function becomes

$$H(z) = \frac{393.9264z^{-1}\left[\frac{1}{1.28\ 10^3}\right]}{1 - 1.0308z^{-1} + 0.3530z^{-2}} \quad H(z) = \frac{0.3078z^{-1}}{1 - 1.0308z^{-1} + 0.3530z^{-2}}$$

Thus, we have

$$a_0 = 0 \quad b_1 = -1.0308 \quad a_1 = 0.3078 \quad b_2 = 0.3530$$

An alternative method of removing the effect of the sampling frequency on the filter gain is to work with normalized frequencies. Thus, in the last example, we would use *T* = 1 and $\omega_p = 2\pi$ x 150/1.28 kHz = 0.7363.

Using these values in Equation (10.36) leads directly to the desired transfer function which is given above. An important advantage of working with normalized

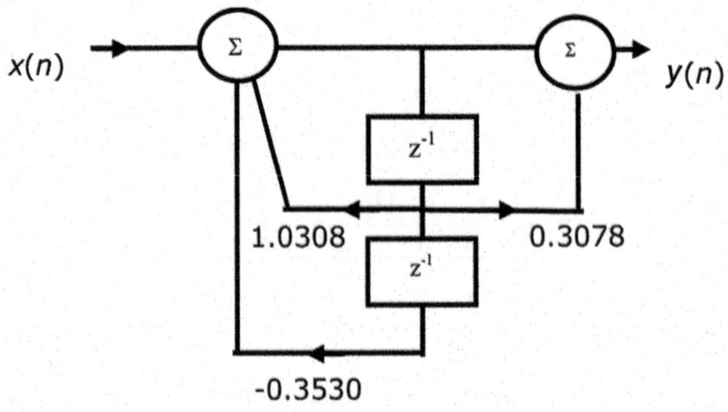

(a) Realization structure.

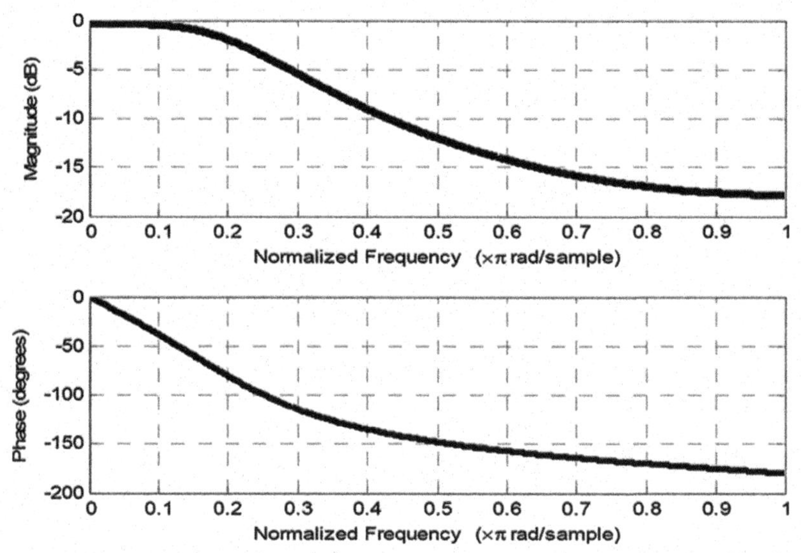

(b) Corresponding magnitude and phase frequency response of a low-pass filter.

frequencies is that the numbers involved are much simpler. It also means that the results can be generalized. The filter is represented in the form of a block diagram in the figure.

10.9.1 Limitation of this Method

The spectrum of the impulse invariant filter corresponding to $H(z)$ would be the same as that of the original analog filter, $H(s)$, but repeats at multiples of the sampling frequency leading to aliasing.

However, if the roll-off of the original analog filter is sufficiently steep or if the analog filter is band-limited before the impulse invariant method is applied, the aliasing will be low. Low aliasing can also be achieved by making the sampling frequency high.

This method may be used for very sharp cutoff low-pass filters with little aliasing, provided that the sampling frequency is reasonably high, but it is unsuitable for high-pass or band-stop filters unless an anti-aliasing filter is used.

10.9.2 Illustration: Impulse Invariant Method

The simple analog filter with the transfer function given below is digitized, using the impulse variant method,

$$H(s) = \frac{C}{s - p} \tag{10.38}$$

The impulse response, $h(t)$, is given by the inverse Laplace transform:

$$h(t) = L^{-1}[H(s)] = L^{-1}\left(\frac{C}{s-p}\right) = Ce^{pt}$$

where L^{-1} symbolizes the inverse Laplace transform. According to the impulse invariant method, the impulse response of the equivalent digital filter, $h(nT)$, is equal to $h(t)$ at the discrete times $t = nT$, $n = 0, 1, 2, \ldots$, i.e.

$$h(nT) = h(t)|_{t=nT} = Ce^{pnT}$$

The transfer function of $H(z)$ is obtained by z-transforming $h(nT)$:

$$H(z) = \sum_{n=0}^{\infty} h(nT)z^{-n} = \sum_{n=0}^{\infty} Ce^{pnT}z^{-n}$$
$$= \frac{C}{1-e^{pT}z^{-1}}$$

Thus, from the result above, it can be written as

$$\frac{C}{s-p} \rightarrow \frac{C}{1-e^{pT}z^{-1}} \tag{10.39}$$

Example 10.15

Consider the first-order transfer function

$$H(s) = \frac{1}{s+a}$$

Assuming that $a = 1$, find the filter transfer function and the difference equation for realizing the structure.

Solution 10.15

$$H(s) = \frac{1}{s+a} \rightarrow h(t) = e^{-at} \quad H(z) = \frac{1}{1-e^{-a}z^{-1}}$$

$$H(z) = \frac{1}{1-0.368\,z^{-1}} \quad H(z) = \frac{1}{1-0.368\,z^{-1}} = \frac{Y(z)}{X(z)}$$

The filter transfer function is expressed in the form of a difference equation as following:

$$y(n) = 0.368y(n-1) + x(n)$$

10.10 Problems and Solutions

Problem 10.1 (Low-Pass filter)

It is required to design a low-pass digital filter to approximate the following analog transfer function:

$$H(s) = \frac{1}{s^2 + s\sqrt{2} + 1}$$

Using the BZT method, obtain the transfer function, $H(z)$, of the digital filter, assuming a 3-dB cutoff frequency of 150 Hz and a sampling frequency of 1.28 kHz.

Solution 10.1

The critical frequency is $\omega_p = 2\pi$ x 150, giving the prewarped analog frequency of

$$\Omega_p = \tan(\omega_p T/2) = 0.3857$$

The prewarped analog filter is given by:

$$H'(s) = H(s)|_{s=s/\Omega_P} = \frac{1}{(s/\Omega_P)^2 + \sqrt{2}s/\Omega_P + 1}$$

$$H'(s) = \frac{\Omega_P^2}{s^2 + \sqrt{2}\Omega_P s + \Omega_P^2} = \frac{0.1488}{s^2 + 0.545\ s\ + 0.1488}$$

Applying the BZT gives

$$H(z) = \frac{0.0878z^2 + 0.1756z + 0.0878}{z^2 + 1.0048z + 0.3561}$$

$$H(z) = 0.0878\frac{1 + 2z^{-1} + z^{-2}}{1 - 1.0048\ z^{-1} + 0.3561\ z^{-1}}$$

Problem 10.2 (Low Pass filter)

Design a unit bandwidth 3-dB digital Butterworth low-pass filter of order 1 having digital frequency ω_p = 1 radian/sec using the conventional bilinear transformation; where in this problem $k = 2/T$, $T = 1$.

Solution 10.2

Prewarp the digital frequency $\omega_p = 1$ radian/sec. requirement to get

$$\Omega_P = 2\tan(\omega_p.T/2)\ = 2\tan(1/2), \quad \Omega_P = 1.0926$$

Use N = 1 analog Butterworth filter as a prototype applying a low pass-to-low pass transformation to get H_P (s), i.e.

$$H_P\ (s)\ = \frac{1}{s+1}$$

Substituting

$$H^{'}(s) = H_P^{'}(s)\,|_{s=\frac{s}{1.0926}} = \frac{1}{\frac{s}{1.0926}+1} = \frac{1}{0.9152s+1}$$

Go through the bilinear transformation

$$H(z) = H^{'}(s)|_{s=(z-1)/(z+1)}$$

$$s = k\frac{1-z^{-1}}{1+z^{-1}} = 2\frac{1-z^{-1}}{1+z^{-1}}$$

$$H(z) = \frac{1}{0.9152438.\quad [2(1-z^{-1})/(1+z^{-1})]+1}$$

$$H(z) = \frac{1+z^{-1}}{2.8305-0.83052z^{-1}} = 0.3533\frac{1+z^{-1}}{1-0.2934z^{-1}}$$

Problem 10.3 (Low Pass filter)

The transfer function of the simple RC low-pass filter is given by

$$H(s) = \frac{V(s)}{I(s)} = \frac{1}{s+1}$$

Use the BZT method to design a corresponding Discrete-Time low-pass filter whose bandwidth is 20 Hz at a sampling frequency of 60 Hz. Plot the magnitude and phase responses of $H(z)$.

Solution 10.3

The critical frequency here is the filter bandwidth. Thus,

$$\omega_p = 2\pi(20) \text{ radians/second}$$

Next, we follow the three design steps associated with the BZT method.

$$\Omega_p = \tan\left(\frac{\omega_p T}{2}\right) \quad \Omega_p = \tan\left(\frac{2\pi(20)}{2(60)}\right)$$

Since $T = 1/60$ sec, Ω_p yields

$$\Omega_p = \tan\frac{\pi}{3} = \sqrt{3}$$

$H(s)$ is known to have a bandwidth of 1 radian/second. We thus use frequency scaling, $H^{'}(s)$, which has a bandwidth of $\Omega_p = \sqrt{3}$; i.e.

$$H^{'}(s) = H(s)|_{s=s/\sqrt{3}} = \frac{\sqrt{3}}{s+\sqrt{3}}$$

Thus, the desired transfer functions are

$$H(z) = \left[\frac{\sqrt{3}}{s + \sqrt{3}}\right]_{s=(z-1)/(z+1)}$$

which yield

$$H(z) = \frac{\sqrt{3}z + \sqrt{3}}{(1 + \sqrt{3})z + (\sqrt{3} - 1)} = \frac{\sqrt{3}z + \sqrt{3}}{2.7321z + 0.7321}$$

Thus, it is

$$H(z) = 0.634\frac{1 + z^{-1}}{1 + 0.268z^{-1}}$$

The magnitude and phase responses are computed using the MATLAB. The plots so obtained are displayed in figure.

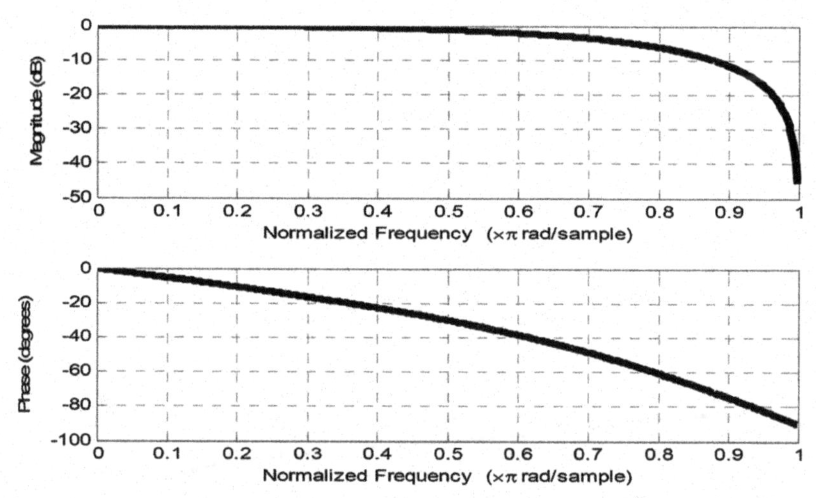

Magnitude and phase response of the filter.

Problem 10.4 (Low Pass Notch Filter)

The transfer function of the circuit is shown below

$$H(s) = \frac{V(s)}{I(s)} = \frac{s^2 + 1}{s^2 + s + 1}$$

Design a discrete-time notch (low-pass) filter with the following specification

1. Notch frequency = 60 Hz.
2. Sampling frequency = 960 Hz.

Plot the corresponding magnitude and phase responses.

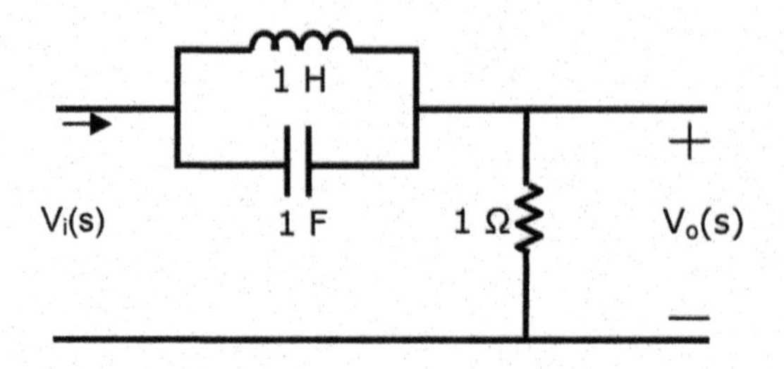

Solution 10.4

The critical frequency is 60 Hz, which corresponds to $\omega_p = 2\pi(60)$ radian/second. Next, we proceed with the three design steps.

$$\Omega_p = \tan\left\{\frac{\omega_p T}{2}\right\} = \tan\left\{\frac{2\pi \times 60}{2 \times 960}\right\} = 0.1989$$

We obtain the scaled transfer function

$$H'(s) = H(s)|_{s=s/\Omega_p}$$

where $H(s)$ and Ω_p are given above. This computation leads to

$$H'(s) = \frac{s^2 + 0.0396}{s^2 + 0.1989s + 0.0396}$$

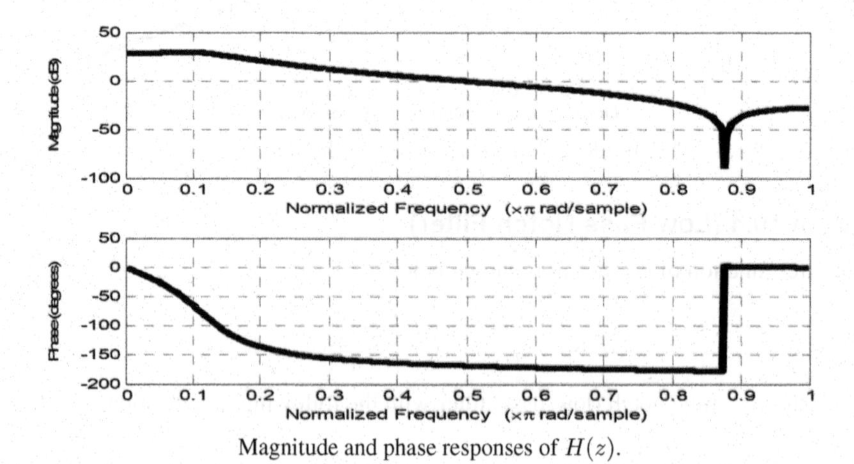

Magnitude and phase responses of $H(z)$.

Using $H(s)$, we evaluate

$$H(z) = H^{'}(s)|_{s=(z-1)/(z+1)}$$

to obtain

$$H(z) = \frac{1.0396 + 1.9208z^{-1} + 1.0396z^{-2}}{1.2385 - 1.9208z^{-1} + 0.8407z^{-2}}$$

whose magnitude and phase responses are plotted in the figure. It is evident that the notch frequency does occur at 60 Hz, as specified.

Problem 10.5 (High Pass Filter)

The normalized transfer function of a sample analog low-pass resistance-capacitance (RC) high-pass filter is given by

$$H(s) = \frac{1}{s+1}$$

Starting from the s plane equation, determine, using the BZT method, the transfer function of an equivalent discrete-time high-pass filter. Assume a sampling frequency of 150 Hz and a cutoff frequency of 30 Hz.

Solution 10.5

The normalized transfer function for the RC filter is $H(s) = \frac{1}{s+1}$

The critical frequency for the digital filter is $\omega_p = 2\pi$ x 30 rad/sec.

The analog frequency, after prewarping, is $\Omega_P = \tan(\omega_p\ T/2) = \tan(\pi/5)$ $= 0.7265$.

The denormalized analog filter transfer function is obtained from $H(s)$ as

$$H'(s) = H(s)|_{s=\Omega_p/s} = \frac{1}{\frac{\Omega_p}{s} + 1} = \frac{s}{s + 0.7265}$$

$$H(z) = H'(s)|_{s=(z-1)/(z+1)} = \frac{(z-1)/(z+1)}{(z-1)/(z+1) + 0.7265} = 0.5792\frac{(1-z^{-1})}{1-0.1584z^{-1}}$$

The coefficients of the discrete-time filter are:

$$a_0 = 0.5792, a_1 = -0.5792 \text{ and } b_1 = -0.1584$$

The difference equation is: $y(n) = 0.1584\ y(n-1) + 0.5792[x(n) + x(n-1)]$

The frequency response for a high-pass filter and its block diagram representation are shown in the figure.

The block diagram representation is shown in the figure.

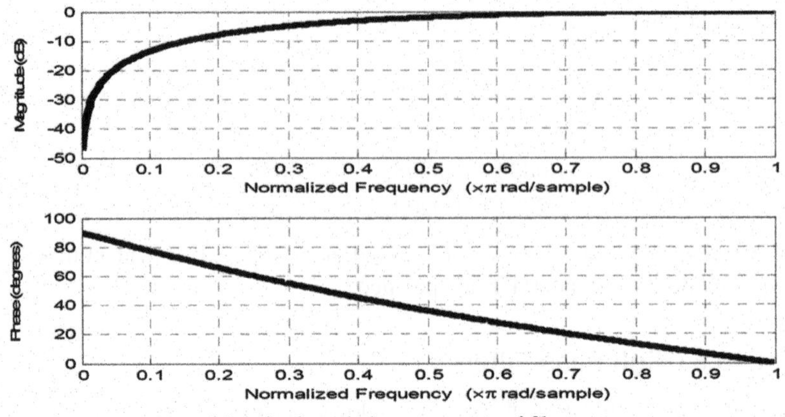

Magnitude and phase response of filter.

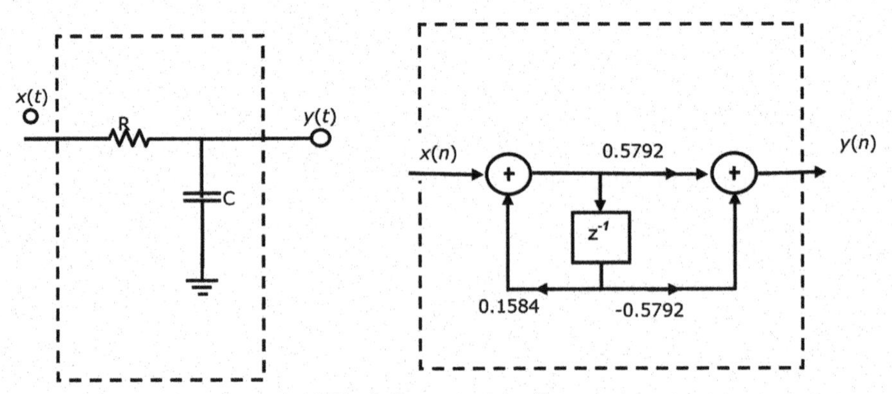

A block diagram of the filter.

Problem 10.6 (Band-Pass Filter)

A discrete-time band-pass filter with Butterworth characteristics meeting the specification given below is required. Obtain the coefficients of the filter using the BZT method.

Pass-band	200–300 Hz
Sampling frequency	2 kHz

Solution 10.6

A first-order, normalized analog low-pass filter is required

$$H(s) = \frac{1}{s+1}$$

The prewarped critical frequencies are

$$\Omega_{P1} = \left(\frac{\omega_{p1}T}{2}\right) = \tan\left(\frac{2\pi \times 200}{2 \times 2000}\right) = 0.3249$$

$$\Omega_{P2} = \left(\frac{\omega_{p2}T}{2}\right) = \tan\left(\frac{2\pi \times 300}{2 \times 2000}\right) = 0.5095$$

$$\Omega_0^2 = \Omega_{P1}.\Omega_{P2} = 0.1655 \; W = \Omega_{P2} - \Omega_{P1} = 0.1846$$

Using the low pass-to-band pass transformation, we have

$$H'(s) = H(s)\Big|_{s=\frac{s^2+\omega_0^2}{Ws}} = \frac{1}{\frac{s^2+\Omega_0^2}{Ws}+1}$$

$$= \frac{Ws}{s^2 + Ws + \Omega_0^2}$$

Applying the BZT gives

$$H'(s) = H(s)\Big|_{s=\frac{z-1}{z+1}} = \frac{1}{\frac{s^2+\Omega_0^2}{Ws}+1}$$

$$= \frac{W\left(\frac{z-1}{z+1}\right)}{\left(\frac{z-1}{z+1}\right)^2 + W\left(\frac{z-1}{z+1}\right) + \Omega_0^2}$$

Substituting the values of Ω_0^2 and W, and simplifying, we have

$$H(z) = 0.1367\frac{1 - z^{-2}}{1 - 1.2362z^{-1} + 0.7265z^{-2}}$$

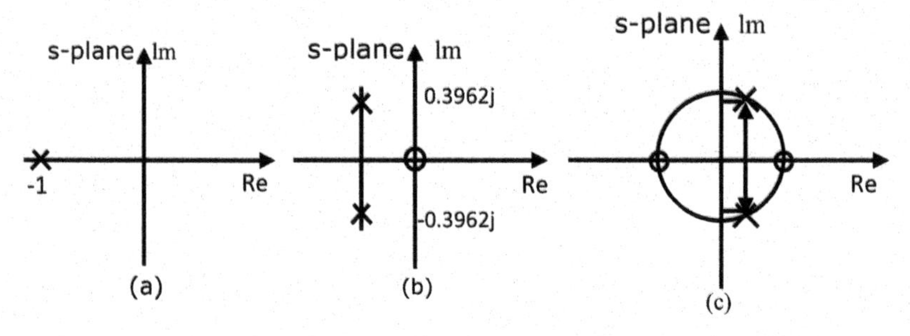

Pole-zero diagrams for (a) prototype low-pass filter and those of (b) intermediate analog band-pass filter, and (c) discrete band-pass filters obtained by band transformation.

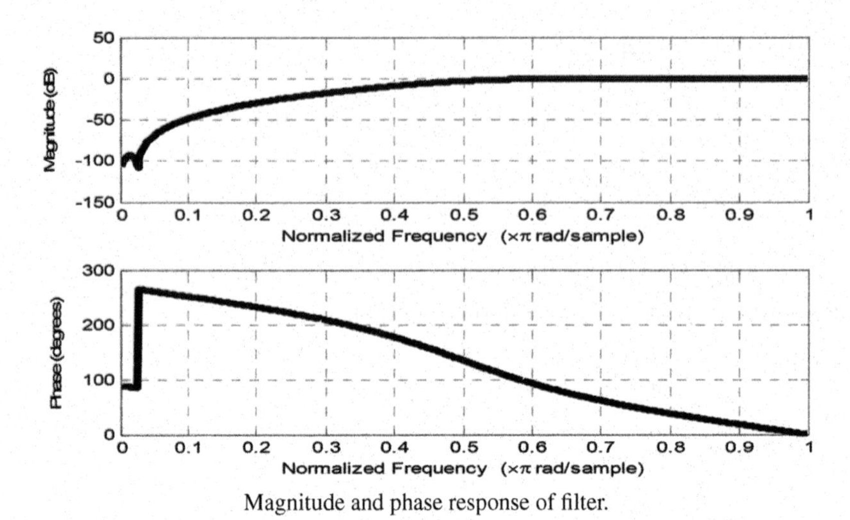

Magnitude and phase response of filter.

The pole-zero diagrams of the normalized prototype low-pass filter (LPF), the analog band-pass filter, and the discrete-time band-pass filter are depicted in the figure. Note that the low pass-to-band pass transformation has introduced a single zero at the origin of the s-plane and a pole at infinity. The BZT method then maps the zeros to $z = \pm 1$. Its poles are at $z = 0.6040 \pm 0.6015j$. The analog band-pass zeros are at $s = 0$ and infinity (not shown) and the poles at $s = -0.0923 \pm 0.3962j$.

Problem 10.7

Design a digital low-pass filter using the bilinear transformation method to satisfy the following characteristics using $N = 2$ and Butterworth approximation with cutoff frequency $\Omega_P = 2$ rad/sec.

Solution 10.7

The design procedure is that of using the bilinear transformation on an analog prototype and consists of the following three steps:

Because a low pass filter has to be designed, frequency transformation $s \to \frac{s}{\Omega_p} = \frac{s}{2}$

Therefore, the required prewarped analog filter using the Butterworth, and the low pass-to-low pass transformation is

$$H_a(s) = \frac{1}{s^2 + \sqrt{2}s + 1}|_{s \to s/2} = \frac{4}{s^2 + 2\sqrt{2}s + 4}$$

Applying the bilinear transformation ($T = 1$) to $H_a(s)$ will take the prewarped analog filter to a digital filter with system function $H(z)$ that will satisfy the given digital requirements:

$$H(z) = H_a(s)|_{s \to [(1-z^{-1})/(1+z^{-1})]}$$

$$H(z) = \frac{4}{\left[\frac{2(1-z^{-1})}{(1+z^{-1})}\right]^2 + 2\sqrt{2}\left[\frac{2(1-z^{-1})}{(1+z^{-1})}\right] + 4}$$

$$H(z) = \frac{1 + 2z^{-1} + z^{-2}}{3.4142135 + 0.5857865\ z^{-2}}$$

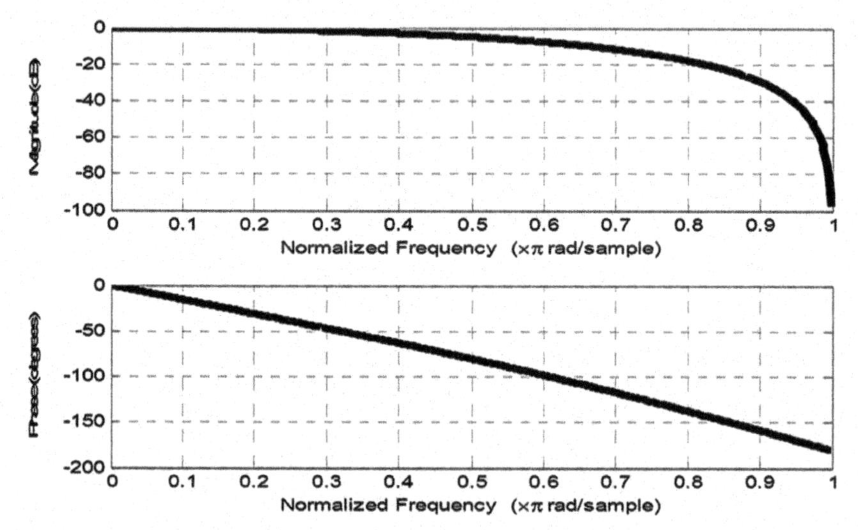

Magnitude and phase response of the filter.

Problem 10.8

Obtain the transfer function and difference equation of the digital low-pass filter. Assuming a sampling frequency of 500 Hz and a cutoff frequency of 50 Hz, Draw direct form-I for the calculated filter transfer function,

$$H(s) = \frac{1}{s+1}$$

Solution 10.8

$$\Omega_P = \tan\left[\omega_P \frac{T}{2}\right] = \tan\left[(314.15)\left[\frac{2 \times 10^{-3}}{2}\right]\right] = 0.3249$$

Now,

$$H'(s) = \frac{1}{\dfrac{s}{0.3249} + 1} = \frac{0.3249}{s + 0.3249}$$

$$s = \frac{z-1}{z+1} \quad H(z) = \frac{0.3249}{\dfrac{z-1}{z+1} + 0.3249}$$

$$H(z) = \frac{0.3249}{\dfrac{(z-1) + 0.3249(z+1)}{(z+1)}} = \frac{0.3249(z+1)}{(z-1) + 0.3249(z+1)}$$

$$H'(s) = \frac{0.3249(z+1)}{(z-1) + 0.3249z + 0.3249}$$

$$H'(s) = \frac{0.3249\ (z+1)}{1.3249\ z - 0.6751}$$

Divide the numerator and denominator by 1.3249.

$$H'(s) = \frac{0.2452(1 + z^{-1})}{1 - 0.5059\ z^{-1}}$$

The filter coefficients are as below:

$$a_0 = 0.2452 \quad b_1 = -0.5059$$
$$a_1 = 0.2452$$

The filter difference equation is described as follows:

$$y(n) = 0.2452\ x(n) + 0.2452\ x(n-1) + 0.5059\ y(n-1)$$

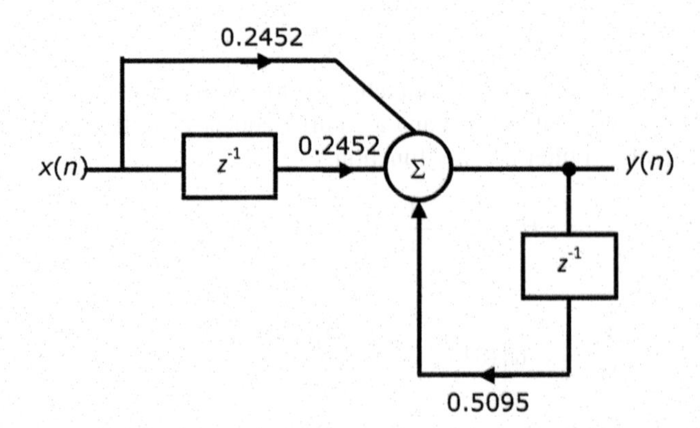

Problem 10.9

A low-pass digital filter meeting the following specifications is required; assume a Butterworth characteristic for the filter.

Pass-band frequency	1000 Hz
Stop-band frequency	3000 Hz
Pass-band ripple	3 dB
Stop-band attenuation	20 dB
Sampling frequency	8 kHz

Determine the following:

(1) Pass- and stop-band edge frequencies for a suitable analog prototype low-pass filter;
(2) The order, N, of the prototype low-pass filter;
(3) Filter coefficients, and hence the transfer function, of the discrete-time filters using the bilinear z-transform.

Solution 10.9

From the specifications, the prewarped frequencies are

$$\Omega_p = \tan\left\{\frac{2\pi \times 1000}{2 \times 8000}\right\} = 0.4142 \quad \Omega_s = \tan\left\{\frac{2\pi \times 3000}{2 \times 8000}\right\} = 2.4142$$

$$\Omega_S^P = \frac{\Omega_s}{\Omega_p} = \frac{2.4142}{0.4142} = 5.8284$$

Thus, the prewarped pass- and stop-band edge frequencies for the prototype LP filter are: 0, 1, 5.8284.

Next, using Equation (10.3) and the values of the parameters given above, the order of the filter is determined.

$$10^{A_s/10} - 1 = 10^{20/10} - 1 = 99; \quad 10^{A_P/10} - 1 = 10^{3/10} - 1 = 0.9952$$
$$\log\left(\tfrac{99}{0.9952}\right) = 1.9976$$

For the prototype LPF,

$$N \geq \frac{\log\left(\frac{10^{\frac{A_s}{10}} - 1}{10^{\frac{A_p}{10}} - 1}\right)}{2\log\left(\frac{\Omega_s^P}{\Omega_p^P}\right)}$$

$$\Omega_P^P = 1.0 \quad \Omega_S^P = 5.8284 \quad 2\log\left(\frac{\Omega_S^P}{\Omega_P^P}\right) = 2\log\left(\frac{5.8284}{1}\right) = 1.5310$$

$$N \geq \frac{1.9976}{1.5310} = 1.3047 \cong N = 2$$

The poles of the prototype filter are (from Equation (9.2))

$$s_{P1} = \cos\left[\frac{(2+2-1)\pi}{4}\right] + j\,\sin\left[\frac{(2+2-1)\pi}{4}\right] = -\frac{\sqrt{2}}{2} + j\frac{\sqrt{2}}{2}$$

$$s_{P2} = -\frac{\sqrt{2}}{2} - j\frac{\sqrt{2}}{2}$$

The s-plane transfer function, $H(s)$, is

$$H(s) = \frac{1}{s^2 + \sqrt{2}\,s + 1}$$

The frequency scaled s-plane transfer function is

$$H^{'}(s) = H(s)\Big|_{s=\frac{s}{\Omega_p}} = \frac{1}{\left(\frac{s}{\Omega_p}\right)^2 + \sqrt{2}\left(\frac{s}{\Omega_p}\right) + 1}$$

$$H^{'}(s) = \frac{\Omega_p^2}{s^2 + \sqrt{2}\,s\,\Omega_p + \Omega_p^2}$$

Applying the BZT method,

$$H(z) = H^{'}(s)\Big|_{s=\frac{z-1}{z+1}} = \frac{\Omega_p^2}{\left(\frac{z-1}{z+1}\right)^2 + \Omega_p\sqrt{2}\left(\frac{z-1}{z+1}\right) + \Omega_p^2}$$

$$H(z) = \frac{\Omega_p^2(z+1)^2}{(z-1)^2 + \Omega_p\sqrt{2}\,(z-1)(z+1) + \Omega_p^2(z+1)^2}$$

After simplification and dividing top and bottom by z^2, we have

$$H(z) = \frac{\Omega_p^2}{1+\sqrt{2}\;\Omega_p + \Omega_p^2} \times \frac{1 + 2\,z^{-1} + z^{-2}}{1 + \frac{2(\Omega_p^2-1)\,z^{-1}}{1+\Omega_p\sqrt{2}+\Omega_p^2} + \frac{(1-\Omega_p\sqrt{2}+\Omega_p^2)\,z^{-2}}{1-\Omega_p\sqrt{2}+\Omega_p^2}}$$

Using the values of the parameters and substituting in the equation above and simplifying, we have

$$H(z) = 0.0976\frac{(1 + 2z^{-1} + z^{-2})}{1 - 1.4714z^{-1} + 0.1380z^{-2}}$$

Problem 10.10

A high-pass digital filter meeting the following specifications is required:

Pass-band frequency	3000 Hz
Stop-band frequency	1000 Hz
Pass-band ripple	3 dB
Stop-band attenuation	20 dB
Sampling frequency	8 kHz

Determine the following:

(1) Pass- and stop-band edge frequencies for a suitable analog prototype low-pass filter;
(2) Order, *N*, of the prototype low-pass filter;
(3) Filters coefficients, and hence the transfer function, of the discrete-time filter using the bilinear z-transform.

Assume a Butterworth characteristic for the filter.

Solution 10.10

From the specifications, the prewarped frequencies are

$$\Omega_{\mathrm{p}} = \tan\left\{\frac{2\pi \times 3000}{2 \times 8000}\right\} = 2.4142 \qquad \Omega_s = \tan\left\{\frac{2\pi \times 1000}{2 \times 8000}\right\} = 0.4142$$

$$\Omega_S^P = \frac{\Omega_p}{\Omega_s} = \frac{2.4142}{0.4142} = 5.8284$$

Thus, the pass- and stop-band edge frequencies for the prototype LP filter are: 0, 1, 5.8284.

(2) Next, we use (10.3) and the values of the parameters above to determine the order of the filter.

Now

$$10^{A_s/10} - 1 = 10^{20/10} - 1 = 99; \quad 10^{A_P/10} - 1 = 10^{3/10} - 1 = 0.9952$$
$$\log\left(\tfrac{99}{0.9952}\right) = 1.9976$$

For the prototype LPF,

$$N \geq \frac{\log\left(\frac{10^{\frac{A_s}{10}}-1}{10^{\frac{A_p}{10}}-1}\right)}{2\log\left(\frac{\omega_s^P}{\omega_p^P}\right)}$$

$$\Omega_P^P = 1.0 \quad \Omega_S^P = 5.8284 \quad 2\log\left(\frac{\Omega_S^P}{\Omega_P^P}\right) = 2\log\left(\frac{\Omega_S^P}{\Omega_P^P}\right) = 1.4206$$

$$N \geq \frac{1.9976}{1.5310} = 1.3047 \cong N = 2$$

The poles of the prototype filter are from Equation (10.2)

$$s_{P1} = \cos\left[\frac{(2+2-1)\pi}{4}\right] + j\,\sin\left[\frac{(2+2-1)\pi}{4}\right] = -\frac{\sqrt{2}}{2} + j\frac{\sqrt{2}}{2}$$

$$s_{P2} = -\frac{\sqrt{2}}{2} - j\frac{\sqrt{2}}{2}$$

The s-plane transfer function, $H(s)$, is

$$H(s) = \frac{1}{(s - s_{P1})(s - s_{P2})} = \frac{1}{s^2 + \sqrt{2}\,s + 1}$$

The frequency scaled s-plane transfer function is

$$H'(s) = H(s)\Big|_{s=\frac{\Omega_p}{s}} = \frac{1}{\left(\frac{\Omega_p}{s}\right)^2 + \sqrt{2}\left(\frac{\Omega_p}{s}\right) + 1}$$

$$H'(s) = \frac{s^2}{s^2 + \sqrt{2}\,s\,\Omega_p + \Omega_p^2}$$

Applying the BZT method,

$$H(z) = H'(s)\Big|_{s=\frac{z-1}{z+1}} = \frac{\left(\frac{z-1}{z+1}\right)^2}{\left(\frac{z-1}{z+1}\right)^2 + \Omega_p\sqrt{2}\left(\frac{z-1}{z+1}\right) + \Omega_p^2}$$

$$H(z) = \frac{(z-1)^2}{(z-1)^2 + \Omega_p\sqrt{2}\,(z-1)(z+1) + \Omega_p^2(z+1)^2}$$

After simplification and dividing top and bottom by z^2, we have

$$H(z) = \frac{1}{1 + \sqrt{2}\;\Omega_p + \Omega_p^2} \times \frac{1 - 2\,z^{-1} + z^{-2}}{1 + \frac{2(\Omega_p^2 - 1)\,z^{-1}}{1 + \Omega_p\sqrt{2} + \Omega_p^2} + \frac{(1 - \Omega_p\sqrt{2} + \Omega_p^2)\,z^{-2}}{(1 + \Omega_p\sqrt{2} + \Omega_p^2}}$$

Using the values of the parameters

$$1 + \sqrt{2}\;\Omega_p + \Omega_p^2 = 10.24263;$$

$$\Omega_p^2 - 1 = 0 \quad 1 - \sqrt{2}\;\Omega_p + \Omega_p^2 = -8.2426 \quad \Omega_p^2 = 5.8284$$

and substituting in the equation above and simplifying, we have

$$H(z) = \frac{5.8284}{10.2426} \frac{(1 - 2z^{-1} + z^{-2})}{1 + \frac{2(4.8284)}{10.2426}z^{-1} - \frac{(-8.2426)}{10.2426}z^{-2}}$$

$$H(z) = 0.5690\frac{(1 - 2z^{-1} + z^{-2})}{1 + 0.9428z^{-2} - 0.8047z^{-2}}$$

Problem 10.11

A requirement exists for a band-pass digital filter, with a Butterworth magnitude-frequency response, that satisfies the following specification:

Lower pass-band edge frequency	500 Hz
Upper pass-band edge frequency	800 Hz
Lower stop-band edge frequency	2000 Hz
Upper stop-band edge frequency	3000 Hz
Pass-band ripple	3 dB
Stop-band attenuation	20 dB
Sampling frequency	6000 Hz

Determine the following:

(1) Pass- and stop-band edge frequencies of a suitable prototype low-pass filter;
(2) Order, *N*, of the prototype low-pass filter;
(3) Filter coefficients, and hence the transfer function, of the discrete-time filter using the BZT method.

Solution 10.11

The prewarped critical frequencies for the band-pass filter are

$$\Omega_{\mathrm{p1}} = \tan\left\{\frac{2\pi \times 500}{2 \times 6000}\right\} = 0.2679 \qquad \Omega_{p2} = \tan\left\{\frac{2\pi \times 800}{2 \times 6000}\right\} = 0.4452$$

$$\Omega_{\mathrm{s1}} = \tan\left\{\frac{2\pi \times 2000}{2 \times 6000}\right\} = 1.732 \qquad \Omega_{s2} = \tan\left\{\frac{2\pi \times 450}{2 \times 1000}\right\} = 6.3138$$

$$\Omega_0 = \sqrt{\Omega_{p2}.\Omega_{p1}} = \sqrt{(0.4452)(0.2679)} = 0.3453$$

$$W = \Omega_{p2}. - \Omega_{p1} = 0.4452 - 0.2679 = 0.1773$$

Thus, the band edge frequencies for the prototype LP filter are (using the relationships above)

$$\Omega_{s1}^{p} = \frac{\Omega_{s1}^2 - \Omega_{\mathrm{o}}^2}{W\Omega_{s1}} = \frac{\Omega_{s1}^2 - \Omega_{\mathrm{o}}^2}{(\Omega_{p2} - \Omega_{p1})\Omega_{s1}} = \frac{(1.732)^2 - 0.1193}{(0.1773)(1.732)} = \frac{2.8805}{(0.3071)} = 9.4379$$

$$\Omega_{s2}^{p} = \frac{\Omega_{s2}^2 - \Omega_{\mathrm{o}}^2}{W\Omega_{s2}} = \frac{\Omega_{s2}^2 - \Omega_{\mathrm{o}}^2}{(\Omega_{p2} - \Omega_{p1})\Omega_{s2}} = \frac{(0.1584)^2 - 1}{(1.3763 - 0.7265)(0.1584)}$$

$$= \frac{-0.9749}{(0.6498)(0.1584)} = \frac{-0.9749}{0.1029} = -9.4705$$

$$\Omega_p^p = 1, \omega_{\mathrm{s}}^{\mathrm{P}} = \min(|\omega_{\mathrm{s1}}^{\mathrm{P}}|, |\omega_{\mathrm{s2}}^{\mathrm{P}}|)$$

$$\Omega_{\mathrm{s}}^{\mathrm{P}} = 9.4705$$

Thus, we require a prototype LPF with $\Omega_p^p = 1, \Omega_s^p = 9.4705$, $A_p = 3$ dB; $A_s = 20$ dB. From Equation, (10.3) the order of the prototype LPF is obtained as

Now

$$10^{A_s/10} - 1 = 10^{20/10} - 1 = 99;$$

$$10^{A_P/10} - 1 = 10^{3/10} - 1 = 0.9952 \quad \log\left(\frac{99}{0.9952}\right) = 1.9976$$

For the prototype LPF,

$$N \geq \frac{\log\left(\frac{10^{\frac{A_s}{10}} - 1}{10^{\frac{A_p}{10}} - 1}\right)}{2\log\left(\frac{\Omega_s^P}{\Omega_p^P}\right)}$$

$$\Omega_P^P = 1.0 \quad \Omega_S^P = 5.3792 \quad 2\log\left(\frac{\Omega_S^P}{\Omega_P^P}\right) = 2\log\left(\frac{\Omega_S^P}{\Omega_P^P}\right) = 1.9443$$

$$N \geq \frac{1.9976}{1.9443} = 1.027 \cong N = 1$$

N must be an integer, and this time for simplicity, we will use $N = 1$. The s-plane transfer function for a first-order prototype LP filter is given by

$$H(s) = \frac{1}{s+1}$$

The frequency scaled s-plane transfer function is

$$H^{'}(s) = H(s)\bigg|_{s=\frac{s^2+\Omega_o^2}{W\Omega}} = \frac{1}{\left(\frac{s^2+\Omega_o^2}{Ws}\right) + 1} = \frac{Ws}{s^2 + Ws + \Omega_0^2}$$

Applying the BZT method,

$$H(z) = H^{'}(s)\bigg|_{s=\frac{z-1}{z+1}} = \frac{W\left(\frac{z-1}{z+1}\right)}{\left(\frac{z-1}{z+1}\right)^2 + W\left(\frac{z-1}{z+1}\right) + \Omega_0^2}$$

$$= \frac{(z-1)(z+1)}{(z-1)^2 + W(z-1)(z+1) + \Omega_0^2(z+1)^2}$$

After simplification, and substituting the values for Ω_0^2 and W,

$$H(z) = \frac{(z^2-1)}{z^2 - 2z + 1 + 0.1773(z^2-1) + 0.119(z^2+2z+1)}$$

$$H(z) = \frac{0.1372(1 - z^{-2})}{z^2 - 2z + 1 + 0.1773(z^2 - 1) + 0.119(z^2 + 2z + 1)}$$

$$H(z) = 0.1372 \frac{(1 - z^{-2})}{1 + 1.3464\ z^{-1} + 0.728\ z^{-2}}$$

Problem 10.12

A requirement exists for a band-stop digital IIR filter, with a Butterworth magnitude-frequency response, that meets the following specifications:

Pass-band frequency	50–500 Hz
Stop-band frequency	200–400 Hz
Pass-band ripple	3 dB
Stop-band attenuation	18 dB
Sampling frequency	1100 Hz

Determine the following:

(1) Pass- and stop-band edge frequencies of a suitable prototype low-pass filter;
(2) Order, *N*, of the prototype low-pass filter;

Solution 10.12

The prewarped critical frequencies for the band-pass filter are

$$\Omega_{p1} = \tan\left(\frac{\omega_{p1}T}{2}\right) = \tan\left(\frac{2\ \pi\ 50}{(2)(1100)}\right) = 0.1437$$

$$\Omega_{p2} = \tan\left(\frac{\omega_{p2}T}{2}\right) = \tan\left(\frac{2\ \pi\ 500}{(2)(1100)}\right) = 6.955$$

$$\Omega_{s1} = \tan\left(\frac{\omega_{s1}T}{2}\right) = \tan\left(\frac{2\ \pi\ 200}{(2)(6000)}\right) = 0.6426$$

$$\Omega_{s2} = \tan\left(\frac{\omega_{s2}T}{2}\right) = \tan\left(\frac{2\ \pi\ 400}{(2)(1100)}\right) = 2.189$$

$$\Omega_0 = \sqrt{\Omega_{p2}.\Omega_{p1}} = \sqrt{(6.955)(0.143)} = 1$$

$$W = \Omega_{p2} - \Omega_{p1} = 6.955 - 0.1437 = 6.811$$

$$\begin{aligned}\Omega_{\mathrm{s}}^{\mathrm{p}(1)} &= \frac{W\Omega_{\mathrm{s1}}}{\Omega_{\mathrm{o}}^2 - \Omega_{\mathrm{s1}}^2} = \frac{(\Omega_{\mathrm{p2}} - \Omega_{\mathrm{p1}})\Omega_{\mathrm{s1}}}{\Omega_{\mathrm{o}}^2 - \Omega_{\mathrm{s1}}^2} \\ &= \frac{(6.811)(0.6426)}{(1) - (0.6426)^2} = \frac{(6.811)(0.6426)}{1 - 0.4129} = \frac{4.3767}{0.5871} = 7.4547\end{aligned}$$

$$\Omega_{\rm s}^{\rm p(2)} = \frac{W\Omega_{\rm s2}}{\Omega_{\rm o}^2 - \Omega_{\rm s2}^2} = \frac{(\Omega_{\rm p2}-\Omega_{\rm p1})\Omega_{\rm s2}}{\Omega_{\rm o}^2 - \Omega_{\rm s2}^2}$$

$$= \frac{(6.811)(2.189)}{(1)^2 - (2.189)^2} = \frac{14.90}{-3.791} = -3.932$$

$\Omega_s^p = \min\left(\left|\Omega_s^{p(1)}\right|, \left|\Omega_s^{p(2)}\right|\right)$ Thus, the band edge frequencies for the prototype LP filter are (using the relationships above)

$$\Omega_p^p = 1; \Omega_s^p = 3.932$$

We require a prototype LPF with $\Omega_p^p = 1$; $\Omega_s^p = 3.932$, $A_p = 3$ dB; $A_s = 20$ dB. From Equation (10.3), the order of the prototype LPF is obtained as

$$10^{A_s/10} - 1 = 10^{20/10} - 1 = 99;$$

$$10^{A_P/10} - 1 = 10^{3/10} - 1 = 0.9952 \quad \log\left(\frac{99}{0.9952}\right) = 1.9976$$

For the prototype LPF,

$$N \geq \frac{\log\left(\frac{10^{\frac{A_s}{10}}-1}{10^{\frac{A_p}{10}}-1}\right)}{2\log\left(\frac{\Omega_s^p}{\Omega_p^p}\right)}$$

$$\Omega_P^P = 1.0 \quad \Omega_S^P = 9.4705 \quad 2\log\left(\frac{\Omega_S^P}{\Omega_P^P}\right) = 2\log(3.932) = 1.189$$

$$N \geq \frac{1.997697}{1.189} = 1.679 \cong N = 2$$

N must be an integer, and for simplicity, we will use $N = 2$. The s-plane transfer function for a first-order prototype LP filter is given by

$$H(s) = \frac{1}{s^2 + \sqrt{2}\, s + 1}$$

Problem 10.13

A band-pass notch digital filter is required to meet the following specification:

(i) Complete signal rejection at 125 Hz.
(ii) A narrow pass band centered at 50 Hz.
(iii) A 3-dB bandwidth of 7.5 Hz.

Assuming a sampling frequency of 500 Hz, obtain the transfer function of the filter, by suitably placing z-plane poles and zeros and its difference equation. The radius (r) of the pole is determined using the following bandwidth relation $r = 1 - (\mathrm{bw}/F_s)\pi$.

Solution 10.13

Zeros

Using the signal rejection frequency, we find zeros.

$$Z_1 = 360^\circ \times \frac{125}{500} = \pm 90^\circ,$$

Poles at the location

Using pass–band-centered frequency.

$$p_{1,2} = \pm\, 360^\circ \times \frac{50}{500} = \pm\, 36^\circ$$

Radius is to be calculated using $r = 1 - (bw/F_s)\pi$.

To ensure that the coefficients are real, it is necessary to have a complex conjugate pole pair. The radius, *r*, of the poles is determined by the desired bandwidth. An approximate relationship between *r*, for $r > 0.9$, and bandwidth (bw), is given by

$$r = 1 - (\text{bw}/F_s)\pi.$$

The radius (r) of the pole is determined by using the following bandwidth relation.

$$r = 1 - (\text{bw}/F_s)\pi., r = 1 - (7.5/500)p, \quad r = 0.9528$$

Complete rejection is required at 125 Hz, which gives the information that where to place the poles and zeros on the z-plane. These are at angles $\pm 90^\circ$ on the unit circle. The pass-band centered at 50 Hz gives the information to place poles at $\pm 36^\circ$.

The pole-zero diagrams are given in the figure below. From the pole-zero diagrams, the transfer function can be written down by inspection:

From the pole-zero diagram,

$$H(z) = \frac{(z - e^{j\pi/2})(z - e^{-j\pi/2})}{(z - re^{j36^\circ})(z - re^{-j36^\circ})}$$

$$H(z) = \frac{(z - j)(z + j)}{(z^2 - re^{-j\pi/2} - zre^{j\pi/2} + r^2 e^{j\pi/2 - j\pi/2})}$$

$$H(z) = \frac{(z^2 + 1)}{(z^2 - z\, r[e^{-j36} + e^{j36}] + r^2 e^{\circ})}$$

$$H(z) = \frac{(z^2 + 1)}{(z^2 - 2z\, r[\cos 36^\circ] + r^2} \qquad H(z) = \frac{(z^2 + 1)}{z^2 - 1.618z + 0.9078}$$

$$H(z) = \frac{(1 + z^{-2})}{1 - 1.618z^{-1} + 0.9078z^{-2}}$$

Problem 10.14

Using the pole-zero placement method, calculate coefficients of a notch filter. Obtain transfer function and the difference equation of a simple digital notch filer that meets the following specification:

Notch frequency	25 Hz
3-dB width of notch	± 10 Hz
Sampling frequency	400 Hz

Solution 10.14

To reject the component at 25 Hz, we place a pair of complex zeros at points on the unit circle corresponding to 25 Hz, that is at angles of $360° \times 25/400 = \pm 22.5°$.

To achieve a sharp notch filter and improved amplitude response on either side of the notch frequency, a pair of complex conjugate poles are placed at a radius $r < 1$. The width of the notch is determined by the locations of the poles. The relationship between the bandwidth and the radius of the poles is given by

$$r \approx 1 - (\text{bw}/F_s)\pi$$

[bw = 20 Hz and $F_s = 400$ Hz, giving $r = (1 - 20/400)\pi = 0.8429$]

$$H(z) = \frac{[z - e^{j22.5°}][z - e^{-j22.5°}]}{[z - 0.8429\, e^{j22.5°})][z - 0.8429\ e^{-j22.5°}]}$$

$$= \frac{z^2 - 2z\cos 22.5° + 1}{z^2 - 1.6858z + 0.7104} = \frac{1 - 1.8477z^{-1} + z^{-2}}{1 - 1.6858z^{-1} + 0.7104z^{-2}}$$

Problem 10.15

Calculate coefficients of a notch filter and the transfer function and the difference equation of a simple digital notch filter that meets the following specifications:

1) Notch frequency 63 Hz.
2) 3-dB width of notch ± 5 Hz.
3) Sampling frequency 500 Hz.

Solution 10.15

Zeros

$$Z_{1,2} = \pm 360° \times \frac{63}{500} = \pm 45.03° \cong 45°$$

Poles

For finding the poles, we use a bandwidth with a noted frequency.

$$Z_{1,2} = \pm 360° \times \frac{(63 + 5)}{500} = \pm\ 48.96° \cong 49°$$

Complete rejection is required at 0 and 250 Hz, which gives the information that where to place the poles and zeros on the z-plane. These are at angles of 0° and $360° \times 250/500 = 180°$ on the unit circle.

The pass band centered at 125 Hz gives information about the placement the poles at $\pm 360° \times 125/500 = \pm 90°$.

To ensure that the coefficients are real, it is necessary to have a complex conjugate pole pair. The radius, *r*, of the poles is determined by the desired bandwidth. An approximate relationship between *r*, for $r > 0.9$, and bandwidth, bw, is given by

$$r \approx 1 - (\mathrm{bw}/F_s)\pi$$

bw = 5 Hz and $F_s = 500$ Hz, giving $r = 1 - (5/500)\pi = 0.968$ rad/sec.

The pole-zero diagrams are given in the figure. From the pole-zero diagrams, the transfer function can be written down by inspection:

$$H(z) = \frac{(z - e^{-j\pi/4})(z + e^{+j\pi/4})}{(z - re^{-J49°})(z - re^{J49°})}$$

$$H(z) = \frac{z^2 - ze^{j\pi/4} - ze^{-j\pi/4} + e^{-j\pi/4 + j\pi/4}}{z^2 - zre^{j49°} - ze^{-j49°} + r^2 e^{-j49° + j49°}}$$

$$H(z) = \frac{z^2 - z[\cos\pi/4 - i\sin\pi/4 + \cos\pi/4 + i\sin\pi/4] + 1}{z^2 - z\,r[\cos 49° - i\sin 49° + \cos 49° + i\sin 49°] + r^2}$$

$$H(z) = \frac{z^2 + 1.050\,z + 1}{z^2 - z(0.968)(0.6011) + 0.9370}$$

$$H(z) = \frac{1 + 1.050\,z^{-1} + z^{-2}}{1 - 0.5818\,z^{-1} + 0.9370\,z^{-2}}$$

Comparing the transfer function, $H(z)$, the following coefficients have been found out.

$$\begin{array}{ll} a_0 = 1 & b_1 = -0.5818 \\ a_1 = 1.05 & b_2 = 0.9370 \\ a_2 = 1 & \end{array}$$

The filter difference equation is given by

$$y(n) = x(n) + 1.050\,x(n-1) + x(n-2) + 0.5818\,y(n-1) - 0.9370\,y(n-2)$$

Problem 10.16

Digitize, using the impulse invariant method, the analog filter with the transfer function.

$$H(s) = \frac{\alpha}{s(s+\alpha)}$$

Assuming a sampling frequency of 1 (normalized) and $\alpha = 0.5$. Draw direct form-I for the calculated filter transfer function.

Solution 10.16

$$H(s) = \frac{\alpha}{s(s+\alpha)} \rightarrow \frac{\alpha}{s^2 + \alpha s}$$

$$H(s) = \frac{C_1}{s - p_1} + \frac{C_2}{s - p_2}$$

$$H(s) = \frac{C_1}{s - p_1} + \frac{C_2}{s - p_2}$$

There are two roots of the transfer function, we use, to residue theorem to find out the residues values of C_1 and C_2

$$\begin{aligned} &C_1 = 1, C_2 = -1; \\ &e^{p1T} = 1; \; e^{p2T} = 0.6065 \\ &p_1 = 0; \; p_2 = -0.5; \; e^{(p1+p2)T} = 0.6065 \end{aligned}$$

Put the values above in Equation (8.22) which are for non-conjugate values.

$$H(z) = \frac{C_1 + C_2 - (C_1 e^{p_2 T} + C_2 e^{p_1 T}) z^{-1}}{1 - (e^{p_1 T} + e^{p_2 T}) z^{-1} + e^{(p_1+p_2)T} z^{-2}}$$

$$H(z) = \frac{1 - 1 - [0.6065 - 1] z^{-1}}{1 - (1.6065) z^{-1} + 0.6065 z^{-2}}$$

$$H(z) = \frac{0.3935 z^{-1}}{1 - 1.6065 z^{-1} + 0.6065 z^{-2}}$$

The filter coefficients are as below:

$$\begin{aligned} a_0 &= 0 & b_1 &= -1.6065 \\ a_1 &= 0.3935 & b_2 &= 0.6065 \end{aligned}$$

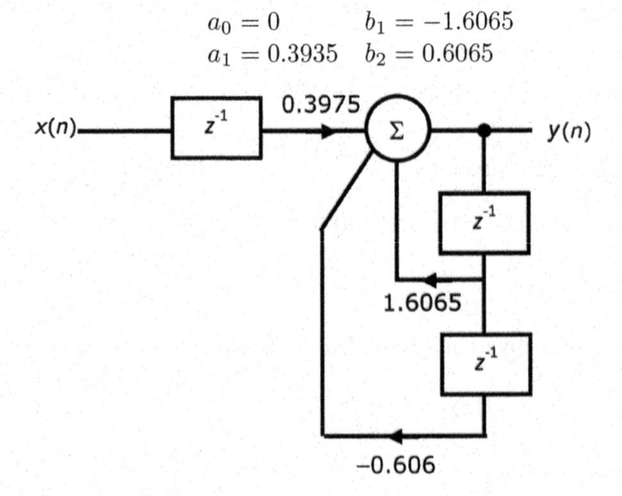

The filter difference equation is described as follows:

$$Y(n) = 0.3935\ x(n-1) + 1.6065\ y(n-1) - 0.6065\ y(n-2)$$

Problem 10.17

Obtain the digital filter transfer function for given

$$H(s) = \frac{1}{(s+a)(s+b)}$$

Assuming that $a = 0.3$ and $b = 0.5$, find the filter transfer function and the difference equation for realizing the structure.

Solution 10.17

$$H(s) = \frac{A}{s+a} + \frac{B}{s+b}$$

This can be written in partial fraction form, after the calculation of the residues A and B as,

$$A = \frac{1}{b-a} \quad B = \frac{1}{a-b} \quad h(t) = \frac{1}{b-a}e^{-at} - \frac{1}{b-a}e^{-bt}$$

$$H(z) = \frac{\frac{1}{b-a}}{1-e^{-a}z^{-1}} - \frac{\frac{1}{b-a}}{1-e^{-b}z^{-1}} = \frac{1/0.2}{1-e^{-0.3}z^{-1}} - \frac{1/0.2}{1-e^{-0.5}z^{-1}}$$

$$H(z) = \frac{5}{1-0.741z^{-1}} - \frac{5}{1-0.607z^{-1}}$$

$$H(z) = \frac{5(1-0.607z^{-1})}{1-0.741z^{-1}} - \frac{5(1-0.741z^{-1})}{1-0.607z^{-1}} \quad H(z) = \frac{0.67z^{-1}}{1-1.348z^{-1}+0.45z^{-1}}$$

$$y(n) - 1.348\ y(n-1) + 0.45\ y(n-2) = 0.67\ x(n-1)$$

Rearranging to get $y(n)$; $y(n) = 1.348\ y(n-1) - 0.45\ y(n-2) + 0.67\ x(n-1)$

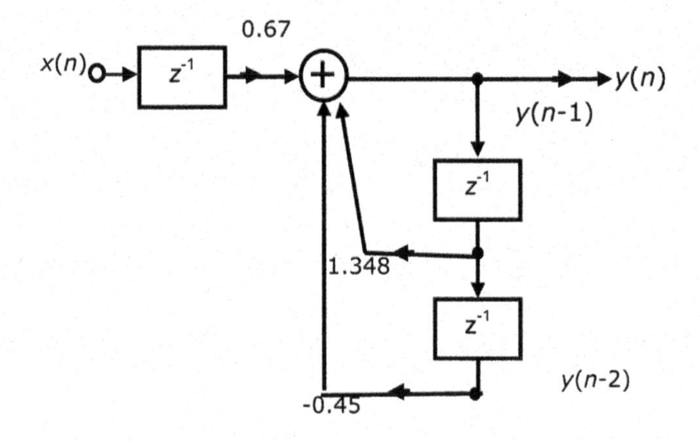

Problem 10.18

Obtain the digital filter transfer function for given

$$H(s) = \frac{s+0.5}{(s+0.5)^2+(s+0.25)^2}$$

Convert this analog filter into a digital filter by making use of the impulse invariant technique, by assuming $T = 0.1$ second.

Solution 10.18

The z-transform of the following transfer function is given by

$$H(s) = \frac{s+a}{(s+a)^2+b^2}$$

In z domain, it is given by

$$H(z) = \frac{1-e^{-aT}(\cos\ bT)\ z^{-1}}{1-2e^{-aT}(\cos\ bT)\ z^{-1}+e^{-2aT}\ z^{-2}}$$

Here, $a = 0.5$ and $b = 0.25$,

$$H(z) = \frac{1-e^{-(0.5)(0.1)}\cos[(0.25)(0.1)]z^{-1}}{1-2e^{-(0.5)(0.1)}\cos[(0.25)(0.1)]z^{-1}+e^{-2(0.5)(0.1)}z^{-2}}$$

$$H(z) = \frac{1-(0.952)(0.1)]z^{-1}}{1-2(0.951)(0.999))z^{-1}+0.904z^{-2}}$$

$$H(z) = \frac{1-(0.951)z^{-1}}{1-1.902z^{-1}+0.904z^{-2}} = \frac{Y(z)}{X(z)}$$

$$y(n)[1-1.902\ y(n-1)+0.904\ y(n-2) = x(n)-0.951\ x(n-1)$$

$$y(n) = 1.902\ y(n-1)-0.904\ y(n-2)+x(n)-0.951\ x(n-1)$$

Problem 10.19

Applying the impulse invariant method to design a digital filter to approximate the following normalized analog transfer function,

$$H(s) = \frac{1}{s^2+\sqrt{2}s+1}$$

Using the impulse invariant method, the transfer function, $H(z)$, of the digital filter assuming a 3-dB cutoff frequency of 1 kHz and a sampling frequency of 5 kHz is obtained.

Solution 10.19

The critical frequency is $\omega_p = 2\pi$ x 1000 = 2000 π, giving the prewarped analog frequency of $\Omega_p = \tan(\omega_p\ T/2)$

$$\Omega_p = \tan\left(\frac{\omega_p T}{2}\right) = \tan\left(\frac{2\ \pi\ 1000}{(2)(5000)}\right) = 0.7265$$

Now replace s by (s/Ω_p). The prewarped analog filter is given by:

$$H'(s) = H(s)|_{s=s/\Omega_p} = \frac{1}{(s/\Omega_p)^2 + \sqrt{2}s/\Omega_p + 1}$$

$$H'(s) = \frac{\Omega_p^2}{s^2 + \sqrt{2}\Omega_p s + \Omega_p^2} = \frac{(0.7265)^2}{s^2 + 1.0267\ s + 0.5271}$$

$$H'(s) = \frac{0.5271}{s^2 + 1.0267\ s + 0.5271}$$

Applying the BZT method gives

$$s = \left[\frac{z-1}{z+1}\right]$$

$$H(z) = \frac{0.5271}{\frac{(z-1)^2}{(z+1)^2} + 1.0267\frac{(z-1)}{(z+1)} + 0.5271}$$

$$H(z) = \frac{0.5271\ (z+1)}{\frac{(z-1)^2}{(z+1)^2} + 1.0267\frac{(z-1)}{(z+1)} + 0.5271}$$

$$H(z) = \frac{0.5271(z + 2z + 1)}{z^2 - 2z + 1 + 1.0267z^2 - 1.0267 + 0.5271\ z^2 + 1.054z + 0.5271}$$

$$H(z) = \frac{0.206 + 0.4127z^{-1} + 0.206z^{-2}}{1 - 0.370z^{-1} + 0.1957z^{-2}}$$

The filter coefficients are as below:

$$a_0 = 0.206 \quad b_1 = -0.370$$
$$a_1 = 0.4127 \quad b_2 = 0.1957$$
$$a_2 = 0.206$$

The filter difference equation is described as follows:

$$y(n) = 0.206\ x(n) + 0.412\ x(n-1) + 0.206\ x(n-2)$$
$$+ 0.37\ y(n-1) - 0.1957\ y(n-2)$$

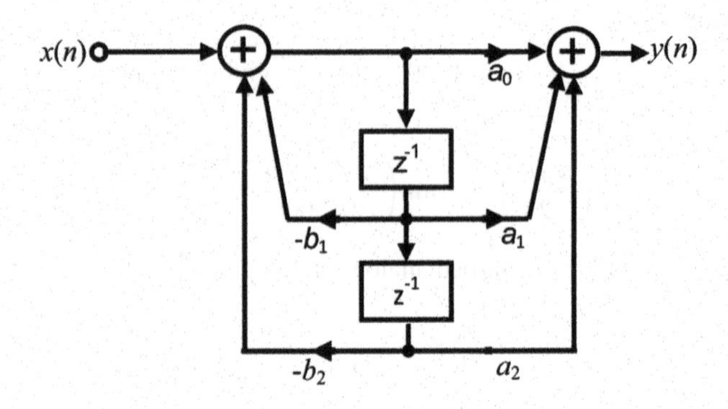

(a) Realization structure.

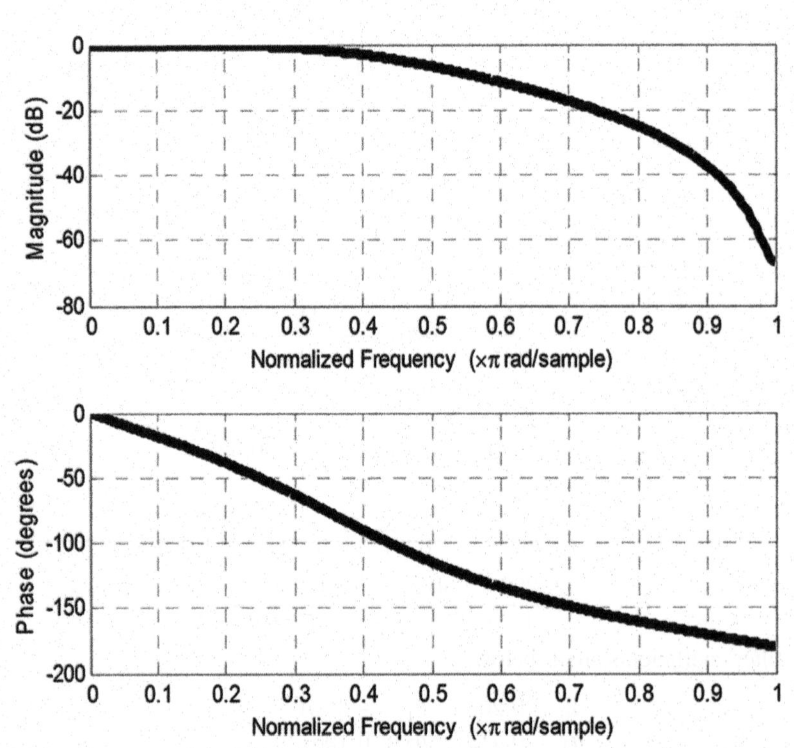

(b) Frequency response of filter.

Problem 10.20

The system function of an analog filter is expressed as follows:

$$H(s) = \frac{s + 0.2}{(s + 0.2)^2 + 9}$$

Convert this analog filter into a digital filter by making use of the impulse invariant technique, by assuming $T = 1$ second.

Solution 10.20

$$H(s) = \frac{s+a}{(s+a)^2 + b^2}$$

where we are given that $a = 0.2$ and $b = 3$, and we can obtain the system response of the digital filter as follows:

$$H(z) = \frac{1 - e^{-aT}(\cos\ bT)\ z^{-1}}{1 - 2e^{-aT}(\cos\ bT)\ z^{-1} + e^{-2aT}\ z^{-2}}$$

Putting the value of $a = 0.2$ and $b = 3$, we obtain

$$H(z) = \frac{1 - e^{-0.2T}(\cos\ 3T)\ z^{-1}}{1 - 2e^{-aT}(\cos\ 3T)\ z^{-1} + e^{-0.4T}\ z^{-2}}$$

Taking $T = 1$ second

$$H(z) = \frac{1 - (0.8187)(-0.99)\ z^{-1}}{1 - 2(0.8187)(-0.99)\ z^{-1} + 0.6703\ z^{-2}}$$

$$H(z) = \frac{1 - 0.8105\ z^{-1}}{1 - 1.6210\ z^{-1} + 0.6703\ z^{-2}}$$

11

Finite Word-Length Effects in Digital Filters

This chapter begins with the introduction to finite word length effects in digital filter, representation of binary numbers, floating-point number types of representation of numbers, quantization error, coefficient quantization error, effects in FIR digital filters, and problems and solutions.

11.1 Introduction

While the digital filters are implemented on digital hardware, there are some effects of using finite lengths to represent all relevant filter parameters. The algorithms for digital signal processing can be realized either with special-purpose digital hardware or as programs for a general-purpose digital computer. However, in both cases, the numbers and coefficients are stored in finite-length registers. Therefore, coefficients and numbers must be quantized by truncation of rounding before they can be stored. Before we proceed to analyze the quantization noise, we here present the refresher for arithmetic used in digital system.

11.2 Methods of Representation of Binary Numbers

We can represent a number either decimal or binary N (say) to any desired accuracy by the following finite series:

$$N = \sum_{i=n_1}^{n_2} c_i r^i$$

where r is known as the radix. For decimal number, the radix = 10, and for binary number, it is 2.

Example 11.1

Represent the decimal number 20.275 into radix $r = 10$ format.

Solution 11.1

For decimal representation having numbers from 0 to 9; in decimal presentation, $r = 10$. The representation of the number can be represented as follows:

$$20.275 = \sum_{i=-3}^{1} c_i 10^i \text{ or } 20.275 = 2 \times 10^1 + 0 \times 10^0 + 2 \times 10^{-1} + 7 \times 10^{-2} + 5 \times 10^{-3}$$

Example 11.2

Represent the binary number 10001.10000011 into radix $r = 2$ format.

Solution 11.2

The representation is called binary representation with two numbers 0 and 1; in this case, $r = 2$. The representation of the binary number is as follows:

$$10001.10000011 = \sum_{i=-8}^{4} c_i 2^i = 1 \times 2^4 + 0 \times 2^3 + 0 \times 2^2 + 0 \times 2^1 + 1 \times 2^0 + 1 \times 2^{-1} + 0 \times 2^{-2} + 0 \times 2^{-3} + 0 \times 2^{-4} + 0 \times 2^{-5} + 0 \times 2^{-6} + 1 \times 2^{-7} + 1 \times 2^{-8}$$

To convert from decimal to binary, we divide the integer part of the number (left to the decimal point) repeatedly by 2 and arrange the remainder in reverse order. The fractional part (right to the decimal point) is repeatedly multiplied by 2, each time removing the integer part, and writing in normal order.

Example 11.3

This example gives a revision of conversion of decimal number into binary number. Convert the given decimal number 15.375 to binary form.

Solution 11.3

We have the following procedure

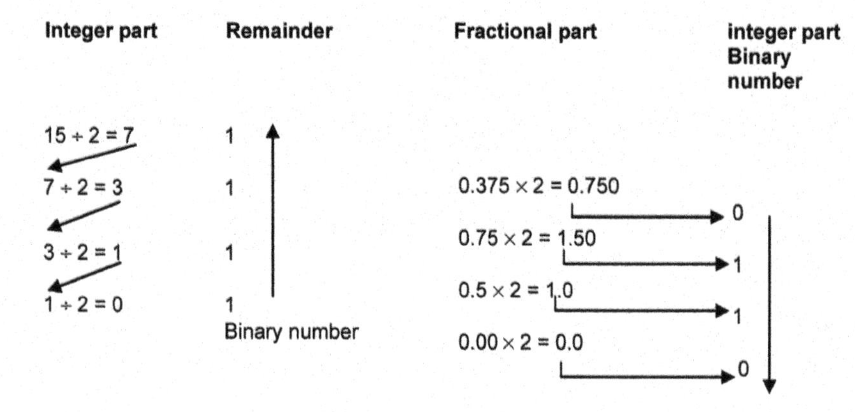

Therefore, from above, we get

$$(15.375)_{10} = (1111.0110\ldots)_2$$

Example 11.4

Convert the given decimal number 20.5 to binary form.

Solution 11.4

We have the following procedure

Integer part	Remainder
20 ÷ 2 = 10	0
10 ÷ 2 = 5	0
5 ÷ 2 = 2	1
2 ÷ 2 = 1	0
1 ÷ 2 = 0	1

Binary number

Fractional part	integer part Binary number
0.5 × 2 = 1.0	1
0.0 × 2 = 0.0	0

Therefore, from above, we get

$$(20.5)_{10} = (10100.10)_2$$

$$10100.10 = 1 \times 2^4 + 0 \times 2^3 + 1 \times 2^2 + 0 \times 2^1 + 0 \times 2^0 + 1 \times 2^{-1} + 0 \times 2^{-2}$$

11.3 Fixed-Point Arithmetic for Binary Number

Binary numbers are represented in digital systems using one of the two formats: fixed or floating point.

11.3.1 Representation of Fixed-Point Number

Let us assume that a word length of b bits is chosen to represent the numbers in a digital filter. Therefore, with a bit words, 2^b different numbers may be represented. The position of binary point is fixed in the fixed-point representation; the position of the binary point is fixed. The bit to the right represents the fractional part of the number and that to the left represents the integer part.

Example 11.5

Represent the binary number 11.10011 to decimal number.

Solution 11.5

$$10.1001101 = \sum_{i=-6}^{1} c_i 2^i = 1\times 2^1 + 0\times 2^0 + 1\times 2^{-1} + 0\times 2^{-2}$$
$$+ 0\times 2^{-3} + 1\times 2^{-4} + 0\times 2^{-5} + 1\times 2^{-6}$$

There are different ways to represent a positive or negative number for fixed-point arithmetic:

(a) Sign-magnitude form
(b) One's complement
(c) Two's complement form

11.3.1.1 Sign-magnitude form

There are certain advantages or disadvantages of each representation, but the most likely used is two's complement form.

In sign-magnitude representation, the most significant bit is set to 1 to represent the negative sign.

In sign-magnitude form, the number 0 has two representations, i.e., 00.000000 or 10.000000. The magnitude of the negative number is given with b bits only as

$$(2^b - 1) \tag{11.1}$$

Example 11.6

Convert the decimal number – 2.25 into binary number using sign-magnitude form.

Solution 11.6

2.25 is represented as 10.01 and −1.75 is represented as 111.101

11.3.1.2 One's complement form

In one's complement form, the positive number is represented as in the sign-magnitude notation. However, the negative number can be obtained by complementing all the bits of the positive number.

In one's complement form, the magnitude of the negative number is given as

$$1 - \sum_{i=1}^{b} c_i 2^{-i} - 2^{-b} \quad (11.2)$$

Example 11.7

Convert 0.775 decimal number into binary and represent the decimal number –0.775 using one's complement.

Solution 11.7

$$(0.775)_{10} = (0.1100011)_2$$

$$(-0.775)_{10} = (1.0011100)_1$$

$(0.775)_{10} = (0.1100011)_2$
↓↓ ↓↓↓↓↓
complementing
1.0011100 each bit
$(-0.775)_{10} = (1.0011100)_1$

Note that this is same as subtracting the magnitude from $2 - 2^{-b}$, where b is the number of bits (without sign bit)

$$2 - 2^{-b} = 10.000000 - 0.0000001 = 1.1111111$$

Now, subtract

$$0.775 = (0.1100011)_2$$

$$\begin{array}{r} 1.1111111 \\ \underline{0.1100011} \\ 1.0011100 \\ = (-0.775)_{10} \end{array}$$

in one's complement form

In one's complement form, the magnitude of the negative number is represented as

$$1-\sum_{i=1}^{b} c_i 2^{-i} - 2^{-b} \quad (1)$$

$$\because 1-(2^{-3}+2^{-4}+2^{-5})-2^{-7}=0.775$$

11.3.1.3 Two's complement form

In two's complements representation, positive numbers are represented as in sign-magnitude form and one's complement. However, the negative number can be obtained by complementing all the bits of the positive number and adding one to the least-significant bit.

Let the given number is A; then, 2's complement of A is $A' = \bar{A}+1$ where $\bar{A}$ is the complement of A.

The magnitude of the negative number is given as follows:

$$x = 1-\sum_{i=1}^{b} c_i 2^{-i} \quad (11.3)$$

Example 11.8

Convert $(5)_{10}$ into binary form; then, using two's complement, find the binary equivalent of $(-5)_{10}$.

Solution 11.8

$A = (+5)_{10} = (0101)_2$
Then,
$\bar{A} = 1010$
$A' = \bar{A}+1 = 1011$
which is two's complement of A and is used for representing negative number
$(-5)_{10} = (1011)_2$

$(5)_{10} = (0101)_2$
↓↓↓↓
complementing
1010 each bit
0001 add 1 to LSB
$(-5)_{10} = 1011$

Example 11.9

Convert $(0.975)_{10}$ into binary form; then, find two's complement of $(-0.975)_{10}$.

Solution 11.9

$$(+0.975)_{10} = (0.11111001100)_2$$

$$(-0.875)_{10} = (1.00000110100)_2$$

$0.975_{10} = (0.11111001100)_2$
↓↓↓↓↓↓↓↓↓↓↓↓↓↓
complementing
1.00000110011 each bit
+0.00000000001 Add 1 to LSB

$(-0.975)_{10} = 1.00000110100$

Here,

$$A = (+0.975)_{10} = (0.11111001100)_2$$

Then, $\bar{A}= 1.00000110011$

$$A' = \bar{A} + 1 = 1.00000110100$$

Before proceeding further, the few operations such as addition and multiplication of fixed-point number are explained in the following paragraph.

11.3.2 Addition

The addition of two fixed-point numbers is quite simple. In fact, the two numbers are added bit by bit starting from right, with carry bit being added to the next bit.

Example 11.10

Add $(0.5)_{10}$ and $(0.125)_{10}$

Solution 11.10

Assuming the total number of bits $b + 1 = 4$ (including sign bit)

Now, we obtain

$$\begin{aligned} (0.5)_{10} &= 0.100_2 \\ (0.125)_{10} &= 0.001_2 \\ & \overline{0.101_2} = (0.625)_{10} \end{aligned}$$

↳sign bit

When two numbers of b bits are added and the sum cannot be represented by b bits, an overflow is said to occur.

Example 11.11

Add $(0.5)_{10}$ and $(0.625)_{10}$

Solution 11.11

If $b = 3$, then the addition of the two numbers is as follows:

$(0.5)_{10} = 0.100$
$(0.625)_{10} = 0.101$
$\underline{1.001} = (-0.125)_{10}$ in sign magnitude
sign bit

The addition of $(0.5)_{10}$ and $(0.625)_{10}$ obtained $(1.125)_{10}$.

Overflow occurs in the above result because $(1.125)_{10}$ cannot be represented by the three-bit number system. Hence, in general, the addition of fixed-point numbers causes an overflow. The subtraction of two fixed-point numbers can be performed easily using two's complement representation.

Example 11.12

Subtract (i) 0.25 from 0.5 and (ii) 0.5 from 0.25

Solution 11.12

Decimal two's complement
(i) $0.5 = 0.100$
add
$-0.25 = 1.110$
$\underline{10.010} = 0.25$
neglect carry bit

Two's complement representation of –0.25
$(0.25)_{10} = (0.010)_2$
↓↓↓↓ complementing each bit
1.101
+0.001 Add 1 to LSB
$= \underline{1.110}$
i.e., $(-0.25)_{10} = (1.110)_2$

Carry is generated after the addition. Therefore, the result is positive. Neglect the carry bit to get the result in decimal.

i.e., $(0.010)_2 = (0.25)_{10}$

(ii) Decimal two's complement
$0.25 = 0.010$
add
$-0.5 = 1.100$
$\underline{1.110}$

Two's complement representation of –0.5
$(0.5)_{10} = (0.100)_2$
↓↓↓↓ complementing each bit
1.011
+0.001 add 1 to LSB
$= \underline{1.100}$
i.e., $(-0.5)_{10} = (1.100)_2$

Carry is not generated after the addition. Therefore, the result is negative. To obtain the decimal output, find the two's complement of 1.110

$$\begin{array}{l} 0.001 \\ \underline{1} \\ \underline{0.010} = (0.25)_{10} \\ \text{i.e., } (-0.25)_{10} \end{array}$$

11.3.3 Multiplication

In the multiplication of two fixed-point numbers, we first separate the sign and magnitude components. The magnitudes of the numbers are multiplied first, and then, the sign of the product is determined and applied to the result. When a b-bit number is multiplied with another b-bit number, the product may contain $2b$ bits.

Example 11.13
Multiply

$$(11)_2 \times (11)_2$$

Solution 11.13

$$(11)_2 \times (11)_2 = (1001)_2$$

If the b bits are organized into $b = b_i + b_f$, where b_i represents integer part and b_f represents the fraction, then the product may contain $2b_i + 2b_f$ bits. In fixed-point arithmetic, multiplication of two fractions results in a fraction. Multiplication occurs in a very similar way as ordinary multiplication. Note that for multiplication of fractions, overflow can never occur.

For example,

$$\begin{array}{lll} 0.1001 \times 0.0011 & = & 0.00011011 \\ \text{(4 bits)} \quad \text{(4 bits)} & & \text{(8 bits)} \end{array}$$

11.4 Floating-Point Number Representation

In floating-point representation, a positive number is represented as $F = 2^c$. M, where M, called mantissa, is a fraction such that $1/2 \leq M \leq 1$ and c, the exponent can be either positive or negative.

In decimal numbers, 4.5, 1.5, 6.5, and 0.625 have floating-point representations as $2^3 \times 0.5625$, $2^1 \times 0.75$, $2^3 \times 0.8125$, and $2^0 \times 0.625$, respectively.

Equivalently, we have

$$2^3 \times 0.5625 = 2^{011} \times 0.1001$$
$$2^1 \times 0.75 = 2^{001} \times 0.1100$$
$$2^3 \times 0.8125 = 2^{011} \times 0.1101$$
$$2^0 \times 0.625 = 2^{000} \times 0.1010$$

$\because (3)_{10} =$	$(011)_2$
$(0.5625)_{10} =$	$(0.1001)_2$
$(1)_{10} =$	$(001)_2$
$(0.75)_{10} =$	$(0.1100)_2$
$(0.8125)_{10} =$	$(0.1101)_2$
$(0.625)_{10} =$	$(0.1010)_2$

Negative floating-point numbers are generally represented by considering the mantissa as a fixed-point number. The sign of the floating-point number is obtained from the first bit of mantissa. In floating-point arithmetic, multiplications are carried out as follows:

$$\text{Let} \quad F_1 = 2^{c_1} \times M_1 \quad \text{and}$$
$$F_2 = 2^{c_2} \times M_2$$
$$\text{Then, the product} \quad F_3 = F_1 \times F_2 = (M_1 \times M_2)2^{c_1+c_2}$$

This means that the mantissas are multiplied using fixed-point arithmetic and exponents are added.

The product $(M_1 \times M_2)$ must be in the range of 0.25 to 1.0. To correct this problem, the exponent $(c_1 + c_2)$ must be altered.

Example 11.14

Multiply $(1.5)_{10}$ and $(1.25)_{10}$ number using block floating-point technique.

Solution 11.14

Consider the multiplication of two numbers

$(1.5)_{10}$ and $(1.25)_{10}$

$$(1.5)_{10} = 2^1 \times 0.75 = 2^{001} \times 0.1100$$
$$(1.25)_{10} = 2^1 \times 0.625 = 2^{001} \times 0.1010$$

$(0.75)_{10} = (0.1100)_2$
$(0.625)_{10} = (0.1010)_2$
0.1100×0.1010
$\overline{0.0111100}$

Now,

$(1.5)_{10}$ and $(1.25)_{10}$

$$= (2^{001} \times 0.1100) \times (2^{001} \times 0.1010)$$
$$= 2^{010} \times 0.01111$$

It may be noted that the addition and subtraction of two floating-point numbers are more difficult than the addition and subtraction of two fixed-point numbers.

To carry out addition, first adjust the exponent of the smaller number until it matches the exponent of larger number. The mantissas are then added or subtracted. Finally, the resulting representation is rescaled so that its mantissa lies in the range 0.5−1.

Suppose we are adding 3.0 and 0.125

$$\begin{aligned} 3.0 &= 2^{010} \times 0.110000 \\ 0.125 &= 2^{000} \times 0.001000 \end{aligned}$$

Now, we adjust the exponent of smaller number so that the both exponents are equal.

$$0.125\ ; = 2^{010\prime}0.0000100$$

Now, the sum is equal to $2^{010} \times 0.110010$

11.5 Comparison of Fixed- and Floating-Point Arithmetic

In this subsection, we are comparing the fixed-point arithmetic and floating-point arithmetic as under Table 11.1.

11.6 Block Floating-Point Numbers

A compromise between fixed- and floating-point systems is known as the block floating-point arithmetic. In this case, the set of signals to be handled is divided into blocks. Each block has the same value for the exponent. The arithmetic operations within the block use fixed-point arithmetic, and only one exponent per block is stored, thus saving memory. This representation of numbers is most suitable in certain FFT flow graphs and in digital audio applications, which has not been discussed here.

11.7 The Quantization Noise

In digital signal processing, signal is to be converted into digital signal using ADC. The process of converting an analog signal into a digital is shown in

Table 11.1 Comparison of fixed- and floating-point arithmetic

S. No	Fixed-Point Arithmetic	Floating-Point Arithmetic
1.	Fast operation	Slow operation
2.	Relatively economical	More expensive because of costlier hardware
3.	Small dynamic range	Increased dynamic range
4.	Roundoff errors occur only for addition	Roundoff errors can occur with both addition and multiplication
5.	Overflow occur in addition	Overflow does not arise
6.	Used in small computers	Used in larger, general-purpose computers

Figure 11.1. At first, the signal $x(t)$ is sampled at regular intervals $t = nT$ where $n = 0, 1, 2$, etc., to create a sequence $x(n)$. This is done by a sampler. The numeric equivalent of each sample $x(n)$ is expressed by a finite number of bits giving the sequence $x_q(n)$. The difference signal $e(n) = x_q(n) - x(n)$ is known as the quantization noise or A/D conversion noise.

Let us assume a sinusoidal signal varying between +1 and –1 having a dynamic range.

If the ADC used to convert the sinusoidal signal employs $(b + 1)$ bits including sign bit, the number of levels available for quantizing $x(n)$ is 2^{b+1}. Therefore, the interval between successive levels will be

$$q = \frac{2}{2^{b+1}} = 2^{-b} \tag{11.4}$$

where q is known as quantization step size. If $b = 3$ bits, then $q = 2^{-3} = 0.125$.

The common methods of quantization are as follows:

(i) Truncation
(ii) Rounding.

11.7.1 Quantization Error Due to Truncation and Rounding

The input to a digital filter is represented by a finite word length sequence, and the result of processing generally leads to a filter variable that requires additional bits for accurate representation; otherwise, the noise occurs in the filter.

11.7.2 Truncation

Truncation is a process of discarding all bits less significant than least-significant bit that is retained. Suppose, if we truncate the following binary number from 8 bits to 4 bits, we obtain the following.

$$\underset{8\text{ bits}}{0.00110011} \quad \text{to} \quad \underset{4\text{ bits}}{0.0011}$$

$$\underset{8\text{ bits}}{1.01001001} \quad \text{to} \quad \underset{4\text{ bits}}{1.0100}$$

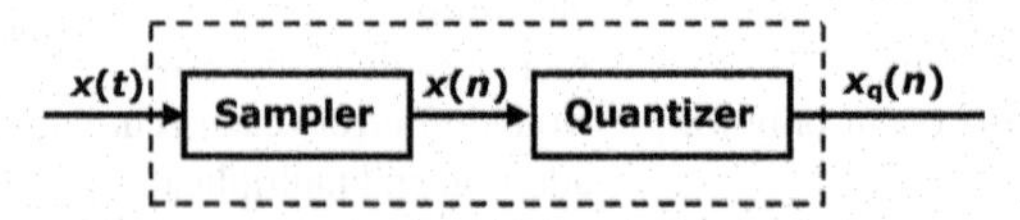

Figure 11.1 Block diagram of an A/D converter.

When we truncate the number, the signal value is approximated by the highest quantization level which is not greater than the signal. Or it is done by discarding all bits less significant than the least-significant bit that is retained.

If the quantization method is that of truncation, then the number is approximated by the nearest level that does not exceed it. In this case, the error $x_T - x$ is negative or zero where x_T is truncated value of x and assumed $|x| \leq 0$. The error made by truncating a number to b bits following the binary point satisfies the following inequality:

$$0 \geq x_T - x > -2^{-b} \tag{11.5}$$

Example 11.15

Consider the decimal number 0.12890625. Perform the truncation operation to 4 bits.

Find also the truncation error.

Solution 11.15

Its binary equivalent is 0.00100001. If we truncate the binary number to 4 bits, we have $x_T = (0.0010)_2$ whose decimal value is 0.125.

Now, the error $(x_T - x) = -0.00390625$, which is greater than $-2^{-b} = 2^{-4} = -0.0625$ satisfying the inequality given in Equation (11.4), which holds for two's complements if $x > 0$. If $x < 0$, we have to find whether the Equation (11.5) holds good for all types of representations.

11.7.2.1 Truncation using two's complement representation

From Equation (11.4), in two's complement representation, the magnitude of the negative number is given as follows

$$x = 1 - \sum_{i=1}^{b} c_i 2^{-i}$$

If we truncate the number to N bits, then we have

$$x_T = 1 - \sum_{i=1}^{N} c_i 2^{-i}$$

The change in magnitude due to quantization will be

$$x_T - x = \left(1 - \sum_{i=1}^{b} c_i 2^{-i}\right) - \left(1 - \sum_{i=1}^{N} c_i 2^{-i}\right)$$

$$x_T - x = \left(1 - \sum_{i=1}^{b} c_i 2^{-i} - 1 + \sum_{i=1}^{N} c_i 2^{-i}\right) = \left(\sum_{i=1}^{N} c_i 2^{-i} - \sum_{i=1}^{b} c_i 2^{-i}\right) \tag{11.6}$$

$$x_T - x = \sum_{i=N}^{b} c_i 2^{-i}$$

$$\text{or} \quad x_T - x \geq 0$$

From the Equation (11.6), it may be observed that the truncation increases the magnitude, which implies that the error is negative and satisfies the following inequality:

$$0 \geq x_T - x \geq -2^{-b} \tag{11.7}$$

Using Equation (11.6), with two's complement representation of mantissa, we have

$$0 \geq M_T - M > -2^{-b} \tag{11.8}$$

$$0 \geq e > -2^{-b} 2^{c} \tag{11.9}$$

We define relative error as

$$\varepsilon = \frac{x_T - x}{x} = \frac{e}{x} \quad \text{or} \quad e = \varepsilon x \tag{11.10}$$

Now, Equation (11.7) can be written as follows:

$$0 \geq \varepsilon x > -2^{-b}.2^{c} \tag{11.11}$$

$$0 \geq \varepsilon 2^{c} M > -2^{-b}.2^{c}$$

$$0 \geq \varepsilon M > -2^{-b} \tag{11.12}$$

If $M = 1/2$, the relative error is maximum.

Therefore, we have

$$0 \geq \varepsilon > -2.2^{-b} \tag{11.13}$$

If $M = -1/2$, the relative error range will be

$$0 \leq \varepsilon < 2.2^{-b} \tag{11.14}$$

11.7.2.2 Truncation using one's complement representation

The magnitude of negative number with b bits is given as in one's complement representation and is given below:

$$x = 1 - \sum_{i=1}^{b} c_i 2^{-i} - 2^{-b} \tag{11.15}$$

When the number is truncated to N bias, then we have

$$x_T = 1 - \sum_{i=1}^{N} c_i 2^{-i} - 2^{-N} \tag{11.16}$$

The change in magnitude due to truncation will be

$$\begin{aligned} x_T - x &= \sum_{i=N}^{b} c_i 2^{-i} - (2^{-N} - 2^{-b}) \\ x_T - x &< 0 \end{aligned} \tag{11.17}$$

Therefore, the magnitude decreases with truncation which implies that error is positive and satisfies the following inequality

$$0 \leq x_T - x < 2^{-b} \tag{11.18}$$

In floating-point system, the effect of truncation is visible only in the mantissa. Let the mantissa is truncated to N bits.

If $x = 2^c.\ M$, then we have

$$\begin{aligned} x_T &= 2^c.\ M_T \\ \text{Error} \qquad e = x_T - x &= 2^c(M_T - M) \end{aligned} \tag{11.19}$$

In one's complement representation, the error for truncation of positive values of the mantissa is as follows

$$0 \geq M_T - M > -2^{-b} \tag{11.20}$$

$$\text{or} \quad 0 \geq e > -2^{-b}.2^c \tag{11.21}$$

$$\text{with} \quad e = \varepsilon x = \varepsilon 2^c.M \tag{11.22}$$

and $M = -½$, we obtain the maximum range of the relative error for positive M as

$$0 \geq \varepsilon > -2.2^{-b} \tag{11.23}$$

For negative mantissa values, the error will be

$$0 \leq M_T - M < 2^{-b} \tag{11.24}$$

$$\text{or} \quad 0 \leq e < 2^c 2^{-b} \tag{11.25}$$

with $M = -½$. The maximum range of the relative error for negative M is

$$0 \geq \varepsilon > -2.2^{-b} \tag{11.26}$$

which is the same as positive M (11.22).

The probability density function $P(e)$ for truncation of fixed-point and floating-point numbers is shown in Figure 11.2.

In fixed-point arithmetic, the error due to rounding a number to b bits produce and error e = x_T – x which satisfies the following inequality:

$$\frac{-2^{-b}}{2} \leq x_T - x \leq \frac{2^{-b}}{2} \tag{11.27}$$

This is because of the fact that with rounding if the value lies halfway between two levels, it can be approximated to either nearest higher level or nearest lower level. For fixed-point numbers, (11.27) satisfied regardless of whether sign-magnitude one's complement or two's complement is used for negative numbers.

$$\text{and} \quad x_T = M_T 2^c \tag{11.28}$$

$$\text{Then} \quad e = x_T - x = (M_T - M)2^c. \tag{11.29}$$

But for rounding, we write

$$\frac{-2^{-b}}{2} \leq M_T - M \leq \frac{2^{-b}}{2} \tag{11.30}$$

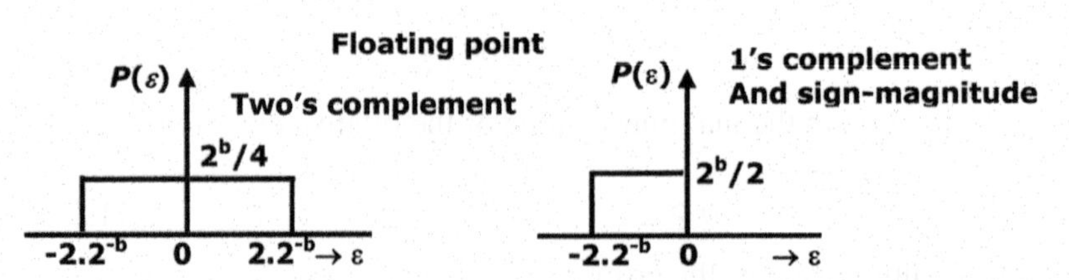

Figure 11.2 Probability density functions $P(e)$ for truncation for fixed point and floating point.

Using Equations (11.13) and (11.29), the Equation (11.30) can be written as

$$-2^c \frac{2^{-b}}{2} \leq x_T - x \leq 2^c \frac{2^{-b}}{2} \tag{11.31}$$

$$\text{or} \quad -2^c \frac{2^{-b}}{2} \leq \varepsilon x \leq 2^c \frac{2^{-b}}{2} \tag{11.32}$$

$$\text{We have} \quad x = 2^c.M \tag{11.33}$$

$$\text{and then} -2^c.\frac{2^{-b}}{2} \leq \varepsilon 2^c.M \leq 2^c.\frac{2^{-b}}{2} \tag{11.34}$$

$$\text{which gives} \quad \frac{-2^{-b}}{2} \leq \varepsilon.M \leq \frac{2^{-b}}{2} \tag{11.35}$$

The mantissa satisfies $1/2 \leq M < 1$

If $M = 1/2$, then we get the maximum range of relative error, i.e.,

$$-2^{-b} \leq \varepsilon < 2^{-b} \tag{11.36}$$

11.7.3 Rounding

Rounding of a number of b bits is accomplished by choosing the rounded result as the b-bit number closest to the original number unrounded. When the unrounded number lies midway between the adjacent b-bit numbers, a random choice is made to which of these numbers to round to.

The probability density function for rounding is shown in Figure 11.3

Table 11.2 Quantization error ranges due to truncation and rounding

Type of Quantization	Type of Arithmetic	Fixed-Point Number Range	Floating-Point Number Relative Error Range
Truncation	Two's complement	$-2^{-b} < e \leq 0$	$-2.2^{-b} < \varepsilon \leq 0,\ M>0$ $0 \leq \varepsilon < 2.2^{-b},\ M<0$
Sign-magnitude Truncation	One's complement Sign-magnitude	$-2^{-b} < e \leq 0, x > 0$ $0 \leq e < 2^{-b}, x < 0$	$-2.2^{-b} < \varepsilon \leq 0$
Rounding	Sign-magnitude One's complement Two's complement	$\frac{-2^{-b}}{2} \leq e \leq \frac{2^{-b}}{2}$	$-2^{-b} \leq \varepsilon \leq 2^{-b}$

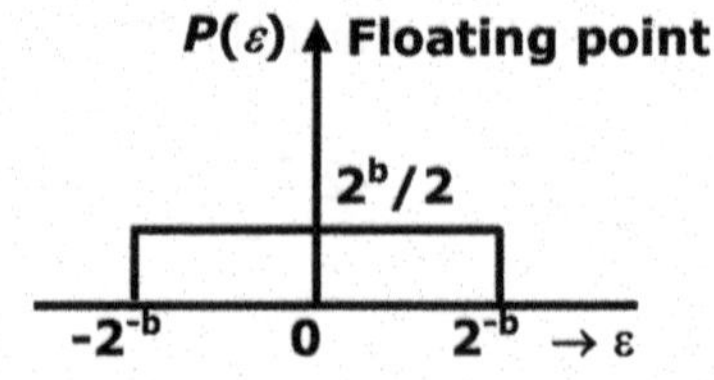

Figure 11.3 Probability density functions $P(\varepsilon)$ for rounding.

Example 11.16

Round the binary number 0.11010 rounded to three bits and binary number 0.110111111 to 8 bits.

Solution 11.16

When 0.11010 rounded to three bits is either 0.110 or 0.111 and the number 0.110111111 is rounded to 8 bits, then the result may be 0.11011111 or 0.11100000. Rounding up or down will have negligible effect on accuracy of computation.

11.8 The Input Quantization Error

We have observed that the quantization error occurs whenever a continuous signal is converted into a digital signal. Thus, the quantization error is given as

$$e(n) = x_q(n) - x(n) \tag{11.37}$$

here $x_q(n)$ = sampled quantized value of signal and
$x(n)$ = sampled unquantized value of signal.
Depending on the manner in which $x(n)$ is quantized, different distributions of quantization noise can be obtained. If rounding of a number is used to obtain $x_q(n)$, then the error signal satisfies the following relation.

$$-\frac{q}{2} \leq e(n) \leq \frac{q}{2} \tag{11.38}$$

Since the quantized signal may be greater or less than the actual signal.
As an example,

Let $x(n) = (0.70)_{10} = 0.10110011\ldots)_2$
↑┘
add
After rounding $x(n)$ to 3 bits, we get

$$x_q(n) = 0.101 \Big\} \text{1 add}$$

$$\overline{0.110}$$

$$=(0.75)_{10}$$

Now, the error will be

$$e(n) = x_q(n) - x(n) = 0.05$$

which satisfies the inequality.

The probability density function $P(e)$ for roundoff error and quantization characteristics with round are shown in Figure 11.4 (a) and (b).

Note that the other type of quantization can be obtained by truncation. In truncation, the signal is represented by the highest quantization level which is not greater than the signal. Here, in two's complement truncation, the error $e(n)$ is always negative and satisfies the following inequality.

$$-q \leq e(n) < 0 \tag{11.39}$$

The quantizer characteristics for truncation and probability density function $P(e)$ for two's complement truncation are shown in Figure 11.5(a) and (b), respectively.

From Figures 11.4 and 11.5, it is obvious that the quantization error mean value is 0 for rounding and $-q/2$ for two's complement truncation.

11.9 The Coefficient Quantization Error

As a matter of fact, in design of a digital filter, the coefficients are evaluated with infinite precision. However, when they are quantized, the frequency

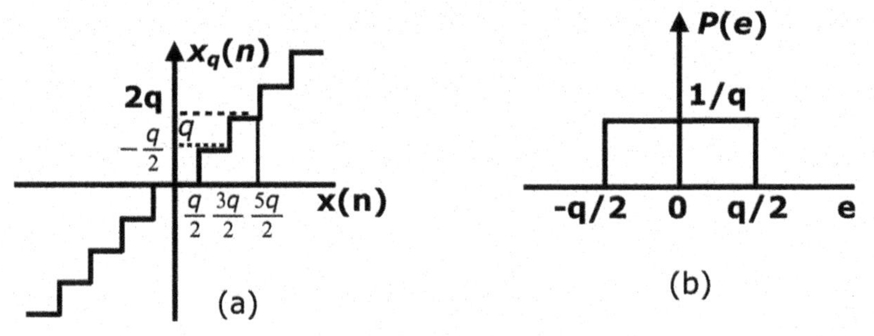

Figure 11.4 (a) Quantizer characteristics with rounding. (b) Probability density functions for roundoff error.

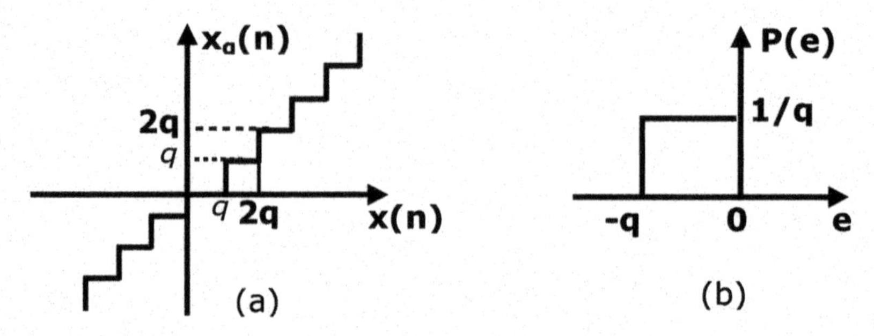

Figure 11.5 (a) Quantizer characteristics with two's complement truncation. (b) Probability density function of roundoff error.

response of the actual filter deviates from that which would have been obtained with an infinite word length representation and the filter may actually fail to meet the desired specifications. If the poles of the desired filter are closed to the unit circle, then those of the filter with quantized coefficients may lie outside the unit circle.

Example 11.17

Given a second-order IIR filter with transfer function

$$H(z) = \frac{1.0}{(1 - 0.5z^{-1})(1 - 0.45z^{-1})}$$

Determine the effect on quantization on pole locations of the given system function in direct form and in cascade form. Take $b = 3$ bits.

Solution 11.17

Direct Form I

We can write

$$H(z) = \frac{1}{1 - 0.95z^{-1} + 0.225z^{-2}}$$

$$(0.95)_{10} = (0.1111001\ldots)_2 \quad (-0.95)_{10} = (1.1111001\ldots)_2$$

After truncation, we have $(1.111)_2 = -0.875$

Similarly, we have

$$(0.225)_{10} = (0.001110\ldots)_2$$

After truncation, we have $(0.001)_2 = 0.125$

Therefore, $H(z) = \frac{1}{1-0.875z^{-1}+0.125z^{-2}}$ cascade form $H(z) = \frac{1}{(1-0.5z^{-1})(1-0.45z^{-1})}$

$$(-0.5)_{10} = (1.100)_2 \quad (-0.45)_{10} = (1.01110\ldots)_2$$

After truncation, we have

$$(1.011)_2 = (-0.375)_{10}$$

Thus, $H(z) = \frac{1}{(1-0.5z^{-1})(1-0.375z^{-1})}$.

11.10 Effects in FIR Digital Filters

In case of FIR filters, there are no limit cycle oscillations, if the filter is realized in direct form or cascade form, because these structures have no feedback. However, recursive realizations of FIR system such as the frequency sampling structures are subject to the above problems such as limit cycle oscillation.

Let us consider a linear shift invariant system with unit-sample response $h(n)$, which is nonzero over the interval $0 \le n \le N-1$. The direct form realization of such a system may be obtained using convolution sum as follows:

$$y(n) = \sum_{k=0}^{N-1} h(k)\, x(n-k) \tag{11.40}$$

The direct form realizations of the system and the roundoff noise model are shown in Figure 11.6, respectively. If rounding is used, the noise value at each multiplier can be assumed to be uniformly distributed between $\pm\frac{2^{-b}}{2}$ with zero mean value and variance of $\pm\frac{2^{-2b}}{12}$.

Now, let us assume the following

(i) The sources $e_k(n)$ are white sources.
(ii) The errors are uniformly distributed.
(iii) The error samples are uncorrelated with the input and each other.

Since the noise sources are assumed independent, the variance of the output noise will be

$$\sigma_\varepsilon^2 = \frac{N2^{-2b}}{12} \tag{11.41}$$

Because the noise power increases with N, N is also equal to the duration of the FIR filter unit-sample response $h(n)$. In fact, this is one of the reasons to make the duration of $h(n)$ as small as possible.

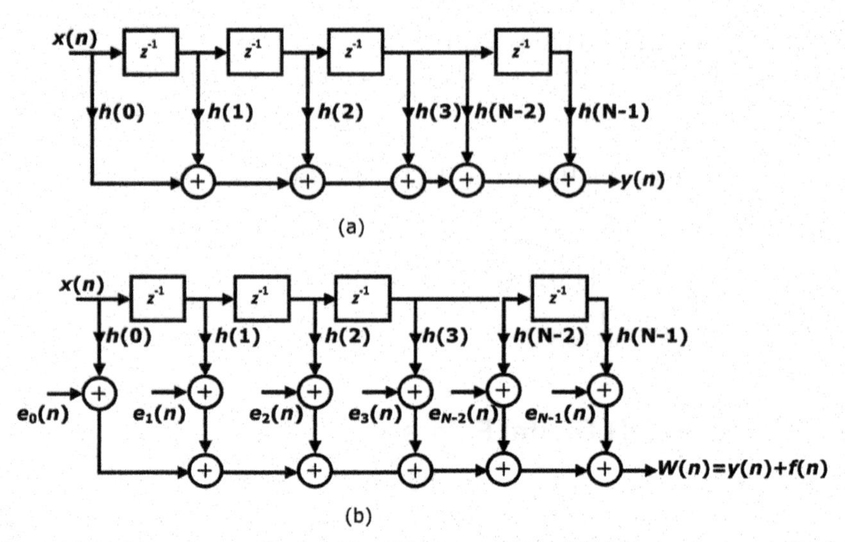

Figure 11.6 Direct form realization of an FIR system (a) Ideal (b) Fixed-point roundoff noise mode.

However, the main disadvantage of fixed-point arithmetic is its dynamic range limitation which necessitates scaling of the input so that no overflow occurs.

For an LTI system, we can find the least upper bound of the output sequence $y(n)$ as follows:

$$|y(n)| \leq x_{\max} \sum_{n=0}^{N-1} |h(n)| \tag{11.42}$$

where $x_{\max}$ is the maximum value of the input sequence. In order to avoid the overflow, we have $S_0\ |y(n)\ | < 1$ and 0 for all n. This means that the scaling factor must satisfy the following inequality.

$$S_0 < \frac{1}{x_{\max} \sum\limits_{n=0}^{N-1} |h(n)|} \tag{11.43}$$

This estimate is too conservative for a narrow band signal such as sine wave. Hence, the input should be scaled in terms of the peak of the frequency response of the system. This means that

$$S_0 < \frac{1.0}{x_{\max} \cdot \max\limits_{0 \leq \omega \leq \pi} [|H(e^{j\omega})|]} \tag{11.44}$$

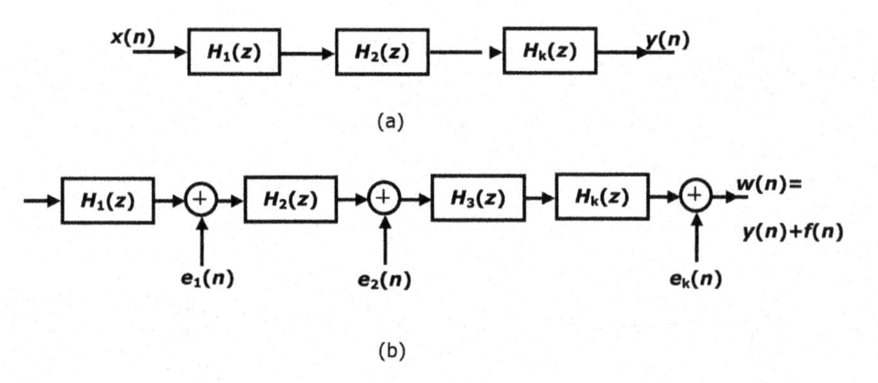

Figure 11.7 Cascade realization of an FIR system (a) Ideal (b) Fixed-point roundoff noise mode.

Further, an FIR filter can also be realized as a cascade of second-order sections as shown in Figure 11.7, where each second-order section $H_k(z)$ is realized in direct form. For our convenience, let N is odd so that $M = \frac{N-1}{2}$. Because each second-order section contains three independent white noise, the variance of each source $e_i(n)$ at the output of each section will be equal to $3\left(\frac{2^{-2b}}{12}\right) = \frac{2^{-2b}}{4}$.

In this case, noise source $e_k(n)$ is filtered by succeeding sections, so that the output noise variance would be dependent upon the order of the second-order sections in the chain. If we define $g_i(n)$ to be the unit-sample response from noise source $e_i(n)$ to the output, we can have

$$\sigma_{e_i}^2 = \frac{2^{-2b}}{4}\left[\sum_{n=0}^{N-2i} g_i^2(n)\right] \tag{11.45}$$

and the total output noise variance will be given by

$$\sigma_\varepsilon^2 = \sum_{i=1}^{M} \sigma_{ei}^2 \quad \text{or} \quad \sigma_\varepsilon^2 = \frac{2^{-2b}}{4}\left[\sum_{i=1}^{M}\sum_{n=0}^{N-2i} g_i^2(n)\right] \tag{11.46}$$

11.11 Problems and Solutions

Problem 11.1

Convert the given decimal number 12.25 to binary form.

Solution 11.1

We have the following procedure

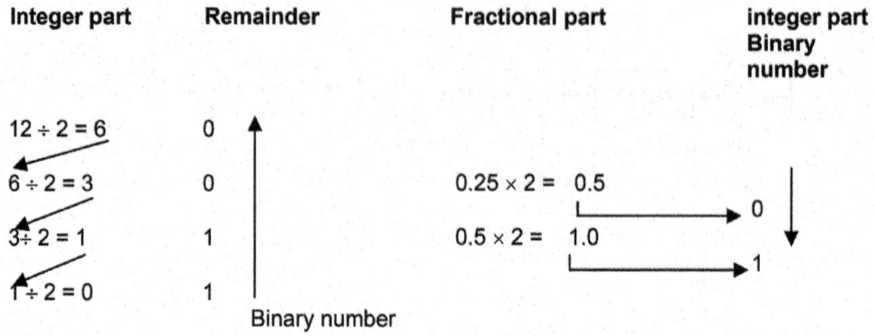

Integer part	Remainder	Fractional part	integer part Binary number
12 ÷ 2 = 6	0		
6 ÷ 2 = 3	0	0.25 × 2 = 0.5	0
3 ÷ 2 = 1	1	0.5 × 2 = 1.0	1
1 ÷ 2 = 0	1		

Binary number

Therefore, from above, we get

$$(12.25)_{10} = (1100.01)_2$$

$$1100.01 = 1 \times 2^3 + 1 \times 2^2 + 0 \times 2^1 + 0 \times 2^0 + 0 \times 2^{-1} + 1 \times 2^{-2}$$

Problem 11.2

Convert the following numbers into decimal

(a) $(1110.01)_2$
(b) $(11011.1110)_2$

Solution 11.2

$$\begin{aligned}\text{(a) } (1110.01)_2 &= (2^3 \times 1 + 2^2 \times 1 + 2^1 \times 1 + 2^0 \times 0) \cdot (0 \times 2^{-1} + 0 \times 2^{-2} \\ &= (8 + 4 + 2 + 0) \cdot (0 + 0.25) = (14.25)_{10}\end{aligned}$$

$$\begin{aligned}\text{(b) } (11011.1110)_2 &= (2^4 \times 1 + 2^3 \times 1 + 2^2 \times 0 + (2^1 \times 1 + 2^0 \times 1) \\ &\cdot (2^{-1} \times 1 + 2^{-2} \times 1 + 2^{-3} \times 1 + 2^{-4} \times 0) \\ &= (16 + 8 + 1) \cdot (0.5 + 0.25 + 0.125) \\ &= (27.875)_{10}\end{aligned}$$

Problem 11.3

Convert the following decimal numbers into binary

(a) $(20.675)_{10}$
(b) $(120.75)_{10}$

Solution 11.3

(a) $(20.675)_{10}$

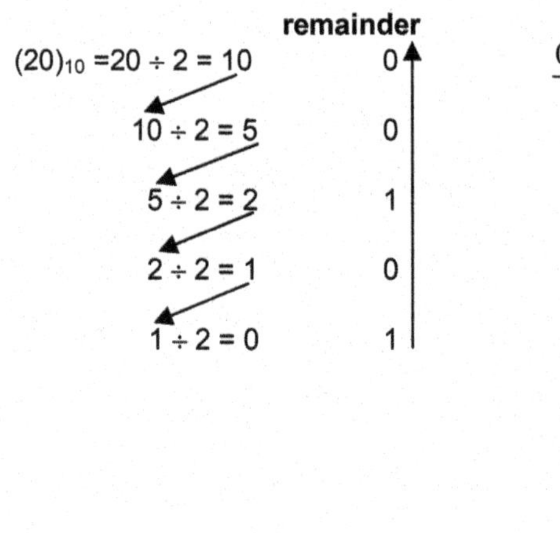

integer part

$\frac{0.675 \times 2}{1.35} \rightarrow 1$

$\frac{0.35 \times 2}{1.70} \rightarrow 0$

$\frac{0.7 \times 2}{1.4} \rightarrow 1$

$\frac{0.4 \times 2}{1.8} \rightarrow 0$

$\frac{0.8 \times 2}{1.6} \rightarrow 1$

$\frac{0.6 \times 2}{1.2} \rightarrow 1$

$\frac{0.2 \times 4}{0.8} \rightarrow 0$

Thus, $(20.675)_{10} = (10100.1010110\ldots)_2$

(b)$(120.75)_{10}$

	remainder
$(120)_{10} = 120 \div 2 = 60$	0
$60 \div 2 = 30$	0
$30 \div 2 = 15$	0
$15 \div 2 = 7$	1
$7 \div 2 = 3$	1
$3 \div 2 = 1$	1
$1 \div 2 = 0$	1

integer part

$\frac{0.75 \times 2}{1.5} \rightarrow 1$

$\frac{0.5 \times 2}{1.0} \rightarrow 1$

Thus, we have

$$(120.75)_{10} = (1111000.11)_2$$

Problem 11.4

Multiply $(2.5)_{10}$ and $(3.25)_{10}$ number using block floating-point technique.

Solution 11.4

Consider the multiplication of two numbers

$(2.5)_{10}$ and $(3.25)_{10}$
$(2.5)_{10} = 2^2 \times 0.625 = 2^{010} \times 0.1010$
$(7.25)_{10} = 2^3 \times 0.90625 = 2^{011} \times 0.111010$
Now,
$(2.5)_{10}(7.25)_{10}$
$= (2^{010} \times 0.1010) \times (2^{011} \times 0.111010)$
$= 2^{101} \times 0.10010001$

$(0.625)_{10} = (0.1010)_2$
$(0.90625)_{10} = (0.111010)_2$

Problem 11.5

The input to the system

$$y(n) = 0.999y(n-1) + x(n) \text{ is applied to an ADC.}$$

What will be the power produced by the quantization noise at the output of the filter if the input is quantized to (a) 8 bits (b) 16 bits.

Solution 11.5

It is given that

$$y(n) = 0.999y(n-1) + x(n)$$

Taking z-transform on both sides in equation (i), we get

$$Y(z) = 0.999z^{-1}Y(z) + X(z)$$

Thus, $H(z) = \frac{Y(z)}{X(z)} = \frac{1}{1-0.999z^{-1}}$.

Taking inverse z-transform, we get

$$h(n) = (0.999)^n u(n)$$

The quantization noise power at the output of the digital filter is given by

$$\sigma_\varepsilon^2 = \sigma_e^2 \sum_{k=0}^{\infty} h^2(k)$$

$$\text{or} \quad \sigma_\varepsilon^2 = \sigma_e^2 \sum_{k=0}^{\infty} (0.999)^{2k} \quad \text{or}$$

$$\sigma_\varepsilon^2 = \sigma_e^2 \frac{1}{1-(0.999)^2} = \sigma_e^2(500.25)$$

$$\text{or } \sigma_\varepsilon^2 = \frac{2^{-2b}}{12}(500.25)$$

(a) Given that $b + 1 = 8$ bits (assuming including sign bit).
Then, $\sigma_\varepsilon^2 = \frac{2^{-14}}{12}(500.25) = 2.544 \times 10^{-3}$
(b) Given that $b + 1 = 16$ bits.
Then, $\sigma_\varepsilon^2 = \frac{2^{-30}}{12}(500.25) = 3.882 \times 10^{-8}$

Problem 11.6
The input to the system

$$y(n) = 0.888y(n-1) + x(n) \text{ is applied to an ADC.}$$

What will be the power produced by the quantization noise at the output of the filter if the input is quantized to (a) 2 bits (b) 4 bits.

Solution 11.6
It is given that

$$y(n) = 0.888y(n-1) + x(n)$$

Taking z-transform on both sides in equation (i), we get

$$Y(z) = 0.888z^{-1}\, Y(z) + X(z)$$

Thus,

$$H(z) = \frac{Y(z)}{X(z)} = \frac{1}{1-0.888z^{-1}}$$

Taking inverse z-transform, we get

$$h(n) = (0.888)^n u(n)$$

The quantization noise power at the output of the digital filter is given by

$$\sigma_\varepsilon^2 = \sigma_e^2 \sum_{k=0}^{\infty} h^2(k)$$

$$\text{or} \quad \sigma_\varepsilon^2 = \sigma_e^2 \sum_{k=0}^{\infty} (0.888)^{2k} \quad \text{or}$$

$$\sigma_\varepsilon^2 = \sigma_e^2 \frac{1}{1-(0.888)^2} = \sigma_e^2(4.73)$$

$$\text{or} \quad \sigma_\varepsilon^2 = \frac{2^{-2b}}{12}(500.25)$$

(a) Given that $b + 1 = 2$ bits (assuming including sign bit).

Then, $\sigma_\varepsilon^2 = \frac{2^{-2}}{12}(4.73) = 95.94 \times 10^{-3}$

(b) Given that $b + 1 = 4$ bits.

Then, $\sigma_\varepsilon^2 = \frac{2^{-20}}{12}(4.73) = 6.5 \times 10^{-3}$

Problem 11.7

Consider the recursive filter shown in the figure. The input $x(n)$ has the range of values of ± 100 V, represented by 8 bits. Calculate the variance of output due to A/D conversion process.

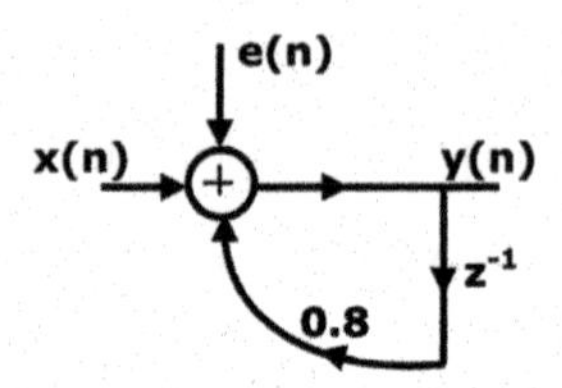

Solution 11.7

Given the range ± 100 V

The differential equation of the system is given by $y(n) = 0.8y(n-1) + x(n)$, whose impulse response $h(n)$ can be obtained as follows:

$$h(n) = (0.8)^n u(n)$$

$$\text{quantization step size} = \frac{\text{Range of the signal}}{\text{Number of Qunatization Levels}} = \frac{200}{2^8} = 0.78725$$

Variance of the error signal is given by

$$\sigma_e^2 = \frac{q^2}{12} = \frac{(0.78125)^2}{12} = 0.05086$$

Variance of the output will be

$$\sigma_\varepsilon^2 = \sigma_e^2 \sum_{k=0}^{\infty} h^2(n) = (0.05086) \sum_{n=0}^{\infty} (0.8)^{2n}$$

$$\text{or} \quad \sigma_\varepsilon^2 = \frac{0.05086}{1 - (0.8)^2} = 0.14128$$

Problem 11.8

Consider the recursive filter shown in the figure. The input $x(n)$ has the range of values of ± 50 V, represented by 16 bits. Calculate the variance of output due to A/D conversion process.

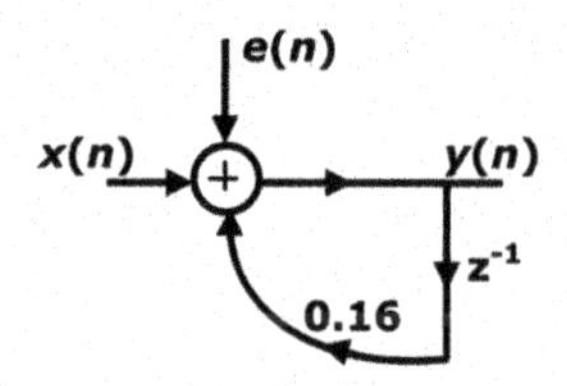

Solution 11.8

Given the range ± 50 V

The differential equation of the system is given by $y(n) = 0.16\, y(n - 1) + x(n)$, whose impulse response $h(n)$ can be obtained as follows:

$$Y(z)[1 - 0.16z^{-1}] = X(z) \quad \frac{Y(z)}{X(z)} = \frac{1}{[1 - 0.16z^{-1}]}$$

$$H(z) = \frac{1}{[1 - 0.16z^{-1}]} \quad h(n) = (0.16)^n u(n)$$

$$\text{quantization step size} = \frac{\text{Range of the signal}}{\text{Number of Qunatization Levels}}$$

$$= \frac{100}{2^{16}} = 1.52 \times 10^{-3}$$

Variance of the error signal is given by

$$\sigma_e^2 = \frac{q^2}{12} = \frac{(1.52 \times 10^{-3})^2}{12} = 1.9253 \times 10^{-7}$$

Variance of the output will be

$$\sigma_\varepsilon^2 = \sigma_e^2 \sum_{k=0}^{\infty} h^2(n) = (1.9253 \times 10^{-7}) \sum_{n=0}^{\infty} (0.16)^{2n}$$

$$\text{or} \quad \sigma_\varepsilon^2 = \frac{1.9253 \times 10^{-7}}{1 - (0.16)^2} = 1.976 \times 10^{-7}$$

Problem 11.9

Find the effect of coefficient quantization on pole locations of the given second-order IIR system, when it is realized in cascade form. Assume a word length of 4 bits through truncation.

$$H(z) = \frac{1}{1 - 0.9z^{-1} + 0.2z^{-2}}$$

Solution 11.9

Direct Form I

Let 4 bits include a sign bit, and then, we have

$$(0.9)_{10} = (0.111011\ldots)_2$$

	Integer part
0.9 × 2 = 1.8	1
0.8 × 2 = 1.6	1
0.6 × 2 = 1.2	1
0.2 × 2 = 0.4	0
0.4 × 2 = 1.8	1
0.8 × 2 = 1.6	1

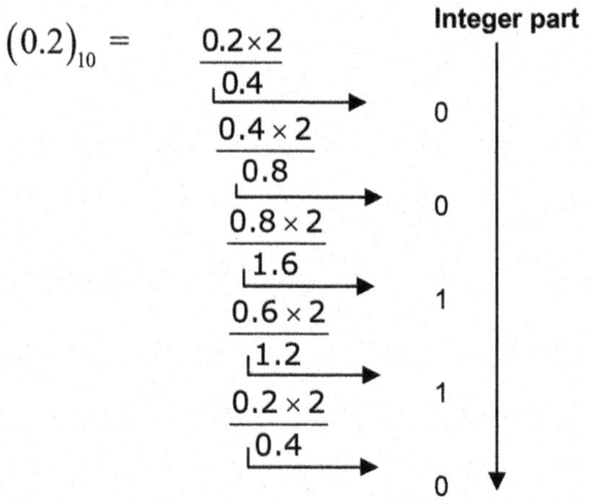

$(0.2)_{10} =$

	Integer part
$0.2 \times 2 = 0.4$	0
$0.4 \times 2 = 0.8$	0
$0.8 \times 2 = 1.6$	1
$0.6 \times 2 = 1.2$	1
$0.2 \times 2 = 0.4$	0

After truncation, we obtain

$$(0.111)_2 = (0.875)_{10}$$

$$(0.2)_{10} = (0.00110\ldots)_2$$

After truncation, we obtain

$$(0.001)_2 = (0.125)_{10}$$

The system function after coefficient quantization will be

$$H(z) = \frac{1}{(1 - 0.875z^{-1} + 0.125z^{-2})}$$

Now, the pole locations are given by

$$z_1 = 0.695, \qquad z_2 = 0.1798$$

Cascade form $H(z) = \frac{1}{1-0.9z^{-1}+0.2z^{-2}}$ $\quad H(z) = \frac{1}{(1-0.5z^{-1})(1-0.4z^{-1})}$

$$(0.5)_{10} = (0.1000)_2$$

After truncation, we obtain

$$(0.100)_2 = (0.5)_{10}$$
$$(0.4)_{10} = (0.01100\ldots\ldots)_2$$

After truncation, we obtain

$$(0.011)_2 = (0.375)_{10}$$

The system function after coefficient quantization will be

$$H(z) = \frac{1}{(1 - 0.5z^{-1})(1 - 0.375z^{-1})}$$

The pole locations are given by

$$z_1 = 0.5$$
$$z_2 = 0.375$$

Problem 11.10

Find the effect of coefficient quantization on pole locations of the given second-order IIR system, when it is realized in direct form I. Assume a word length of 4 bits through truncation.

$$H(z) = \frac{1}{1 - 0.95z^{-1} + 0.2z^{-2}}$$

Solution 11.10

Direct Form I

Let 4 bits include a sign bit; then, we have

$$(0.95)_{10} = (0.11110\ldots)_2$$

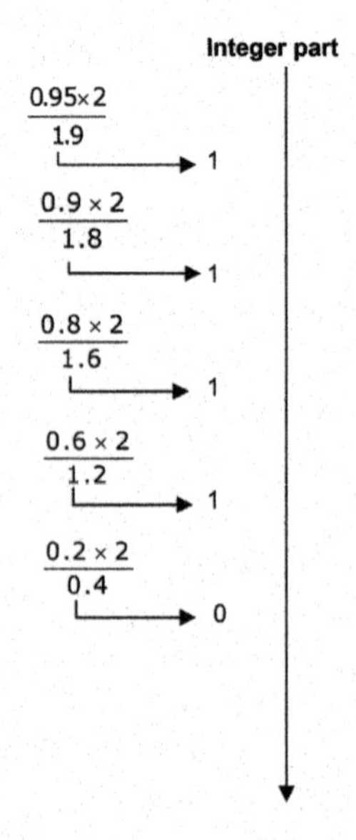

After truncation, we obtain

$$(0.111)_2 = (0.875)_{10}$$
$$(0.2)_{10} = (0.00110\ldots)_2$$

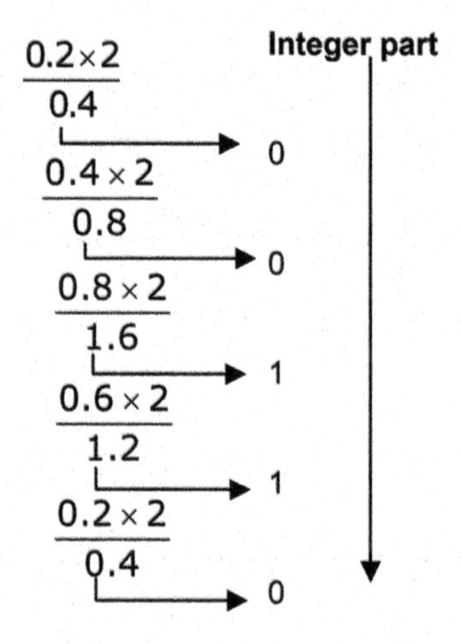

After truncation, we obtain

$$(0.001)_2 = (0.125)_{10}$$

The system function after coefficient quantization will be

$$H(z) = \frac{1}{(1 - 0.875z^{-1} + 0.125z^{-2})}.$$

12

Review Question with Answers and Multiple Choice Questions

12.1 Review Questions with Answers

1. What do you mean by signal processing?
   Any operation that changes the characteristic of a signal is called signal processing.
2. Define a signal.
   Any physical quantity that varies with time, space, or any other independent variable is called a signal.
3. What do you mean by a system?
   A physical device that performs an operation on a signal is defined as a system.
4. Mention a few advantages of DSP.
   (a) Good accuracy.
   (b) Data storage becomes easier.
   (c) Cheaper.
   (d) Easy implementation of algorithms.
5. What are the limitations of DSP?
   (a) Power consumption is more in DSP.
   (b) Bandwidth is limited by sampling rate.
6. Mention some of the applications of DSP.
   (a) Speech processing.
   (b) Image processing.
   (c) Telecommunication.
   (d) Military applications.
7. Mention some of the Military applications of DSP
   (a) Sonar Signal processing.

(b) Radar signal processing.

8. How are signals classified?

(a) Continuous time signal.
(b) Discrete-time signal.

9. What do you mean by continuous-time signal?
A signal which varies continuously with time is called continuous time signal.
10. What do you mean by discrete-time signal?
A signal which has values only at discrete instants of time is called a discrete-time signal.
11. Sketch a continuous time signal.

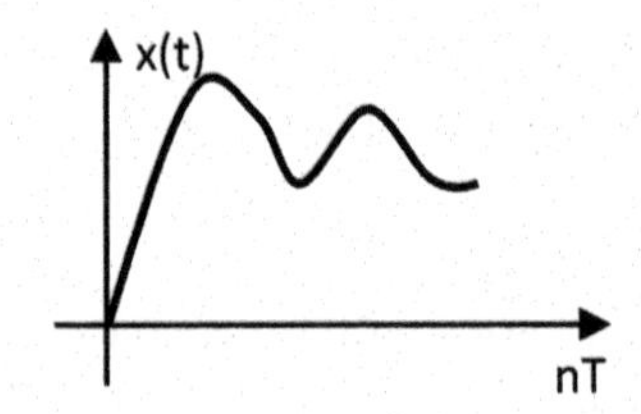

12. Sketch a sample discrete-time signal.

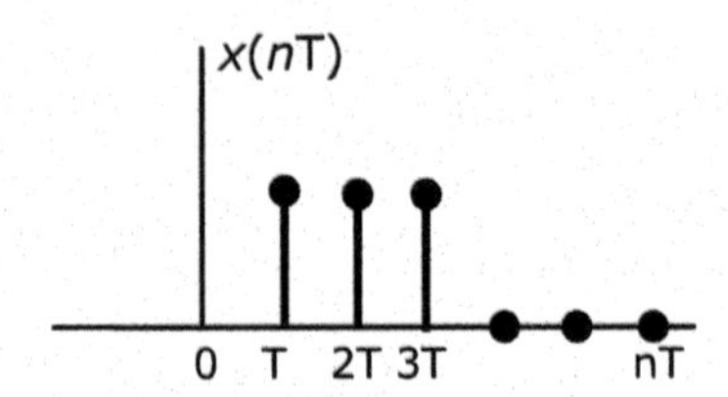

13. What are energy signals?
The energy of a signal is defined as

$$E = \sum_{n=-\infty}^{\infty} |x(n)|^2$$

If E is finite, then $x(n)$ is an energy signal.
14. What are power signals?
The average power of a signal (discrete-time) $x(n)$ is

$$P = Lt_{N \to \infty} \frac{1}{2N+1} \sum_{n=-N}^{N} |x(n)|^2$$

If P is non-zero and finite, the signal is called a power signal.

15. Differentiate between DSPs and Microprocessors.
    (a) Microprocessors are built for a range of general purpose functions, and large blocks of software, such as operating systems like UNIX.
    (b) Microprocessors are not called upon for real-time applications.
    (c) DSPs are employed as attached processors, assisting a general-purpose host microprocessor.
16. What do you mean by aliasing?
    Aliasing is the result of sampling, which means that we cannot distinguish between high and low frequencies.
17. What is Quantization error?
    It results due to limited precision (word length) while converting between analog and digital forms, when sorting data, or when performing arithmetic.
18. What are the effects of the limitations of DSP?
    (a) Aliasing.
    (b) Quantization Error.
19. Define Sampling
    It is a process of converting a continuous-time signal into a discrete-time signal.
20. Sketch the block diagram of an A/D converter.

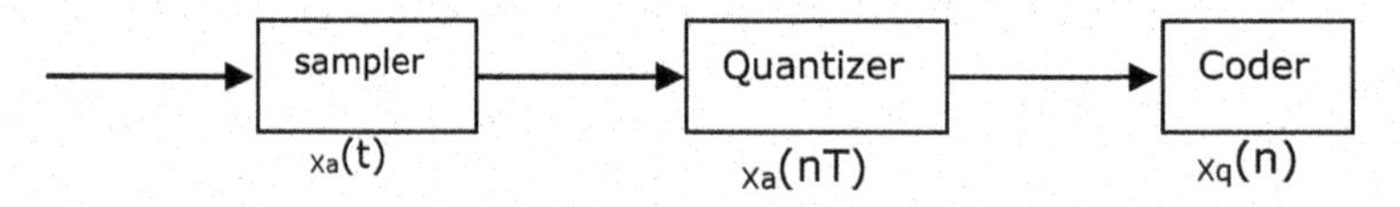

21. What are the different representations of a signal?
    (a) Functional representation.
    (b) Sequence representation.
    (c) Tabular representation.
    (d) Graphical representation.

    Functional representation

$$\begin{aligned} x(n) &= 1, \text{ for } n = 0, 1, 2 \\ &= 0, \text{for other } n. \end{aligned}$$

Sequence representation

$$x(n) = \{0,\ 0,\ 0,\ 1,\ 1,\ 1,\ 0,\ 0,\ \ldots.\}$$

Tabular representation

N	0	1	2	3	4
$x(n)$	1	1	1	0	0

Graphical representation

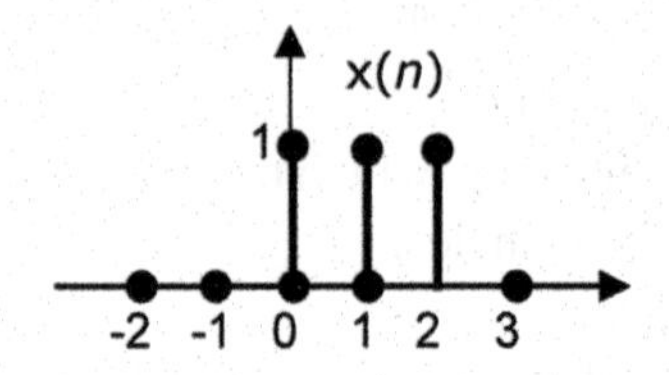

22. What are the different operations performed on a discrete-time signal?
    (a) Delay.
    (b) Advance.
    (c) Folding.
    (d) Time scaling.
    (e) Amplitude scaling.
    (f) Addition.
    (g) Multiplication.
23. What is a stable system?
    A system is said to be bounded-input bounded-output (BIBO) stable if and only if every bounded input produces a bounded output.

    $$|x(n)| \leq M_x < \infty \text{ and } |y(n)| < M_y < \infty \text{ if those exists,}$$

    $$M_x \text{ and } M_y.$$
24. What is linear/discrete convolution?
    The output $y(n)$ obtained by convolving the Impulse response $h(n)$ with the inputs signal $x(n)$.
    $$y(n) = \sum_{k=-\infty}^{\infty} x(k)h(n-k) = \sum_{k=-\infty}^{\infty} h(k)x(n-k)$$
25. What are the properties of convolution?
    (a) Commutative
    $x(n) \otimes h(n) = h(n) \otimes x(n)$
    (b) Associative
    $[x(n) \otimes h_1(n)] \otimes h_2(n) = x(n) \otimes [h_1(n) \otimes h_2(n)]$

(c) Distributive

$$x(n) \otimes [h_1(n) + h_2(n)] = x(n) \otimes h_1(n) + x(n) \otimes h_2(n)$$

26. Evaluate

$$y(n) = x(n) \otimes h(n) \text{ where } \quad x(n) = h(n) = \{\underset{\uparrow}{1}, 2, -1\}$$

By overlap and add method, we have $y(n) = \{1, 4, 2, -4, 1\}$

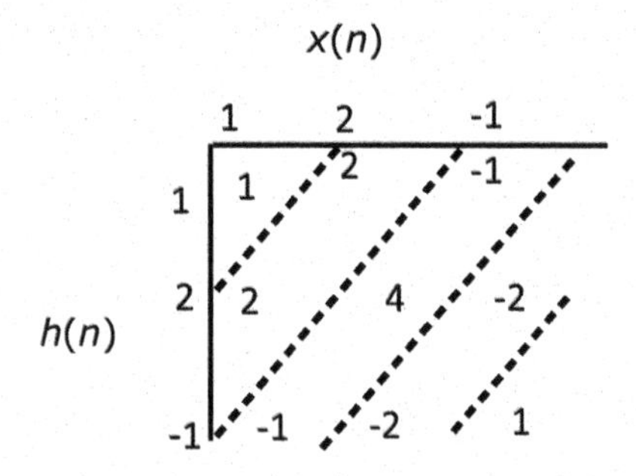

27. How do you find the step response if impulse response is known?

$$y(n) = x(n) \otimes h(n)$$

For $x(n) = u(n)$, $y(n) = u(n) \otimes h(n)$

$$= \sum_{k=-\infty}^{\infty} u(n-k)h(k) = \sum_{k=-\infty}^{\infty} u(k) \;\; v(n-k) = 0, for\ k > n.$$

28. Define Fourier Transform of a sequence.
The F.T. of a discrete-time signal $x(n)$ is

$$X(\omega) = \sum_{n=-\infty}^{\infty} x(n)e^{-j\omega n}$$

29. Write down the sufficient condition for the existence of Discrete-Time Fourier Transform (DTFT)?
For a sequence $x(n)$.

$$\sum_{n=-\infty}^{\infty} |x(n)| < \infty$$

30. State Parseval's is the theorem for discrete-time signals.

$$E = \sum_{n=-\infty}^{\infty} |x(n)|^2 = \frac{1}{2\pi}\int_{-\pi}^{\pi} |X(\omega)|^2 d\omega$$

31. What is DTFT pair?
The F.T. pair of discrete-time signal is

$$x(n)| = \frac{1}{2\pi}\int_{2\pi} |X(\omega).e^{j\omega n} d\omega$$

$$X(\omega) = \sum_{n=-\infty}^{\infty} x(n).e^{-j\omega n}$$

32. State the properties of Fourier Transform of a discrete-time a periodic sequence.
    (a) The Fourier spectrum of an a periodic sequence is continuous.
    (b) The Fourier spectrum is periodic with period 2π.
33. Find the discrete Fourier Transform of a sequence given by

$$\begin{aligned} x(n) &= 1, -2 \leq n \leq 2 \\ &= 0, \text{ otherwise.} \end{aligned}$$

$$X(\omega) = \sum_{n=-\infty}^{\infty} x(n)e^{-j\omega n} = \sum_{n=-2}^{2} e^{-j\omega n}$$

$$\begin{aligned} X(\omega) &= e^{2j\omega} + e^{j\omega} + 1 + e^{-j\omega} + e^{-j2\omega} \\ X(w) &= 1 + 2\ \cos\omega + \ 2\ \cos 2\omega \end{aligned}$$

34. Find the Fourier Transform of

$$x(n) = (0.8)^n,\ n = 0, \pm 1, \pm 2, \ldots.$$

For negative values of n, $x(n)$ is not absolutely summable. Therefore, the Fourier Transform does not exist.

35. What is the frequency shifting property of Discrete-Time Fourier Transform?
If $\text{DTFT}[x(n)] = X(e^{j\omega})$
Then $\text{DTFT}[e^{j\omega_0 n}x(n)] = X[e^{j(\omega-\omega_0)}]$

36. What is the time-shifting property of discrete-Time Fourier Transform?
If $\text{DTFT}[x(n)] = X(e^{j\omega})$ then
Then $\text{DTFT}[x(n-k)] = e^{-j\omega k}X(\omega)$
37. What is the linear property of DTFT?
If $\text{DTFT}[x_1(n)] = X_1(\omega)$ and $\text{DTFT}[x_2(n)] = X_2(\omega)$
Then $\text{DTFT}[a_1x_1(n) + a_2x_2(n)] = a_1X_1(\omega) + a_2X_2(\omega)$
38. Find the transfer function of the 3-sample average:

$$\begin{aligned} h(n) &= \tfrac{1}{3}, -1 \leq n \leq 1 \\ &= 0, \text{ otherwise} \end{aligned}$$

$$H(\omega) = \sum_{n=-\infty}^{\infty} h(n)e^{-j\omega n} = \sum_{n=-1}^{1} h(n)e^{-j\omega n} = \sum_{n=-1}^{1} \frac{1}{3}e^{-j\omega n}$$

39. Write mathematical expression of the *z*-transform.
The *z*-Transform of a discrete-time signal *x*(*n*) is

$$X(z) = \sum_{n=-\infty}^{\infty} x(n)z^{-n}$$

40. What do you mean by ROC?
ROC means Region of Convergence, i.e., the ROC of *X*(*z*) is the set of all values of *z* for which *X*(*z*) attains a finite value.
41. Explain the linearity property of *z*-Transform.

$$\begin{aligned} &\text{If} \quad Z[x_1(n)] = X_1(z) \text{ and } Z[x_2(n)] = X_2(z) \text{ then} \\ &Z[a_1x_1(n) + a_2x_2(n)] = a_1X_1(z) + a_2X_2(z). \end{aligned}$$

42. State the time-shifting property of *z*-Transform.

$$\begin{aligned} &\text{If} \quad Z[x(n)] = X(z), \text{ then} \\ &Z[x(n-k)] = z^{-k}X(z). \end{aligned}$$

43. State the scaling property of *z*-Transform.

$$\begin{aligned} &\text{If} \quad Z[x(n)] = X(z), \text{ then} \\ &Z[a^n x(n)] = X(a^{-1}z) \end{aligned}$$

44. State the time reversal property of *z*-Transform.

$$\begin{aligned} &\text{If} \quad Z[x(n)] = X(z), \text{ then} \\ &Z[x(-n)] = X(z^{-1}) \end{aligned}$$

45. State the convolution property of z-Transform.

$$\text{If } Z[x_1(n)] = X_1(z) \text{ and } Z[x_2(n)] = X_2(z),$$
$$\text{then } Z[x_1(n) * x_2(n)] = X_1(z)X_2(z)$$

46. State the multiplication property of z-Transform.

$$\text{If } Z[x_1(n)] = X_1(z), \text{ and } Z[x_2(n)] = X_2(z) \text{ then}$$

$$Z[x_1(n)x_2(n)] = \frac{1}{2\pi j}\oint_C X_1(v)X_2\left(\frac{z}{v}\right)v^{-1}dv$$

47. State the Parseval's Theorem in z-Transform.
If $x_1(n)$ and $x_2(n)$ are complex-valued functions, then

$$\sum_{n=-\infty}^{\infty} x_1(n) * x_2(n) = \frac{1}{2\pi j}\oint_C X_1(v)X_2 * \left(\frac{1}{v}\right)v^{-1}dv$$

48. State the initial value theorem of z-Transform.
If $x(n)$ is causal, then

$$x(0) = Lt_{z\to\infty} X(z)$$

49. State the final value theorem of z-Transform.
If $x(n)$ is causal, then

$$x(\infty) = Lt_{z\to 1}(z-1)X(z)$$

50. Determine the z-Transform of a digital impulse.
Since $x(n)$ magnitude is zero except for $n = 0$, where $x(n)$ is 1 we have $X(z) = 1$.
51. What are discrete-time signals?
Signals represented as

$$h(n), \quad N_1 \leq n \leq N_2$$
$$h(nT), \quad N_1 \leq n \leq N_2$$

The first equation represents a non-uniformly spaced samples and the second equation represents uniformly spaced samples.
52. Define a unit sample sequence.

$$u_0(n) = \begin{cases} 1, & n = 0 \\ 0, & n \neq 0 \end{cases}$$

53. Define an impulse delayed by n_0 samples.

$$u_0(n - n_0) = \begin{cases} 1, & n = n_0 \\ 0, & n \neq n_0 \end{cases}$$

54. What is a unit step sequence?

$$u(n) = \begin{cases} 1, & n \geq n_0 \\ 0, & n < 0 \end{cases}$$

55. What is decaying exponential?

$$f(n) = \begin{cases} a^n, & n \geq 0 \\ 0, & n < 0 \end{cases}$$

56. Define a sinusoid.

$$x(n) = \cos\left(\frac{2\pi\, n}{n_0}\right), \quad \text{for all } n.$$

57. What is a linear time invariant system?
If $x_1(n)$ and $x_2(n)$ are inputs to a system with $y_1(n)$ and $y_2(n)$ as their corresponding outputs, the system is said to be linear if the sequence $ax_1(n) + bx_2(n)$ applied at the input produces a sequence $ay_1(n) + ay_2(n)$ as the output.
In a time invariant system, if the input sequence $x(n)$ produces an output sequence $y(n)$, then the input sequence $x(n - n_0)$ produces an output sequence $y(n - n_0)$, for all n_0.
58. When is an LTI system said to be causal?
An LTI system is said to be casual or realizable if the output at $n = n_0$ is dependent only on the values of the input for $n = n_0$. This means the impulse response $h(n) = 0$, for $n < 0$.
59. When is an LTI system said to be stable?
An LTI system is said to be stable if every bounded input produces a bounded output.
60. State the necessary and sufficient condition for stability of an LTI system.
For an LTI system with impulse response $h(n)$, the condition for stability is

$$\sum_{n=-\infty}^{\infty} |h(n)| < \infty$$

61. Define DTFT (Discrete-Time Fourier Transform) of a signal.

    The DTFT $[x(n)] = X[w]$ $X[\omega] = \sum_{n=0}^{\infty} x(n)e^{-j\omega n}$

    where $X[\omega]$ is the frequency response of the system with impulse response $x(n)$.

62. Define IDTFT (Inverse Discrete-Time Fourier Transform) of a spectrum.

$$\text{IDTFT} \quad [H(\omega)] = h(n) = \frac{1}{2\pi}\int_{2\pi} H(\omega)\, e^{jwn}\, d\omega$$

    represents the IDTFT, where $H(\omega)$ is periodic with period 2π.

63. Write the DTFT of (a) $x(n)$ (b) x*(n) (c) x*(-n)

    (a) DTFT $[x(n)] = X[w]$
    (b) DTFT $[x*\ (n)] = X*\ [-w]$
    (c) DTFT $[x*(-n)] = X*\ [w]$

64. Write the time shift property of a discrete signal.

$$\text{DTFT}[x(n - k)] = e^{-jwk}X[w]$$

65. State the time reversal property of discrete-time signal.

$$\text{DTFT}[x(-n)] = X[-w]$$

66. State the convolution property of discrete-time signals

$$\text{DTFT}\ [x_1(n) \otimes x_2(n)] = X_1[w]\ X_2[w]$$

    Convolution of discrete-time signals in the time domain equals the frequency-domain multiplication.

67. State the frequency-shifting property of discrete-time signals.

$$\text{DTFT}\ [e^{jw0n}x(n)] = X[w - w_0]$$

68. State the modulation theorem of discrete-time signals.

$$\text{DTFT}[x(n)\cos\omega_0 n] = \frac{1}{2}[X(\omega+\omega_0) + X(\omega-\omega_0)]$$

69. What is the DTFT of $x_1(n)\ x_2(n)$?

$$\text{DTFT}\ [x_1(n)\ x_2(n)] = \frac{1}{2\pi}\int_{-\pi}^{\pi} X_1(\lambda)X_2(\omega-\lambda)d\lambda$$

70. Define Discrete Fourier Transform (DFT).

$$\mathrm{DFT}[x(n)] = \sum_{k=0}^{N-1} x(n)W_N^{kn}$$

71. Define Inverse Discrete Fourier Transform (IDFT).

$$\mathrm{DFT}[X_p(k)] = x_p(nT) = \frac{1}{N}\sum_{k=0}^{N-1} X_p(k)W_N^{kn}$$

72. For a given signal, $x(n) = e^{j\omega n}$, $-\infty < n < \infty$ obtain the spectrum $H(\omega)$.

$$y(n) = \sum_{m=-0}^{\infty} h(m)e^{j\omega(n-m)} = e^{j\omega m}\sum_{m=-0}^{\infty} h(m).e^{-j\omega n}$$

$$y(n) = x(n)H(\omega)$$
$$H(w) = \frac{y(n)}{x(n)}$$

73. For an LSI system impulse response $h(n) = a^n\, u(n)$, $|a| < 1$, find the frequency response.

$$H(\omega) = \sum_{n=0}^{\infty} a^n e^{-j\omega n} = \sum_{n=0}^{\infty}(ae^{-j\omega})^n$$

$$H(\omega) = \frac{1}{1 - ae^{-j\omega}}$$

74. What do you mean by recursive realization?
Recursive realization is obtained by using the functional relationship between input sequence $\{x(n)\}$ and output sequence $[y(n)]$ defined by

$$y(n) = F[y(n-1), y(n-2), \ldots., x(n), x(n-1), \ldots]$$

75. What do you mean by non-recursive realization?
Non-recursive realization is obtained by using the functional relationship between input sequence $\{x(n)\}$ and output sequence $[y(n)]$ defined by $y(n) = F[x(n), x(n-1), \ldots.]$, i.e., current output sample $y(n)$ is a function only of past and present inputs.

76. What are the different realization structures?
    (a) Direct form-I
    (b) Direct from-II
    (c) cascade form
    (d) parallel form
    (e) Lattice form
77. Compare Direct Form-I and direct From-II realization structures.
    The number of memory location necessary is more in the case of direct form-I, then in direct form-II.
78. What is a discrete-time system?
    A discrete-time system is a device that operates on a discrete-time input signal $x(n)$, to produce a discrete-time output signal $y(n)$.
79. Sketch the direct form-I realization structure of a second-order system.

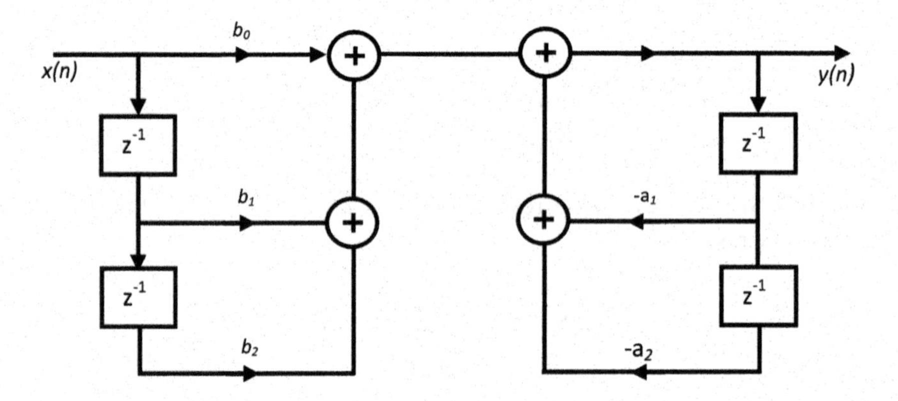

80. How are the discrete-time systems classified?
    (a) Static and dynamic systems
    (b) Time-variant and time-invariant systems.
    (c) Linear and non-linear systems.
    (d) Stable and unstable systems
    (e) Causal and non-causal systems.
81. What do you mean by a time-invariant system?
    If the input-output relation of a system does not vary with time, the system is a time-invariant system.
82. What is causal system?
    A system is said to be causal if the output of the system at any time is dependent only on present and past inputs, but not on future inputs.

83. What is an LTI system?
An LTI system is one which possesses the linearity and time invariance.
84. Define impulse response of a system.
The response or output signal $b(x)$, obtained from a discrete time system when the input signal is a unit sample sequence (unit impulse), is known as the unit sample response or impulse response.
85. State the causality condition for an LTI system.
Unit sample response $h(n) = 0$, for negative values of n. $h(n) = 0, n < 0$.
86. State the condition for system stability.

$$\sum_{k=-\infty}^{\infty} |h(k)| < \infty$$

87. What is the unit step response for the LTI system with impulse response $h(n) = \mathrm{a}^n\, u(n)$, $|\mathrm{a}| < 1$.

$$y(n) = \sum_{k=-\infty}^{\infty} h(k) = \sum_{k=0}^{n} a^k = \frac{1 - a^{n+1}}{1 - a}.$$

88. Find the impulse response of the Low-pass filter defined by

$$H(\omega) = 0, \omega_0 \leq\ |\omega|\ \leq \pi$$
$$h(n) = \tfrac{1}{2\pi} \int_{-\pi}^{\pi} H(\omega)\ e^{j\omega n} d\omega$$

$$h(n) = \frac{1}{2\pi} \int_{-\omega_0}^{\omega_0} e^{j\omega n} dw = \frac{1}{2\pi jn} e^{j\omega n}\Big|_{-\omega_0}^{\omega_0}$$

$$h(n) = \frac{1}{\pi\, n(2j)} \left[e^{j\omega_0 n} - e^{-j\omega_0 n}\right] = \frac{\sin \omega_0\, n}{\pi\, n}.$$

89. Define the frequency response of a discrete-time system.
Let $x(n) = \mathrm{e}^{j\omega n}$

$$\text{Then}\quad y(n) = \sum_{k=-\infty}^{\infty} h(k) x(n-k) = \sum_{k=0}^{\infty} h(k) e^{j\omega(n-k)}$$

$$= e^{j\omega n} \sum_{k=0}^{\infty} h(k) e^{-j\omega k} \qquad y(n)\ = e^{j\omega n} H(\omega)$$

$H(\omega)$ is called the frequency response of the LTI system.

$$H(\omega) = \sum_{k=0}^{\infty} (ae^{-j\omega})^n$$
$$H(\omega) = \frac{1}{1-ae^{-j\omega}}$$

90. State the properties of frequency response $H(\omega)$ of an LTI system.
   (a) $H(\omega)$ is a continuous function of w, and with period 2π.
   (b) $|H(\omega)|$ is even symmetric w.r.t, $\omega = \pi$.
   (c) $\angle H(\omega)$ is anti-symmetric w.r.t, $\omega = \pi$.
91. Find the transfer function of the system.
   $y(n) - \frac{1}{2}y(n-1) = x + \frac{1}{3}x(n-1)$. Transfer function is defined as $H(\omega) = \frac{Y(\omega)}{X(\omega)}$.
   Fourier Transforming the given equation, we have

$$Y(\omega) - \frac{1}{2}e^{-j\omega}Y(\omega) = X(\omega) + \frac{1}{3}e^{-j\omega}X(\omega)\ Y(\omega)\left[1 - \frac{1}{2}e^{-j\omega}\right]$$

$$= X(\omega)\left[1 + \frac{1}{3}e^{-j\omega}\right]$$

$$H(\omega) = \frac{Y(\omega)}{X(\omega)} = \frac{(1 + \frac{1}{3}e^{-j\omega})}{(1 - \frac{1}{3}e^{-j\omega})}.$$

92. Find the transfer function of a first-order recursive filter.

$$y(n) = a\ y(n-1) + x(n).$$

   Fourier Transforming the equation.

$$Y(\omega) = ae^{-j\omega}\ Y(\omega) + X(\omega)$$

   Therefore, $H(\omega) = \frac{Y(\omega)}{X(\omega)} = \frac{1}{1-ae^{-j\omega}}$
93. What are the types of discrete convolution?
   (a) Circular convolution
   (b) Linear convolution
94. What do you mean by circular convolution?
   $y_p(n) = \sum_{l=0}^{N-1} x_p(l)h_p(n-l)$ is defined as the circular/periodic convolution of $x_p(n)$ and $h_p(n)$.

95. Obtain the N-point DFT of the circular convolution between $x(n)$ and $h(n)$.

$$y_p(n) = \sum_{l=0}^{N-1} x_p(l).h_p(n-l)$$

Taking N-point DFT of the above equation, we have $Y_p(k) = H_p(k).X_p(k)$

96. Why is the circular convolution called so?
Circular convolution represents the convolution of two sequences defined on a "circle".

97. What is linear convolution?
Linear or a periodic convolution of $x(n)$ and $h(n)$ yields the sequence $y(n)$ given by

$$y(n) = \sum_{m=0}^{n} h(m)x(n-m)$$

98. Can you realize linear convolution using circular convolution?
Yes, we can realize linear convolution using circular convolution.

99. What are the two methods of sectioned convolution?

    (a) Overlap and add method
    (b) Overlap and save method.

100. Define DFT (Discrete Fourier Transform),
The DFT of a sequence $x_p(n)$ is defined as

$$X_p(K) = \sum_{n=0}^{N-1} x_p(n)e^{-j(2\pi/N)nk}$$

101. Define Inverse Discrete Fourier Transform (IDFT).

$$\text{IDFT}[X_p(K)] = x_p(n) = \frac{1}{N}\sum_{n=0}^{N-1} X_p(K)e^{j(2\pi/N)kn}$$

102. State the linearity property of DFT.
If $x_p(n)$ and $y_P(n)$ are periodic sequence, both of period N samples, with DFTs $X_P(k)$ and $Y_P(k)$, then the DFT of $x_p(n) + y_P(n)$ is $X_P(k) + Y_P(k)$.

103. State the shifting property of DFT.
If $\text{DFT}[x_p(n)] = x_p(k)$, then

$$\text{DFT}[x_p(n-n_0] = X_p(K)e^{-j\left(\frac{2\pi}{N}\right)n_0k}$$

104. State the relationship between z-Transform and DFT.

$$X(z)|z = e^{j(2\pi/N)K} = X\left[e^{j\left(\frac{2\pi}{N}\right)k}\right] = \sum_{n=0}^{N-1} x(n)e^{-j(2\pi/N)k}$$

105. What are two types of FFT?
    (a) Decimation in Time.
    (b) Decimation in frequency.
106. How are filters classified as per frequency response?
    (a) High-pass filter
    (b) Low-pass filter
    (c) Band-pass filter
    (d) Band-elimination filter
107. How are filters classified based on impulse response?
    (a) Infinite Impulse response filter
    (b) Finite Impulse response filter
108. What are IIR filters?
    IIR filters are those which have the present output samples depending on the present input, past input sample, and output sample.
109. What are FIR filters?
    Filters whose present output samples depend on the present and previous input samples are called FIR filters.
110. Give the advantages of FIR filters.
    (a) FIR filters are very stable
    (b) Both recursive and non-recursive realization of FIR filters is applicable
    (c) Irrespective of the magnitude level, FIR filters can be realized.
111. State any low disadvantages of FIR filters.
    (a) Memory space required for storage is large
    (b) The order of filter realized for a given set of specifications is very high compared with that of IIR filter.
112. What do you understand by the statement "FIR filters have linear phase"?
    The output signal shape is not altered by the FIR filter for a linear-phase filter and $\theta(\omega)$ is directly proportional to ω.

113. State the methods available to design FIR filters.

    (a) Frequency sampling method
    (b) Windows method
    (c) Optimal design method.

114. State why FIR filter is always stable?
    As all the poles of the FIR filter transfer function are at the origin, the filter is stable.
115. State the condition for stability of a filter.

$$\sum_{n=\infty}^{\infty} |h(n)| < \infty$$

    where $h(n)$ is the impulse response of the filter.
116. State the condition for causality of a filter.
    $h(n) = 0, \text{ for } n < 0.$ is the condition for causality of the filter, where $h(n)$ is the impulse response of the filter.
117. Compare and contrast FIR and IIR of a filter.

FIR filter	IIR filter
1. Linear-phase filters	1. Non-linear-phase filters
2. Both recursive and non-recursive realization is present.	2. Only recursive realization is possible.

118. State the conditions for IR filter where they act as linear filters

    (a) symmetric condition $h(n) = h(N - 1 - n)$
    (b) anti-symmetric condition $h(n) = -h(N - 1 - n)$

119. What are the characteristics of a linear-phase filter?

    (a) The filter should have constant group delay
    (b) The filter has to have constant phase delay.

120. Define 'group delay' of a filter.
    Group delay of a filter is defined as the derivative of the phase w.r.t frequency.
121. Define 'phase delay' of a filter.
    Phase delay is defined as the ratio of phase to frequency.
122. How do you represent the frequency response of linear-phase filters?
    $H(\omega) = H_1(\omega)e^{j(\beta-\alpha)}$, where $H_1(\omega)$ is purely real and α and β are given by $\alpha = \left(\frac{N-1}{2}\right)$ and $\beta = \pm\frac{\pi}{2}$

123. How will you design an FIR filter using the Fourier series method?
$h(n)$ is obtained from

$$h(n) = \frac{1}{2\pi} \int_{-\pi}^{\pi} H(\omega) e^{j\omega} d\omega$$

Truncate $h(n)$ at $n = \pm\left(\frac{N-1}{2}\right)$ to obtain the finite duration sequence of $h(n)$. We then proceed to find $H(z)$ using

$$H(z) = z^{-(N-1)/2} \left[h(0) + \sum_{n=1}^{\frac{N-1}{2}} h(n)(z^n + z^{-n}) \right]$$

124. What are the difficulties in the representation as

$$H(\omega) = \sum_{n=-\infty}^{\infty} h(n) e^{-j\omega n}$$

(a) The impulse response $h(n)$ is infinite in duration since the summation extends from $-\infty$ to $+\infty$.
(b) The filter is unrealizable as the impulse response begins at $-\infty$, i.e. no finite amount of delay can make the impulse response realizable.

125. Explain Gibbs phenomenon.
In order to obtain an FIR filter approximating $H(w)$, we need to truncate the infinite Fourier series at $n = \pm((N-1)/2)$. This leads to oscillations being produced in stop band as well as pass band. This is referred to as the Gibbs Phenomenon.

126. State the disadvantages of the Fourier series method of designing FIR filter.
Truncation of the Fourier series of the infinite impulse response at $n = \pm((N-1)/2)$ leads to oscillations in the stop band and pass band, which is a disadvantage.

127. State the procedure to design FIR filter using the windows method.
It is required to find the impulse response $h_d(n)$ using

$$h_D(n) = \frac{1}{2\pi} \int_{-\pi}^{\pi} H_D(\omega) e^{j\omega n} d\omega$$

The infinite impulse response is multiplied with a chosen window sequence w(n) to get filter coefficients

$$\begin{aligned} h(n) &= h_D(n)\omega(n),\ for\ |n| \leq \tfrac{N-1}{2} \\ &= 0,\ \text{otherwise} \end{aligned}$$

Next step is to find the transfer function of the filter

$$H(z) = z^{-(N-1)/2}\left[h(0) + \sum_{n=1}^{\frac{N-1}{2}} h(n)(z^n + z^{-n})\right]$$

128. How is the windowing technique better than the Fourier series method of designing FIR filters?
Fourier series method leads to Gibbs oscillation, which is avoided using the window technique.
129. How is the Blackman window defined?

$$\omega(n) = 0.42 + 0.5\ \cos\left(\tfrac{2\pi n}{N}\right) + 0.08\cos\left(\tfrac{4\pi n}{N}\right)$$
$$\text{for } \frac{-(N-1)}{2} \leq n \leq \frac{(N-1)}{2}$$

130. How is the rectangular window defined?
The weighting function of the rectangular window is

$$\omega_R(n) = \begin{cases} 1 & -\left(\frac{N-1}{2}\right) \leq n \leq \left(\frac{N-1}{2}\right) \\ 0 & \text{otherwise} \end{cases}$$

131. What are the major requirements of window characteristics?
    (a) Small width of the main lobe of the frequency response of the window containing as much of the total energy as possible.
    (b) Side lobes of the frequency response that decrease in energy rapidly as w tends to π.
132. How are the Hamming window and Hanning window defined?
The generalized weighting function of the window is

$$\omega_H(n) = \{\alpha - (1-\alpha)\cos\left(\tfrac{2\pi n}{N}\right), \quad -\left(\tfrac{N-1}{2}\right) \leq n \leq \left(\tfrac{N-1}{2}\right)$$

Hamming window has a value of $\alpha = 0.54$ and the Hanning window has a value of $\alpha = 0.5$ in the above-mentioned weighting function.
133. Compare a Hamming window with a rectangular window.
    (a) The main lobe of the frequency response of the Hanning window is twice that of the rectangular window.
    (b) For $\alpha = 0.54$, (Hamming window), 99.96% of the spectral energy is in the main lobe and the peak side lobe ripple is down about 40 dB from the main lobe peak. In rectangular window, the spectral side lobes are down by only about 14 dB from the main lobe peak.

134. Write down the weighting function of the Kaiser window.

$$\omega(n) = \frac{I_0[\sqrt{1 - [2n/(N-1)]^2}]}{I_0[\beta]}, \quad \left(\frac{N-1}{2}\right) \leq n \leq \left(\frac{N-1}{2}\right)$$

where β is a constant that specifies a frequency response tradeoff between the peak height of the side lobe ripples and the width or energy of the main lobe and $I_0(x)$ is the modified width-order Bessel function.

135. Why is the Kaiser window an optimum window?
It is a finite duration sequence that has the minimum spectral energy beyond some specified frequency.

136. Sketch the frequency response of a Hamming window.

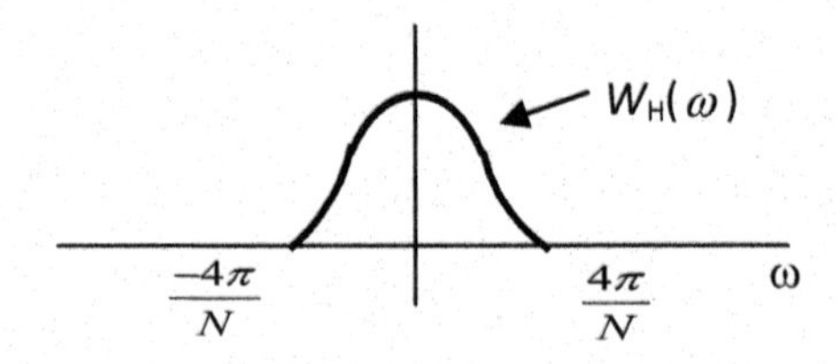

137. What is the weighting function of a Bartlett window?

$$\begin{aligned}\omega(n) &= 1 - \tfrac{2|n|}{N-1} \quad \tfrac{-(N-1)}{2} \leq n \leq \tfrac{(N-1)}{2} \\ &= 0 \quad otherwise\end{aligned}$$

138. Sketch the frequency response of a rectangular window.

$$W_R(\omega) = \frac{\sin\left(\frac{\omega N}{2}\right)}{\sin\left(\frac{\omega}{2}\right)}$$

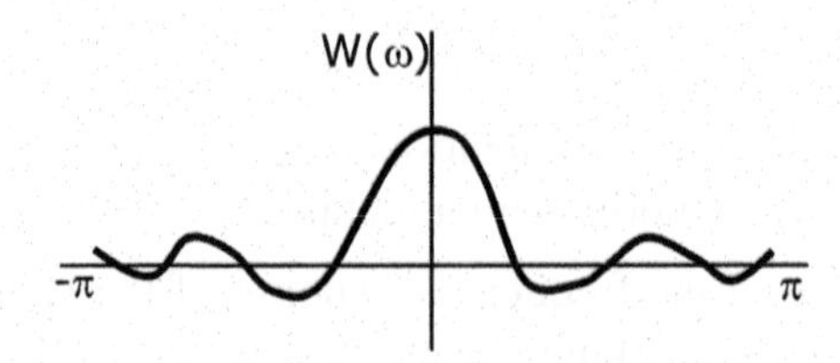

139. State the advantage of the Kaiser window.
Designer is allowed to choose the side lobe level and *N*. The side lobe level can be varied from the low value in the Blackman window to a high value in the rectangular window.

140. How is an FIR filter designed using the Frequency sampling method?
The desired magnitude response is sampled and a linear-phase response is specified. The samples of the required frequency response are the DFT coefficients. The filter coefficients are now determined by getting the IDGT of the DFT coefficients.
141. What is an IIR filter?
When a filter produces a unit-sample response that has an infinite duration, it is called IIR filter.
142. Sate the properties of the Butterworth filter.
    (a) Magnitude response decreases monotonically as the frequency is increased from 0 to 8.
    (b) Magnitude response is normally that about $\omega = 0$, in that all the derivatives up to order N are equal to zero at $\omega = 0$.
    (c) The phase response curve approaches $-N(\pi/2)$ for large ω, where N is the number of poles on the Butterworth circle in the left-half s-plane.
143. Sate the properties of Chebyshev filter.
    (a) The magnitude response has a ripple in either the pass band or in stop band
    (b) The poles of the Chebyshev filter lie on an ellipse.
144. Sketch the Butterworth pole locations for $N = 3$.

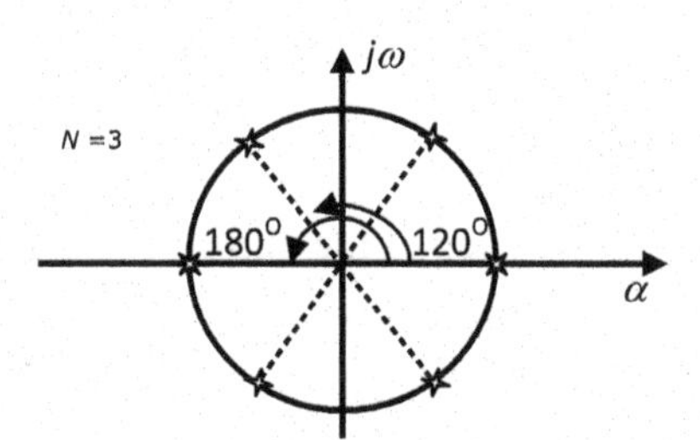

145. What is the power transfer function of the Butterworth filter?

$$|H_B(\omega)|^2 = \frac{1}{1 + \left[\frac{\omega}{\omega_c}\right]^{2N}}$$

146. What is the power function of the Chebyshev filter?

$$|H_c(\omega)|^2 = [1 + \mu^2 C_N^2(\omega/\omega_p)]^{-1}$$

Where C_N are Chebyshev's polynomials.

147. State the design parameters of the Chebyshev filter.
    (a) Pass-band frequency
    (b) Allowable deviation in pass band.
148. Write down Chebyshev's polynomials $C_N(x)$ for $N = 0, 1, 2$
    (a) For $N = 0,\ C_N(x) = 1$
    (b) For $N = 1,\ C_N(x) = x$
    (c) For $N = 2,\ C_N(x) = 2x^2 - 1$
149. How are Chebyshev's polynomials defined?

$$C_{N+1}(x) = 2x\, C_N(x) - C_{N-1}(x),\ \text{for } N \geq 1$$

150. Why are the Chebyshev filters said to be optimum?
    The Chebyshev filters are optimum in that for the given pass-band and stop-band levels, it has the smallest transition region, $(w_s - w_p)$, of any filter that consists only of finite poles. Such "all pole" filters have all then zero at $s = \infty$.
151. State the important features of the Chebyshev filters.
    (a) $|C_N(x)| \leq 1$, for $|x| \leq 1$
    (b) For $|x| >> 1$, $|C_N(x)|$ increases as the N-th power of x
152. How does a Butterworth filter differ from a Chebyshev filter?
    The poles of the Butterworth filter lie on a circle, whereas those of the Chebyshev filters lie on an ellipse.
153. Differentiate between a causal and stable IIR filter from that of an FIR.
    FIR filter can have linear phase, whereas an IIR filter cannot follow linear phase.
154. Define the major and minor axes of the ellipse of a Chebyshev filter.
    Minor axis, radius $r = \frac{w_\rho[\rho^{1/N} - \rho^{-1/N}]}{2}$
    Minor axis, radius $R = \frac{w_\rho[\rho^{1/N} - \rho^{-1/N}]}{2}$;
    where $\rho = \mu^{-1} + \sqrt{1 + \mu^{-2}}$
155. How will you obtain digital filters from analog filters?
    (a) The given analog filter specifications are transformed into digital filter specifications.
    (b) Arrive at the analog transfer function from the analog prototype.
    (c) Transform the transfer function of the analog properties into its equivalent digital filter transfer function.

156. What are the methods of obtaining digital filters from analog filters?
    (a) Impulse invariant method
    (b) Bilinear transform method
157. What is the pole mapping procedure in the Impulse invariant method?
    A pole located at $s = s_p$ in the s-plane is transformed into a pole in the z-plane located at $z = e^{s_P T_S}$.
158. What is the disadvantage of the Impulse invariant method?
    The method is unsuccessful for implementing digital filters for which $|H_A(j\Omega)|$does not approach zero for large values of Ω. Hence "aliasing" occurs.
159. How is the disadvantage in Impulse invariant method overcome?
    The Bilinear Transform method is used to prevent 'aliasing' which is a disadvantage found with the impulse invariant method. The entire $j\Omega$ axis, for $-\infty < \Omega < \infty$, maps uniformly onto a unit circle,

$$\frac{-\pi}{T_s} < \frac{\omega}{T_s} < \frac{\pi}{T_s}.$$

160. Give an example of a simple Bilinear transformation.

$$s = \frac{2}{T_s}\left[\frac{z-1}{z+1}\right]$$

161. Stage the relation between continuous-time and discrete-time frequencies.

$$\Omega = 2\ arc\ \tan\ [\omega\ T_s/2]$$

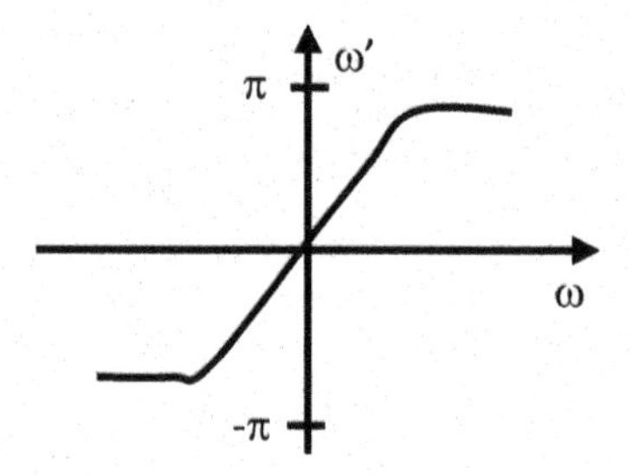

162. What do you mean by "prewarping"?
    By using the formula

$$\Omega = \frac{2}{T}\tan\left(\frac{\omega T}{2}\right)$$

The compression in the magnitude response is compensated by using the prewarping.

163. Stage the advantage of bilinear transformation.
    (a) No aliasing effect.
    (b) Provides one-to-one mapping procedure.
164. Stage the disadvantage of Bilinear transformation.
    (a) Impulse response as well as phase response of the analog filter is not preserved
    (b) Provides highly non-linear frequency compression at high frequencies.
165. What are the effects of using finite register lengths while implementing digital filters in the digital hardware?
    (a) A/D conversion noise
    (b) Round-off noise (uncorrelated)
    (c) Coefficient inaccuracy
    (d) Round-off noise or limit cycles (correlated)
166. What are the types of arithmetic used in filter algorithms?
    (a) Fixed point arithmetic
    (b) Floating point arithmetic
    (c) Block floating point arithmetic
167. What is meant by diced point arithmetic?
    The position of the binary point is fixed. The bit to the right represents the fractional part of the number and that to the left represents the integer parts.
168. What are the types of fixed point arithmetic?
    (a) Sign-magnitude representation
    (b) 1's – complement
    (c) 2's – complement
169. What is sign-magnitude representation?
    The leading binary digit is used to represent the sign to correspond to + and 1 corresponds to –. For a b bit word, (b – 1) bits are used to represent the magnitude.
170. How many numbers are represented in sign-magnitude form for a b bit word?
    $(2^b - 1)$ numbers are represented exactly

171. Covert (01.01100) into a base-10 number.

$$(01.0100)_2 = [0 \times 2^1] + [1 \times 2^0] + [0 \times 2^{-1}] + [1 \times 2^{-2}] + [1 \times 2^{-3}] + [0 \times 2^{-4}] + [0 \times 2^{-5}] = (1.375)_{10}$$

172. What are the representations of 0 in sign-magnitude form?

$$(00.00000)_2 \text{ and } (10.00000)_2$$

173. What do you mean by 2's – complement representation?
To obtain the negative of a positive number, we complement all the bits of the positive number and odd one unit in the position of the least significant bit.
174. Obtain the negative of $(01.01100)_2$ using 2's complement notation

$$-(01.01100)_2 = (10.10011)_2 + (00.00001)_2 = (10.10100)_2$$

175. What do you mean by 1's complement form?
The negative of a positive number is obtained by complementing all the bits of the positive number which is the 1's complement.
176. Obtain the negative of 01.01100 using 1's complement notation
Negative of $(01.01100)_2 = -(01.01100)_2 = (10.10011)$
177. Represent 0 in 1's complement form

$$(00.00000)_2 \text{ and } (11.11111)_2$$

178. What do you mean by quantization?
Quantization is a technique by which the numerical equivalent of each sample of $s(n)$ is expressed by a finite number of the bits giving the sequence $s_Q(n)$.
179. What do you mean by floating point arithmetic?
A number is represented as $f = 2^e m$, where m is called mantissa and f is a fraction such that $0.5 \leq m \leq 1$, and c is the exponent.
180. Compare fixed point and floating point arithmetic.
Fixed point Arithmetic Floating point Arithmetic
    (a) Overflow occurs in the addition operation
    (b) There is no overflow
    (c) Roundoff error occurs in addition
    (d) Roundoff error occurs in both addition and multiplication.
181. Mention the types of quantization.
    (a) Truncation
    (b) Rounding

182. What do you mean by Truncation?
The process of discarding all bits less significant than the least significant bit that is retained is called truncation.
183. Mention the different types of quantization errors, occurring due to finite word length registers in digital filters.
    (a) Coefficient quantization error
    (b) Input quantization error
    (c) Product quantization error
184. What do you mean by rounding?
Rounding of a number of 'b' bits is done by choosing the rounded result as the b bit number closest to the original un-rounded quantity.
185. What is meant by coefficient in accuracy?
The coefficients of a digital filter are obtained by some theoretical design procedure that essentially assumes infinite precision representation of the filter coefficients. For practical realizations, the coefficients should be quantized to a fixed number of bits. As a result of this, the frequency response of the actual filter which is realized deviates from that which has been obtained with an infinite word-length representation. This is called coefficient inaccuracy.
186. State the two approaches to analyze/synthesize digital filters with finite precision coefficients.
    (a) Treat the coefficient quantization errors as statistical quantities
    (b) Study each individual filter separately so as to optimize between ideal and actual responses.
187. State the inequalities satisfied by truncation error where x_T is a truncated value of x.

$$\text{(i)} \quad \text{for } x > 0,\ 0 \geq x_T - x > -2^{-b}$$

$$\text{(ii)} \quad \text{for } x < 0,\ 0 \leq x_T - x < -2^{-b}$$

188. State the inequality satisfied by truncation error for a floating point word $x = 2^e.\, m$ in 2's complement representation of mantissa.
$x_T - x = (1+ \in)x$ where x_T = truncated value of x.
The inequality is $0 \leq \in x \leq -2^{-b}.2^c$.

$$\text{For } x > 0, \quad 2^{c-1} < x \leq 2^c.$$

$$\text{Where by} \quad 0 \geq \in > -2^{-b}.2, x > 0 \text{ and } 0 \leq \in < 2^{-b}.2, x < 0$$

189. State the inequality satisfied by truncation error for a floating point word $x = 2^e. m$, in 1's complement notation of mantissa.

$$0 < \in < -2.\ 2^{-b},\ \text{for all } x.$$

190. State the inequality satisfied by fixed-point arithmetic, the error being made by rounding a number to 'b' bits following the binary point.

$$\frac{-2b^{-2}}{2} \leq x_T - x \leq \frac{2^{-2}}{2}.$$

for all the three types of number systems (two's complement, one's complement, and sign-magnitude)

191. State the inequality satisfied by floating arithmetic, the error being due to rounding.

$$-2^c.\frac{2^{-b}}{2} \leq x_T - x < -2^c.\frac{2^{-b}}{2}.$$

$$\text{If} \quad x_T - x = \in x, \quad \text{then} \quad -2^{-b} \leq \in \leq 2^{-b}.$$

192. Sketch the probability density function for rounding

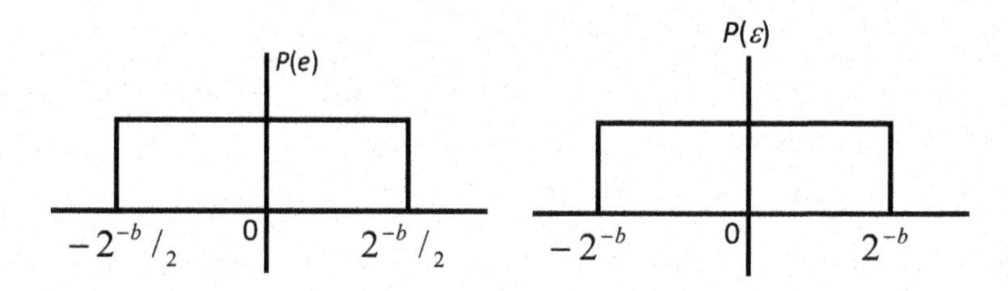

193. Sketch the probability density function Truncation with 2's complement.

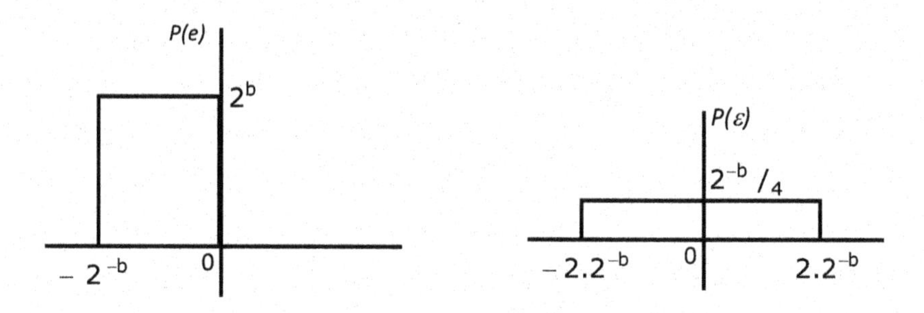

194. Sketch the probability density function of Truncation with 2's complement/sign-magnitude.

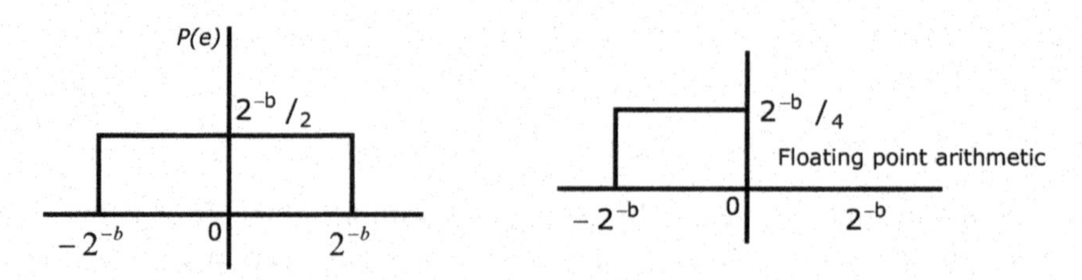

195. What are limit cycle oscillations?
In recursive systems, the non-linearities due to the finite precision arithmetic operations cause periodic oscillations to occur in the output, even when the input sequence is zero or some non-zero constant value. Such oscillations are called "limit cycle oscillations".
196. What is 'dead band'?
The amplitudes of the output of recursive systems during a limit cycle are confined to a range of values called 'dead band'.
197. Stage the methods of avoiding overflow oscillations.
   (a) Saturation arithmetic
   (b) Scaling
198. What do you mean by saturation arithmetic?
Whenever an overflow is obtained, the sum of the adder is set equal to the maximum value. This is known as saturation arithmetic, useful in avoiding overflow oscillations.
199. What is the disadvantage of saturation arithmetic?
Saturation arithmetic causes undesirable signal distortion due to the non-linearity of the clipper.
200. What do you mean by scaling in digital filters?
In order to prevent overflow, the signal level at some points in the digital filters have to be scaled in order that no overflow occurs in the header. This is meant by scaling in digital filters.
201. State the necessary and sufficient condition to prevent overflow by scaling.

$$A_x < \frac{1}{\sum\limits_{m=-\infty}^{\infty} |h_k(m)|}$$

where the signal $x(n)$ is upper bounded by A_x.
202. Which is preferred in digital filter implementation-rounding or truncation? Why?
Rounding is preferred to truncation, because

(a) The variance of the rounding error signal is low.
(b) Mean of rounding error is zero.

203. Give an expression for steady-stage noise power due to quantization of 'b' bits of binary sequence.

$$\text{Noise Power } \sigma_e^2 = \frac{2^{-2b}}{12}$$

204. Write the different types of arithmetic in digital systems?
There are three types of arithmetic used in digital systems as follows:

(a) Fixed point arithmetic
(b) Floating point arithmetic
(c) Block floating arithmetic

205. What do you mean by a fixed-point number?
In fixed-point arithmetic, the position of the binary point is fixed. The bit to the right represent the fractional part of the number and those to the left represent the integer part. As an example, the binary number 01.1100 has the value 1.75 in decimal.
206. What are the different quantization methods?
The common methods of quantization are a under:

(a) Truncation
(b) Rounding

207. Write the different types in fixed-point number representation?
Depending on the way, negative numbers are represented, there are there different forms of fixed point arithmetic, following

(a) sign-magnitude
(b) 1's complement
(c) 2's complement.

208. What do you mean by sign-magnitude representation?
For sign-magnitude representation, the leading binary digit is used to represent the sign. If it is equal to 1, the number is negative, otherwise the number is positive. For example, the decimal number –1.75 is represented as 11.110000 and 1.75 is represented as 01.110000. In sign-magnitude from, the number 0 has two representations, i.e., 00.000000 or 10.000000. With b bits only, $2^b - 1$ numbers may be represented.
209. What is 1's complement representation?
In one's complement form, the positive number may be represented as in the sign-magnitude system. To obtain the negative of a positive

number, one simply complements all the bits of the positive number. As an example, the negative of 01.110000 will be represented as under:

$$-(01.110000)_2 = (10.001111)_2$$

The number 0 has two representations, i.e., 00.00000 and 11.11111 in a 1's complement representation.

210. What do you mean by 2's complement representation?

In two's complement representation, positive numbers are represented as in sign-magnitude and one's complement. The negative number can be obtained by complementing all the bits of the positive number and adding one to the least significant bit.

For example, we have

$$\begin{aligned} (0.5625)_{10} &= (0.100100)_2 \\ (-0.5625)_{10} &= \underline{1.011011} \\ &\quad \underline{0.000001} \\ &\quad (1.011100)_2 \end{aligned}$$

211. Write a short note on floating point arithmetic.

In floating point representation, a positive number is represented as $F = 2^C . M$, where M, called mantissa, is a fraction such that $\frac{1}{2} \leq M < 1$ and c, the exponent can be either positive or negative.

The decimal numbers 2.25 and 0.75 have floating point representations as

$$2.25 = 2^2 \times 0.5625 = 2^{010} \times 0.1001$$

and $2.75 = 2^0 \times 0.75 = 2^{000} \times 0.1100$ repetitively

Negative floating point numbers are generally represented by considering the mantissa as a fixed point number. The sign of the floating point number is obtained from the first bit of mantissa.

212. What is do you mean by block floating point representation? What are its advantages?

In block floating point arithmetic, the set of signals to be handled is divided into blocks. Each block has the same value for the exponent. The arithmetic operations with in the block use fixed-point arithmetic and only one exponent per blocks. Each block uses fixed-point arithmetic and only one exponent per block is stored hence saving memory. This representation of numbers is most suitable in certain FFT flow graphs and in digital applications.

213. Write the advantages of floating point arithmetic?

(a) Larger dynamic range.
(b) Overflow in floating point representation is unlikely.

214. Express the fractions $\frac{7}{8}$ and $\frac{-7}{8}$ in sign magnitude, 2's complement, and 1's complement.
Fraction $\frac{7}{8} = (0.111)_2$ in sign magnitude is complement and 2's complement.
Fraction $\left(\frac{7}{8}\right) = (1.111)_2$ in sign magnitude

$$= (1.000)_2 \text{ in 1's complement}$$

$$= (1.001)_2 \text{ in 2's complement}$$

215. Compare the fixed-point arithmetic and floating-point arithmetic.

S. No.	Fixed-Point Arithmetic	Floating-Point Arithmetic
1.	Fast Operation	Slow Operation
2.	Relatively Economical	More expensive due to costlier hard-ware
3.	Small dynamic range	Increased dynamic range
4.	Roundoff errors occur only for addition	Roundoff errors can occur with both addition and multiplication
5.	Overflow occurs in addition	Overflow does not arise
6.	Used in small computers	Used in larger, general-purpose computers

216. Write the three quantization errors due to finite word length registers in digital filters?

(a) Input quantization error.
(b) Coefficient quantization error.
(c) Product quantization error.

217. How the multiplication and addition are carried out in floating point arithmetic? In floating point arithmetic, multiplications are carried out as under:

$$\text{Let } f_1 = M_1 \times 2^{c1} \text{ and } f_2 = M_2 \times 2^{c2}.$$

$$\text{The } f_3 = f_1 \times f_2 = (M_1 \times M_2)2^{(c1+c2)}$$

This means that the mantissas are multiplied using fixed-point arithmetic and the exponents are added.

The sum of two floating-point numbers is carried out by shifting the bits of the mantissa of the smaller number to the right until the exponents of the two numbers are equal and then adding the mantissas.

218. Write a short note on coefficient inaccuracy? What is coefficients quantization error? What is its effect?
The filter coefficients are computed to infinite precision in theory. However, in digital computation, the filter coefficients are represented in binary and are stored in registers. If a b bit register is used, then the filter coefficients must be rounded or truncated to b bits, which produces an error.
Because of quantization of coefficients, the frequency response of the filter can differ appreciably from the desired response and some times the filter may actually fail to meet the desired specifications. If the poles of desired filter are close to the unit circle, then those of the filter with quantized coefficients may be just outside the unit circle, and hence leading to instability.
219. What do you mean by product quantization error (or) What is product roundoff error in digital signal processing (DSP)?
Product quantization errors arise at the output of a multiplier. Multiplication of a b bit data with a b bit coefficient results in a product having 2b bits. Since a b bit register is used, the multiplier output must be rounded or truncated to b bits, which produces an error. This error is called product quantization error.
220. What do you mean by input quantization error?
In digital signal processing (DSP), the continuous-time input signals are converted into digital using a b bit ADC. The representation of continuous signal amplitude by a fixed digit produce an error is called input quantization error.
221. What do you mean by truncation? What is the error that arose due to truncation in floating point numbers?
Truncation is a process of discarding all bits less significant than least significant bit that is retained.
Suppose we truncate the following number from 7 bits to 4 bits, then we obtain 0.0011001 to 0.0011 and 0.0100100 to 0.0100
For truncation in floating point systems, the effect is seen only in mantissa. If the mantissa is truncated to b bits, then the error ε satisfies

$$0 \geq \varepsilon > -2.2^{-b} \quad \text{for } x > 0$$

and $0 \leq \varepsilon < -2.2^{-b} \quad \text{for } x < 0$

If the mantissa is represented by 1's complement or sign magnitude, then the error satisfies

$$-2.2^{-b} < \varepsilon \leq 0 \text{ for all } x$$

Here, $\varepsilon = \frac{x_T - x}{x}$

where x_T is the truncated value of x.

222. What is the relationship between truncation error e and the bits b for representing a decimal into binary?

For a 2's-complement representation, the error due to truncation for both positive and negative values of x is $0 \geq x_T - x > -2^{-b}$

where b is the number of bits and x_T is the truncated value of x.

This equation holds good for both sign-magnitude and 1's-complement, the truncation error being satisfied.

$$0 \leq x_T - x < 2^{-b}$$

223. What do you mean by rounding? Discuss its effect on all types of number representations.

Rounding a number to b bits is accomplished by choosing the rounded result as the b bit number closets to the original number unrounded.

For fixed-point arithmetic, the error made by rounding a number to b bits satisfies the following inequality:

$$\frac{-2^{-b}}{2} \leq x_T - x \leq \frac{2^{-b}}{2}$$

for all three types of numbers system, i.e., two's-complement, one's-complement, and sign-magnitude.

For floating point numbers, the error made by rounding a number to b bits satisfies the following inequality:

$$-2^{-b} \leq \varepsilon \leq 2^{-b} \quad \text{where} \quad \varepsilon = \frac{x_T - x}{x}$$

224. What is meant by A/D conversion noise?

A digital signal processor contains a device, A/D converter, which operates on the analog input $x(t)$ to produce $x_q(n)$ that is a binary sequence of 0's and 1s.

Firstly, the signal $x(t)$ is sampled at regular intervals to produce a sequence $x(n)$ is of infinite precision. Each sample $x(n)$ is expressed in terms of a finite number of bits giving the sequence $x_q(n)$. The difference signal $e(n) = x_q(n) - x(n)$ is known as A/D conversion noise.

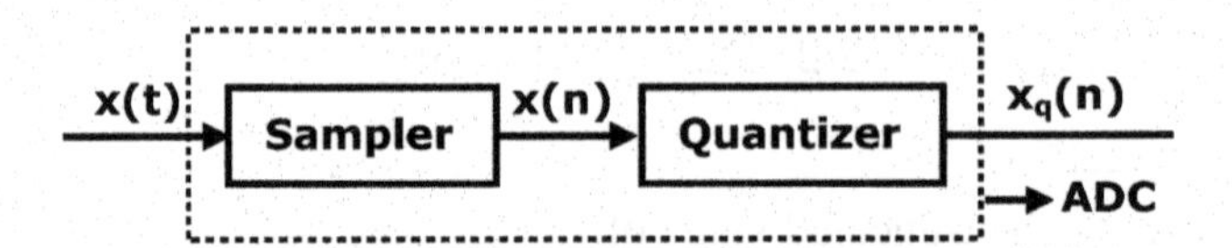

225. What is the effect of quantization on pole locations?
Quantization of coefficients in digital filters leads to slight changes in their value. This change in value of filter coefficients modifies the pole-zero locations. Sometimes, the pole locations would be changed in such a manner that the system may be driven into instability conditions.
226. Which type of realization is less sensitive to the process of quantization?
Cascade form
227. What do you mean by quantization step size?
Let us assume a sinusoidal signal varying between +1 and −1 having a dynamic range of 2. If the ADC used to convert the sinusoidal signal employs b^{+1} bits including sign bit, then the number of levels available for quantizing $x(n)$ is 2^{b+1}. Thus, the interval between successive levels will be

$$q = \frac{2}{2^{b+1}} = 2^{-b}$$

where q is called the quantization step size.
228. Draw the following characteristics:
(a) Quantizer characteristic with rounding
(b) Quantizer characteristic with truncation.

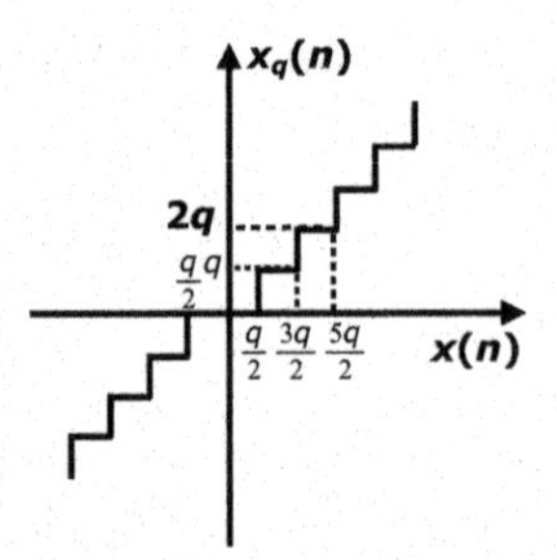

(a) Quantizer characteristics with rounding.

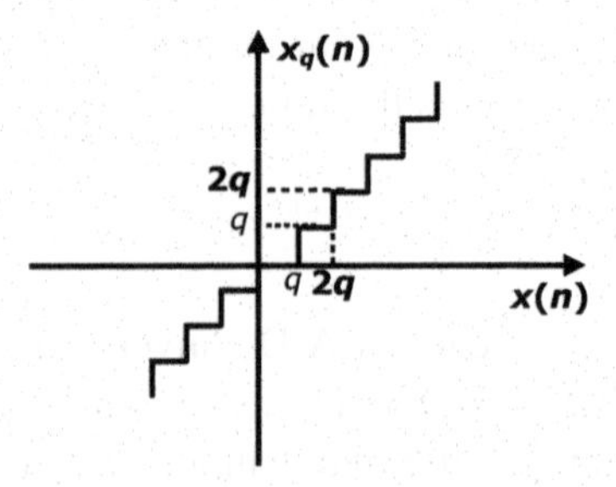

(b) Quantizer characteristics with truncation.

229. Draw the probability density function (PDF) for rounding.
Figure illustrates the probability density functions (PDF) for rounding for two cases:

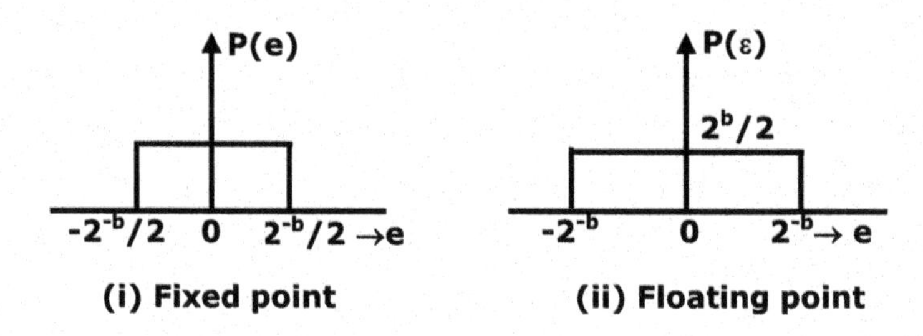

230. What are the assumptions made connecting the statistical impedance of various noise sources, which occur in realizing the filter?
Following are the assumptions:

(a) For any n, the error sequence $e(n)$ is uniformly distributed over the range $\frac{-q}{2}$ and $\frac{q}{2}$. This implies that mean value of $e(n)$ is zero and its variance will be $\frac{2^{-b}}{12}$.
(b) The error sequence $e(n)$ is a stationary white noise source.
(c) The error sequence $e(n)$ is uncorrelated with the signal sequence $x(n)$.

231. How will you relate the steady-state noise power due to quantization and the b bits representing the binary sequence?
Steady-state noise power will be

$$\sigma_e^2 = \frac{2^{-2b}}{12}$$

where b is the number of bits excluding sign bit.

12.2 Multiple Choice Questions

1. Which of the following terms would be used to describe a signal that is restricted to 256 voltage levels?

(a) Analog
(b) Linear
(c) Digital.
(d) None of the above.

2. The resolution of a digital signal can be increased by
   (a) Increasing the number of bits.
   (b) Decreasing the number of bits.
   (c) Sampling the signal less often
   (d) Using a faster computer.
3. The process of converting from the continuous-time domain to the discrete-time domain is called
   (a) Sampling.
   (b) Quantization.
   (c) Fourier analysis.
   (d) All of the above.
4. How would you convert an analog signal to a digital signal?
   (a) Sampling only.
   (b) Quantization only.
   (c) Both a and b.
   (d) None of the above.
5. The electrical signal from a microphone is best described as
   (a) A quantized signal.
   (b) A discrete signal.
   (c) An aliased signal.
   (d) An analog or continuous signal.
6. We have discussed transducers that produce electrical signals. Which of the following transducers does the opposite-takes in an electrical signal and converts it to some other type of signal?
   (a) Loudspeaker
   (b) Thermistor
   (c) Thermocouple
   (d) Strain gauge
7. Which of the following is an example of a bit?
   (a) 1
   (b) 5
   (c) 63
   (d) 101

8. Match the following.

(a) Bit	(i) $f_s/2$
(b) Byte	(ii) Low/pass
(c) Anti-aliasing filter	(iii) 0
(d) Nyquist frequency	(iv) 0011 1001

9. Match the waveforms in Figures 12.2 to 12.7 with the locations A.B.C.D. and F in the DSP system shown in Figure 12.1

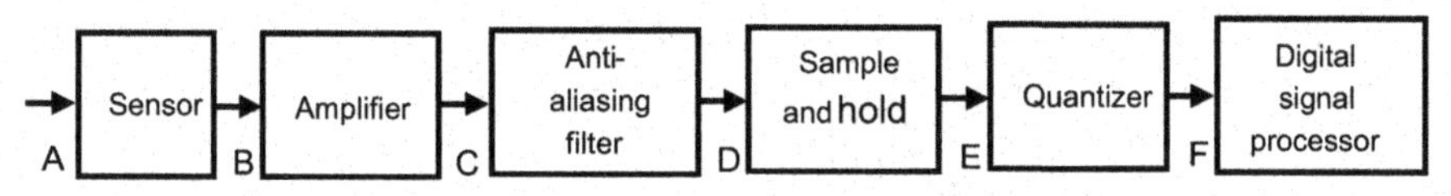

Figure 12.1 A typical DSP system.

For your correct answer, write the sequence of figure number in the answer sheet.

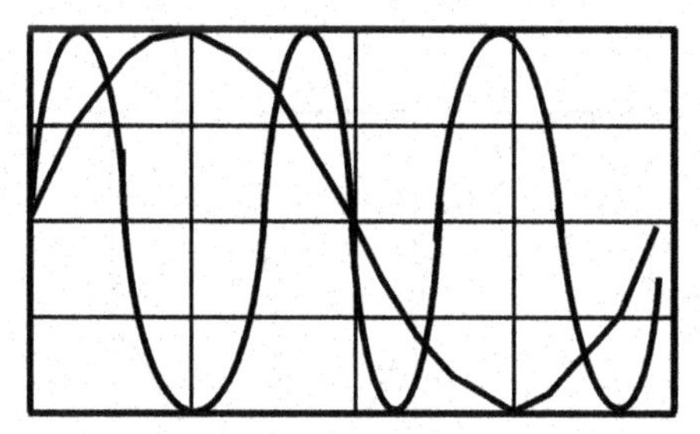

Figure 12.2 Two sine waves of frequencies 2 and 6 kHz, each of peak amplitude 1 V.

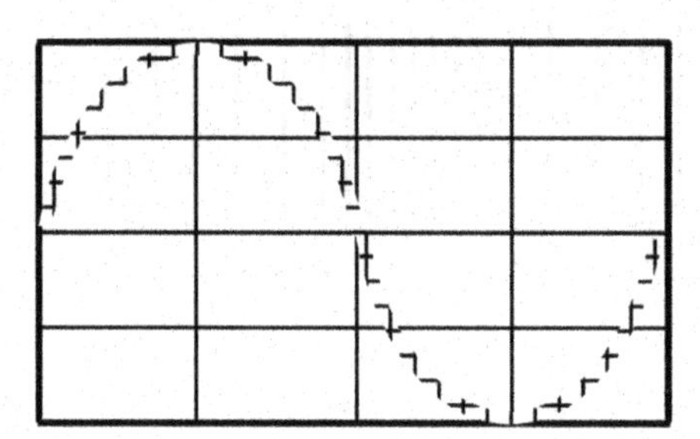

Figure 12.3 A waveform in one part of a typical DSP system.

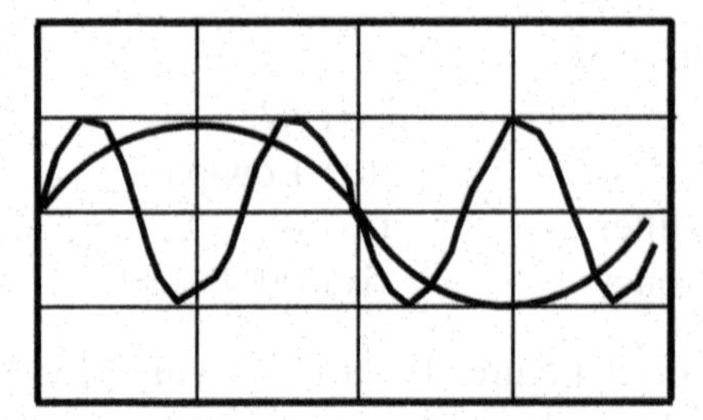

Figure 12.4 Two sine waves of frequencies 2 and 6 kHz, each of peak amplitude 0.5 V.

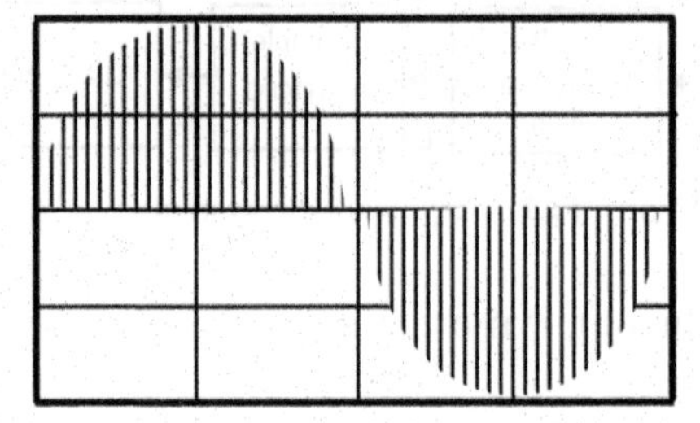

Figure 12.5 Samplers of a sine wave.

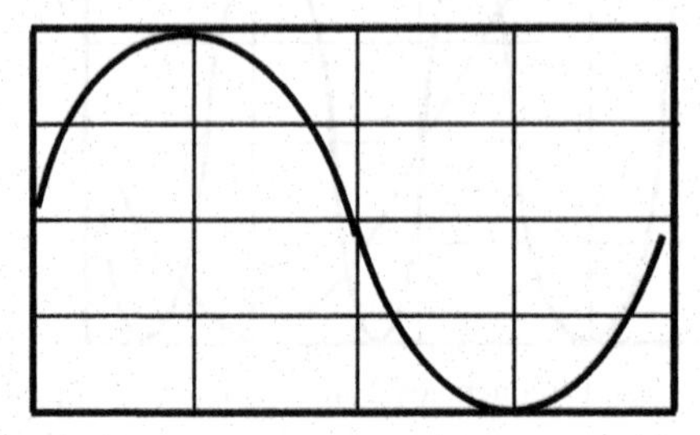

Figure 12.6 Sine wave of frequency 2 kHz.

Figure 12.7 Pressure wave with Alternate compressions and rarefactions.

10. Which of the following binary numbers is equal to the decimal number 65?

    (a) 1100 1000
    (b) 1000 0000

(c) 0100 0001
(d) 0000 1110

11. What is the name given to the frequency that is exactly half of the sampling frequency?
    (a) Anti-aliasing frequency
    (b) Nyquist frequency
    (c) Shannon frequency
    (d) Sample-and-hold frequency
12. When a sampled high-frequency signal produces the same discrete values as a low-frequency signal, the problem is called
    (a) Aliasing
    (b) Quantization
    (c) Component error
    (d) Sample and hold
13. The anti-aliasing filter
    (a) Is a low-pass analog filter.
    (b) Is a low-pass digital filter.
    (c) Was invented by Nyquist.
    (d) Allow frequencies higher than the sampling frequency to pass through.
14. What is the rms quantization noise voltage for a 10-bit system with a signal range of 0 to 3 V?

$$V_{\text{noise}}(\text{rms}) = \frac{V_{\text{full scale}}\,(0.289)}{2^n}$$

    (a) 0.847 mV.
    (b) 1.19 mV.
    (c) 3.76 mV.
    (d) None of the above.
15. What is the maximum signal-to-noise ratio for the system in Question 14? Max signal to noise ratio (dB) $= 6.02\ \mathrm{n} + 1.76$
    (a) 38 dB
    (b) 44 dB
    (c) 53 dB
    (d) 62 dB

16. A DSP system similar to the one shown in Figure 12.1 is used in a home theater system that produces surround sound. The sampling frequency is 44,100 Hz. Which of the following statements is definitely false?
    (a) The Nyquist frequency is 20 kHz.
    (b) The anti-aliasing filter will not pass a signal with a frequency of 40 kHz.
    (c) The sampling interval is 1/44100 s.
    (d) The microprocessor processes the quantized samples.
17. A dB value of −40 dB corresponds to a voltage ratio of
    (a) 100
    (b) 1
    (c) 0.1
    (d) 0.01
18. Which of the following is the symbol for convolution?
    (a) X
    (b) *
    (c) Θ
    (d) .
19. Which of the following is not required for correlation?
    (a) Accumulation
    (b) Multiplication
    (c) Flipping
    (d) Shifting
20. Random noise
    (a) Is a deterministic signal
    (b) Can be either Normal or Uniform.
    (c) Is always predictable.
    (d) Has an autocorrelation function that is periodic.
21. Match the following:

(a) Impulse	(i) Sample values are unpredictable.
(b) Step	(ii) Has only one nonzero value.
(c) Random noise	(iii) Amplitude decreases as time increase.
(d) Decaying exponential	(iv) Has only two possible values.

22. Which of the following is false?
    (a) The autocorrelation of random noise is an impulse
    (b) The autocorrelation of a pulse is a triangular wave.
    (c) The autocorrelation of a periodic signal is also periodic.
    (d) The autocorrelation of a triangular waveform is a square wave.
23. DSP commercial products came into market in the year
    (a) 1920.
    (b) 1960.
    (c) 1980.
    (d) 1990.
24. Sine waves, cosine waves, rectangular waves, and triangular waves are all examples of
    (a) Single-frequency functions.
    (b) Multiple-frequency functions.
    (c) Periodic functions.
    (d) All of the above.
25. What name is given to the lowest frequency in a Fourier series?
    (a) Fundamental.
    (b) First harmonic.
    (c) Both a and b.
    (d) None of the above.
26. A signal is composed entirely of these frequencies: 60, 180, 300, 420, 540, and 660 Hz. The signal is made up of
    (a) The fundamental and both even and odd harmonics.
    (b) The fundamental and only even harmonics.
    (c) The fundamental and only odd harmonics.
    (d) None of the above.
27. What is the bandwidth of an ideal 1-kHz square wave?
    (a) 1 kHz
    (b) 10 kHz
    (c) 1 MHz
    (d) Infinite
28. Spectrum analyzers display signals in the
    (a) Phasor domain
    (b) Time domain

(c) Frequency domain
(d) Propagation delay domain

29. A periodic signal displays on a spectrum analyzer as a single vertical line. What would the signal look like on an oscilloscope?

    (a) Square wave
    (b) Sawtooth wave
    (c) Pulse wave
    (d) Sine wave

30. Fourier synthesis of periodic signals with discontinuities produces aberrations due to

    (a) Laplace's phenomenon
    (b) Blackman's phenomenon
    (c) Gibbs' phenomenon
    (d) Component tolerance

31. A sawtooth waveform drops instantaneously form some positive value to zero. That falling edge represents a

    (a) Discontinuity
    (b) Fourier series
    (c) Missing harmonic
    (d) Limited bandwidth

32. The sum of a long Fourier series for a sawtooth function will show distortion called

    (a) Crossover
    (b) Gibbs' phenomenon
    (c) Clipping.
    (d) All of the above.

33. Conversion from the frequency domain to the time domain is accomplished by the

    (a) Fourier transform
    (b) Inverse Fourier transform
    (c) Gibbs' transform
    (d) Logarithmic transform

34. The samples of a cosine wave at zero frequency are equivalent to samples of

    (a) A sine wave at a frequency of 10 Hz.

(b) A dc signal.
(c) An aperiodic signal.
(d) An unknown signal.

35. FFT is

(a) Acronyms for fast Frequency transform.
(b) A transform for determining content of continuous-time a periodic signal.
(c) An efficient algorithm for calculating the DFT when N is a power of 2.
(d) An algorithm for calculating the length of the diagonal of a right-angled triangle.

36. The fastest way to compute the frequency content of a signal is to

(a) Choose the number of samples as a power of 2, and use an FFT algorithm.
(b) Choose any reasonable number of samples and use a generic DFT algorithm
(c) Perform the calculations using pencil and paper.
(d) Look up the answer in a book.

37. The underlying principle behind the DFT is

(a) Convolution
(b) Correlation
(c) Aliasing
(d) Quantization

38. The sines and cosines chosen by the DFT for correlation are

(a) Orthogonal
(b) Aperiodic
(c) Random
(d) Rectangular

39. The length of one side of a right-angled triangle is 5, and the length of the diagonal is 6. The length of the third side of the triangle is thus

(a) $\sqrt{5^2 + 6^2}$
(b) $\sqrt{6^2 - 5^2}$
(c) $5^2 + 6^2$
(d) $6^2 + 5^2$

40. The DC value of the signal having the following samples, 3.5, 3.5, −2.0, −3.5, 2.0 −1.5, 3.0, 3.0, is

    (a) 1.0
    (b) 3.5
    (c) 3.0
    (d) 1.5

41. Approximately how much faster is the FFT compared with the DFT if the number of samples is 4096?

    (a) 0.003
    (b) 256
    (c) 341
    (d) 396

42. To convert a frequency-domain representation to a time-domain representation, one uses the

    (a) Inverse FFT
    (b) DFT
    (c) FFT
    (d) Decibel unit

43. When two sine waves of frequencies 50 and 30 Hz are multiplied together, the resulting signal consists of the following frequencies (in Hertz):

    (a) 50 and 30
    (b) 80 and 20
    (c) −50 and 30
    (d) 100 and 60

44. If f_s is the sampling frequency and N is the number of samples, the formula for the frequency resolution Δf is given by

    (a) N/f_s
    (b) f_s/N
    (c) kf_s/N where $k = (N - 1)/2$
    (d) $0.5 - 0.5 \cos (2\pi\ n/N)$

45. If a signal is sampled for 2 s at a sampling frequency of 10,000 Hz, what is the frequency resolution of the FFT (in Hz)?

    (a) 500
    (b) 20.000
    (c) 1.0
    (d) 0.5

46. The frequency resolution Δf can be increased by
    (a) Keeping N fixed and reducing the sampling frequency.
    (b) Keeping the sampling frequency fixed and increasing N.
    (c) Both (a) and (b).
    (d) Neither (a) and (b).
47. Which of the following windows is best suited for separating two frequencies that are close to each other and have almost the same amplitudes?
    (a) Rectangular
    (b) Hamming
    (c) Hanning
    (d) All of the above are equal for separating frequencies.
48. The equation for the causal Hamming window is
    (a) $\omega[n] = 0.54 - 0.46 \cos\,(2\pi n/N - 1)$
    (b) $\omega[n] = 0.54 \;-\; 0.46 \sin\,(2\pi n/N - 1)$
    (c) $\omega[n] = 0.5 \;-\; 0.5 \cos\,(2\pi n/N - 1)$
    (d) $\omega[n] = 1.0$
49. Spectral leakage occurs when
    (a) Performing autocorrelation.
    (b) A signal frequency is not an exact integer multiple of Δf.
    (c) N is a power of 2.
    (d) All of the above.
50. A signal consists of a pure tone of frequency 100 Hz. It is sampled at f_s = 1024 Hz for 0.5 s. The FFT of the sampled signal using a rectangular window
    (a) Will show a single peak a t 100 Hz.
    (b) Will exhibit spectral leakage.
    (c) Is not possible because the number of samples is not a power of 2.
    (d) Will be twice as fast as a DFT on the same number of samples.
51. The window that has been widely used in speech-processing applications is the
    (a) Hanning window
    (b) Flat-top window
    (c) Hamming window
    (d) Rectangular window

52. Match the following

(a) Rectangular	(i) Amplitude accuracy
(b) Hamming	(ii) Uniform
(c) Hanning	(iii) Speech-processing applications
(d) Flat-top	(iv) Raised cosine

53. Another name for an ideal filter is
    (a) Bessel
    (b) Brick wall
    (c) Linear
    (d) Digital
54. Which of the following filter designs has ripple in the pass band?
    (a) Bessel
    (b) Tschebyscheff
    (c) Inverse Tschebyscheff
    (d) Butterworth
55. Suppose a filter has a linear-phase response and the delay at 1 kHz is 1 ms. What is the delay at 3 kHz?
    (a) 1 ms
    (b) 2 ms
    (c) 3 ms
    (d) None of the above
56. Which filter has the best pulse response?
    (a) Butterworth
    (b) Tschebyscheff
    (c) Elliptic
    (d) Bessel
57. The transition bandwidth of an ideal filter is
    (a) 0 Hz.
    (b) Half of the pass band.
    (c) Half of the stop band.
    (d) None of the above.
58. Suppose a low-pass moving-average filter uses the coefficients 0.25, 0.25, 0.25, and 0. 25. What will the output sequence be with an input signal at a constant value of 2?

(a) 0.1, 0.1, 0.1, 0.1, 0.1, 0.1, and so on.
(b) 0.5, 1.0, 1.5, 2.0, 2.0, 2.0, and so on.
(c) 1.0, 2.0, 3.0, 4.0, 4.0, 4.0, and so on
(d) None of the above.

59. A high-pass moving-average filter uses the coefficients −0.25, −0.25, 1.0, −0.25, and −0.25. What will the output sequence be with an input signal at a constant value of 2?

(a) −0.5, −1.0, 1.0, 0.5, 0, 0, 0, and so on
(b) −0.25, −1.25, 1.25, 1.75, 2.0, 2.0, 2.0, and so on
(c) 0.5, 1.5, 1.0, 0, −1.5, −1.5, −1.5, and so on
(d) None of the above.

60. What is the formal name given to the process of shift, multiply, add . . . shift, multiply, add . . . that goes on inside all DSP processors?

(a) Quantization
(b) Low-pass filtering
(c) High-pass filtering
(d) Convolution

61. What kinds of systems can change their characteristic, on the fly, as environmental factors change?

(a) Tschebyscheff systems
(b) RC systems
(c) Differential op-amp systems
(d) Adaptive systems

62. The horizontal axis for the frequency-response graph of a digital filter ends at

(a) $2f_s$
(b) f_s
(c) $f_s/2$
(d) $f_s/4$

63. The fact that sampling and amplitude modulation produces pairs of sidebands supports the concept of

(a) Negative frequencies.
(b) Fourier square waves.
(c) Gibbs' distortions.
(d) All of the above.

64. The coefficients for FIR filters are almost always arranged to be symmetrical to obtain
    (a) A sharper transition.
    (b) A linear phase response.
    (c) Elimination of Gibbs' distortion.
    (d) All of the above.
65. Windows are used in FIR filter designs to
    (a) Reduce the Gibbs' phenomenon.
    (b) Convert low-pass to high-pass.
    (c) Convert low-pass to band-pass.
    (d) All of the above.
66. Which of the following filters has an impulse response that is a sine wave that decays exponentially?
    (a) 11-tap FIR
    (b) Sixth-order IIR.
    (c) 49-tap FIR band-pass
    (d) Windowed sine type.
67. When a signal passes through two FIR filters in cascade, the same effect can be obtained by
    (a) Adding the filter coefficients.
    (b) Multiplying the filter coefficients.
    (c) Dividing the filter coefficients.
    (d) Convolving the filter coefficients.
68. The b coefficients are sometimes called the
    (a) FIR coefficients
    (b) Feedback coefficients
    (c) Feed-forward coefficients
    (d) Blackman coefficients
69. The filter structure having the minimum possible number of delay elements is known as
    (a) Direct form I
    (b) Canonic form
    (c) Second-order section
    (d) Cascaded form

70. The number of possible filter structures is
    (a) 2
    (b) 5
    (c) 100
    (d) Infinite
71. Figure is a
    (a) First-order IIR filter
    (b) First-order FIR filter
    (c) Second-order IIR filter
    (d) Second-order FIR filter

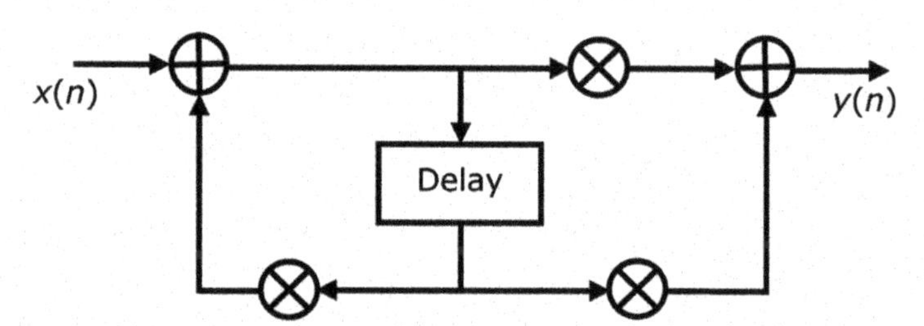

Figure 12.8 A filter with one delay, two adders, and three multipliers.

72. The filter structure in Figure 12.8 is
    (a) A direct form I structure
    (b) A higher-order filter.
    (c) An FIR filter.
    (d) A direct canonic form structure.
73. The choice of a filter structure depends on
    (a) Software complexity.
    (b) Hardware requirement
    (c) Finite word length effects.
    (d) All of the above.
74. Using finite-precision arithmetic, one needs to be careful about
    (a) Coefficient quantization errors.
    (b) Overflow and saturation.
    (c) Limit cycles
    (d) All of the above.

75. Which of the following would represent a negative number in a DSP system that uses two's complement representation?
    - (a) 0101 0001 0100 0111
    - (b) 1001 1111 0001 0101
    - (c) 0001 1001 1111 1011
    - (d) 0010 0100 1001 1111
76. DSP chips that use the type of numbers shown in question 75 are called
    - (a) Fixed-point processors.
    - (b) Floating-point processors.
    - (c) Adaptive processors.
    - (d) None of the above.
77. DSP chips that store numbers as a 24-bit mantissa and an 8-bit exponent are called
    - (a) Fixed-point processors.
    - (b) Floating-point processors.
    - (c) Adaptive processors.
    - (d) None of the above.
78. The binary number 1000 000 is in unsigned binary form. What is its decimal equivalent?
    - (a) +129
    - (b) −127
    - (c) +127
    - (d) −129
79. The binary number 1000 0001 is in two's complement signaled binary form. What is its decimal equivalent?
    - (a) +129
    - (b) −127
    - (c) +127
    - (d) −129
80. The number 1011 corresponds to
    - (a) Decimal one thousand and eleven.
    - (b) Decimal eleven.
    - (c) Decimal negative five.
    - (d) It depends. We need to know if the number is in base 10 or base 2 notations. If base 2, then we need to know if it is unsigned integer or signed two's complement integer.

81. A 16-bit number in two's complement notation can represent decimal numbers from

    (a) 0 to 65536
    (b) –32,768 to 32,767
    (c) 0 to −65.536
    (d) None of the above

82. What is the binary equivalent of the hex number 4A9E?

    (a) 0100 1010 1000 1110
    (b) 1000 1010 1001 1110
    (c) 0100 1010 1001 1110
    (d) 1110 1001 1010 0010

83. What is the hex equivalent of the binary number 1010 1010?

    (a) AA
    (b) AA0
    (c) 55
    (d) 550

84. Floating-point numbers are specified by

    (a) A mantissa
    (b) An exponent
    (c) A sign bit
    (d) All of the above

85. For an application where it is imperative that limit cycles should not occur, the filter use is

    (a) An FIR filter
    (b) An IIR filter
    (c) A filter with feedback
    (d) A direct from 1 IIR filter

13

Examination Question Papers

13.1 Practice Question Paper 1

QP # 1.1

(a) What is DSP? List the key operations of DSP. Discuss in brief any one of them.

(b) (i) List the steps for the conversion of analog signal into digital signal.
(ii) Discuss in brief the quantization error. How we can reduce it?

(c) A discrete time signal $x(n)$ is shown in the figure.

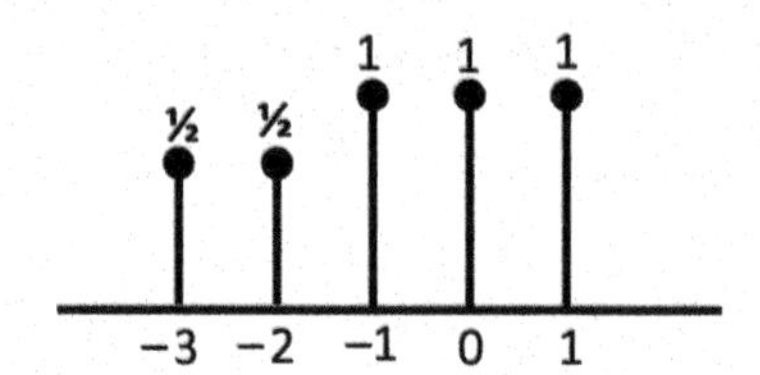

Sketch

- $x(n^2)$
- odd part of $x(n)$

QP # 1.2

(a) Define cross- and autocorrelation?

(b) Define cross-correlation coefficient and find $\rho_{x1x2}(1)$ for

$$x_1(n) = \{4, 1, -2, 3, 0\}$$
$$x_2(n) = \{-3,\ 2,\ 3,\ 5,\ 0\}$$

[Hint: at lag "1"]

(c) Determine convolution $y(n)$ of the signals

$$x(n) = \begin{cases} n/2, & 0 \leq n \leq 3 \\ 0, & \text{elsewhere.} \end{cases}$$

$$h(n) = \begin{cases} 2, & -1 \le n \le 2 \\ 0, & \text{elsewhere} \end{cases}$$

QP # 1.3

(a) Determine the causal signal $x(n)$ if its Z-transform is given as

(i) $X(z) = \frac{z^{-6} + z^{-7}}{1 - z^{-1}}$ (ii) $X(z) = \frac{0.1}{z - 0.9}$

(b) Determine the zero input response of the system described by the block diagram

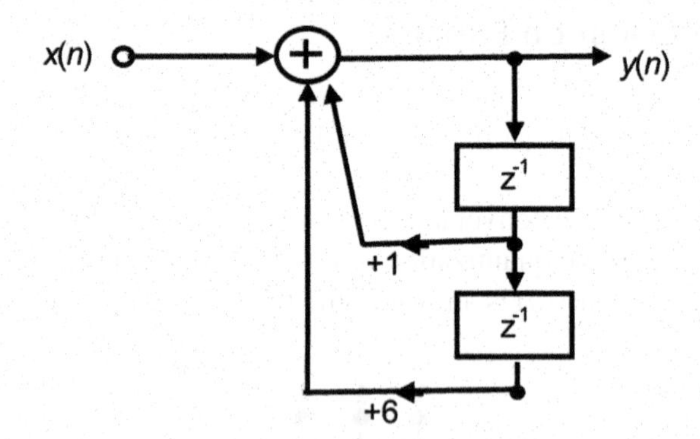

QP # 1.4

(a) (i) A discrete signal is given by $x(n) = \{\underset{\uparrow}{1}, \ 0, \ 3, \ 4\}$

Determine numerically its DFT, $X(k)$ over four points ($N = 4$)
(ii) Find the magnitude $X(k)$ and phase $\Phi(k)$

(b) Find IDFT of

$$X(K) = \{2, \ 1 + j, \ 0, \ 1 - j\}$$

(c) Find Fourier transform using FFT approach when

$$x(n) = \{1, \ 1, \ 1, \ 1\}$$

QP # 1.5

(a) Realize the following transfer function into four forms

$$H(z) = \frac{1 + z^{-1}}{1 + 3z^{-1}}$$

(i) Direct form II. (ii) Cascade. (iii) Parallel.

(b) Sketch the cascade form of the following transfer function
$H(z) = \frac{z+1}{z^2-3z+2}$

13.2 Practice Question Paper 2

QP # 2.1

(a) Give short and precise answers for the followings:

(i) Name the three steps involved in analog to digital conversion of a signal.
(ii) Any physical quantity, which varies with time or with any other independent variable, is called ____.
(iii) If sampling frequency is increased, number of quantization level ____.

(b) List the key operations of DSP.
(c) If $x(n) = \{1, \underset{\uparrow}{2}, 3\}$, then find out $y(n)$ while $y(n) = x(n)$.
(d) Define the following signals with graphical representation of each:

(i) Unit impulse.
(ii) Unit ramp signal.
(iii) Exponential signal.
(iv) Unit step signal.

QP # 2.2

(a) Give brief and precise answers for the followings:

(i) Constant multiplier is a system ____ memory. (With/Without)
(ii) Unit delay element is a system ____ memory. (With/Without)
(iii) Discuss static and dynamic systems in comparison with each other.

(b) Confirm whether the followings are time variant or time invariant signals:

(i) $y(n) = nx(n)$
(ii) $y(n) = x(n)\cos\omega n$

(c) Briefly discuss the followings:

(i) Casual and noncasual systems.
(ii) Stable and unstable systems.

(d) Define and differentiate linear and nonlinear systems.

QP # 2.3

(a) Short and smart answers are required for the followings:

(i) Define correlation and convolution. Also differentiate both these terminologies with respect to each other.
(ii) List down the four processes of convolution.

(b) For the following table, find out $\rho_{12}(1)$.

N	0	1	2	3	4	5	6	7	8
$x_1(n)$	1	3	4	4	2	4	3	1	2
$x_2(n)$	1	1	1	0	0	0	0	0	1

QP # 2.4

(a) Draw and specify the butterfly diagram of 4-point DFT, first finding the equation for $X(0)$, $X(1)$, $X(2)$, $X(3)$.

(b) Find the DFT of signal $x(n)$ and sketch the phase and magnitude plot if the signal is sampled at 10 KHz frequency.

n	0	1	2	3
$x(n)$	3	0	1	5

QP # 2.5

(a) Find the complete solution of the equation given below:

$$y(n) - \left(\frac{5}{6}\right) y(n-1) + \left(\frac{1}{6}\right) y(n-2) = 5^{-n} \quad \text{for} \quad n \geq 0, \text{ where}$$

$$y(-2) = 25, \quad y(-1) = 6$$

(b) Find inverse z-transform for the following:

$$Y(z) = \frac{0.2}{(z - 0.3 - j0.6)(z + 0.3 + j0.6)}$$

QP # 2.6

(a) Consider an LTI system function defined by

$$H(z) = \frac{9 + 4\,Z^{-1}}{7 + 2\,Z^{-1}}$$

Write down its difference equation and draw direct form I realization block diagram.

(b) Consider the following LTI system function:

$$H(z) = \frac{1}{1 - \frac{2}{3}Z^{-1}}$$

Find the difference equation and also draw its block diagram.

QP # 2.7

(a) Mention the 5 steps of digital FIR filter design with advantages/ disadvantages of using window method.
(b) Obtain coefficients of an FIR low-pass filter to meet the specifications given on the next page using the window method.

Stop band attenuation	>60 dbs
Sampling frequency	= 12 KHz
Transition width	= 0.8 KHz
Passband edge freq	= 1.8 KHz

13.3 Practice Question Paper 3

QP # 3.1

(a) Determine and make a labeled sketch of the odd and even components of signal shown in set. Verify that the addition of the components produces the original signal. $x(n) = \{-1,\ 1, -1,\ \underset{\uparrow}{1},\ 1\}$ and $y(n) = \{1, 2, \underset{\uparrow}{3}\}$
(b) Define the followings:
   (i) Linear and nonlinear
   (ii) Time invariant and time variant
   (iii) Static and dynamic
   (iv) casual and noncasual
(c) Determine whether the system $y(n) = n\ x(5n)$ given is
   (i) Linear or nonlinear
   (ii) Time invariant or time variant
   (iii) Static or dynamic
   (iv) Casual or noncasual

QP # 3.2

(a) (i) Write down major advantages and disadvantages of DSP?
   (ii) List down the key operations of DSP and discuss briefly any two of them

(b) Write note on any two of the following

(i) Audio applications of DSP
(ii) Telecommunication application of DSP
(iii) Biomedical application of DSP

QP # 3.3

(a) What is correlation and list down the types of correlation?
(b) Find cross-correlation r_{XY} by lag 3.

$$x(n) = \{\underset{\uparrow}{1}, 3, 7, -5, -9, -2, 3, 7\}$$

$$y(n) = \{\underset{\uparrow}{2}, 5, -1, 3, -4, 0, 5, 6\}$$

(c) Find the cross-correlation coefficient of data given below

$$x(n) = \{\underset{\uparrow}{3}, -1, 5, -3, -1, 4, 7\}$$

$$y(n) = \{\underset{\uparrow}{7}, 9, 3, -1, 0, 2, 4\}$$

QP # 3.4

(a) Find DFT of the following signal

$$x(n) = \{\underset{\uparrow}{3}, 6, 9, 12, 15\}$$

(b) Find FFT of the following signal

$$x(n) = \{\underset{\uparrow}{2}, 4, 8, 10\}$$

QP # 3.5

(a) Find the z-transform of the following signal

$$x(n) = \{5, 7, 3, \underset{\uparrow}{2}, 1, 8, 9, 4\}$$

(b) Find the inverse z-transform of the following transfer function

$$Y(z) = \frac{-25Z^{-1}}{1 - 7Z^{-1} + 12Z^{-2}}$$

(c) Determine the zero input response of system

$$x(n) = 15y(n-1) + 37y(n-2)$$

QP # 3.6

(a) Draw a structure by the help of the following transfer function

$$H(z) = \frac{15 + 19\, z^{-1} + 12\, z^{-2}}{5 + 17\, z^{-1} + 30\, z^{-2}}$$

(i) Direct form I
(ii) Direct form II

(b) Draw a structure by the help of following transfer function

$$H(z) = \frac{11 + 41\, z^{-1} + 30\, z^{-2}}{1 + 4\, z^{-1} + 4\, z^{-2}}$$

(i) Cascade form
(ii) Parallel form

QP # 3.7

(a) List down the main advantages and disadvantages of digital filters
(b) What are the types of digital filters and list down the difference between them
(c) Write down the difference equation and filtering operation in block diagram of the following IIR filters.

$$H(z) = \frac{0.7 + 0.89\, z^{-1} + 0.579\, z^{-2}}{1 - 0.35\, z^{-1} - 0.89\, z^{-2}}$$

QP # 3.8

(a) Obtain the coefficients $h(0)$, $h(1)$, $h(2)$ of an FIR low-pass filter to meet the specification given below using the window method

Passband edge frequency	4 kHz
Transition width	1 kHz
$30\,\text{dB} < A_s < 40\,\text{dB}$	
Sampling frequency	10 kHz

(b) Satisfy the symmetry conditions of the following FIR digital filters.

$$h(n) = h(N - n - 1)$$
$$\text{where } N = 7$$

13.4 Practice Question Paper 4

QP # 4.1

(a) Given the graphical representation of $x(n)$:

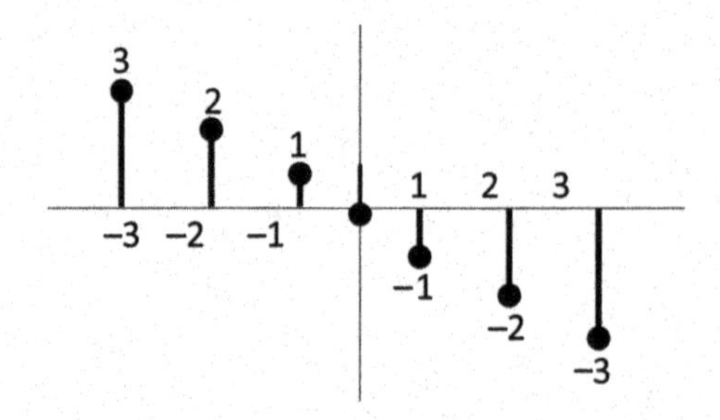

Represent $x(n)$ as:

(i) Function
(ii) Sequence
(iii) Table
(iv) Show whether the signal is symmetric or asymmetric

(b) Prove that $x(n) = x_e(n) + x_o(n)$ where $x(n) = \{-1, 2, 1, \underset{\uparrow}{4}, 3, -2, 1\}$.

(c) For the given $x(n)$:

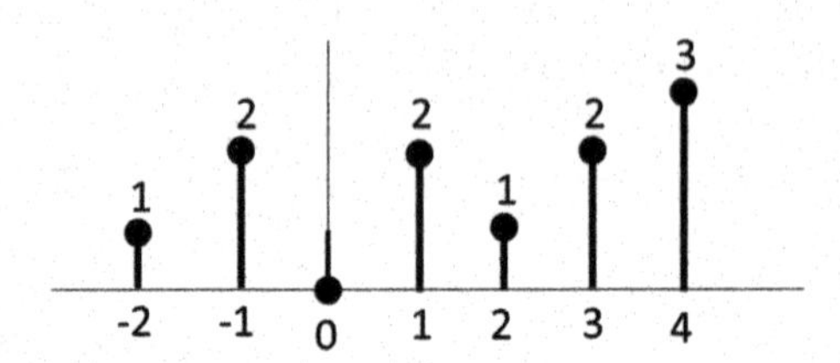

Find:

(i) $x(2n)$
(ii) $x(n^2)$

(d) What are the main types of signals?

QP # 4.2

(a) Define DSP. Write down the four main areas of its application. Also list down the key DSP operations.
(b) What are the main advantages of digital signal processing over analog signal processing.
(c) Check the given signals for any three of these properties and state reasons:

- Static/dynamic
- Time variant/invariant

- Linear/nonlinear
- Causal/noncausal
- Stable/unstable

(i) $y(n) = x(n) + 7x(n-1) + 2x(n-2)$
(ii) $y(n) = nx(n) + x(2n)$

QP # 4.3

(a) Convolve $x(n)$ and $h(n)$ using $x(n) \otimes h(n) = \sum_{k=0}^{n} x(k).h(n-k)$

$$\text{where } x(n) = \begin{cases} 2n+1 & 0 \le n \le 4 \\ 0 & \text{elsewhere} \end{cases},$$

$$h(n) = \begin{cases} n+2 & -1 \le n \le 1 \\ 0 & \text{elsewhere} \end{cases},$$

(b) Let

$$x_1(n) = \{1, 0, 1, 1, 1, 0, 1, 1, 0, 1\}$$
$$x_2(n) = \{0, 2, 3, 0, 1, 2, 9, 0.75, 1, 2\}$$

(i) Correlate to find the true value of $r_{12}(3)$.
(ii) Calculate the normalizing factor.

QP # 4.4

(a) Butterfly diagram, calculate FFT of $x(n) = \{6, 1, 5, 9\}$
(b) Apply DFT to find X(K) for the sequence in part a where

$$x(k) = \sum_{n=0}^{N-1} x(n)e^{-jKn\Omega T} \quad \text{and} \quad \Omega = \frac{2\pi}{NT}$$

(c) What are the advantages of FFT over DFT?

QP # 4.5

(a) Solve the following difference equation given that $y(-2) = 0$ and $y(-1) = 1$

$$y(n-2) - \frac{3}{2}y(n-1) + \frac{1}{2}y(n) = 1 + 3^{-n}$$

(b) Find Inverse Z Transform of the following:

$$\text{(i) } E(z) = \frac{1}{z^3 - 1.9z^2 + 0.98z - 0.08}$$

$$\text{(ii) } E(z) = \frac{-0.5995z}{(z^2 + 0.6160)(z - 0.6159)}$$

QP # 4.6

(a) Draw for the given transfer function:

$$H(z) = \frac{1 + \frac{1}{2}z^{-1}}{(1 - z^{-1} + \frac{1}{4}z^{-2})(1 - z^{-1} + \frac{1}{2}z^{-2})}$$

(i) Cascade Realization
(ii) Parallel Realization

(b) Write the transfer function for the following signal flow diagram.

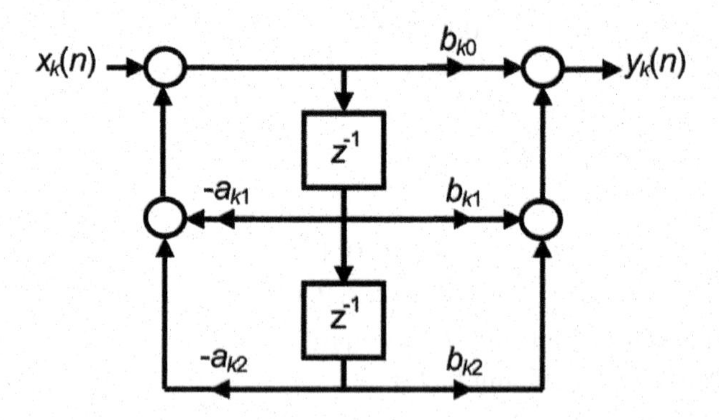

QP # 4.7

(a) A bandpass filter in the IF stage of Nav–Star receiver has the following characteristics:

Passband	160–240 Hz
Transition width	55 Hz
Passband ripple	0.2 dB
Stop band attenuation	55 dB
Sampling frequency	1.5 KHz

Obtain the ripple parameters for this FIR filter using KAISER window.

(b) Obtain *h*(0), *h*(1), and *h*(2) for an FIR low-pass filter to meet the specification given below using Hamming window:

Passband edge frequency	1.9 KHz
Transition width	0.8 KHz
Stop band attenuation	>50 dB
Sampling frequency	10 KHz

(c) Using Hamming window, obtain filter coefficients for a filter with the following specifications:

Passband edge frequency	1.5 KHz
Transition width	0.7 KHz
Stop band attenuation	>30 dB
Sampling frequency	6 KHz

13.5 Practice Question Paper 5

QP # 5.1

(a) A discrete time signal $x(n)$ is shown in the figure. Sketch the signals

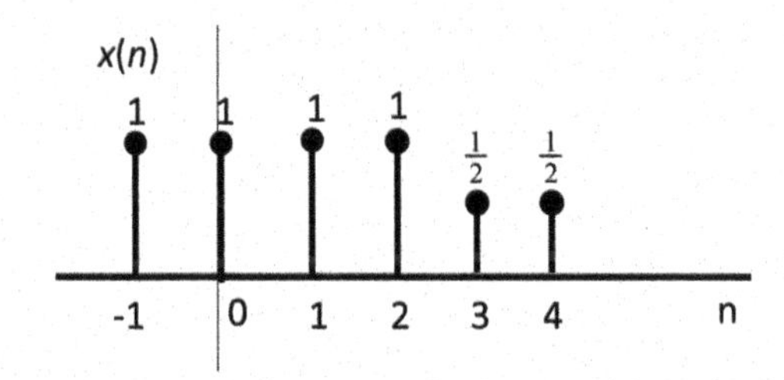

Express the signal in functional representation.

(i) $x(-n+4)$.
(ii) First delay $x(n)$ by 4 and then fold.
(iii) Even component of $x(n)$.
(iv) $x(n).u(2-n)$.

(b) A discrete time signal can be

(i) Static or dynamic.
(ii) Linear or nonlinear.
(iii) Time invariant or time variant.
(iv) Causal or noncausal.
(v) Stable or unstable.

Examine the following system and apply any two properties on the systems given

i) $y(n) = |x(n)|$
ii) $y(n) = x(2n)$

QP # 5.2

(a) What is DSP? Write down its applications.
(b) Write down any five advantages and disadvantages of DSP.
(c) What are key operations of DSP? Explain any two.
(d) What are the advantages of DSP over analog signal processing?

QP # 5.3

(a) Determine the convolution $y(n)$ of the signals

$$x(n) = \begin{cases} 1/2n, & 0 \leq n \leq 4 \\ 0, & \text{elsewhere} \end{cases}$$

$$h(n) = \begin{cases} 1, & -2 \leq n \leq 2 \\ 0, & \text{elsewhere} \end{cases}$$

(b) Discuss briefly auto and cross-correlation. Find cross-correlation $r_{xy}(j)$ of the following sequence up to 1 lag

$$\begin{aligned} x(n) &= \{2,\ 1, -1, -2, -3\} \\ y(n) &= \{-2,\ 2,\ 1,\ 5, -3\} \end{aligned}$$

Also find the cross-correlation coefficient $r_{xy}(1)$.

QP # 5.4

(a) A discrete time signal $x(n)$ takes the following value

$$x(0) = 1,\ x(1) = 0,\ x(2) = 1,\ x(3) = 0$$

Determine numerically discrete Fourier transform $x(k)$ over 4 points.

(b) Verify part (*a*) using FFT method.

QP # 5.5

(a) Determine the response $y(n), n = 0$ of the system described by the second-order difference equation

$$y(n) - 3y(n-1) - 4y(n-1) = x(n) + 2x(n-1)$$

when the input sequence is $x(n) = 4^n u(n)$.

(b) Determine the causal signal $x(n)$ if its Z-transform $X(z)$ is given by

$$X(z) = \frac{z^{-1}}{1 - 0.5z^{-1} - 0.25z^{-2}}$$

QP # 5.6

(a) Determine the difference equation for the following figure.

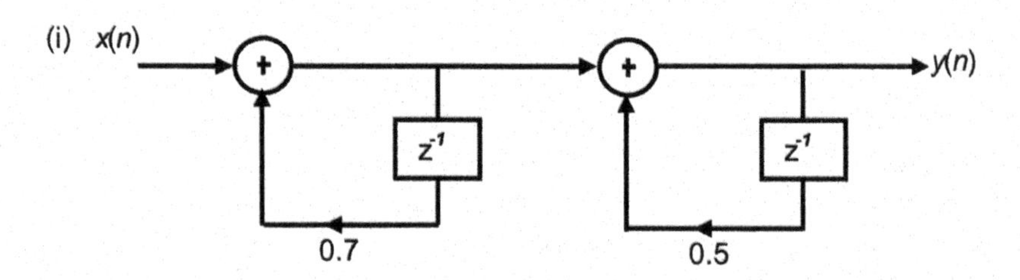

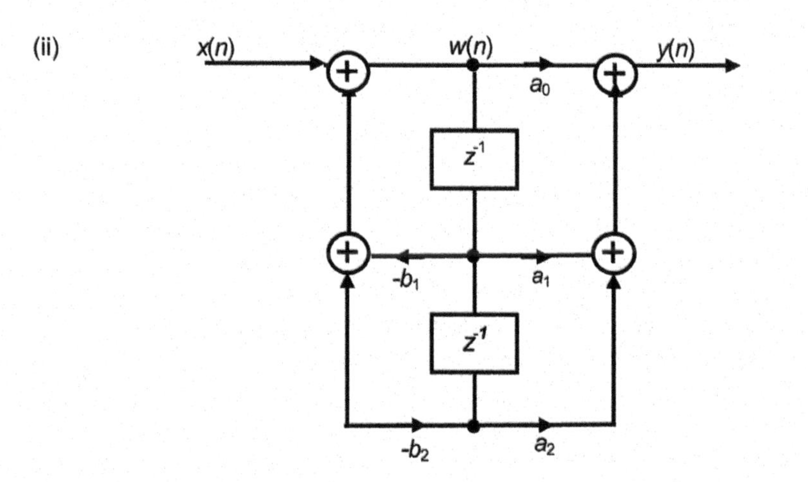

(b) Given the transfer function

$$H(z) = \frac{0.5(1 + 0.4z^{-1})(1 - 0.2z^{-1})}{(1 + 0.8z^{-1})(1 - 0.6z^{-1})}$$

Obtain the following realizations:

(i) Cascade form
(ii) Parallel form.

QP # 5.7

(a) Discuss briefly the advantages and disadvantages of digital filters.
(b) Summarize the key features of FIR and IIR filters.
(c) FIR digital filter has impulse response $h(n)$, defined over the interval; $0 \leq n \leq N - 1$

Show that whether $N = 9$ and $h(n)$ satisfy the symmetry condition $h(n) = h(N - n - 1)$, the filter has a linear phase characteristic.

QP # 5.8

(a) Obtain the coefficient $h_d(0)$, $h_d(1)$, and $h_d(2)$ of an FIR low-pass filter to meet the specifications given below using window method:

Passband edge frequency	= 1.5 KHz
Transition width	= 0.6 KHz
Stop band attenuation	>50 db
Sampling Frequency	= 11 KH

(b) Using BZT method, design a digital filter to approximate the following analog transfer function

$$H(s) = \frac{1}{s^2 + 4s + 1}$$

Assume cutoff frequency of 160 Hz and sampling frequency of 1.20 kHz.

$$H(s) = \frac{2.6978}{s^2 + 6.57s + 2.6978}$$

$$H(z) = \frac{0.263(1 + 2z^{-1} + 2z^{-2})}{10.2678(1 + 0.331z^{-1} - 0.2797z^{-2})}$$

13.6 Practice Question Paper 6

QP # 6.1

(a) A discrete time signal $x(n)$ is defined as

$$x(n) = \begin{cases} 2 + n/6 & -4 \leq n \geq -1 \\ 3n & 0 \leq n \geq 4 \\ 0 & \text{elsewhere} \end{cases}$$

Determine its value and sketch the signal $x(n)$.

(b) A discrete time signal $x(n)$ is shown in figure

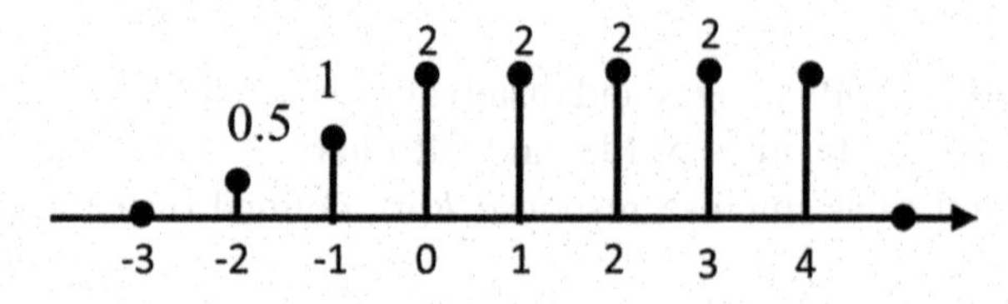

Sketch (i) $x(2n)$ (ii) even component of $x(n)$ (iii) odd component of $x(n)$

QP # 6.2

(a) What are the advantages and disadvantages of DSP, compare with analog signal processing design.
(b) List down the key operation of DSP.
(c) A discrete time signal can be

(i) Static or dynamic (ii) linear or nonlinear (iii) time invariant or variant
(ii) casual or noncasual (v) stable or unstable.

Examine the following system with respect to the properties above.
(i) $y(n) = x(n)\,u(n)$
(ii) $y(n) = x(-n+4)$
(iii) $y(n) = |x(n)|$
(iv) $y(n) = \sin[x(n)]$
(v) $y(n) = x(n) + nx(n+1)$

Answer: (c)

(i) static, linear, time invariant, causal, stable.
(ii) static, linear, time invariant, causal, stable.
(iii) dynamic, linear, time invariant, noncausal, unstable.
(iv) static, nonlinear, time invariant, causal, stable.
(v) dynamic, nonlinear, time variant, linear, stable.

QP # 6.3

(a) Determine the convolution $y(n)$ of the signal given below by the matrix method.

$$x(n)\begin{cases} 2n/3 & 0 \leq n \leq 6 \\ 0 & \text{elsewhere} \end{cases}$$

$$h(n)\begin{cases} 3 & -2 \leq n \leq 2 \\ 0 & \text{elsewhere} \end{cases}$$

(b) What is correlation and what are its types?
(c) The data values are as follows:

x	0	1	2	3	4	5	6	7	8
$x_1(n)$	0	3	5	6	5	2	0.5	0.25	0
$x_2(n)$	1	0	1	1	1	0	0	0	1
$x_3(n)$	0	9	7	6	5	6	1.5	0.75	0
$x_4(n)$	2	3	2	4	2	3	2	4	2

Determine the following:

(i) $r_{12}(1)$ (ii) $r_{34}(3)$ (iii) $\rho_{12}(1)$ (iv)$\rho_{34}(3)$

QP # 6.4

A discrete signal $x(n)$ takes the following values:

$$x(n)\begin{cases} 1+2n & 0 \le n \le 3 \\ 0 & \text{elsewhere} \end{cases}$$

(a) Determine numerically the discrete Fourier transform $x(k)$ over 4 points ($N = 4$)
(b) Verify part (a) using inverse discrete Fourier transform method.
(c) Verify part (a) using fast Fourier transform method.

QP # 6.5

(a) Determine the total solution $y(n)$ to the difference equation.

$$y(n) - 3y(n-1) + 2y(n-2) = 6^n \text{ with initial condition}$$

$$y(-2) = 5 \text{ and } y(-1) = 6$$

(b) Solve the difference equation using Z-transform approach.

$$f(n) + b^2 f(n-2) = 0$$

where $|b| < 1$ and initial condition $f(-1) = 0$ and $f(-2) = -1$, show that $f(n) = b^{n+2} \cos(n\pi/2)$

QP # 6.6

(a) What are the advantages and disadvantages of digital filters, as compared with analog filters.
(b) The following transfer function represents two different filters:

$$\text{(i) } H(z) = \frac{0.8(z^2 - 0.28)}{z^2 + 0.7\,z - 0.91} \qquad \text{(ii) } H(z) = \sum_{k=0}^{4} h(k) z^{-k}$$

(i) State whether it is a FIR or IIR filter.
(ii) Represent the filtering operation in a block diagram form and write down the difference equation.

QP # 6.7

(a) What is the difference between phase delay and group delay and what are the conditions for an FIR filter to have linear phase response?
(b) Obtain $h(6)$, $h(7)$, $h(8)$, $h(9)$, and $h(10)$. Coefficients of an FIR filter using Hamming window method. The specifications are as follows:

Pass edge frequency	2.5 KHz
Transition width	0.5 KHz
Stop band attenuation	>50 dB
Sampling frequency	50 KHz

And also $h_D(n)$ for low-pass filter is given by

$$h_D(n) = \frac{2f_c \sin(n\omega_c)}{n\omega_c} \qquad n \neq 0$$

And window function, $\omega(n), |n| \leq (N-1)/2 = 0.5 + 0.5\cos(2\pi n/N)$
Hamming window information:

Transition width (Hz) normalized = $3.1/N$,
Stop band attenuation (db) = 44 db_{max}
Passband ripple (db) = 0.0546
Main lobe relative to side lobe (db) = 31

QP # 6.8

(a) What are the methods used for the calculation of IIR filters coefficient?
(b) What are the design stages for digital IIR filters?
(c) The normalized transfer function for an IIR filter is $H(s) = 1/(s+1)$.

Using bilinear Z-transform method, determine the transfer function and difference equation. Assume a sampling frequency of 250 Hz and cutoff frequency 50 Hz.

$$\text{Ans : } \quad (c) \quad \Omega_p = \tan\left(\frac{100\pi}{250 \times 2}\right) = \tan\left(\frac{\pi}{5}\right) = 0.726542$$

$$H'(s) = \frac{0.726542}{s + 0.726542}$$

$$H(z) = \frac{0.42082(1 + z^{-1})}{1 - 0.15841z^{-1}}$$

$$y(n) = 0.15841\, y(n-1) + 0.42082\,[x(n) + x(n-1)]$$

13.7 Practice Question Paper 7

QP # 7.1

(a) Describe key operations of DSP and explain one of them.
(b) Write down the advantages and disadvantages of DSP.

(c) Make labeled sketches of signals $x(n)$ and $y(n)$(3 + 3 + 2 + 2)

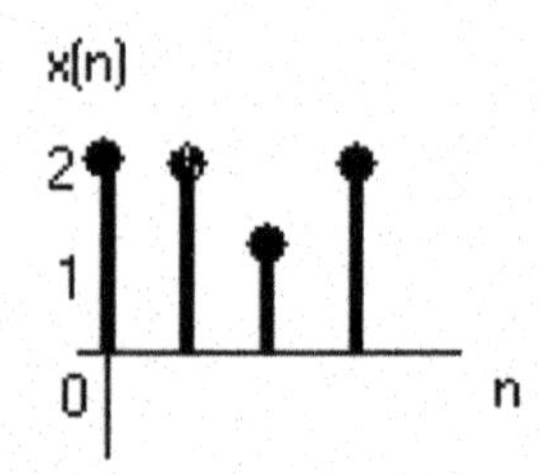

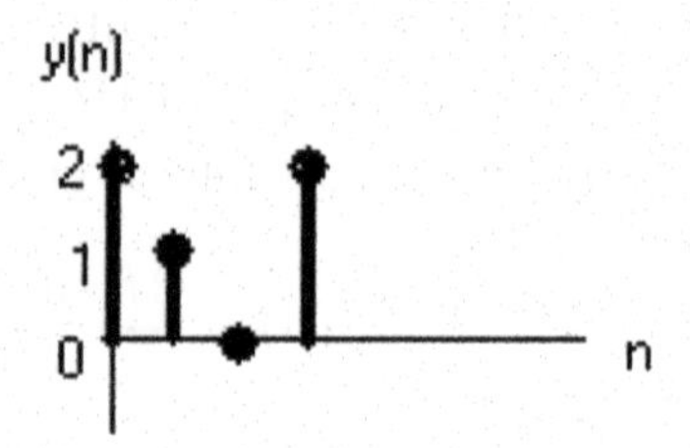

(i) $0.5x(2n) - y(n)$
(ii) $x(n-2) + y(n-4)$
(iii) Illustrate the following signals with labeled sketches:

$$x(n) = u(n) - u(n-1) - 2u(n-2)$$
$$y(n) = u(n) - 3u(n-2) + 2u(n-3)$$

QP # 7.2

(a) Define the following with example:

(i) Static and dynamic system
(ii) Linear versus nonlinear system
(iii) Time invariant versus time variant system.
(iv) Stable and unstable system.

(b) A system is given as

$$y(n) = x(n) - bx(n-1)$$

Show that the system is

(i) Time invariant or not
(ii) Linear or nonlinear.
(iii) If $y(n) = x(n) + 3x(n+4)$, then this system is causal or not?

QP # 7.3

(a) Find cross-correlation of two sequences and $x_1(n)$ and $x_2(n)$ at lag $j(2)$ that is $r_{12}(2)$

Also find $\rho_{12}(2)$ when sequences are

n	1	2	3	4	5	6
x_1	4	−1	2	3	−2	4
x_2	4	2	3	−5	1	1

(b) (i) If two signals are

$$x(n) = \{3,\ 1,\ 2\}$$
$$h(n) = \{1,\ 2,\ 3,\ 2\}$$

What will be the result of convolution using matrix method?

(ii) If two signals are

$$x(n) = \{1,\ 1,\ 1\}$$
$$h(n) = \{2,\ 1,\ 6\}$$

What will be the result of convolution using analytical formula method?

QP # 7.4

(a) Find discrete Fourier transform of the sequence:

$$x(n) = \cos\left(\frac{n2\pi}{N}\right)$$

(b) Find fast Fourier transform of the sequence using radix four:

$$x(n) = \{1,\ 0,\ 2,\ 1\}$$

(c) Find inverse Fourier transform of the following sequence

$$X(k) = \{2, 1 + j,\ 0,\ 1 - j\}$$

QP # 7.5

(a) Find z-transform of the sequence:

(i) $x(n) = a^n un$
(ii) $x(n) = \sin(n\omega t)$

(b) Find inverse Z-transform.

$$H(z) = \frac{z^2}{(z - 0.5 - j0.5)\,(z - 0.5 + j0.5)}$$

(c) Find total solution of the given homogenous equation:

$$y(n)\ -\ 0.5y(n-1)\ =\ 0$$

QP # 7.6

(a) Calculate $h(0)$, $h(1)$, and $h(2)$ for FIR low-pass filter to meet the specifications given below using the window method.

Passband edge frequency	2 kHz
Transition width	0.7 kHz
Stop band attenuation	>45 db
Sampling frequency	11 kHz

(b) Determine the difference equation for the IIR filter using BZT (bilinear z-transform) method while

Sampling frequency	300 Hz
Cutoff frequency	45 Hz
Normalized transfer function	$= 2/(s+3)$

(c) What is the advantage of Kaiser Window over Blackman Window?

13.8 Practice Question Paper 8

QP # 8.1

The signals in figure a, are zero except as shown.

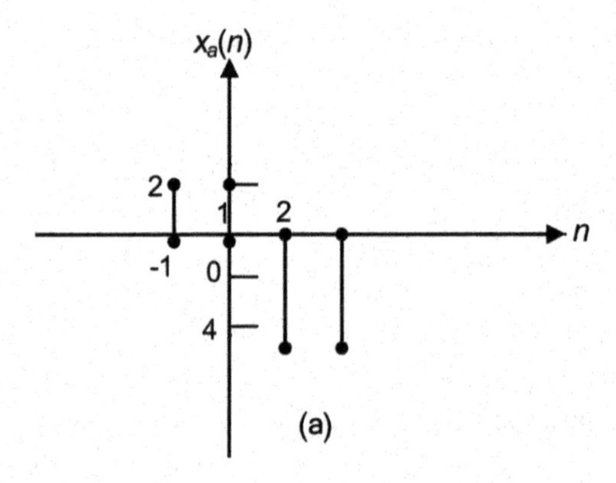

(a) For the signal $x_a(n)$ of figure a, plot the following:

(i) $x_a(n)\,\delta(n-1)$ (ii) $x_a(n) + u(2-n)$
(iii) $-4x_a(n-3)$ (iv) $2 + 2x_a(n-3)$
(v) $4x_a(-n) - 2$

(b) Find the even and odd parts of $X_b(n)$ in figure b.

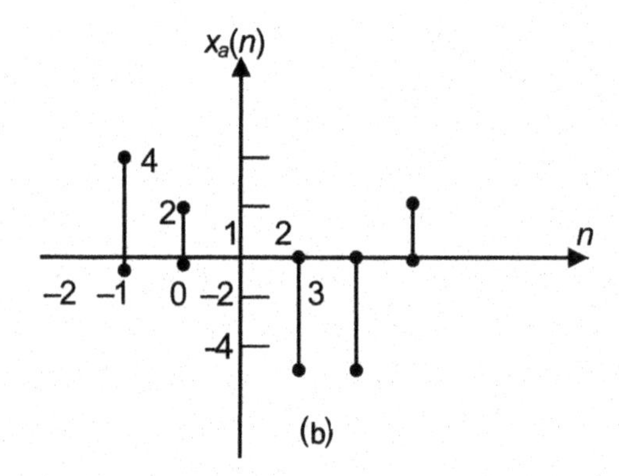

(b)

(c) Give proofs of the following statements.

(i) The sum of two even functions is even.
(ii) The sum of two odd functions is odd.
(iii) The product of two odd functions is odd.
(iv) The product of two even functions is even.
(v) The product of an even and odd function is odd.

QP # 8.2

(a) Give the advantages of digital signal processing over analog signal processing.

(b) Classify the following signals according to whether they are

- one or multidimensional.
- single or Multichannel.
- continuous time or discrete time.
- analog or digital (in amplitude).

1. Closing prices of utility stocks on the New York Stock Exchange.
2. A color movie.
3. Position of the steering wheel of a car in motion relative to car's reference frame.
4. Position of the steering wheel of a car in motion relative to ground reference frame.
5. Weight and height measurements of a child taken every month.

(c) A discrete time system can be

- static or dynamic
- linear or nonlinear.

- time invariant or time variant.
- casual or noncasual.
- stable or unstable

Examine the following system with respect to the properties above.

(i) $y(n) = \sin(x[n-1])$
(ii) $y(n) = \ln x(n)$

QP # 8.3

(a) Compute the convolution $y(n)$ of the signals analytically.

$$x(n) = \begin{cases} \alpha^n, & -3 \leq n \leq 5 \\ 0, & \text{elsewhere} \end{cases}$$

$$h(n) = \begin{cases} 1, & -2 \leq n \leq 2 \\ 0, & \text{elsewhere} \end{cases} \quad \text{where,} \quad \alpha = \frac{1}{3}$$

(b) Determine the cross-correlation of the sequence $r_{xy}(l)$ of the sequences

$$x(n) = \{\ldots, 0, 0, 2, 3, 1, \underset{\uparrow}{5}, 2, -1, 0, 0, \ldots\}$$
$$y(n) = \{\ldots, 0, 0, 1, -1, 2, -2, \underset{\uparrow}{4}, 1, -2, 5, 0, 0, \ldots\}$$

Also find the value of $r_{xy}(-3)$ and $\rho_{xy}(-3)$

QP # 8.4

(a) Determine the DFT and IDFT of the following sample sequences:

$$x(n) = \{1, \underset{\uparrow}{1}, 1, 1\}$$

$$y(n) = \{\underset{\uparrow}{0}, 1, 1, 0\}$$

Also draw the corresponding amplitude and phase spectra.
(b) Verify the DFT calculated in part a using fast Fourier method.

QP # 8.5

(a) Determine the inverse Z-transform of the following transfer function

$$H(z) = \frac{0.178\, z^3 + 0.342\, z^2}{z^3 - 1.702\, z^2 + 0.837\, z - 0.135}$$

$$h(k) = ? \quad \text{For } -2 \leq k \leq 2$$

(b) Using residue theorem, determine $f(n)$ for the following:

$$F(z) = \frac{z^2 + 4z}{(z - 0.735)^3}$$

(c) Solve the following difference equation:
$y(n) - 3y(n-1) - 4y(n-2) = x(n) + 2x(n-1)$ where $x(n) = 4^n u(n)$

QP # 8.6

(a) Consider an FIR filter with system function

$$H(z) = 1 + 2.88z^{-1} + 3.40z^{-2}$$

Sketch the direct form and lattice realization of the filter and determine in detail the corresponding input–output equations.

(b) Obtain the direct form I, direct form II, and cascade and parallel structures for the system given below.

$$y(n) = 0.1y(n-1) + 0.2y(n-2) + x(n) + \frac{3}{4}x(n-1)$$

QP # 8.7

(a) Briefly describe the advantages of FIR filters over IIR filters.
(b) List down the various steps involved in designing a digital filter.
(c) An FIR digital filter has an impulse response, $h(n)$, defined over the interval

$$0 \le n \le N - 1.$$

Show that whether N = 6 and $h(n)$satisfy the symmetry condition $h(n) = h(N - n - 1)$

The filter has linear phase characteristic.

QP # 8.8

(a) Using the pole–zero replacement method, calculate the coefficient of a bandpass filter having the following specifications:

- Complete signal rejection at dc and 500 Hz.
- A narrow passband filter centered at 250 Hz.
- A 3 dB bandwidth of 10 Hz.
- Sampling frequency of 750 Hz.

Obtain transfer function and hence the difference equation.

(b) Convert the analog filter with system function into a digital IIR filter assuming a 3 dB cutoff frequency of 250 Hz and sampling frequency of 2.28 kHz. Also obtain a suitable realization.

$$H(s) = \frac{1}{s^2 + 4s + 4}$$

13.9 Practice Question Paper 9

QP # 9.1

(a) Define the following

(i) DSP
(ii) Signal
(iii) System
(iv) List the key operations of DSP

(b) Examine the following system with the given properties.
$y(n) = x(n + 2)$

(i) Static or dynamic
(ii) Linear or nonlinear
(iii) Casual or noncausal
(iv) Stable or unstable
(v) Time invariant or time variant

QP # 9.2

(a) (i) Represent the following signal in

$$x(n) = \begin{cases} n+1 & -2 \leq n \leq 0 \\ 1 & 1 \leq n \leq 4 \\ 0 & \text{elsewhere} \end{cases}$$

i. Graphical
ii. Sequential
iii. Tabular

(ii)

i. $x_e(\text{n}) = ?$
ii. $x_o(\text{n}) = ?$
iii. $x(2\text{n}) = ?$

(b) Mathematically brief about the following:

(i) Unit impulse
(ii) Unit step
(iii) Unit ramp
(iv) Exponential

QP # 9.3

(a) (i) Write main difference between convolution and correlation.
(ii) List the four processes of convolution.
(iii) Find the $\rho_{12}(1)$ of the given data.

N	1	2	3	4	5	6	7	8	9
$x_1(n)$	0	3	5	5	5	2	0.5	0.25	0
$x_2(n)$	1	1	1	1	1	0	0	0	1

(b) Determine analytically the convolution of the signal and prove it by matrix method.

$$x(n) = \{\underset{\uparrow}{1},\ 2,\ 4\}$$

$$h(n) = \{\underset{\uparrow}{1},\ 1,\ 1,\ 1,\ 1\}$$

QP # 9.4

(a) Draw and label the butterfly diagram of 4-point FFT (decimation in time).
(b) Find the DFT of the signal $x(n)$ and sketch the phasor plot and the magnitude plot, if the signal is sampled at 10 kHz.

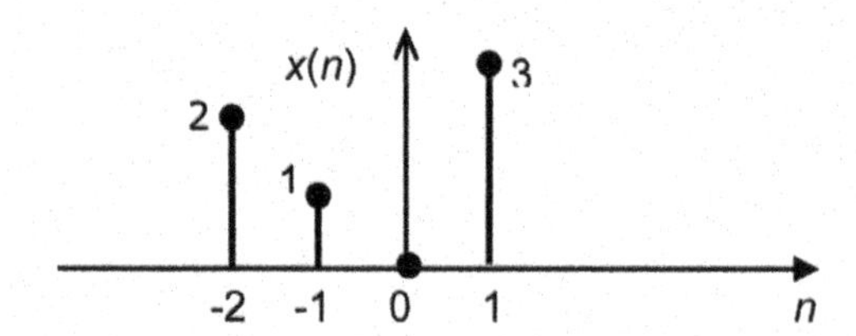

QP # 9.5

(a) Solve the following difference equation using ZT approach
$2f(n-2) - 3f(n-1) + f(n = 3^{n-2}$, where,
$f(-2) = -4/9,\ f(-1) = -1/3$

$$\text{(i) } E(z) = \frac{1}{z(z-1)(z-0.8)}$$

$$\text{(ii) } Y(z) = \frac{-3.894z}{z^2 + 0.6065}$$

QP # 9.6

(a) List the methods of finding FIR filter coefficients and describe the advantages and disadvantages of window method.

(b) Cutoff frequency 3 kHz
Transmission width 1 kHz
Stop band attenuation >45 db
Sampling frequency 10 kHZ

Obtain the coefficients of FIR filter $h(0)$, $h(1)$, and $h(2)$.

QP # 9.7

(a) Briefly explain the design stages of digital IIR filters

(b) Using BZT method, determine the difference equation and draw the block diagram of RC filter whose transfer function is

$$H(s) = \frac{8}{2s + 1}$$

Assume
sampling frequency = 300 HZ
cutoff frequency = 60 HZ

13.10 Practice Question Paper 10

QP # 10.1

(a) Determine $h(n)$ the inverse Z-transform of the signal

$$H(z) = \frac{1 + {}^1\!/_2\, z^{-1}}{1 + 2z^{-1} + z^{-2}}$$

(b) Determine the casual signal $X(z)$ if $x(n)$ is given by

$$x(n) = 5.25\, u(n) + 3.5\, n(1)^{n-1} + 0.75\, (-1)^n$$

QP # 10.2

The difference equation of the second-order system is given by

$$y(n) + 0.2\,y(n-1) - 0.48\,y(n-2) = x(n)$$

Use one–sided Z-transform to find its impulse response when

(i) the initial conditions are zero

$$y(n) = 0.571\,(-0.8)^n\,u(n) + 0.429\,(0.6)^n\,u(n)$$

(ii) the initial conditions are as follows:

$$\begin{aligned} y(-1) &= -1.25 \\ y(-2) &= -0.52 \\ y(n) &= (-0.8)^n\,u(n) \end{aligned}$$

QP # 10.3

Determine the impulse response and the step response of the causal system given as follows:

$$y(n) = \frac{1}{2}\,[x(n) + x(n-1)] + y(n-1)\,n \geq 0$$

$$h(0) = 0.5\,h(n) = 1\ n \geq 1\ y(n) = (2n+1),\ n \geq 0\ (n+0.5)\ u(n)$$

QP # 10.4

(a) Determine the Fourier transform of the signal by evaluating its Z-transform on the unit circle

$$x(n) = (\cos\omega_0\,n)\,u(n)$$

$$X(\omega) = \frac{1 - e^{-j\omega}\cos\omega_0}{1 - 2\,e^{-j\omega}\cos\omega_0 + e^{-2j\omega}}$$

(b) A linear time invariant system is described by the difference equation.

$$y(n) = a\,y(n-1) + b\,x(n)\ 0 < a < 1$$

Determine (i) $H(w)$ of the system $\frac{b}{1-ae^{-jw}}$
The output of the system to the input signal

$$x(n) = 5 + 12\sin\frac{\pi}{2}n$$

QP # 10.5

(a) For the general linear time invariant recursive system described by the difference equation

$$y(n) = -\sum_{K=1}^{N} a_k\, y(n-k) + \sum_{K=0}^{N} b_k\, y(n-k)$$

Illustrate how you will obtain direct form II structure for.

(b) Find the direct form I and direct form II structure of the discrete time system

$$H(z) = \frac{8z^3 - 4z^2 + 11z + 2}{(z - 1/4)\,(z^2 - z + 1/2)}$$

QP # 10.6

The transfer function of an analog, third-order, Butter worth low-pass filter with a cutoff frequency of 1 rad/sec is

$$H(s) = \frac{1}{(s+1)\,(s + 0.5 + j0.866)\,(s + 0.5 - j0.866)}$$

Design an impulse invariant digital equivalent based on a sampling interval of 0.5 sec.

$$H(z) = \frac{z}{z - 0.65} + \frac{z\,(0.8956 - z)}{z^2 + 1.4142 + 6065}$$

QP # 10.7

Determine the convolution of the following pairs of signals by means of the Z-transform

$$x_1(n) = (1/2)^n\, u(n);\; x_2(n) = \cos \pi n\, u(n)$$
$$y(n) = [2/3 \cos \pi/n + 1/3\,(1/2)^n]\, u(n)$$

Write a short note on biomedical signals and their analysis using digital signal processing.

13.11 Practice Question Paper 11

QP # 11.1

(a) Determine the Z-transform of the signals.

$$\text{(i) } x(n) = \frac{1}{2}(n^2 + n)\left(\frac{1}{3}\right)^{n-1} u(n-1)$$

$$(ii)\ x(n) = \left(\frac{1}{3}\right)^n u(n) - \left(\frac{1}{2}\right)^n u(-n-1)$$

(b) Determine the causal signal $x(n)$ if its Z-transform $X(z)$ is given by

$$x(z) = \frac{1}{(1-2z^{-1})(1-z^{-1})^2}$$

$$X(z) = \frac{A}{1-2z^{-1}} + \frac{B}{1-z^{-1}} + \frac{C\,z^{-1}}{(1-z^{-1})^2}$$

$$A = 4; B = -3; C = -1$$

$$x(n) = [4(2^n) - 3 - n]\,u(n)$$

QP # 11.2

(a) Determine the convolution of the following pair of signal by means of Z-transform

$$x_1(n) = n\,u(n),\ x_2(n) = \cos \pi n\ u(n)$$

(b) Determine the impulse response of the following causal system.

$$y(n) = \frac{1}{6}\,y(n-1) - \frac{1}{6}\,y(n-2) = x(n)$$

QP # 11.3

Use one-sided Z-transform to determine $y(n)$, $n > 0$ with the case as indicated

$$\frac{1}{\sqrt{5}}\left(\frac{1}{2}\right)^{n+1}\left[\left(1+\sqrt{5}\right)^{n+1} - \left(1-\sqrt{5}\right)^{n+1}\right]u(n)$$

$y(n) = y(n-1) + y(n-2)$ initial conditions are $y(-2) = 0$ $y(-2) = 1$

$$x(n)\ \frac{1}{\sqrt{5}}\left(\frac{1}{2}\right)^{n+1}\left[\left(1+\sqrt{5}\right)^{n+1} - \left(1-\sqrt{5}\right)^{n+1}\right]u(n)$$

QP # 11.4

Consider the system described by the difference equation
Determine

(i) its frequency response
(ii) its response to the input

$$x(n) = 5 + 6\cos\left(\frac{\pi}{2}n + \frac{\pi}{4}\right)$$

QP # 11.5

Obtain the direct form I and direct form II for the following systems.

$$y(n) = -0.1\,y(n-1) + 0.72\,y(n-2) + 0.7\,x(n) - 0.25\,x(n-2)$$

QP # 11.6

The transfer function of an analog third-order Butter worth loss-pass filter is given by

$$H(s) = \frac{1}{(s+1)\,(s^2 + s + 1)}$$

Design impulse invariant digital filters based on a sampling interval of
(i) 50 m sec. (ii) 0.5 sec.

QP # 11.7

Design a low-pass filter using bilinear transformation with the following specification:

$$\begin{aligned} \alpha\,\text{max} &= 2\,\text{dB} \\ \alpha\,\text{min} &= 14\,\text{dB} \\ \omega_p &= 0.65\,\text{kHz} \\ \omega_s &= 2.75\,\text{kHz} \end{aligned}$$

Assume any necessary data if required.

13.12 Practice Question Paper 12

QP # 12.1

(a) Determine the causal signal $x(n)$ if its z-transform $X(z)$ is given by

$$X(z) = \frac{6z^3 + 2z^2 - z}{z^3 - z^2 - z + 1}$$

$$x(n) = 5.25\,u(n) + 3.5\,n\,(1)^{n-1} + 0.75\,(-1)^n$$

(b) Determine the convolution of the signals by means of the Z-transform

$$x_1(n) = n\,u(n),\; x_2(n) = 2^n\,u(n-1)$$

$$y(n) = x_1(n) + x_2(n) = \left[-2\,(n+1) + 2^{n+1}\right] u(n)$$

QP # 12.2

Consider the causal system

$$y(n) = 0.48\,y(n-2) - 0.2\,y(n-1) + x(n)$$

Determine

(a) its impulse response when initial conditions are zero
(b) its impulse response when initial conditions are as follows

$y(-1) = -1.25,\ y(-2) = -0.52$

(c) The zero-state step response

Answer:

(a) $y(n) = 0.571\,(-0.8)^n\,u(n) + 0.429\,(06)^n\,u(n)$
(b) $y(n) = (-0.8)^n\,u(n)$
(c) $y(n) = -0.64(0.6)^n\,u(n) + 0.25\,(-0.8)^n\,u(n) + 1.39\,u(n)$

QP # 12.3

(a) State with justification two major advantages and disadvantages of DSP compared with analog signal processing design.
(b) What are the key operations of DSP? List them and discuss in brief any one of them.

QP # 12.4

(a) A discrete time signal $x(n)$ is defined as

$$x(n) = \begin{cases} 1 + \frac{n}{3} & -3 \le n \le -1 \\ 1, & 0 \le n \le 3 \\ 0, & \text{elsewhere} \end{cases}$$

Determine its values and sketch the signal $x(n)$.
(b) A discrete time signal $x(n)$ is shown in the figure.

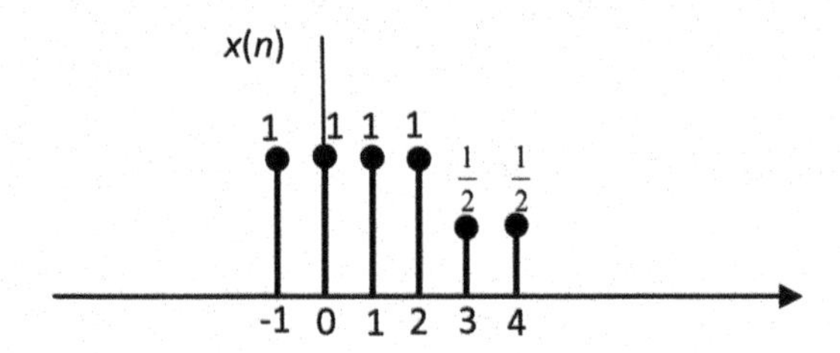

Sketch (i) $x(n^2)$ (ii) even part of $x(n)$

QP # 12.5

A discrete time system can be
(a) static or dynamic (b) linear or nonlinear (c) time invariant or time variant
(d) causal or noncausal (e) stable or unstable

Examine the following systems with respect to the properties above.
(a) $(n) = |x(n)|$
(b) $y(n)$ $y(n) = \cos[x(n)]$
(c) $y(n) = x(-n+2)$
(d) $y = x(2n)$

QP # 12.6

Determine analytically the convolution $y(n)$ of the signals

$$x(n) = \begin{cases} \frac{1}{3}n, & 0 \leq n \leq 6 \\ 0, & \text{elsewhere} \end{cases}$$

$$h(n) = \begin{cases} 1, & -2 \leq n \leq 2 \\ 0, & \text{elsewhere} \end{cases}$$

13.13 Practice Question Paper 13

QP # 13.1

(a) Using direct z-transform, determine the z-transform of the signal $x(n) = e^{-an}$ and hence for the signal $x(n) = e^{-an} \sin bn$.

$$X(z) = \frac{e^{-an} z \sin bn}{z^2 - 2e^{-an} z \cos bn + e^{-2an}}; \quad |z| > 1$$

Using z-transform properties.
(b) Determine the casual signal $x(n)$ if its Z-transform $X(z)$ is given by

$$X(z) = \frac{(z^2-5)}{(z-1)(z-2)^2}$$

$$x(n) = \frac{5}{4}\delta(n) - \left[4 - \frac{1}{2n}(2)^{n-1} + \frac{11}{4}2^n\right] u(n)$$

QP # 13.2

A causal LTI system has impulse response $h|n|$, for which the z-transform is

$$H(z) = \frac{1 + z^{-1}}{\left(1 - \frac{1}{2}z^{-1}\right)\left(1 + \frac{1}{4}z^{-1}\right)}$$

(a) What is the region of convergence of $H(z)$? $|z| > \frac{1}{2}$

(c) Find the impulse response *x*(*n*) of the system

$$x(n) = 2\left(\frac{1}{2}\right)^n u(n) - \left(-\frac{1}{4}\right)^n u(n)$$

QP # 13.3

(a) Compute the zero-state response for the system and the input signal given below.

$$\begin{aligned} y(z) &= \frac{10}{1+z^{-2}}\,x(z) \\ y(n) &= -y(n-2) + 10\,x(n) \\ y(z) & \frac{100}{(1+z^{-2})^2} \end{aligned}$$

$$x(z) = \frac{10}{1+z^{-2}}\,x(n) = 10\,(\cos n\,\pi/2\,)\,u(n)$$

$$y(z)\,\frac{50}{1+j\,z^{-1}} + \frac{50}{1-j\,z^{-1}} + \frac{-25\,jz^{-1}}{(1-j\,z^{-1})^2} + \frac{25\,jz^{-1}}{(1-j\,z^{-1})^2}$$

(b) A linear time invariant system is described by the following difference equation.

$$y(n) = 0.9\,y(n-1) + x(n)$$

Determine the output of the system to the input signal $\cos\frac{\pi n}{2}\,u(n)$

$$x(n) = 5 + 12\,\sin n\,\pi/2 - 20\,\cos(\pi n + \pi/4)$$

QP # 13.4

(a) Find out the direct realizations (form I and form II) of the discrete time system using summers, multipliers, and delay elements. Compare the delay elements of both the realizations.

$$y(n) = -0.1\,y(n-1) + 0.2\,y(n-2) + 3x(n) + 3.6\,x(n-1) + 0.6\,x(n-2)$$

$$Y(z) = \frac{3 + 3.6z^{-1} + 0.6z^{-2}}{1 + 0.1z^{-1} - 0.2z^{-2}}$$

(b) Compute the convolution $y(n) \otimes h(n)$ of the following signals

$$\begin{aligned} x(n) &= (1/3)\,n\;\; 0 \le n \le 4 \\ h(n) &= (1,\,-2,\,-3,\,4) \end{aligned}$$

$$y(n) = \left[0,\,\frac{1}{3},\,0,\,-\frac{4}{3},\,-\frac{4}{3},\,-3,\,0,\,\frac{16}{3}\right]$$

QP # 13.5

Determine the sequence $x_3(n)$ corresponding to the circular convolution of sequences

$$x_3(n) = \left\{ \underset{\uparrow}{1}, 2, 2, 1 \right\}$$

$$x_2(n) = \left\{ \underset{\uparrow}{2}, 1, 1, 2 \right\} \quad y(n) = \left\{ \underset{\uparrow}{9}, 10, 9, 8 \right\}$$

QP # 13.6

Design using bilinear transformation a second-order low-pass digital Butter worth filter without off frequency 1 kHz at a sampling rate of 10^4 sample per seconds.

The analog low-pass filter specifications are

Passband cutoff = 4.828 rads/sec
Stop band cutoff = 2 rads/sec
Passband ripple = 3 dB
Stop band ripple = 15 dB

$$H(z) = \frac{0.068\left[1 + 2z^{-1} + z^{-2}\right]}{1 - 1.142^{-1} + 0.412z^{-2}}$$

$$\Omega_c = 0.65 \times 10^4$$

13.14 Practice Question Paper 14

QP # 14.1

(a) Determine the z-transform of the signal

$$x(n) = (\cos \omega_0 n)\, u(n)$$

[Hint] $a^n (\cos \omega_0 n)\, u(n) \leftrightarrow \frac{1-a\,z^{-1}\cos \omega_0}{1-a\,z^{-1}\cos \omega_0 + a^2 z^{-2}}$

(b) Determine the z-transform of the signal

$$x(n) = n^2 u(n)$$

QP # 14.2

(a) Determine the casual signal *x*(*n*), if its z-transform *x*(*z*) is given by

$$X(z) = 1 - \frac{2 - 1.5\,z^{-1}}{1.5\,z^{-1} + 0.5\,z^{-2}}$$

$$x(n) = \left(\frac{1}{2}\right)^n u(n) + u(n)$$

$$x(n) = [2,\,1.5,\,1.25,\,1.1\ldots.]$$

QP # 14.3

The difference equation of the second-order system is given by

$$y(n) = \tfrac{1}{4}\,y(n-2) + x(n) \qquad \begin{aligned} y(-1) &= 0 \\ y(-2) &= 1 \end{aligned}$$

Use one-sided Z-transform to find its

(i) impulse response

$$y(n) = \frac{5}{8}\left[\left(\frac{1}{2}\right)^n + \left(-\frac{1}{2}\right)^n\right] u(n)$$

(ii) step response

$$y(n) = \left[\frac{4}{3}\left(\frac{1}{2}\right)^n + \left(-\frac{1}{2}\right)^n\right] u(n)$$

$$y(-1) = 1;$$

$$y(-2) = 0$$

$$H(z) = \frac{2.48}{1 - 0.64z^{-1}} + \frac{1.51}{1 + 0.39z^{-1}}$$

$$h(n) = 2.48\,(0.64)^n\,u(n) + 1.51\,(-0.39)^n\,u(n)$$

$$y(n) = \left[\frac{4}{3} - \frac{3}{8}\left(\frac{1}{2}\right)^n + \frac{7}{24}\left(-\frac{1}{2}\right)^n\right] u(n)$$

QP # 14.4

(a) Compute the convolution $y(n) = x(n) * h(n)$ of the following signals

$$x(n) = (1, 1, 0, 1, 1)$$
$$h(n) = (1, \underset{\uparrow}{-2}, -3, 4)$$

(b) Compute the convolution of the pair of signals using Z-transform.

$$x_1(n) = u(n); \quad x_2(n) = \delta(n) + \left(\frac{1}{2}\right)^n u(n)$$

QP # 14.5

A linear time invariant system is described by the difference equation

$$y(n) + \frac{1}{2}y(n-1) + x(n) + \frac{1}{2}x(n-1)$$

Determine

(i) magnitude $H(\omega)$ of the system

(ii) the output of the system to the input signal

QP # 14.6

(a) Examine the system shown below with respect to its linearity, time variant, and causality.

$y(n) = x(n) + n\,x(n+1)$ linear, time variant, and noncausal.

(b) Find out the direct realizations (form I and form II) of the discrete time system using summers, multipliers, and delay elements. Compare the delay elements of both the realizations.

$$H(z) = \frac{8z^3 - 4z^2 + 11z - 2}{\left(z - \frac{1}{4}\right)\left(z^2 - z + \frac{1}{2}\right)}$$

QP # 14.7

Find the transfer function of an analog low-pass filter given

$$H(s) = \frac{s + 0.1}{(s - 0.1)^2 + 9}$$

13.15 Practice Question Paper 15

QP # 15.1

(a) Determine the z-transform of the signal

$$x(n) = (-1)^n(\cos \pi/3\,n)u(n)$$

(b) Determine the casual signal $x(n)$ if its Z-transform $X(z)$is given by

$$X(z) = \frac{6z^3 + 2z^2 - z}{z^3 - z^2 - z + 1}$$

QP # 15.2

The difference equation of the second-order system is given by

$$y(n) + 0.2\,y(n-1) - 0.48\,y(n-2) = x(n)$$

and use one-sided Z-transform to find its impulse response when

QP # 15.3

Determine the impulse response and the step response of the causal system given below:

$$y(n) = \frac{1}{2}\,[x(n) + x(n-1)] + y(n-1)\,n \geq 0$$

QP # 15.4

(a) Determine the Fourier transform of the signal

$$x(n) = (\cos\,\omega_0\,n)\,u(n)$$

by evaluating its Z-transform on the unit circle

(b) A linear time invariant system is described by the difference equation.

$$y(n) = a\,y(n-1) + b\,x(n),\; 0 < a < 1$$

Determine the output $y(n)$ of the system to the input signal $x(n) = 5 + 12\,\sin\,\frac{\pi}{2}\,n$

QP # 15.5

(a) For the general linear time invariant recursive system described by the difference equation

$$y(n) = -\sum_{K=1}^{N} a_k\,y(n-k) + \sum_{K=0}^{N} b_k\,y(n-k)$$

illustrate how you will obtain direct form II structure.

(b) Find the direct form I and direct form II structure of the discrete time system

$$H(z) = \frac{8z^3 - 4z^2 + 11z + 2}{(z - 1/4)\,(z^2 - z + 1/2)}$$

QP # 15.6

The transfer function of an analog, third-order, Butter worth low-pass filter with a cutoff frequency of 1 rads/sec is

$$H(s) = \frac{1}{(s+1)\,(s+0.5+j0.866)\,(s+0.5-j0.866)}$$

(a) Design an impulse invariant digital equivalent based on a sampling interval of 0.5 sec.

$$H(z) = \frac{z}{z - 0.65} + \frac{z\,(0.8956 - z)}{z^2 + 1.414z + 6065}$$

(b) Determine the poles and zeros of the filter function obtained in part (a)
Note: work the question for three decimal places only

QP # 15.7

(a) Determine the convolution of the following pairs of signals by means of the z-transform

$$x_1(n) = (1/2)^n\, u(n) \;\; x_2(n) = \cos \pi n\; u(n)$$

(b) Write a short note on biomedical signals and their analysis using digital signal processing.

13.16 Practice Question Paper 16

QP # 16.1

(a) Determine the inverse z-transform by partial fraction expansion method

$$X(z) = \frac{1 - 1/2\, z^{-1}}{1 + 3/4\, z^{-1} + 1/8\, z^{-2}}$$

(b) By long division method, find *h*(*n*).

$$H(z) = \frac{z}{3\, z^2 - 4z + 1}$$

QP # 16.2

A causal LTI system has impulse response *h*(*n*), for which the z-transform is

$$H(z) = \frac{1 + z^{-1}}{(1 - 1/2\, z^{-1})\,(1 + 1/4z^{-1})}$$

(a) What is the region of convergence of $H(z)$?
(b) Find the z-transform $Y(z)$ of the input $y(n)$ shown below

$$y(n) = -1/3\,(1/4)^n u(n) - 4/3\,(2)^n\, u(n-1)$$

(c) Find the impulse response $H(z)$ of the system

$$2\,(1/2)^n\, u(n) - (-1/4)^n\, u(n)$$

QP # 16.3

The difference equation satisfied by the input and output of a system is

$$y(n) = a\,y(n-1) + x(n)$$

Consider the input $x(n) = k\,\delta(n)$
where k is an arbitrary number and the auxiliary condition $y(-1) = c$.

(a) Determine recursively
(i) the output for $n \geq 0$
(ii) the output for $n \leq -1$
(iii) the output for all n

(b) If it is known that a linear system requires that output be zero for all time where the input is zero for all time, discuss about the linearity of the system.

(c) State whether the system is time variant or time invariant and causal or noncausal.

QP # 16.4

Find the step response $y(n)$, using one-sided z-transform for the system where $x(n)$ is given below

$$\begin{aligned} y(n) &= -y(n-2) + 10\,x(n) \\ x(n) &= 10(\cos n\,\pi/2)\,u(n) \end{aligned}$$

QP # 16.5

(a) A linear time invariant system is described by the following difference equation.

$$y(n) = 0.9\,y(n-1) + x(n)$$

Determine the following:

(i) The magnitude and phase of the frequency response $H(\omega)$ of the system. Plot the responses.

(ii) The output of the system to the input signal

$$x(n) = 5 + 12\sin\left(\frac{n\pi}{2}\right) - 20\cos(\pi n + \pi/4)$$

(iii) The output of the system to the input signal

$$x(n) = 50 + 8.916\sin\left(\frac{n\pi}{2} - 42^o\right) - 10.7\cos(n\,\pi + \pi/4)$$

(b) Realize using direct form I and direct form II for the system

$$y(n) = -0.1y(n-1) + 0.2\,y(n-2) + 3.6x(n-1) + 0.6x(n-2)$$

References

[1] Gold, B., and Rader, C. M. (1969). *Digital Processing of Signals.* New York, NY: McGraw-Hill.

[2] Oppenheim, A. V., and Schafer, R. W. (1989). *Discrete-Time Signal Processing.* EnglewoodCliffs, NJ: Prentice Hall.

[3] Oppenheim, A., and Schafer, R. W. V. (1975). *Digital Signal Processing.* Englewood Cliffs, NJ: Prentice Hall.

[4] Rabiner, L., and Gold, B. R. (1975). *Theory and Application of Digital Signal Processing.* Englewood Cliffs, NJ: Prentice Hall.

[5] Mitra, S., and Kaiser, J. F. K. (eds) (1993). *Handbook of Digital Signal Processing.* New York, NY: Wiley.

[6] Mitra, S. K. (2015). *Digital Signal Processing: A Computer-Based Approach.* New York, NY: McGraw Hill.

[7] Parks, T., and Burrus, C. S. W. (1987). *Digital Filter Design.* New York, NY: Wiley.

[8] Antoniou, A. (1993). *Digital Filters: Analysis and Design, 2nd Edn.* New York, NY: McGraw-Hill.

[9] Elliott, D. F., (ed). (1987). *Handbook of Digital Signal Processing.* New York, NY: Academic Press.

[10] Rabiner, L., and Rader, C. M. R. (eds) (1972). *Digital Signal Processing.* New York, NY: IEEE Press.

[11] Selected Papers in Digital Signal Processing, II, edited by the Digital Signal Processing Committee and IEEE ASSP, IEEE Press, New York, 1976.

[12] Roberts, R., and Mullis, C. T. A. (1987). *Digital Signal Processing.* Reading, MA: Addison-Wesley.

[13] Lynn, P., and Fuerst, W. A. (1989). *Introductory Digital Signal Processing with Computer Applications.* New York, NY: Wiley.

[14] Proakis, J., and Manolakis, D. G. G. (1988). *Introduction to Digital Signal Processing, 2nd Edn.* New York, NY: Macmillan.

[15] Ifeachor, E., and Jervis, B. W. C. (1993). *Digital Signal Processing: A Practical Approach.* Reading, MA: Addison-Wesley.

[16] Haddad, R., and Parsons, T. W. A. (1991). *Digital Signal Processing: Theory, Applications, and Hardware.* New York, NY: Freeman, W. H.

[17] DeFatta, D. J., Lucas, J. G., and Hodgkiss, W. S. (1988). *Digital Signal Processing: A System Design Approach.* New York, NY: Wiley.

[18] Robinson, E., and Treitel, S. (1980). *Geophysical Signal Analysis.* Englewood Cliffs, NJ: Prentice Hall.

[19] Kay, S. M. (1988). *Modern Spectral Estimation: Theory and Application.* Englewood Cliffs, NJ: Prentice Hall.

[20] Marple, S. L. (1987). *Digital Spectral Analysis with Applications.* Englewood Cliffs, NJ: Prentice Hall.

[21] Widrow, B., and Stearns, S. D. (1985). *Adaptive Signal Processing.* Englewood Cliffs, NJ: Prentice Hall.

[22] Chillingworth, H. R. (1973). *Complex Variables.* Oxford: Pergamon.

[23] Scholfield, P. H. (1958). *The Theory of Proportion in Architecture.* London: Cambridge Univ. Press.

[24] Kappraff, J. (1990). *Connections: The Geometric Bridge Between Art and Science.* NewYork, NY: McGraw-Hill.

[25] Linden, D. A. (1959). A discussion of sampling theorems. Proc. IRE, 47, 1219–1226.

[26] Jerri, A. J. (1977). The Shannon Sampling Theorem—Its Various Extensions and Applications: A Tutorial Review. *Proc. IEEE* 65, 1565–1596.

[27] Butzer, P., and Strauss, R. L. L., (1992). Sampling theory for not necessarily band-limited functions: a historical overview. *SIAM Rev.* 34, 40–53.

[28] Sheingold, D. H. (ed). (1986). *Analog-Digital Conversion Handbook, 3rd Edn.* EnglewoodCliffs, NJ: Prentice Hall.

[29] VanDoren, A. (1982). *Data Acquisition Systems.* Reston, VA: Reston Publishing.

[30] Bennett, W. R. (1948). Spectra of quantized Signals. *Bell Syst. Tech. J.* 27, 446–472.

[31] Widrow, B. (1961). "Statistical analysis of amplitude-quantized sampled-data systems," in *Proceeding of the Transactions of the American Institute of Electrical Engineers, Part II: Applications and Industry,* Vol. 79 (Rome: IEEE), 555–568.

[32] Swaszek, P. F. (ed). (1985). *Quantization.* New York, NY: Van Nostrand Reinhold.

[33] Sripad, A., and Snyder, D. L. B. (1977). "A necessary and sufficient condition for quantization errorsto be uniform and white," *IEEE Transactions on Acoustics, Speech, and Signal Processing*, Vol. 25 (Rome: IEEE), 442–448.

[34] Barnes, C. W., Tran, B., Leung, S. (1985). "On the statistics of fixed-point roundoff error," in *IEEE Transactions on Acoustics, Speech, and Signal Processing*, Vol. 33 (Rome: IEEE), 595–606.

[35] Gray, R. M. (1990). "Quantization noise spectra," in *IEEE Transactions on Information Theory*, Vol. 36, 1220–1244 and earlier references therein. Reprinted in Ref. [276], p. 81.

[36] Schuchman, L. (1964). "Dither Signals and Their Effect on Quantization Noise," in *IEEE Transactions on Communication Technology* Vol. 12, 162–165.

[37] Jayant, N., and Noll, P. S. (1984). *Digital Coding of Waveforms.* Englewood Cliffs, NJ: Prentice Hall.

[38] Blinn, J. F. (1994). "Quantization Error and Dithering," *IEEE Computer Graphics and Applications* Vol. 14, 78–82.

[39] Lipshitz, S. P., Wannamaker, R. A., and Vanderkooy, J. (1992). "Quantization and dither: a theoretical Survey. *J. Audio Eng. Soc.* 40, 355–375.

[40] Vanderkooy, J., and Lipshitz, S. P. (1984). Resolution below the least significant bit in digitalsystems with dither. *J. Audio Eng. Soc.* 32, 106–113.

[41] Vanderkooy, J., and Lipshitz, S. P. (1987). Dither in digital audio. *J. Audio Eng. Soc.* 35, 966 (1987).

[42] Vanderkooy, J. and Lipshitz, S. P. (1989). "Digital Dither: Signal Processing with Resolution Far Below the Least Significant Bit," in *Proceedings of the 7th International Conference: Audio in Digital Times*, Toronto, 87.

[43] Wannamaker, R. A. (1992). Psychoacoustically optimal noise shaping audio. *J. Eng. Soc.* 40, 611–620.

[44] Gerzon, M. A., Graven, P. G., Stuart, J. R., and Wilson, R. J. (1993). Psychoacoustic Noise Shaped Improvements in CD and Other Linear Digital Media," *Presented at 94th Convention of the AES*, Berlin, AES Preprint no. 3501.

[45] van der Waal, R., Oomen, A., and Griffiths, F. (1994). "Performance Comparison of CD, Noise-Shaped CD and DCC," *Presented at 96th Convention of the AES*, Amsterdam, AES Preprint no. 3845. Moorer, J. A., and Wen, J. C. (1993). Whither dither: experience with high-order

dithering algorithms in the studio. *Paper Presented at 95th Convention of the AES* (New York, NY: AES).

[46] Moorer, J. A., and Wen, J. C. (1993). Whither dither: experience with high-order dithering algorithms in the studio. *Paper Presented at 95th Convention of the AES* (New York, NY: AES).

[47] Wannamaker, R. A. (1994). Subtractive and nonsubtractive dithering: a comparative analysis. *Paper Presented at 97th Convention of the AES* (San Francisco, CA: AES).

[48] Ranada, D. (1994). Super CD's: do they deliver the goods? *Stereo Rev.* p. 61.

[49] Mullis, C. T., and Roberts, R. A. (1982). "An interpretation of error spectrum shaping in digital filters," in *Proceedings of the IEEE Transactions on Acoustics, Speech, and Signal Processing (ASSP-30)* (Rome: IEEE).

[50] Dattoro, J. (1989). "The implementation of digital filters for high-fidelity audio," in *Proceedings of the AES 7th International Conference Audio in Digital Times*, Toronto, 165.

[51] Wilson, R. (1993). Filter topologies. *J. Audio Eng. Soc.* 41, 455.

[52] Zolzer, U. (1994). Roundoff error analysis of digital filters. *J. Audio Eng. Soc.* 42, 232.

[53] Chen, W. (1994). Performance of the cascade and parallel IIR filters. *Paper Presented at 97th Conventionof the AES* (San Francisco, CA: AES).

[54] Horning, D. W., and Chassaing, R. (1991). IIR filter scaling for real-time signal processing. *IEEE Trans. Educ.* 34, 108.

[55] Baudendistel, K. (1994). Am improved method of scaling for real-time signal processing applications. *IEEE Trans. Educ.* 37, 281.

[56] Chassaing, R. (1992). *Digital Signal Processing with C and the TMS320C30.* New York, NY: Wiley.

[57] El-Sharkawy, M. (1990). *Real Time Digital Signal Processing Applications with Motorola's DSP56000 Family.* Englewood Cliffs, NJ: Prentice Hall.

[58] El-Sharkawy, M. (1994). *Signal Processing, Image Processing and Graphics Applications with Motorola's DSP96002 Processor*, Vol. I. Englewood Cliffs, NJ: Prentice Hall.

[59] Ingle, V. K., and Proakis, J. G. (1991). *Digital Signal Processing Laboratory Using the ADSP-2101 Microcomputer.* Englewood Cliffs, NJ: Prentice Hall.

[60] Tow, J. (1988). "Implementation of digital filters with the WE_R DSP32 digital signal processor," in Proceedings of the AT&T Microelectronics: Application Note, Allentown, PA.

[61] Freeny, S. L., Kaiser, J. F., and McDonald, H. S., (1978). "Some applications of digital signal processing in telecommunications," in *Applications of Digital Signal Processing*, ed. A. V. Oppenheim (Engle wood Cliffs, NJ: Prentice Hall).

[62] Boddie, J. R., Sachs, N., and Tow, J. (1981). Receiver for TOUCH-TONE service. *Bell Syst. Tech. J.* 60, 1573.

[63] Mar, A. (ed.) (1990). *Digital Signal Processing Applications Using the ADSP-2100 Family*. Englewood Cliffs, NJ: Prentice Hall.

[64] Special Issue on Digital Audio (1985). *IEEE ASSP Mag*. 2(4).

[65] Ando, Y. (1985). *Concert Hall Acoustics*. New York, NY: Springer-Verlag.

[66] Begault, D. (1993). The evolution of 3-D audio. *MIX* 17, 42–46.

[67] Blesser, B., and Kates, J. M., (1978). "Digital processing of audio signals," in *Applications of Digital Signal Processing*, ed. A. Oppenheim (Englewood, NJ: Prentice-Hall), 29–116.

[68] Brighton, N., and Molenda, M. (1993). Mixing with delay. *Electronic Musician* 9, 88.

[69] Eargle, J. M. (1992). *Hand book of Recording Engineering*, 2nd Edn, NewYork, NY: Van Nostrand Reinhold.

[70] Freudenberg, P. (1994). All about dynamics processors, parts 1 & 2. *Home Studio Record.* 18, 44.

[71] Griesinger, D. (1989). "Practical processors and programs for digital reverberation," in *Proceedings of the AES 7th International Conference, Audio in Digital Times*, Toronto, 187.

[72] Childers, D. G. (1989). "Biomedical signal processing," in *Selected Topics in Signal Processing*, ed. S. Haykin (Englewood Cliffs, NJ: Prentice Hall).

[73] Cohen, A. (1986). *Biomedical Signal Processing*, Vol. 1 and 2. Boca Raton, FL: CRC Press.

[74] Goovaerts, H. G., and Rompelman, O. (1991). Coherent average technique: a tutorial review. *J. Biomed. Eng.* 13, 275–280.

[75] Horowitz, P., and Hill, W. (1989). *The Art of Electronics*, 2nd Edn. Cambridge: Cambridge University Press.

[76] Rompelman, O., and Ros, H. H. (1986). Coherent averaging technique: a tutorial review, Part1: Noise reduction and the equivalent filter. *Biomed.*

J. Eng. 8, 24; and Part 2: Trigger jitter, overlapping responses, and non-periodic stimulation. *ibid* 30.

[77] Shvartsman, V., Barnes, G., Shvartsman, L., and Flowers, N. (1982). Multichannel signal processing based on logic averaging. *IEEE Trans. Biomed. Eng.* 29, 531–536.

[78] Thomas, C. W., Rzeszotarski, M. S., and Isenstein, B. S. (1982). "Signal averaging by parallel digital filters," in *Proceedings of the IEEE Transactions on Acoustics, Speech, and Signal Processing, ASSP-30* (Rome: IEEE), 338.

[79] Wilmshurst, T. H. (1990). *Signal Recovery from Noise in Electronic Instrumentation*, 2nd Edn. Bristol: Adam Hilger and IOP Publishing.

[80] Bromba, M., and Ziegler, H. (1979). Efficient computation of polynomial smoothing digital filters. *Anal. Chem.* 51, 1760–1762.

[81] Bromba, M., and Ziegler, H. (1980). Explicit formula for filter function of maximally flat non recursive digital filters. *Electronics Lett.* 16, 905–906.

[82] Bromba, M., and Ziegler, H. (1981). Application hints for Savitzky-Golay digital smoothing filters. *Anal. Chem.* 53, 1583–1586.

[83] Edwards, T. H., and Wilson, P. D. (1974). Digital least squares smoothing of spectra. *Appl. Spectrosc.* 28, 541–545.

[84] Edwards, T. H., and Wilson, P. D. (1976). Sampling and smoothing of spectra. *Appl. Spectrosc. Rev.* 12, 1–81.

[85] Enke, C. G., and Nieman, T. A. (1976). Signal-to-noise ratio enhancement by least-squares polynomial smoothing. *Anal. Chem.* 48, 705A–712A.

[86] Ernst, R. R. (1966). "Sensitivity enhancement in magnetic resonance," in *Advances in Magnetic Resonance*, Vol. 2, ed. J. S. Waugh (New York, NY: Academic Press).

[87] Hamming, R. W. (1983). *Digital Filters*, 2nd Edn. Englewood Cliffs, NJ: Prentice Hall.

[88] Kendall, M. (1976). *Time-Series*, 2nd Edn. New York, NY: Hafner Press.

[89] Kendall, M., and Stuart, A. (1968). *Advanced Theory of Statistics*, Vol. 3, 2nd Edn. London: Charles Griffin & Co.

[90] Madden, H. H. (1978). Comments on the Savitzky-Golay convolution method for least-squares fit smoothing and differentiation of digital data. *Anal. Chem.* 50, 1383–1386.

[91] Savitzky, A., and Golay, M. (1964). Smoothing and differentiation of data by simplified least squares procedures. *Anal. Chem.* 36, 1627–1639.

[92] Williams, C. S. (1986). *Designing Digital Filters.* Englewood Cliffs, NJ: Prentice Hall.

[93] Ziegler, H. (1981). Properties of digital smoothing polynomial (DISPO) filters. *Appl. Spectrosc.* 35, 88–92.

[94] Kaiser, J. F., and Reed, W. A. (1977). Data smoothing using lowpass digital filters. *Rev. Sci. Instrum.* 48, 14–47.

[95] Kaiser, J. F. (1975). Hamming, sharpening the response of a symmetric non recursive filter by multiple use of the same filter. *IEEE Trans. Acoust. Speech Signal Process.* 25, 415.

[96] Harris, F. J. (1978). On the use of windows for harmonic analysis with the discrete fourier transform. *Proc. IEEE* 66, 51–83.

[97] Geckinli, N. C., and Yavuz, D. (1978). Some Novel Windows and a Concise Tutorial Comparison of Window Families. *IEEE Trans. Acoust. Speech Signal Process.* 26, 501–507.

[98] Kaiser, J. F., and Schafer, R. W. (1980). On the use of the I0-sinh window for spectrum analysis. *IEEE Trans. Acoust. Speech Signal Process.* 28, 105–107.

[99] Nuttal, A. H. (1981). Some windows with very good sidelobe behavior. *IEEE Trans. Acoust. Speech Signal Process.* 29, 84–91.

[100] Brigham, E. O. (1988). *The Fast Fourier Transform.* Englewood Cliffs, NJ: Prentice Hall.

[101] Ramirez, R. W. (1985). *The FFT, Fundamentals and Concepts.* Englewood Cliffs, NJ: Prentice Hall.

[102] Burrus, C. S., and Parks, T. W. (1985). *DFT/FFT and Convolution Algorithms.* New York, NY: Wiley.

[103] Van Loan, C. (1992). *Computational Frameworks for the Fast Fourier Transform.* Philadelphia: SIAM.

[104] Bergland, G. D. (1969). A guided tour of the fast fourier transform. *IEEE Spectrum.* 6, 41–52.

[105] Cooley, J. W., Lewis, P. A. W., and Welch, P. D. (1970). The fast fourier transform algorithm:programming considerations in the calculation of sine, cosine, and laplace transforms. *J. Sound Vib.*, 12, 315–337.

[106] Harris, F. J. (1982). The discrete fourier transform applied to time domain signal processing. *IEEE Commun. Mag.* 20, 13–22.

[107] Cooley, J. W., Lewis, P. A. W., and Welch, P. D. (1967). Historical notes on the fast fourier transform. *IEEE Trans. Audio Electroacoust.* 15, 76.

[108] Heideman, M. T. Johnson, D. H. and Burrus, C. S. (1984). Gauss and the history of the fast fourier transform. *IEEE ASSP Mag.* 4, 14–21.

[109] Cooley, J. W. (1992). How the FFT gained acceptance. *IEEE Signal Proc. Mag.* 9, 10–13.

[110] Kraniauskas, P. (1994). A plain man's guide to the FFT. *IEEE Signal Proc. Mag.* 11, 24–36.

[111] Deller, J. R. (1994). Tom, dick, and mary discover the DFT. *IEEE Signal Proc. Mag.* 11, 36–50.

[112] Kaiser, J. F. (1963). "Design methods for sampled data filters," in *Proceeding of the 1st Allerton Conference Circuit System Theory*, Rome, 221.

[113] Kaiser, J. F. (1966). "Digital filters," in *System Analysis by Digital Computer*, eds F. F. Kuo and J. F. Kaiser (New York, NY: Wiley), 228.

[114] Kaiser, J. F. (1974). "Nonrecursive digital filter design using the i0-sinh window function," in *Proceeding of the IEEE International Symposium on Circuits and Systems*, Rome, 20.

[115] Helms, H. D. (1968). Nonrecursive digital filters: design methods for achieving specifications on frequency response. *IEEE Trans. Audio Electroacoust.* 16, 336–342.

[116] Helms, H. D. (1971). Digital filters with equiripple or minimax response. *IEEE Trans. AudioElectroacoust.* 19, 87–93.

[117] Dolph, C. L. (1946). A current distribution which optimizes the relationship between beamwidth and side-lobe levels. *Proc. I.R.E.* 34, 335–348.

[118] Barbiere, D. (1952). A method for calculating the current distribution of tschebyscheff arrays. *Proc. I.R.E.* 40, 78–82.

[119] Stegen, R. J. (1953). Excitation coefficients and beamwidths of tschebyscheff arrays. *Proc. I.R.E.* 41, 1671.

[120] Hansen, R. C. (1986). "Linear arrays," in *Handbook of Antenna Design*, eds A. W. Rudge, et al., Vol. 2, 2nd Edn (London: Peregrinus and IEE).

[121] Saramaki, T. Finite Impulse Response Filter Design. 5, 155.

[122] Benson, K. B., and Whitaker, J. (1990). *Television and Audio Handbook*, (New York, NY: McGraw-Hill).

[123] Schuck, P. L. (1989). "Digital FIR Filters for Loudspeaker Crossover Networks II: Implementation Example," in *Proceeding of the AES 7th International Conference, Audio in Digital Times*, Toronto, 181.

[124] Wilson, R. et al. (1989). Application of digital filters to loudspeaker crossover networks. *J. Audio Eng. Soc.* 37, 455.

[125] Steiglitz, K., Parks, T. W., and Kaiser, J. F. (1992). METEOR: a constraint-based FIR filter design program. *IEEE Trans. Acoust. Speech Signal Process.* 40, 1901.

[126] Steiglitz, K., and Parks, T. W. (1986). What is the filter design problem? Proceeding of the Princeton Conference Information Science System, Rome, 604.

[127] Hirano, K., Nishimura, S., and Mitra, S. (1974). Design of digital notch filters. *IEEE Trans. Commun.* 22, 964.

[128] Swami, M. N. S., and Thyagarajan, K. S. (1976). Digital bandpass and band stop filters with variable center frequency and bandwidth. *Proc. IEEE* 64, 1632.

[129] Moorer, J. A. (1983). The manifold joys of conformal mapping: applications to digital filtering in the studio. *J. Audio Eng. Soc.* 31, 826.

[130] White, S. A. (1986). Design of a digital biquadratic peaking or notch filter for digital audio equalization. *J. Audio Eng. Soc.* 34, 479.

[131] Regalia, P. A., and Mitra, S. K. (1987). Tunable digital frequency response equalization filters. *IEEE Trans. Acoust. Speech Signal Process.* 35, 118.

[132] Shpak, D. J. (1992). Analytical design of biquadratic filter sections for parametric filters. *J. Audio Eng. Soc.* 40, 876.

[133] Massie, D. C. (1993). An engineering study of the four-multiply normalized ladder filter. *J. Audio Eng. Soc.* 41, 564.

[134] Harris, F., and Brooking, E. (1993). A versatile parametric filter using imbedded all-pass sub-filterto independently adjust bandwidth, center frequency, and boost or cut. *Paper presented at the 95th Convention of the AES*, (New York, NY: AES Preprint).

[135] Bristow-Johnson, R. (1994). The equivalence of various methods of computing biquad coefficients for audio parametric equalizers. *Paper Presented at the 97th Convention of the* AES, (San Francisco, AES Preprint).

[136] Vaidyanathan, P. P. (1993). *Multirate Systems and Filter Banks*, Englewood-Cliffs, NJ: Prentice Hall.

[137] Candy, J. C., and Temes, G. C., (eds) (1992). *Oversampling Delta-Sigma Data Converters.* Piscataway, NJ: IEEE Press.

[138] Candy, J. C., and Temes, G. C. (). Oversampling Methods for A/D and D/A Conversion," in Ref.[113].

[139] Gray, R. M. (1987). Oversampled sigma-delta modulation. *IEEE Trans. Commun.* 35, 481–485. Reprinted in Ref. [276], p. 73.

[140] Goedhart, D., van de Plassche, R. J., and Stikvoort, E. F. (1982). Digital-to-analog conversion in playing a compact disc. *Philips Tech. Rev.* 40, 174–179.

[141] Hauser, M. W. (1991). Principles of oversampling A/D conversion. *J. Audio Eng. Soc.* 39, 3–26.
[142] Knuth, D. E. (1981). *The Art of Computer Programming*, Vol. 2, 2nd Edn. Reading, MA: Addison-Wesley.
[143] Bratley, P., Fox, B. L., and Schrage, L. (1983). *A Guide to Simulation*, New York, NY: Springer-Verlag.
[144] Press, W. H., Flannery, B. P., Teukolsky, S. A., and Vetterling, W. T. (1992). *Numerical Recipes in C*, 2nd Edn. New York, NY: Cambridge University Press.
[145] Park, S. K., and Miller, K. W. (1988). Random number generators: good ones are hard to find. *Comm. ACM*, 31, 1192–1201.
[146] Ripley, B. D. (1983). Computer generation of random variables: a tutorial. *Int. Stat. Rev.* 51, 301–319.
[147] Best, D. J. (1979). Some easily programmed pseudo-random normal generators. *Aust. Comput. J.* 11, 60–62.
[148] Marsaglia, G. (1990). Toward a universal random number generator. *Stat. Probab. Lett.* 8, 35–39.
[149] Marsaglia, G., and Zaman, A. (1991). A new class of random number generators. *Annals Appl. Prob.* 1, 462–480.
[150] Marsaglia, G., and Zaman, A. (1994). Some portable very-long random number generators. *Comput. Phys.* 8, 117–121.
[151] van der Ziel, A. (1986). *Noise in Solid State Devices and Circuits*, New York, NY: Wiley.
[152] Reichbach, M. (1993). Modeling 1/f Noise by DSP," Graduate Special Problems Report, ECE Department, Rutgers University, Fall 1993.
[153] Lyons, R. G. (1997). *Understanding Digital Signal Processing*. Boston, MA: Addison-Wesley.
[154] Smith, S. W. (2003). *Digital Signal Processing A Practical Guide for Engineers and Scientists*. Oxford: Newnes.
[155] Walpole, R. E., and Myers, R. H. (1985). *Probability and Statistics for Engineers and Scientists*, 3rd Edn. New York, NY: MacMillan Publishing Company.
[156] Ifeachor, E. C., and Jervis, B. W. (2002). *Digital Signal Processing A Practical Approach*, 2nd Edn. Harlow: Prentice-Hall.
[157] Boyer, C. B. (1959). *The History of the Calculus and Its Conceptual Development*. New York, NY: Dover Publications, Inc.
[158] Richard, C. J. (2003). Twelve Greeks and Romans Who Changed The World. Rowman & Little_eld, 2003.

[159] Gonick, L. (1990). *The Cartoon History of the World*, Vol. 1–7. New York, NY: Doubleday.

[160] Geselowitz, M. N. (2002). *Hall of Fame: Heinrich Hertz.* IEEE-USA News & Views, November 2002.

[161] Andrew Bruce, D. D., and Gao, H.-Y. (1996). *Wavelet Analysis.* IEEE Spectrum, 26–35.

[162] Lee, E. A., and Varaiya, P. (2003). *Structure and Interpretation of Signals and Systems.* Reading, MA: Addison-Wesley.

[163] Hamming, R. W. (1998). *Digital Filters*, 3rd Edn. Mineola, NY: Dover Publications.

[164] Sheffield, D. (1980). Equalizers. *Stereo Rev.* 72–77.

[165] Kaiser, G. (1994). *A Friendly Guide to Wavelets.* Boston, MA: Birkhauser.

[166] Thompson, S. P. and Gardner, M. (1998). *Calculus Made Easy.* New York, NY: St. Martin's Press.

[167] Swokowski, E. W. (1980). *Elements of Calculus with Analytic Geometry.* Boston, MA: Prindle, Weber and Schmidt.

[168] Chen, W. K. (ed.). (2005). *The Electrical Engineering Handbook.* Burlington, MA: Elsevier Academic Press.

[169] Ozer, J. (1995). *New Compression Codec Promises Rates Close to MPEG*, "CD-ROM Professional."

[170] Hickman, A., Morris, J., Rupley, C. L. S., and Willmott, D. (1997). Web acceleration," *PC Magazine*," June 10, 1997.

[171] Boles, W. W., and Tieng, Q. M. (1993). "Recognition of 2D objects from the wavelettransform zero-crossing representations," in *Proceedings SPIE, Mathematical Imaging*, Vol. 2034, San Diego, CA, 104–114.

[172] Vishwanath, M., and Chakrabarti, C. (1994). "A VLSI architecture for real-time hierarchical encoding/decoding of video using the wavelet transform," in *Proceedings of the IEEE International Conference on Acoustics, Speech and Signal Processing (ICASSP'94)*, Vol. 2, Adelaide, 401–404.

[173] Weeks, M., and Bayoumi, M. (2003). Discrete wavelet transform: architectures, design and performance issues. *J. VLSI Sig. Process.* 35, 155–178.

[174] Haar, A. (1910). Zur theorie der orthogonalen funktionensysteme. *Mathemat. Ann.* 69, 331–371.

[175] Smith, M. J. T., and Barnwell, T. P. III. (1986). Exact reconstruction techniques fortree-structured subband coders. *IEEE Trans. ASSP* 34, 434–441.

[176] Jaffard, S., Meyer, Y., and Ryan, R. D. (2001). *Wavelets Tools for Science & Technology*. Philadelphia, PA: Society for Industrial and Applied Mathematics (SIAM).

[177] Anton, H. (1991). *Elementary Linear Algebra*, 6th Edn. New York, NY: John Wiley & Sons, Inc.

[178] Chakrabarti, C., and Vishwanath, M. (1995). Efficient realizations of the discrete and continuous wavelet transforms: from single chip implementations to mappingson simd array computers. *IEEE Trans. Sig. Process*. 43, 759–771.

[179] Goldberg, D. (1991). What every computer scientist should know about floating-point arithmetic. *J. ACM Comput. Sur*. 23, 5–48.

[180] Wu, S.-W. (1998). Additive vector decoding of transform coded images. *IEEE Trans. Image Process*. 7, 794–803.

[181] Al-Shaykh, O. K., and Mersereau, R. M. (1998). Lossy compression of noisy images. *IEEE Trans. Image Process*. 7, 1641–1652.

[182] Fgee, E.-B., Phillips, W. J., and Robertson, W. (1999). "Comparing audio compression using wavelets with other audio compression schemes," in *Proceedings of the IEEE Canadian Conference on Electrical and Computer Engineering*, Vol. 2, Alberta, 698–701.

[183] Rabie, T. (2005). Robust estimation approach for blind denoising. *IEEE Trans. Image Process*. 14, 1755–1765.

Index

About the Authors

Muhammad N. Khan received the B.E. degree in electronic engineering from Dawood College of Engineering and Technology, Pakistan in 2003 and the M.Sc. degree in electrical engineering from the Delft University of Technology, Delft, The Netherlands in 2009 and PhD degree in electrical engineering at the Institute for Telecommunications Research (ITR), University of South Australia, Australia. He worked with Pakistan Telecommunication Industries (PCI) Pty. Ltd. from 2003 to 2005 as assistant manager, specialising in telecommunication. He worked with the SUPARCO, SRDC Lahore from 2005 to 2007 as RF communication engineer, specialising in modem and codec design for satellite communications. Currently, he is Associate Professor in Electrical Engineering Department, The University of Lahore, Pakistan. His research interests include communication theory, digital modulation and iterative decoding, and free space optical communications.

Syed K. Hasnain worked in different national level engineering colleges with an excellent teaching record. He worked with the Pakistan Naval Engineering

College as a Professor, specialising in digital signal processing for communications. Currently, he is a Professor in Electrical Engineering Department, Swedish College of Engineering and Technology, Rahim Yar Khan, Pakistan. His research interests include detection schemes and channel coding.

Mohsin Jamil received the B.E. degree in industrial electronic engineering from NED university of Engineering and Technology in 2004, Pakistan and the M.Sc. degree in electrical engineering from National University of Singapore in 2007 and PhD degree in electrical engineering (Control Systems/Power Electronics) University of Southampton in 2011 U.K. Currently, he is an Assistant Professor in Mechanical Engineering Department, National University of Sciences and Technology, Islamabad, Pakistan. His research interests include automation and control.

Lightning Source UK Ltd.
Milton Keynes UK
UKOW06n1148270117
293027UK00002B/47/P

9 788793 379404